Communications
in Computer and Information Science 159

Yuanxu Yu   Zhengtao Yu   Jingying Zhao (Eds.)

Computer Science for Environmental Engineering and EcoInformatics

International Workshop, CSEEE 2011
Kunming, China, July 29-31, 2011
Proceedings, Part II

 Springer

Volume Editors

Yuanxu Yu
Dalian University of Technology, Dalian, China
E-mail: yuanxuyu@gmail.com

Zhengtao Yu
Kunming University of Science and Technology, Kunming, China
E-mail: ztyu@hotmail.com

Jingying Zhao
International Association for Scientific and High Technology
Kunming, China
E-mail: cseee2011@gmail.com

ISSN 1865-0929   e-ISSN 1865-0937
ISBN 978-3-642-22690-8   e-ISBN 978-3-642-22691-5
DOI 10.1007/978-3-642-22691-5
Springer Heidelberg Dordrecht London New York

Library of Congress Control Number: 2011932218

CR Subject Classification (1998): I.2, H.3, H.4, C.2, H.5, I.4

Typesetting: Camera-ready by author, data conversion by Scientific Publishing Services, Chennai, India

Printed on acid-free paper

Springer is part of Springer Science+Business Media (www.springer.com)

Preface

Following the great progress made in the field of computer science for environmental engineering and ecoInformatics, the 2011 International Workshop on Computer Science for Environmental Engineering and EcoInformatics (CSEEE 2011) aimed at providing a forum for the presentation and discussion of state-of-the-art development in computer science and ecoInformatics. Emphasis was given to basic methodologies, scientific development and engineering applications.

The conference was co-sponsored by Kunming University of Science and Technology and the International Association for Scientific and High Technology, and was held in cooperation with Yunnan University, Chongqing Jiaotong University and the International Science and Engineering Research Center. The purpose of CSEEE 2011 was to bring together researchers and practitioners from academia, industry, and government to exchange their research ideas and results in the areas of the conference.

We would like to thank all the participants and the authors for their contributions. We would also like to gratefully acknowledge Stefan Göller and Leonie Kunz, who enthusiastically supported the conference. In particular, we appreciate the full-hearted support of all the reviewers and committee members of the conference. We believe that CSEEE 2011 was successful and enjoyable for all participants and look forward to seeing you at future CSEEE events.

July 2011

Yuanxu Yu
Zhengtao Yu
Jingying Zhao

Organization

Workshop Chair

Dechun Kang   University of Melbourne, Australia

Program Committee

Chih-Chen Chang	Hong Kong University of Science and Technology, China
Yuanxu Yu	Dalian University of Technology, China
Zhengtao Yu	Kunming University of Science and Technology, China
Jian-Zhong Li	Tongji University, China
Li-Xiang Zhang	Kunming University of Science and Technology, China
Liao-Yuan Ye	Yunnan Normal University, China
Man-Chung Tang	T.Y. Lin International Engineering Consulting (China) Co., Ltd., China
Meng-Lin Lou	Tongji University, China
Pei-Yuan Lin	Southwest Jiaotong University, China
Wan-Cheng Yuan	Tongji University, China
Wen Pan	Kunming University of Science and Technology, China
Z. John Ma	University of Tennessee, USA
Qun Ding	Heilongjiang University, China
Zhao Zhiwei	Harbin Institute of Technology, China
Liu Dongmei	Harbin Institute of Technology, China
Guo Hai	Dalian Nationalities University, China
Lv Wei-min	Province Light Industrial Science Research Institute, China
Jiang Yi-feng	Zhejiang University of Technology, China
Gao Xue-feng	Jilin University, China
Li Jun	Zhejiang University of Technology, China
Ze-Min Xu	Kunming University of Science and Technology, China
Jingying Zhao	International Association for Scientific and High Technology, China

Reviewers

Haigang Zhang	Shanghai Ocean University, China
Lei Xiao	South China Agricultural University, China

Jiaoyong Liu	Sichuan University, China
Li Zhonghai	Shenyang Aerospace University, China
Lijuan Wu	Shenyang Normal University, China
Lili Zhai	Harbin University of Science and Technology, China
Linchong Yu	Xiamen University of Technology, China
Qing Ye	North China University of Technology, China
Qingmin Kong	Beihang University, China
Yusheng Tang	Guangxi University, China
Zhijun Wang	Liaoning Technical University, China
Benfu Lv	Chinese Academy of Science, China
Chenglei Zhao	Shanghai Jiao Tong University, China
Chengyu Hu	Shandong University, China
Cunbao Chen	Southeast University, China
Furao Shen	Nanjing University, China
Guanqi Gou	Hunan Institute of Science and Technology, China
Haitao Li	Beihang University, China
Hansuk Sohn	New Mexico State University, USA
Hofei Lin	Dalian University of Technology, China
Jiaohua Cheng	Fujian Agriculture and Forestry University, China
Jing Na	Kunming University of Science and Technology, China
Jinsong Gui	Central South University, China
Junfeng Man	Hunan University of Technology, China
Kai Wan	University of Illinois-Urbana Champaign, USA
Fanyu Kong	Shangdong University, China
Taosheng Li	Wuhan University of Technology, China
Liang Tang	Chongqing University, China
Yimin Mao	Hunan University, China
Maoxin Wang	Chinese Academy of Meteorological Sciences, China
Meng Yu	GuiLin University of Electronic Technology, China
Mingwen Wang	Jiangxi Normal University, China
Mingxing He	Xihua University, China
Na Chen	Beijng Jiaotong University, China
Qiang Li	Tianjin University of Technology, China
Qiguo Duan	Tongji University, China
Quan Rui	Wuhan University of Technology, China
Shaoying Zhu	Jinan University, China
Shun Long	Jinan University, China
Susy H.Y.	Hsiuping Institute of Technology, Taiwan

TingTing Wang	Jiangnan University, China
Weiguang Wang	Hohai University, China
Xin Guan	Liaoning Technical University, China
Xin Yan	Wuhan University of Technology, China
Yi-Chih Hsieh	National Formosa University, Taiwan
Ying Shi	Wuhan University of Technology, China
Yunfei Yin	Chongqing University, China
Yunjie Li	Liaoning Technical University, China
Yunsong Tan	Wuhan Institute of Technology, China
Yuxin Dong	Harbin Engineering University, China
Ze Cao	Changchun Institute of Engineering, China
Zhang Yan	Lanzhou University, China
Zhi Li	Xi'an Jiao Tong University
ZhiZhuang Liu	Hunan University of Science and Engineering, China
Jipeng Zhou	Jinan University, China

Cosponsored by

Kunming University of Science and Technology, China
International Association for Scientific and High Technology, China

In Cooperation with

Yunnan University
Chongqing Jiaotong University
T.Y. Lin International Engineering Consulting (China) Co., Ltd.
International Science & Engineering Research Center

Table of Contents – Part II

Section 1: Artificial Intelligence and Pattern Classification

Section 2: Computer Networks and Web

Section 3: Computer Software, Data Handling and Applications

Section 4: Data Communications

Section 5: Data Mining

Section 6: Data Processing and Simulation

Section 7: Information Systems

Section 8: Knowledge Data Engineering

Section 9: Multimedia Applications

Table of Contents – Part I

Section 3: Computing Practices and Applications

Section 4: EcoInformatics

Section 5: Image Processing and Information Retrieval

Section 6: Pattern Recognition

Section 7: Wireless Communication and Mobile Computing

A Hybrid Collaborative Filtering Algorithm for Patent Domain

Liping Zhi[1, 2] and Hengshan Wang[1]

[1] Business School, University of Shanghai for
Science & Technology, China
[2] School of Computer Science & Information Engineering,
Anyang Normal University, China
zhiliping1216@163.com, wanghs@usst.edu.cn

Abstract. In this paper, we present a hybrid collaborative filtering algorithm based on ontology and item rating prediction. As items can be classified and predicted from the semantic level, this method can efficiently improve the extreme sparsity of user rating data. Experimental results indicate that the algorithmic accuracy, measured by mean absolute error (MAE) has been improved larger compared with traditional algorithms.

Keywords: Collaborative filtering algorithm, Hybrid, Patent.

1 Introduction

The World Intellectual Property Organization revealed that 90-95% of world's inventions were found in patented documents. Along with the explosion of information, it is more and more difficult to acquire knowledge from the Internet. For example, we can retrieve "TV" patents from the website of China's intellectual property rights(http://www.cnipr.com/) for almost 1050 pages (10493 correlative patents in total). Among which , 2728 invention patents, 4178 utility patents, 3888 appearance design patents. Then, how toretrieval useful relevant patents for users from vast intellectual property database and recommend to users automatically so as to realizing personalized service? However, there isn't any intelligent recommended technique that has been applied to patent information research. We firstly introduce the collaborative filtering algorithm into patent domain, which can be applied to domestic patent retrieval system and web, for it could greatly improve the retrieval efficiency of patents. Our research will have a positive action for the intellectual property in the world..

2 Study on Patent-Based Intelligent Recommended Technology

2.1 Introduction of Intelligent Recommended Technology

Designing accurate and efficient intelligent recommended algorithm is the core problem of intelligent recommendation. Currently popular recommended techniques

Y. Yu, Z. Yu, and J. Zhao (Eds.): CSEE 2011, Part II, CCIS 159, pp. 1–6, 2011.

conclude: cooperative filter recommendation, recommendation based on user's statistical information, content-based recommendation, knowledge-based recommendation, rule-based recommendation, etc. Collaborative filtering recommendation put forward by Tapestry in 1992, is the most used method in the research and application of intelligent recommendation, which is widely used in E-commerce sites, digital library, web search, press filtering etc. The biggest advantage of collaborative filtering is no need to analysis object's properties, so as to handle unstructured complex objects, such as music, books, movies, etc. Patent information has typical characteristics of unstructured objects, so it's very suitable for using collaborative filtering algorithm.

However, collaborative filtering recommended technique also exists its deficiencies. With user rating data becoming extremely sparse , traditional collaborative filtering recommendation algorithms calculate similarity between items all have their weakness , and they do not consider the semantic relationship between different items , thus recommendation quality is very poor. For the past few years, effective integration of many technologies cause researchers' attention.

The term "ontology" comes from the field of philosophy that is concerned with the study of being or existence. In 1993, Gruber gave a definition of ontology, namely "ontology is an explicit specification of a conceptualization"[1]. People have made deep research on the application of ontology technology in knowledge engineering, information management and the semantic Web. Literature[2]thought: It has certain advantages to apply ontology into retrieval of patent information. Because the patent information has a standard, highly standardized format, and a unified classification system, it is easy to construct the patent ontology.

Based on the above analysis, the paper proposes a new hybrid collaborative filtering recommended algorithm based on domain ontology, using vector distance algorithm to calculate semantic similarity of items, and synthetically considering the semantic similarity and the score similarity of items, using hybrid similarity to predict the score of the item, then the item similarity will be even more close to human subjective judgments. As items can be classified and predicted from the semantic level, this method can efficiently improve the extreme sparsity of user rating data, and provide better recommended quality.

2.2   Construction of Domain Ontology of Patent

Aiming at the specific circumstances of the patent management & analysis system, this paper makes some improvement on the existing modeling methods and proposes a new ontology modeling method combined UML with OWL [4]. This method absorbs advantages of those two languages, simultaneously, overcomes the shortcomings, so as to provide convenient communication and understanding [5]. Here is the ontology constructing method of patent domain, showed in fig.1 [6].

This paper uses a method based on OWL-DL to establish logic and reasoning layer for patent ontology , supplying formal semanteme of concepts of patent ontology. we choose Protégé 2000 developed by Stanford University to construct domain ontology.

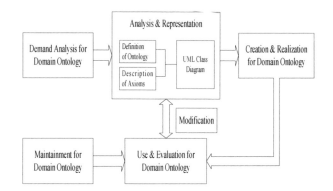

Fig. 1. Flowchart of the hybrid collaborative filtering algorithm

2.3 Vector Distance Algorithm

The vector distance algorithm will generate a central vector to represent each text sets according to arithmetic average distance, then before the arriving of a new text vector, determine the new text vector and compute the distance between the vector and its central vector ,namely similarity[7], specific steps are as follows:

1) Computing the central vector for each text sets, namely the arithmetic average distance value of all text vectors;
2) The new text will be represented as its eigenvector after its arriving;
3) Computing the similarity between the target vector and its central vector, and the formula is as follows:

$$\text{Sim}(d_i, d_j) = \frac{\sum_{k=1}^{M} W_{ik} \times W_{jk}}{\sqrt{(\sum_{k=1}^{M} W_{ik}^2)(\sum_{k=1}^{M} W_{jk}^2)}} \tag{1}$$

Through Eq.(1), d_i is the eigenvector of a new text and d_j is the central vector of the j class. M is the dimensions of the eigenvector, while W_k represents the k dimension of the vector.

1) The total similarity is evaluated in the following formula :

$$\text{SIM}(di, dj) = \sum w_i \text{Sim}(d_i, d_j) \tag{2}$$

Through Eq.(2), w_i is the weight of attribute value of some class in patent, e.g. name, abstract, inventor. $\text{Sim}(d_i, d_j)$ is the similarity of these attributes.
2) Comparing the similarity between the text and its central vector, and divide the text into the most similar category.

2.4 Hybrid Similarity

The hybrid collaborative filtering recommended algorithm we proposed is based on domain ontology, which synthetically considers the semantic similarity of and the score similarity of items, using hybrid similarity to predict the score of the object item, so the item similarity will be even more close to human subjective judgment. It is evaluated as the following formula [7] :

$$Sim_{ij} = w\,Sim_{ij1} + (1-w)Sim_{ij2} \qquad (3)$$

Through Eq.(3), Sim_{ij} is the hybrid similarity and Sim_{ij1} is the semantic similarity. W is the contribution weight of semantic similarity, while Sim_{ij2} is the score similarity and $(1-w)$ is the contribution weight of score similarity.

Supposing if there were n items, the hybrid collaborative filtering recommended algorithm will recommend according to the score of the most similar item. P_{ui} representing the score of item i by user u is evaluated as the following formula :

$$P_{ui} = \sum_{j=1}^{n}(Sim(i,j) \times R_{uj}) \ / \ \sum_{j=1}^{n}Sim(i,j) \qquad (4)$$

Through Eq.(4), R_{uj} represents the score of item j by user u, and $Sim(i,j)$: the hybrid similarity of item i and j.

3   Experimental Results and Analysis

3.1   Data Set

In order to judge the performance of decribed algorithm, the simulation experiment is carried out. We choose the data-set supplied by MovieLens (http://movielens.umn.edu/) , which is a kind of research recommendation system. At present, the number of users of the site has more than 45,000 people , also, the number of the scored films has exceeded 6600.

MovieLens provides three dimensions of different data-set, considering the experimental environment of computer operation ability, this paper chooses the medium scale data-set (including 6040 independent users ,3900 movies, and about 100 million ratings). Randomly extracting rating by 100 users from data-set to judge the performance of decribed algorithm. The 100 users have scored 12976 assessment to 2318 movies. The experimental data sets need to be further divided into two parts. The one contains 90% of the data as training set, and the other one remaining 10% of data as the probe set. In order to compare the proposed algorithm, we select item-based filtering algorithm and hybrid filtering algorithm as a reference [8].

3.2   Measures

The results can be valued by lots of methods. We choose the most commonly used quality measurement method: mean absolute error (MAE), which is easy to understand and value the quality of recommendation system directly. Based on average absolute deviation MAE as metrics. Less the value of MAE is, higher quality recommendation is. Assume that the user rating set for prediction is represented as { p1 ,p2 , ... , pN }, Accordingly, the actual set is {q1 ,q2 , ... , qN }, then the equation of mean absolute error (MAE) is defined as[9]:

$$\text{MAE} = \frac{\sum_{i=1}^{N} |Pi - qi|}{N} \tag{5}$$

3.3 Experimental Results

Considering the influence of the weight parameter(ω) in hybrid similarity calculation, we need to find optimal value range of ω while using it to predict. We put weight parameters from 0 to 1, the interval for 0.1. Experimental results as shown near the scope of 0.4 is optimal.

To test the performance of the proposed algorithm in this paper, we make a comparison of the traditional item-based collaborative filtering algorithm and hybrid collaborative filtering algorithm, The weight parameters ω takes 0.4,while the number of recent neighbor increased to 25 from 5, the interval for 5, experimental results are shown in fig.2:

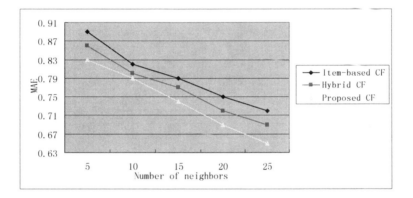

Fig. 2. Comparison of accuracy of recommendation algorithms

4 Conclusion

This paper studies the technologies of collaborative filtering algorithm for patent, using hybrid similarity to predict the score of the item, then the item similarity will be even more close to human subjective judgments. As items can be classified and predicted from the semantic level, this method can efficiently improve the extreme sparsity of user rating data, and provide better recommended quality.

Acknowledgments. We acknowledge GroupLens Research Group for providing us the data set Movielens. The research is financed by Shanghai Leading Academic Discipline Project (No.S30504) & the Innovation Fund Project For Graduate Student of Shanghai (No.JWCXSL1021) & the National Natural Science Foundation of China (No.70901010, 71071098, 10905052) & the Innovation Fund Project For Graduate Student of Shanghai (No.JWCXSL1001).

References

1. Gruber, T.R.: A Translation Approach to Portable Ontologies. Knowledge Acquisition 5(2), 199–220 (1993)
2. Yong, Z., Na, L.: Research on Knowledge Service System of Patent based on Ontology. Modern Information, 93–95 (2005)
3. Jianxia, L.: Study on Intelligent Recommendation Technology and Application. In: E-Commerce. Master Thesis, Chongqing University, pp. 30–34 (2009)
4. Shu-hao, Y., Shou-bao, S., Ren-jin, L.: A Comparative Studies of UML and OWL in Ontology Modeling. Computer Technology and Development 4, 155–157 (2007)
5. Ren, H., Yong-zhong, H., Zhen-lin, L., Xiao-nan, L.: Research on OWL Ontology Building Method. Computer Engineering and Design 4, 1397–1400 (2008)
6. Stader, J.: Results of the Enterprise Project. In: Proeedings of the 16th International Conference of the British Computer Society Specialist Group on Expert Systems, Cambridge, UK, pp. 121–130 (1996)
7. Song-jie, G., Hong-yan, P.: A Collaborative Filtering Algorithm Using Domain Ontology. Applications of the Computer Systems 5, 20–23 (2008)
8. Ai-lin, D., Yang-yong, Z., Bai-le, S.: A Collaborative Filtering Recommendation Algorithm Based on Item Rating Prediction. Journal of Software 8, 1621–1628 (2003)
9. Sarwar, B., Karypis, G., Konstan, J., Riedl, J.: Item-Based Collaborative Filtering Recommendation Algorithms. In: Proceedings of the 10th International World Wide Web Conference, pp. 285–295 (2001)

An Intelligent Application for the WBC Application Enabler Sub-layer: Design and Implementation

Zhanlin Ji[1], Ivan Ganchev[2], and Máirtín O'Droma[3]

[1] CAC Dept, Hebei United University, TangShan, 064300, P.R. China
[2] ECE Dept, University of Limerick, Limerick, Ireland
Zhanlin.ji@ieee.org, Ivan.Ganchev@ul.ie, Mairtin.ODroma@ul.ie

Abstract. This paper describes the design and implementation of a new intelligent application, which runs at the application enabler sub-layer of a wireless billboard channel service provider's (WBC-SP) node. The application is implemented with three tiers: a service discovery and maintenance tier acting as a client-server distributed system for data collection and organization; an intelligent application tier holding all business logic and common application programming interfaces (APIs); and a multi-agent systems (MAS) tier maintaining the advertisement, discovery and association (ADA) agents' lifecycle.

Keywords: Ubiquitous Consumer Wireless World (UCWW); Wireless Billboard Channel (WBC); Advertisement, Discovery and Association (ADA); Software Architecture; Multi-Agent Systems (MAS).

1 Introduction

Wireless billboard channels (WBCs) [1-3], are novel infrastructural components for facilitating the service advertisement, discovery and association (ADA) in the ubiquitous consumer wireless world (UCWW) [4, 5]. In the UCWW, maximizing the consumer wireless transactions, and not the subscriber contracts, is the main business driver for service providers. The mobile user is not constrained to any particular access network provider (ANP) and may use any available service through any available access network, and pay for the use of services through a trusted third-party authentication, authorization and accounting service provider. The consumer-user is free to choose what is 'best' for them, i.e. the service and access network they consider best matches their needs at any time or place.

The newly conceived WBC infrastructural component of the UCWW aims to satisfy this requirement by facilitating the service providers to 'push' advertisements and information about their wireless service offerings to potential consumers. Taking into consideration the potentially large number of wireless services already available, efficient and easy-to-implement mechanisms and applications for wireless services' ADA are needed.

Y. Yu, Z. Yu, and J. Zhao (Eds.): CSEEE 2011, Part II, CCIS 159, pp. 7–12, 2011.
© Springer-Verlag Berlin Heidelberg 2011

The push-based WBC concept is based on a unidirectional communication between a WBC-service provider's (WBC-SP) node and a user node (mobile terminal). Each node contains three logical layers. The physical layer is represented by a transmitting system in the WBC-SP node and a receiving system in the user node. The link layer is concerned with typical frame processing issues. The service layer describes the service discovery model, and data collection, clustering, scheduling, indexing, discovery and association schemes. The service layer consists of two sub-layers: a service enabler sub-layer and an application enabler sub-layer. This paper focuses on the design and implementation of the application enabler sub-layer on the WBC-SP node.

2   WBC Service Layer's Architecture

The WBC service layer's architecture has been proposed in [6, 7]. Its application enabler sub-layer part consists of three tiers:

• The service discovery and maintenance tier is a web tier with two main types of actors: service providers (xSPs) - both ANPs and teleservice providers (TSPs) - who submit/publish and manage the service descriptions (SDs) of their services via the WBC-SP portal application, and the WBC-SP who monitors the status of xSPs' SDs and maintains the WBC center's (WBCC) server;

• The application tier consists of common application programming interfaces (APIs), such as the shared ontology API, common APIs, Drools/Jess API and SDs clustering, scheduling, and indexing APIs. This tier is shared by the other two tiers;

• The multi-agent system (MAS) container tier provides an agent run-time environment. The WBCC controls the life cycle of all agents. A special gateway agent is used for communication with the service discovery and maintenance tier via a shared message channel.

The design and implementation of the WBC application enabler sub-layer follow the personal software process (PSP) methodology [8]. An experience repository database is used to store all development experiences. Test-driven development (TDD) [9] and feature-driven development (FDD) [10] methods were selected when designing this PSP project. With these two methodologies, the three-tier heterogeneous WBC application enabler sub-layer's architecture is plotted into a set of functional unit modules (features). For each unit module, from bottom tier to top tier, unified modeling language (UML) diagrams following the corresponding interface were first designed. Then the interface was fully implemented and a unit testing was performed.

The WBC application enabler sub-layer is built with a number of open-source integrated development environments (IDEs)/APIs/frameworks as been stated in [6]. The benefits of using open-source software include public collaboration, not bounding to a single development company, auditability, flexibility and freedom. The details of each tier's design and implementation are presented in the following sections.

3   Design and Implementation of Service Discovery and Maintenance Tier

To design and implement the service discovery and maintenance tier quickly and efficiently, a project skeleton as well as a number of build/test/deploy schemes was first developed. To describe the requirements for analysis, design, implementation and documentation, a use case diagram was first designed as shown in Figure 1. It includes a number of modules, such as login, SD uploading, search, configuration, etc. For each unit module, these main design and implementation steps were followed:

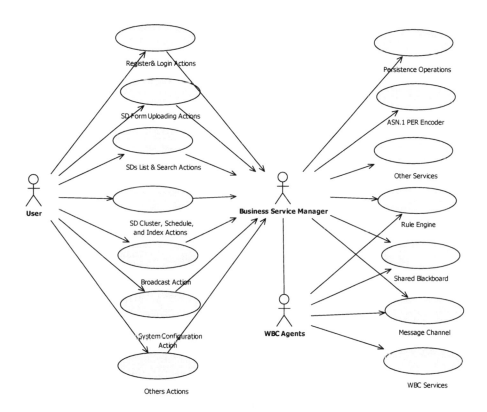

Fig. 1. A Use Case View of the Service Discovery and Maintenance Tier

Step 1: Create the corresponding Plain Old Java Object (POJO) aentity and add Java Persistence API's (JPA) annotations. Implement the toString(), equals() and hashCode() methods. Then run the command mvn test-compile hibernate3:hbm2ddl to create the aentity database table.

Step 2: Create the aentity's data access object (DAO) bean aentityDao for database's Create, Read, Update and Delete (CRUD) operations. Configure the aentityDao with the class org.wbc.dao.hibernate.GenericDaoHibernate in the context file wbcContext.xml. Once the aentityDao is being defined, add the setter method with the parameter of GenericDao in order to use this bean as a dependency of the object.

Step 3: Create the aentityManager to act as a service facade to aentityDao. Configure the aentityManager with the class org.wbc.service.impl. GenericManagerImpl in the context file wbcContext.xml. Similarly to step 2, add the setter method with the parameter of GenericManager in order to use this bean as a dependency of the object.

Step 4: Create the web tier using xwork. Implement an aentity action to process the request and response operations.

4   Design and Implementation of Application Tier

To achieve a loose-coupling system and enable WBC advertise-processing to run in an intelligent way, the WBC APIs, rule engine and database are integrated in this tier. Considering that the programming to an interface is the key principle of the reusable object-oriented design [11], all APIs of the application tier are extract interfaces to other tiers. Figure 2 shows the interface design of the application tier.

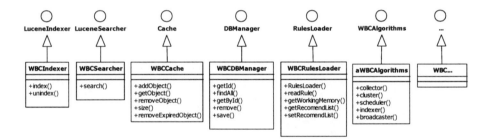

Fig. 2. The Interface Design of the Application Tier

A rule-based expert system operates at this tier for facilitating the data broadcasting on the WBC-SP node. The Drools was selected as the rule engine in WBC [2]. All the .drl configuration files are stored in a database, which can be accessed and updated via the portal application.

5   Design and Implementation of MAS Container Tier

The WBC-JADE is a lightweight MAS. The UI of WBC-JADE has been redesigned based on JADE (the popup UI in Figure 3) [2]. In addition to the WBC collecting agents, clustering agents, scheduling agents, indexing agents, broadcasting agents and personal assistance agent (PAA), a gateway agent, message channel and blackboard were also developed.

The developed UI of the MAS portal is depicted in Figure 3.

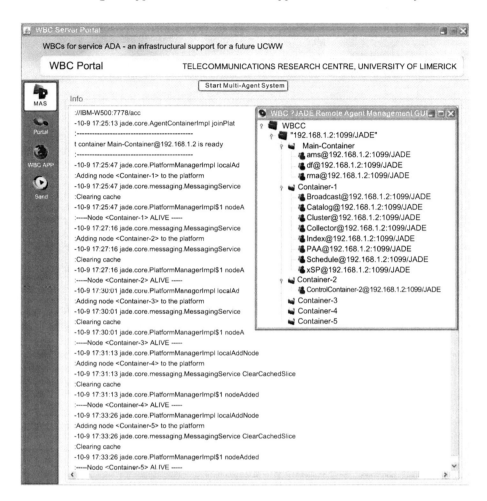

Fig. 3. The WBC-SP MAS Portal

6   Conclusions

This paper has described the design and implementation of a new intelligent application running on the WBC application enabler sub-layer. The application is implemented with three tiers: a service discovery and maintenance tier that maintains the client requests and server responses, supports the service discovery model, and serializes objects to the persistence sub-tier; a rule-based expert system running at the intelligent application tier and providing WBC common services and maintaining ontologies; and a peer-to-peer multi-agent system (MAS) tier running a set of agents to collect, cluster, schedule, index, and broadcast descriptions of (wireless) services.

Acknowledgments. This publication has been supported by the Irish Research Council for Science, Engineering and Technology (IRCSET) and the Telecommunications Research Centre, University of Limerick, Ireland.

References

1. Flynn, P., Ganchev, I., O'Droma, M.: WBCs - ADA vehicle and infrastructural support in a UCWW. In: 10th IEEE International Symposium on Consumer Electronics, pp. 351–356. IEEE Press, New York (2006)
2. Ji, Z., Ganchev, I., O'Droma, M.: 'WBC over DVB-H' Testbed Design, Development and Results. EURASIP Journal on Wireless Communications and Networking 1–18 (2010) ID 769683
3. Ji, Z., Ganchev, I., O'Droma, M.: Performance Evaluation of 'WBC over DVB-H' System. IEEE Transactions on Consumer Electronics 55, 754–762 (2009)
4. O'Droma, M., Ganchev, I.: Toward a ubiquitous consumer wireless world. IEEE Wireless Communications 14, 52–63 (2007)
5. O'Droma, M., Ganchev, I.: The Creation of a Ubiquitous Consumer Wireless World through Strategic ITU-T Standardization. IEEE Communications Magazine 48, 158–165 (2010)
6. Ji, Z., Ganchev, I., O'Droma, M.: Intelligent Software Architecture for the Service Layer of Wireless Billboard Channels. In: 6th Annual IEEE Consumer Communications & Networking Conference, pp. 172–173. IEEE Press, New York (2009)
7. Ji, Z., Ganchev, I., O'Droma, M.: Building a Heterogeneous Software Architecture for the WBC Service Layer. In: China-Ireland International Conference on Information and Communications Technologies, pp. 14–18. IEE Press, London (2008)
8. Humphrey, W.: PSP: A Self-Improvement Process for Software Engineers-Instructor's Guide. Software Engineering Institute, Carnegie Mellon University, Pittsburgh (2005)
9. Crispin, L.: Driving software quality: How test-driven development impacts software quality. IEEE Software. 23, 70–71 (2006)
10. Coad, P., LeFebrve, E., De Luca, J.: Feature-driven development. Prentice Hall PTR, Englewood Cliffs (1999)
11. Hightower, R., Onstine, W., Visan, P.: Professional Java tools for extreme programming: Ant, Xdoclet, JUnit, Cactus, and Maven. Wrox (2004)

Numerical Simulation on Indoor Air Pollution Diffusion and Ventilation

Dong Li[*], Yan Zhao, and Guozhong Wu

The College of Architecture and Civil Engineering, Northeast Petroleum University,
Hei Longjiang Daqing 163318, China
{Lidonglvyan,YanZhao,wgzdq}@126.com

Abstract. Forecasting diffusion regularity of Indoor air pollution is the vital significance to design a reasonable ventilation plan. Analyzed pollution diffusion characteristics of indoor air based on dynamic simulation theory, established the boundary and initial conditions using the simplified air terminal device model, the indoor air pollution diffusion model was built. The influencing results of indoor pollution diffusion were analyzed with different conditions of ventilation, human interference and pollution condition. The results show that (1) the diffusion concentration of carbon monoxide in the diffusion indoor area decrease adopting air hood in the local pollution area. (2)the interference of gas pollution diffusion is not neglected by human in room. (3) optimizing experiment stations and air hood, indoor pollutants concentration need the standard requirement and energy saving could be satisfied at the same time.

Keywords: Pollutants, Diffusion, Dynamic Simulation, Air Terminal Device Model, Ventilation Mode.

1 Introduction

The laboratories of chemistry, petroleum chemical industry produce various pollutants. How to reduce the concentration of the pollutants directly relates to the indoor air quality and laboratory staff's health. At the beginning of ventilation system design process in the lab, engineering design personnel hope that they could detailedly understand the indoor flow field and pollutants concentration distribution, which provide the basis for designing better ventilation system plan.

Numerical simulation is a cost-effective and efficient approach to predict the indoor flow field and pollutants concentration distribution of buildings. CFD simulation provides detailed spatial distributions of air velocity, air pressure, temperature, pollutants concentration distribution and turbulence by numerically solving the governing conservation equations of fluid flows. It is a reliable tool for the evaluation of contaminant distributions. These results can be directly or indirectly used to quantitatively analyze the indoor environment and determine ventilation system performances. However, the application of CFD for the indoor flow field and pollutants

---

[*] Corresponding author.

Y. Yu, Z. Yu, and J. Zhao (Eds.): CSEEE 2011, Part II, CCIS 159, pp. 13–17, 2011.

concentration distribution prediction has been limited due to long computational time and excessive computer resource requirements. CFD simulations require high computation cost, especially when grid size requirements for various computational domains are inconsistent, such as indoor airflow simulation and pollutants concentration distribution simulation at the same time. The basic calculation principle of CFD and its application in HVAC areas was studied by ZhiLi Zhang[1].The distribution of indoor air pollutants was simulated using three-dimension large eddy model by Wu Meng[2], and the flow field and pollutants were obtained according to the different type of room and vent arrangement. The pollutant transport was simulated using COHERENS model by HuaZuLin[3]. This paper is intended to find the appropriate external coupling strategies to improve the accuracy and efficiency in assessing the performance of the indoor flow field and pollutants concentration distribution, and to reveal the detailed pollutants concentration distribution information in some particular states. The computational fluid dynamics method was adopted, and the physics and mathematics model of indoor pollution air transport in laboratory was established, then the concentration distribution of indoor pollution were simulated under different ventilation and air velocity, and compared the exhaust hood with no exhaust hood in the area of local pollutants district.

2 Indoor Pollutants Flow Model

The laboratory model is shown in Fig.1, with the room length 8m in X direction, the room length 6m in Z direction, the room height 4 m in Y direction.

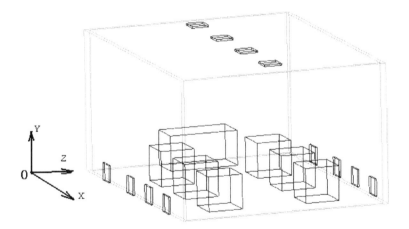

Fig. 1. Scheme of laboratory model

The room ventilation system adopts bilateral air supply, with arranged four air return opening and eight air supply outlet, and the air supply outlet are located in the lower of side wall, the air return opening located in the top of the room center, whose sizes are

0.5m × 0.5m. There are seven experimental tables in the room, and in which are pollution source on the top of experimental table, with the pollution diffusion velocity 0.1m/s.

The control equations include continuity equation, the momentum equation, the energy equation and k-ε equation and turbulent viscosity coefficient [4-6].

The continuity equation

$$\frac{\partial \rho}{\partial t} + \frac{\partial (\rho \bar{v}_j)}{\partial x_j} = 0 \cdot \tag{1}$$

The momentum equation

$$\frac{\partial (\rho \bar{v}_i)}{\partial t} + \frac{\partial (\rho \overline{v_i v_j})}{\partial x_j} = -\frac{\partial p}{\partial x_j} + \frac{\partial}{\partial x_j}\left[u(\frac{\partial \bar{v}_j}{\partial x_j} + \frac{\partial \bar{v}_i}{\partial x_j}) \right] \cdot \tag{2}$$

The energy equation

$$\frac{\partial}{\partial t}(\rho c_p \bar{T}) + \frac{\partial}{\partial x_j}(\rho v_j c_p \bar{T}) = \frac{\partial}{\partial x_j}(\lambda \frac{\partial \bar{T}}{\partial x_j}) - \frac{\partial}{\partial x_j}(\overline{\rho v'_j c_p T}) + \omega_s Q_s \cdot \tag{3}$$

The component equation

$$\frac{\partial}{\partial t}(\rho \bar{Y}_i) + \frac{\partial}{\partial x_j}(\rho v_i \bar{Y}_i) = \frac{\partial}{\partial x_j}(D_\rho \frac{\partial \bar{Y}_s}{\partial x_j}) - \frac{\partial}{\partial x_j}(\overline{\rho v'_j Y'_s}) - \overline{\omega}_s \cdot \tag{4}$$

3 Results and Discussion

3.1 Influence of Air Hood

There are two simulation schemes in the numerical simulation. First, the air hood is urgently adopted in the experimental table, whose pollutant diffusion is obvious. Second, the air hood is not adopted in the experimental table, although those pollutant diffusion are obvious. Comparing and analyzing the pollutant concentration distribution of two conditions with room air distribution and velocity distribution were accomplished, and numerical simulation results are shown in Fig. 2.

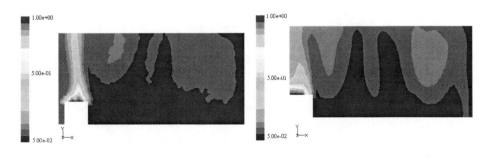

(a) with air hood   (b) without air hood

Fig. 2. Carbon monoxide concentration distribution of Z = 3m section

As shown in Fig. 2, the pollutant diffusion area is more obvious without air hood, and the distribution of pollutant is almost filled with whole room, with the pollutant concentration more high in upper of room rear. However, because of air hood, the pollutant concentration distribution is concentrate in air hood, and the pollutant concentration is lower outside air hood, so the air quality is improved in room.

3.2 Influence of Existing Man Body

For considering the body of man to affect the pollutant concentration distribution, given the body of man is as a column of 0.3m×0.3m×1.6m. The effect of flow field is ignored with excreted pollutant in the body of man self. Comparing and analyzing the pollutant concentration distribution of two conditions with room air distribution and velocity distribution were accomplished, and numerical simulation results are shown in Fig. 3.

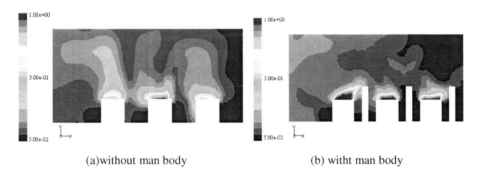

(a)without man body   (b) witht man body

Fig. 3. Carbon monoxide concentration distribution of Z = 4.5m section

As shown in Fig. 3, because of body barrier, indoor air flow field is affected, which cause mass distribution of pollutant, and the local pollutant concentration is more high.

3.3 Influence of Optimizing Experiment Stations and Air Hood

Scheme 1 is as following. Experimental table 1 and 2 are adopted, with air supply outlet 1, 2, 7 and 8 open. Experimental table 1 and 2 are adopted in Scheme 2, with air supply outlet 1, 2, 3, 6, 7 and 8 open. Comparing and analyzing the pollutant concentration distribution of two conditions with room air distribution and velocity distribution were accomplished, and numerical simulation results are shown in Fig. 4.

As shown in Fig. 4, indoor pollutant concentration distribution are similar under two air supply forms, but the pollution concentration of Scheme 2 is significantly lower than one in Scheme 1. Therefore, the numbers of open air supply outlet and the position of opened air supply outlet should be adapted to the position of experimental table.

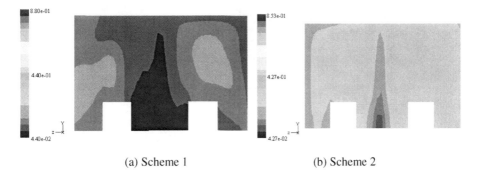

(a) Scheme 1   (b) Scheme 2

Fig. 4. Carbon monoxide concentration distribution of X= 4.5m section

4 Conclusion

The influences that the pollution sources acting on the indoor ventilation were analyzed. The laboratories using air hood in the locally severe pollution areas are smaller pollution diffusion areas than the others not applying, and the concentrations decrease quickly, so the indoor ventilation effect is better. Because of the body barrier, there are more concentrated pollutant distribution and e higher local pollution concentrations, so the body has great influence on the flow distribution, and when designing the indoor air-conditioning organizations, the body should be considered. Optimizing experiment stations and air hood, indoor pollutants concentration need the standard requirement and energy saving could be satisfied at the same time.

References

1. Yang, Z., Ji, X.: Some Aspects of Indoor Air Quality Assessment and Study. Environmental Science and Technology 26(6), 61–63 (2003) (in Chinese)
2. Zhang, Z., Wu, X.: The Basic Methods of CFD Problems and Its Applications in HVAC. Energy Technology 23(1), 8–11 (2002) (in Chinese)
3. Wu, M., Hu, G., Fan, J., Cen, K.: Three-Dimension Large Eddy Simulation of Lndoor Air Pollution. Journal of Zhejiang University (Engineering Science) 40(6), 986–990 (2006) (in Chinese)
4. Hua, Z., Gu, L., Cha, Y., Dong, X.: Numerical Simulation of Three-dimension Flow and Pollutant Transportation of Tidal River Based on COHERENS Model. Environmental Science and Technology 32(6), 14–18 (2009) (in Chinese)
5. Khan, J.A., Feigley, C.E., Lee, E.: Effects of Inlet and Exhaust Loca-tions and Emitted Gas Density on Indoor Air Contaminant Concentrations. Building and Environment 7(41), 851–863 (2006)
6. Luo, S., Heikkinen, J., Roux, B.: Simulation of Air Flow in IEA Annex 20Test Room-validation of a Simplified Model for the Nozzle Diffuser in Isothermal Test Cases. Building and Environment 12(39), 1403–1415 (2004)

Research on Collision Detection of Tree Swaying in Wind

Zhanli Li, Cheng Gao, and Yu Sun

College of Computer Science & Technology, Xi'an University of
Science & Technology Xi'an, P.R. China 710054
lizl@xust.edu.cn, xustgao@163.com, sunyu@xust.edu.cn

Abstract. Simulating the tree swaying in wind field fast and realistically is still a challenge in computer graphics. Collision detection plays a crucial role in enhancing the effect of simulating the swaying of a tree under different wind scale. Based on the branch position relationship analysis, the detection methods in all kinds of conditions were proposed. A novel collision detection algorithm was formed. Experiment results show that the novel algorithm can reduce the computing complexity and improve the efficiency compared to OBB algorithm.

Keywords: Tree swaying, Branches, Collision Detection, Cylinder.

1 Introduction

Many collision detection algorithms have been proposed in recent years within the fields of Computational Geometry, Robotics, and especially Computer Graphics. The purpose of collision detection is to discover the collision and to provide the information for the further collision response. There are five categories of the object collision detection algorithm, they are Spatial Subdivision [1], Bounding Volume Hierarchy method, Stochastic Methods, Distance Field Methods [2], and Image-space Techniques [3]. Of which the most widely used algorithm is Bounding Volume Hierarchy. The typical algorithms of Bounding Volume Hierarchy include Sphere Bounding Box, Axis-Aligned Bounding Box, Oriented Bounding Box [4,5], Fixed Directions Hulls, Discrete Orientation Polytopes, etc.

In tree dynamic simulation, both domestic and foreign scholars have got different research results [6,7,8,9,10], but a few of them focus on collision detection in the dynamic simulation of trees. Two methods for the collision detection of trees simulation were proposed in the references [11]. One is patch-based collision detection, the other is collision detection based on convex polyhedron. Without affecting real-time animation features and providing a better authenticity, a more appropriate method of collision detection is needed.

2 Analysis and Simplification

Tree model is a combination of branches and leaves. Shown in Figure 1, the geometric model of a single branch can be considered to be a cylinder. To simplify the model, tree model can be reduced to the structure of two sub-branches tree, every

Y. Yu, Z. Yu, and J. Zhao (Eds.): CSEEE 2011, Part II, CCIS 159, pp. 18–23, 2011.
© Springer-Verlag Berlin Heidelberg 2011

branch has no more than two sub-branches, this is to say, the tree model has two branches, one sub-branch or no sub-branch, as shown in Figure 2.

Fig. 1. One branch model   **Fig. 2.** Tree model

To simulate the process of the trees swaying in the wind, there are five important problems in the collision detection: the collision between branches and branches, branches and leaves, leaves and leaves, leaves and ground, branches and ground. Due to the particularity and the frequency, we only discuss the collision detection between branches and branches.

When the simulative trees swaying in the weak wind, because of the weak force, the trees will not move heavily and collisions will occur in a relatively small probability, even if it occurs, the collision is too slight to be detected and has little effect on the authenticity of virtual environment. However, when the wind becomes strong, some problems happened. When the simulative trees swaying in strong wind, the movements of branches were enhanced and the collision between branches and branches occurred. Under the strong wind, penetration phenomenon would inevitable happen without the collision detection. In the natural environment, penetration phenomenon is clearly not realistic. So, it is particularly important to design an appropriate collision detection and collision response method, in order to enhance virtual reality and immersive effect.

As the branches model can be simplified into cylinders, collision detection between the branches can be considered as collision detection between cylinders.

3   Collision Detection Algorithms

3.1   Principle of the Collision Detection Algorithm

As the detected object is a model composed of special geometry, we can design corresponding special collision detection algorithm to deal with the problem. In this paper, we use the routine algorithms in computational geometry, to do the intersection test on the detected objects, and analyze the relative position and the shortest distance between the cylinders in the three-dimensional space.

In the three-dimensional space, the distance to a straight line is equal to all the fixed-length points, known as the cylindrical surface, and the straight line is called the axis. By using two planes which are perpendicular to the axis to cut cylindrical surface, the geometry composed of two sections and the cylindrical surface surrounded is called straight cylinder. The relative position relationship of two spatial cylinders Cy_A and Cy_B can be determined by the position of axes. Given the radius of two cylinders for r_A and r_B.

The relative positions of two lines in the three-dimensional space have two cases: skewed and coplanar. While two lines are coplanar, there are also three cases: intersection, parallelism and superposition.

We can assume the equations of two lines l_1 and l_2 are denoted as following:

$$l_1 : \frac{x - x_1}{X_1} = \frac{y - y_1}{Y_1} = \frac{z - z_1}{Z_1} \qquad l_2 : \frac{x - x_2}{X_2} = \frac{y - y_2}{Y_2} = \frac{z - z_2}{Z_2} \tag{1}$$

The related position of the two spatial lines:

1) If $\Delta = \begin{vmatrix} x_2 - x_1 & y_2 - y_1 & z_2 - z_1 \\ X_1 & Y_1 & Z_1 \\ X_2 & Y_2 & Z_2 \end{vmatrix} \neq 0$, the two lines are skewed.

2) If $\Delta = 0$, $X_1 : Y_1 : Z_1 \neq X_2 : Y_2 : Z_2$, the two lines are intersectional.

3) If $X_1 : Y_1 : Z_1 = X_2 : Y_2 : Z_2 \neq (x_2 - x_1) : (y_2 - y_1) : (z_2 - z_1)$, the two lines are parallel.

4) If $X_1 : Y_1 : Z_1 = X_2 : Y_2 : Z_2 = (x_2 - x_1) : (y_2 - y_1) : (z_2 - z_1)$, the two lines are coincident.

3.1.1 Skewed Case

If the two axes L_p and L_q are skewed, it can be determined whether the collision occurred by comparing the distance between the two lines and the sum of two cylinders' radius. The distance between the two lines is calculated as following:

$$D_{skewed} = \frac{\left\| \begin{matrix} x_2 - x_1 & y_2 - y_1 & z_2 - z_1 \\ X_1 & Y_1 & Z_1 \\ X_2 & Y_2 & Z_2 \end{matrix} \right\|}{\sqrt{ \begin{vmatrix} Y_1 & Z_1 \\ Y_2 & Z_2 \end{vmatrix}^2 + \begin{vmatrix} Z_1 & X_1 \\ Z_2 & X_2 \end{vmatrix}^2 + \begin{vmatrix} X_1 & Y_1 \\ X_2 & Y_2 \end{vmatrix}^2 }} \tag{2}$$

When $D_{skewed} > r_A + r_B$, the collision has not occurred; when $D_{skewed} \leq r_A + r_B$, the problem is subdivided by calculating the foot of common perpendicular between the two skewed lines. If the feet both fall within the two cylinders' axial segments, the collision occurs. If the foot just falls within the cylinder Cy_A's axial segment, calculating the minimum distance N_{min} between the centers of the cylinder Cy_B and the axis of cylinder Cy_A, comparing to the radius of cylinder Cy_A, and judging whether the collision occurs. If the feet didn't fall within the two cylinders' axial segments, the collision does not occur.

3.1.2 Parallel Case

If the two axes L_p and L_q in the case of parallelism, take any point $M_0(x_0, y_0, z_0)$ on the line l_2, the linear equation of l_1 has been given, then the distance between the two lines is computed as following:

$$D_{parallel} = \frac{\sqrt{\begin{vmatrix} y_0 - y_1 & z_0 - z_1 \\ Y_1 & Z_1 \end{vmatrix}^2 + \begin{vmatrix} z_0 - z_1 & x_0 - x_1 \\ Z_1 & X_1 \end{vmatrix}^2 + \begin{vmatrix} x_0 - x_1 & y_0 - y_1 \\ X_1 & Y_1 \end{vmatrix}^2}}{\sqrt{X_1^2 + Y_1^2 + Z_1^2}} \quad (3)$$

When the two axes are parallel, one axial segment would be projected onto another axis. If the two axial segments have some part to coincide and $D_{parallel} \le r_A + r_B$, the collision occurs; if the two axial segments coincide and $D_{parallel} \le r_A + r_B$, the collision does not occur; if the two axial segments don't coincide, the collision does not occur.

3.1.3 The Intersectional Case

If the two axes L_p and L_q in the case of intersection, the two cylinders can be projected onto the plane which located the axes. Let us simplify the collision problem between two cylinders in the three-dimensional space into two rectangles in the two-dimensional space. In this condition, we can carry out intersection test in the two-dimensional space to determine whether the collision occurred. In order to determine whether two rectangles intersected under the two-dimensional space, we can respectively determine whether the four vertexes of each rectangle are contained in another rectangle, if there is any vertex contained in another rectangle, the two rectangles are intersected.

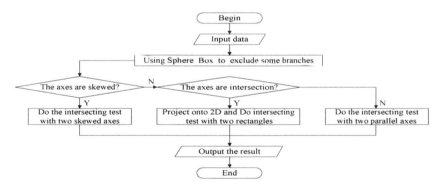

Fig. 3. Flow Chart of the Algorithm

4 Experimental Results and Analysis

This experiment is carried out with VC++6.0 and OpenGL under Duo T6670 2.20GHz CPU, 2G memory PC. The branches in different locations are detected using the algorithm described in this paper and OBB algorithm. The red line in Figure 4 to Figure 6 named X-axis, the green line named Y-axis, the blue line named Z-axisFigure 4(a), Figure 5(a) and Figure 6(a) respectively represent the initial position under different circumstances. At the beginning of the experiment, let us give the right branch a force f in the graph X-axis negative direction and make it move closer to the other branch up until the collision occurs.

When the axes are in different conditions, collisions of two algorithms are shown in Fig. 4 to Fig. 6.

a. Initial Position

b. Paper Algorithm

c. OBB Algorithm

Fig. 4. Skewed Case

a. Initial Position

b. Paper Algorithm

c. OBB Algorithm

Fig. 5. Intersectional Case

a. Initial Position

b. Paper Algorithm

c. OBB Algorithm

Fig. 6. Parallel Case

Compared with the running time in all cases, as shown in Figure 7, the efficiency of this paper's algorithm is higher than that of OBB algorithm.

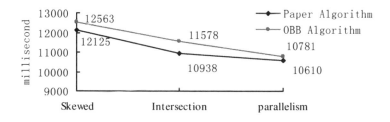

Fig. 7. The comparison of collision detection efficiency of two algorithms

5 Conclusion

In this paper we present a fast collision detection algorithm between the branches to simulate the tree swaying in high efficiency. As the branches model can be simplified into cylinders, collision problems are tested by analyzing the relative positions between cylinders in 3D. Firstly, it analyzes the relative positions of branches in different conditions. Secondly, it adopts different methods to test the intersection between branches. Experimental results show that our algorithm acquired satisfied results and improved the efficiency compared to the OBB algorithm. Our algorithm is not only suitable for the collision detection of the branches, but also suitable for the collision detection of the robot arm in the industrial production, the collision detection of the limbs in 3D character animation and so on.

Acknowledgements

This study was supported by the Scientific Research Program Funded by Shaanxi Provincial Education Commission (Program NO.09JK577).

References

1. Zou Y.-s., Ding G.-f., Xu M.-h., He Y.: Survey on Real-time Collision Detection Algorithms. Application Research of Computers (2008)
2. Zhao, S.: Research on Distance Fields and its Application in Collision Detection. JiLin University (2008).
3. Huang X.-j., Tian H.-w.: Realization of OpenGL for Rigid-body Collision Detection Based on Image Space. Computer Knowledge and Technology (2007)
4. Wang, W., Ma, J., Liu, W.: Research and Application of Collision Detection Based on Oriented Bounding Box. Computer Simulation (2009)
5. Wang, P.: Liu X.-m., Guan Y.: Improved Collision Detection Algorithm Based on OBB. Computer Engineering and Design (2009)
6. Singh, P.A., Zhao, N., Chen, S.-C., Zhang, K.: Tree Animation for a 3D Interactive Visualization System for Hurricane Impacts. In: Proceedings of the IEEE International Conference on Multimedia & Expo., pp. 598–601 (2005)
7. Hiromi, O.: Practical Experience in the Physical Animation and Destruction of Trees. In: Budapest, H. (ed.) Computer Animation and Simulation 1997: Proceedings of the Euro Graphics Workshop, pp. 149–159, 201–203 (1997)
8. Li, F.-q.: A New Modeling Strategy Of Plant Simulation. J. Normal University (Natural Science) 4, 54–59 (2003)
9. Yan, T., Wu, W.-h.: Multiple-image-based Modeling of Trees. J. System Simulation 12, 565–571 (2000)
10. Mao, W.-q., Geng, W.-d., Pan, Y.-h.: Feature-Based Synthesis Reasoning for Plant Modeling. J. Computer Aided Design & Computer Graphics 12, 595–600 (2000)
11. Feng, J.-h.: Going with Wind—Physically-based Animation. Institute of Software Chinese Academy of Sciences 1(2), 9–13 (1999)

FCM Algorithm Based on Improved Automatic Segmentation of MRI Brain Images

Zhuang Miao[1], Chengcheng Liu[2], and Xiaomei Lin[2]

[1] Department of neurosurgery, China-Japan Union Hospital, Jilin University,
Changchun 130033, China
[2] ChangChun University of Technology, No.2055 Yan'an Street,
Changchun , P.R. China
miaozhuang99@163.com, elaine2008study@sina.com,
linxiaomei@mail.ccut.edu.cn

Abstract. Traditional fuzzy C means is widely used in image segmentation. It is a classic method of fuzzy clustering analysis, but the FCM algorithm requires logarithmic or using non-parametric model of the partial field for MR data. Also we can use a continuous spatial distribution of brain tissuepriori constraint model. The computational algorithm have the problems of large amount of calculation、complex parameter estimation、segmentation of noise-sensitive and sensitive to the shortcomings of the initialization algorithm. So we put forward an improved FCM algorithm,which use the bias field parameter model and neighbor to simultaneously complete the pixel domain constraintsimage segmentation and bias field estimates.The experiment can show that the improved algorithm has a good segmentation, while robust to noise.

Keywords: Image segmentation, Parameter model, Domain constraint.

1 Introduction

Because of time-consuming and subjective factors, the artificial segmentation of brain MR images results are not good repeatability. Therefore, automatic segmentation method of brain tissue is needed to complete the Automatic segmentation of MR images.But the morphology of brain tissue structure for the complex maneuver, combined with noise、partial volume effect (PVE) and image bias field (BF) existing the division of their organization with strong pixel ambiguity and uncertainty, which makes the fuzzy clustering compared to other technologies are more widely used in brain MR image segmentation.But fuzzy C means clustering algorithm itself, there are some drawbacks, such as the need to get MR data on the number and the bias field using non-parametric model and a continuous spatial distribution of brain tissue using a priori constraints model.Therefore, adaptive fast automatic improvement of FCM segmentation algorithm is put forward based on gray value of pixel to complete MR brain segmentation that contain partial field. In this algorithm.We propose a new segmentation objective function parameter model to approximate the bias field and similar to the MRF neighborhood priori constraints to simulate the spatial consistency

Y. Yu, Z. Yu, and J. Zhao (Eds.): CSEEE 2011, Part II, CCIS 159, pp. 24–29, 2011.
© Springer-Verlag Berlin Heidelberg 2011

of brain tissue. Results of simulation and clinical brain MR image segmentation show that our algorithm is not sensitive to initial values, a strong noise suppression, effectively overcome the bias field and the segmentation results are accurate and fast.

2 Improved FCM Algorithm Based on Automatic Segmentation of MR

2.1 Improved FCM Algorithm for the Establishment of the Objective Function

The standard FCM algorithm is based on image pixel gray value of N, and the cluster centers V for each pixel.A cluster center of the membership function U, the iterative optimization of the objective function.Objective function

$$J_{FCM}(U,V) = \sum_{j \in \Omega} \sum_{k=1}^{c} u_{jk}^{q} \left\| y_j - v_k \right\|^2 \tag{1}$$

Where: Ω is the spatial location of the image pixel set, and its absolute value $|\Omega| =$ N; C is class number for the cluster and $C \geq 2$; q is weighted index and the fuzzyq> l; U=(ujk) for the fuzzy membership function matrix, which ujk of pixels j are k-class membership,and $\sum_{k=1}^{c} u_{jk} = 1$. A cluster center for the collection of C; yj j for the pixel gray value; $\| . \|$ is the norm, usually the Euclidean norm.When the image pixel gray value of the cluster approach the center of a certain category, the class membership is given high value in the gray value of a class away from the cluster center. The type of membership degrees are given low values.Through the optimization and adjustment of various types of cluster centers, FCM objective function is minimized [1] and obtain the results of fuzzy partition (each pixel value of the membership function U).Fuzzy partition of the results can be transformed into final result through the maximum degree of membership, in other words ,the image pixel value according to its maximum degree of membership is the only one divided into the appropriate class to get the final segmentation results.

In order to simultaneously adjust the bias field and restrain the impact of noise on the segmentation results, we propose the following new clustering objective function:

$$J = \sum_{j=\Omega} \sum_{k=1}^{c} u_{jk}^{q} \left\| y_j - \sum_i c_i \phi_i(y_j) v_{k} \right\|^2 + \lambda \sum_{j=\Omega} \sum_{k=1}^{c} \sum_{i \in N_j} \frac{\left(u_{jk} - u_{ik} \right)}{|Nj|} \tag{2}$$

Where: $\sum_i c_i \phi_i(y_j)$ is the bias field B based on parameter model for the expression of

MR images ; I is the number of basis functions of the model of the parameters; j is the pixel value of the function at the base;c_1 is the model parameters to be estimated, the basis function coefficients; N_j is neighborhood of pixel j; $| N_j |$ is the cardinality of the neighborhood pixel j, j pixel neighborhood that is the number of pixels; λ_1 neighborhood constraint for the weight coefficient.

2.2 Improved FCM Algorithm for the Estimated Cluster Center

In the formula (1-2), we choose the Euclidean norm.Vk on L taking the partial derivative and partial derivative is set equal to zero, that is,

$$\left[\frac{\delta L}{\delta V_K} = \sum_{j \in \Omega} u_{j^k}^2 (y_j - \sum_l c_l \phi_l(y_j) v_k)(\sum_l c_l \phi_l(y_j))\right]_{vk=vk^0} = 0 \qquad (3)$$

Solution of equation (1-3) is estimated to have vk

$$v_k^0 = \frac{\sum\limits_{j=\Omega} u_{j^k}^2 (\sum\limits_l c_l \phi_l(y_j)) y_j}{\sum\limits_{j=\Omega} u_{j^k}^2 (\sum\limits_l c_l \phi_l(y_j))^2} \qquad (4)$$

2.3 Improved FCM Algorithm for the Estimated Bias Field B

Similarly, on taking the partial derivative of q, and set the partial derivative equal to zero, that is,

$$\left[\frac{\delta L}{\delta_{C1}} = \sum_{j=\Omega}^{} \sum_{k=1}^{c} u_{j^k}^2 (y_j - \sum_l c_l \phi_1(y_j) v_k)(\phi(y_j))\right]_{ci=ci^0} = 0 \qquad (5)$$

Solution of equation (1-5) was estimated c_1.In order to express clearly, we express a matrix form, that is,

$$\begin{bmatrix} c_1 * \\ c_2 * \\ . \\ . \\ . \end{bmatrix} = (A^T M A)^{-1} A^T W Y \quad \text{of which} \quad A = \begin{bmatrix} \phi_1(Y_1)\phi_2(Y_2)\phi(Y_1)... \\ \phi1(Y_1)\phi_2(Y_2)\phi(Y_1)... \\ . \\ . \\ . \end{bmatrix}$$

$$Y = \begin{bmatrix} y_1 \\ y_2 \\ . \\ . \\ . \end{bmatrix} \quad M = diag(\sum_{k=1}^{c} u_{j_k}^2 v_k^2) \quad W = diag(\sum_{k=1}^{c} u_{j_k}^2 v_k)$$

$$\forall k$$

In equations (1-5), the number of the equation is greater than the number of estimated parameters q, so the equations (1-5) is off set.Therefore, equations (1-5) the solution is in the sense of least squares optimal solution.As we adopt the low-frequency slowly varying parameter model to approximate a continuous image of the real MR bias field B, and in each step of the recursive algorithm to obtain the corresponding MR image segmentation results, so we use the results of the sampling pixel segmentation to quickly estimate parameter bias field B c_1, so that accurate and rapid bias field is estimated to further increase the speed of the algorithm.

3 Division Result and Appraisal

3.1 Comparison of Different Segmentation Algorithms

In addition to quantitative comparison, from visual observation of Figure 1, you can also see the proposed algorithm in this paper stronger than the other algorithms noise abilities, in Figure 1.2g, most of the noise has been removed, especially in the homogeneous region.

Table 1.

	WM / ρ	GM / ρ	CSF / ρ
FCM	84.53	78.69	75.81
ASFCM	90.03	88.18	87.85
Our method	92.59	90.87	90.09

3.2 Image Comparison Algorithm Segmentation

As the traditional FCM algorithm only divided according to the current pixel is clearly present in the image noise can do nothing, the image still contains a great noise, low SNR, segmentation is not satisfactory.Through improved FCM method Considering the role of neighborhood, segmentation is relatively clear, signal to noise ratio is better, stronger noise suppression.

Comparing the results shown in Figure 1 - Figure 3.

It can be seen to improve the FCM segmented image obviously better than the standard FCM segmentation result. This improved method to overcome the impact of noise on the segmentation is robust and superiority.

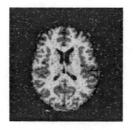

Fig. 1. the original brain image with noise

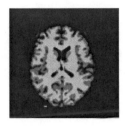

Fig. 2. Standard FCM segmented image

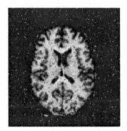

Fig. 3. Improved FCM segmented image

4   Conclusion

In this paper, construct a new fuzzy clustering segmentation objective function. Presents a wide field calibration model using parameters and neighborhood constraint model of FCM neighborhood fast MRI brain tissue segmentation method. Experimental results show that the new method of image segmentation performance is better than the standard FCM method. The new algorithm with FCM method is fast and the MRF model with the airport a higher accuracy and consistency of features. Algorithm basically meet the three-dimensional MR image fast segmentation and the requirements of high-quality segmentation. The improved algorithm also can be a great range of bias field for fast and high-quality MR images segmentation.

References

1. Bezdek, J.: A Convergence Theorem for the Fuzzy ISODATA Clustering Algorithms. IEEE Transactions on Pattern Analysis and Machine Intelligence 2(1), 1–8 (1980)
2. Bezdek, J.C.: Pattem Recogniti on With Fuzzy Objective Function Algorithms. Plenum Press, New York (1981)
3. Hung, M.C., Yang, D.L.: An Efficient Fuzzy C2 Means Clustering Algorithm. In: IEEE International Conference on Data Mining, California, USA, pp. 225–232 (2001)
4. Smith, S., Zhang, Y., Jenkinson, M., et al.: Accurate, Robust, and Automated Longitudinal and Cross-sectional Brain Change Analysis. NeuroImage 17(1), 479–489 (2002)

5. Liew, A.W.C., Yah, H.: Current Methods in the Automatic Tissue Segmentation of 3D Magnetic Resonance Brain Images. Current Medical Imaging Reviews 2, 91–103 (2006)
6. Nikhl, P.R., Bezdek, J.C.: On Cluster Validity for the Fuzzy C- means model. IEEE Transactions on Fuzzy Systems 3, 370–379 (1995)
7. Bezdek, J.: PaRem: Recognition with Fuzzy Objective Function Algorithms (M). Plenum Press, New York (1981)
8. Pham, D.L., Prince, J.L.: Adaptive Fuzzy Segmentation of Magnetic Resonance images. IEEE Transactions on Medical Imaging 8, 737–752 (1999)
9. Ahmed, M.N., Yamany, S.M., Mohamed, N., et al.: A Modified Fuzzy C-Means Algorithm for Bias Field Estimation and Segmentation of MRI data. IEEE Transactions on Medical Imaging 21, 193–199 (2002)

Grey Verhulst Neural Network Model of Development Cost for Torpedo

Qing-wei Liang[1], Min-quan Zhao[1,2], and Pu Yang[1]

[1] The College of Marine, Northwestern Polytechnical University,
Xi'an, Shaanxi, China
[2] No. 92785 Unit, People's Liberation Army, Qinhuangdao, Hebei, China
{liangqingwei,zh8015,xfnonsl}@163.com

Abstract. The cumulative curve of development cost for torpedo shows similar to shape S. Establish the grey Verhulst model of development cost for torpedo, selecting the data for several years. We can get fitted values of its historical data and the predictive value of the next data. To take full advantage of information resource contains in the raw data, we find a mapping between historical data and the predictive value using BP neural networks. Simulation shows that this method has good prediction accuracy.

Keywords: Torpedo, Development Cost, BP Neural Networks, Grey Verhulst Model.

1 Introduction

The development costs, production costs and maintenance costs of torpedo weapon system increasing in recent years. And its various costs substantial growth because of social and economic aspects. It has been a common concern how to use the limited financial resources to obtain the highest efficiency of torpedo. We study the life cycle cost (Referred to as LCC) of torpedo weapon system to solve this problem.

The LCC of torpedo means all expenses paid in the Life cycle, including demonstration costs, development costs, production costs, operation and maintenance costs. The development Cost is an important part of LCC. Although it only accounts for 18% of total costs, it determined the 95% of the total. Substantial increase in consumption funds with the progress of life cycle, however, the impact capacity of LCC is getting less and less. This shows that studying development cost has great significance[1] for reducing LCC. Literature [2-4] analyzed the development cost from different aspects. We found, the cumulative cost curve of torpedo is showing a growth trend, increased slowly initially, and then rapid growth, tends to limit slowly in the final. Verhulst model is mainly used to describe the "S" type of process trends, and it has higher accuracy and reliability. So we create a development cost model for torpedo using grey Verhulst model theory.

Grey model has several advantages, such as need not take into account relevant factors, reduce the randomness of time series, and get a wealth of information from small samples sequence. But it has several drawbacks too, such as prediction accuracy

Y. Yu, Z. Yu, and J. Zhao (Eds.): CSEEE 2011, Part II, CCIS 159, pp. 30–35, 2011.
© Springer-Verlag Berlin Heidelberg 2011

is low, and error accuracy unable be controlled. Because of its powerful processing capabilities of nonlinear problems, neural network has a strong advantage in data fitting and function approximation. In the paper, we take full advantage of the complementarity of the two theories, mining information of data as much as possible.

2 Grey Verhulst Prediction Model

2.1 Modeling

The raw data sequence is known:

$$X^{(0)} = \left(x^{(0)}(1), x^{(0)}(2), \cdots, x^{(0)}(n) \right) . \tag{1}$$

After 1-AGO to the data sequence (1), get the generate sequence:

$$X^{(1)} = \left(x^{(1)}(1), x^{(1)}(2), \cdots, x^{(1)}(n) \right) . \tag{2}$$

Where, $x^{(1)}(k) = \sum_{i=1}^{k} x^{(0)}(i), \quad (k = 1, 2, \cdots, n) .$

From the data sequence $X^{(1)}$, construct background value sequence:

$$Z^{(1)} = \left(z^{(1)}(1), z^{(1)}(2), \cdots, z^{(1)}(n) \right) .$$

Where, $z^{(1)}(k) = 0.5 \left(x^{(1)}(k) + x^{(1)}(k-1) \right), \quad (k = 1, 2, \cdots, n) .$

Then, the grey Verhulst model is

$$X^{(0)} + aZ^{(1)} = b \left(Z^{(1)} \right)^2 .$$

In this formula, a, b are model parameters, a is called the development coefficient, its size reflects the growth rate of the sequence $X^{(0)}$, b is the role of vector. LSE parameters vector of grey differential equation meet:

$$\hat{a} = (a \quad b)^T = \left(B^T B \right)^{-1} B^T Y .$$

$$B = \begin{bmatrix} -z^{(1)}(2) & -z^{(1)}(3) & \cdots & -z^{(1)}(n) \\ \left(z^{(1)}(2) \right)^2 & \left(z^{(1)}(3) \right)^2 & \cdots & \left(z^{(1)}(n) \right)^2 \end{bmatrix}^T , Y = \left[x^{(0)}(2) \quad x^{(0)}(3) \quad \cdots \quad x^{(0)}(n) \right]^T .$$

The time response sequence of grey Verhulst model is:

$$\hat{x}^{(1)}(k+1) = \frac{a x^{(1)}(0)}{b x^{(1)}(0) + \left(a - b x^{(1)}(0) \right) e^{ak}} .$$

1-IAGO to the data sequence, get grey Verhulst model to the sequence $x^{(0)}(i)$:

$$\hat{x}^{(0)}(k+1) = \hat{x}^{(1)}(k+1) - \hat{x}^{(1)}(k), \ k = 0,1,\cdots .$$

Then get grey Verhulst model sequence:

$$\hat{X}^{(0)} = \left(\hat{x}^{(0)}(1), \hat{x}^{(0)}(2), \cdots, \hat{x}^{(0)}(n), \hat{x}^{(0)}(n+1)\right) . \tag{3}$$

2.2 Accuracy Test

Usually the testing method is posterior difference method[5]. Accuracy is determined by MSE ratio and small error probability. The basic method is as follows:

Suppose $x^{(0)}$ is the raw data sequence, $\hat{x}^{(0)}$ is simulation sequence, $\varepsilon^{(0)}$ is residual sequence, then

$$S_1^2 = \frac{1}{n}\sum_{k=1}^{n}\left(x^{(0)}(k) - \overline{x}\right)^2 , \quad S_2^2 = \frac{1}{n}\sum_{k=1}^{n}\left(\varepsilon^{(0)}(k) - \overline{\varepsilon}\right)^2 .$$

In the formula: $\overline{x}$ is the mean of $x^{(0)}$; $\overline{\varepsilon}$ is the mean of residual $\varepsilon^{(0)}(k)$。

According to the formula $C = S_2 / S_1$ and $p = P\{|e(i) - \overline{e}| \prec 0.6745 S_1\}$, we can get the ratio C of posterior difference and small error probability p.

3 BP Neural Networks

BP neural network[6][7] is a directed graph posed by many neurons, its network structure imitate biological nervous system, it has the ability of self-learning, adaptive and nonlinear processing. Network implicit the knowledge get through self-learning into the network structure, the way of dealing with complex nonlinear system has the essential differences with traditional methods.

As a typical learning algorithm of artificial neural networks, BP neural network is multi-layer network using differentiable nonlinear function with weight training. It contains the essential part of the neural networks, and its simple structure, plasticity. BP network is a kind of neural network which has 3-layer or more than 3 layer, including input layer, hidden layer and output layer, All connections between the upper and lower layer, and no contact between each layer neurons. In the actual training, provide a set of training samples, and each training sample consisted of input and desired output. In the network learning process, Neuron activation values spread from the input layer to output layer after the middle layers. Neural networks obtain input response in the output layer. Its output compare with the expected output, return along the original route if error can not meet the requirements. And in order to reduce the error gradually, adjust the connection weights according to certain principles. Repeated study until the error precision meet the requirements. Then we can predict using the trained model.

To avoid falling into local optimal solution, we train BP network using LM algorithm in the paper. This network consists of input layer, hidden layer and output layer. In the training model, activation function uses the tangent function between input layer and hidden layer and uses linear function between hidden layer and output layer. The range of Sigmoid function is [0, 1], we normalized the input sample to [0.1, 0.9] for improving the speed of network convergence.

4   Combination Model of Grey Verhulst Neural Network

Although grey Verhulst model has the superiority of less information and simple, BP neural network has the Characteristic of effective simulation. To take full advantage of the information contained in the original data, we establish combination model of grey Verhulst neural network. Specific practices:

First, establish grey Verhulst model according to original data series (1), and get forecast data series (3).

Literature [8][9] take into account the use of neural network model to compensate for the grey model, in order to amend the error. Comparison of sequence (3) and (1), we get the residual sequence: $\varepsilon^{(0)}(i) = x^{(0)}(i) - \hat{x}^{(0)}(i)$, $i = 1, 2, \cdots, n$. Since residuals may be positive, or negative, therefore, positive residuals first: Plus the absolute value of the minimum negative and plus 1 too for each data. Use $e^{(0)}(i), i = 1, 2, \cdots, n$ to distinguish. Regrouping take advantage of the raw data column (1) and residual sequence, to s consecutive years of practical value as input, and the next year's residual of grey model as output, then can be divided into k groups ($s + k - 1 = n$)。

Access matrix:

$$
\begin{pmatrix}
x^{(0)}(1) & x^{(0)}(2) & \cdots & x^{(0)}(s) & e^{(0)}(s+1) \\
x^{(0)}(2) & x^{(0)}(3) & \cdots & x^{(0)}(s+1) & e^{(0)}(s+2) \\
\vdots & \vdots & \ddots & \vdots & \vdots \\
x^{(0)}(k-1) & x^{(0)}(k) & \cdots & x^{(0)}(n-1) & e^{(0)}(n) \\
x^{(0)}(k) & x^{(0)}(k+1) & \cdots & x^{(0)}(n) & \overline{e}^{(0)}(n+1)
\end{pmatrix}. \tag{4}
$$

Where $\overline{e}^{(0)}(n+1)$ is the unknown value.

Establish neural network model using the matrix mentioned above, the input layer is s, and output layer is 1. For network training in the use of top $k-1$ group of data, and get a series of weights and threshold corresponding to each node. After completing the training, simulation prediction use k sets of data, then get predictive value of residual $\overline{e}^{(0)}(n+1)$ at time n+1.

Predictive value of residual compensation for the predictive value to grey Verhulst model. Ultimately, combination prediction is $\hat{x}^{(0)}(n+1) + \overline{e}^{(0)}(n+1)$.

5   Simulation

The development costs over the years to a model of torpedo as shown.

Table 1. Development Cost Unit:10 000 RMB¥, (first fiscal year)

Year	1	2	3	4	5	6	7	8	9	10
Development Cost	496	779	1187	1025	488	255	157	110	87	79

Assume that the development costs of 10th year is unknown. In order to predict the development costs of 10th year, we establish model using the first 9 years of data, and compared with the actual costs.

The sequence can be modeled by grey Verhulst model, and get time response function:

$$\hat{x}^{(1)}(k+1) = \frac{501.5056}{0.0992 + 0.9119e^{-1.0111k}} .$$

From the function, the predicted value sequence of grey Verhulst model is:

$$\hat{X}^{(1)} = (496, 1163.7, 2280.6, 3504.2, 4354.2, 4775.7, 4950, 5016.6, 5041.3, 5050.3)$$

$$\hat{X}^{(0)} = (496, 667.68, 1116.9, 1223.7, 850, 421.44, 174.31, 66.62, 24.68, 9.04) \qquad (5)$$

From equation (1) and (5), we get grey Verhulst residual sequence:

$$E^{(0)} = (0, 111.32, 70.1, -198.7, -362, -166.44, -17.31, 43.38, 62.32)$$

After treatment, access to positive

$$E^{(0)} = (363, 474.32, 433.1, 164.3, 1, 196.56, 345.69, 406.38, 425.32) \qquad (6)$$

Substituted into the accuracy formula, get:

$$S_1 = 383.72 ; \quad S_2 = 145.72 ; \quad C = 0.38 ; \quad P = 0.89 。$$

According to model prediction accuracy table[10], the accuracy of model is 2.

In order to obtain more accurate model, improve model using BP neural network.

Make use of equation (4), grouped the original data equation (1) and (6), and s is 3, then it can be divided into 7 sets of data(k is 7).

Through multiple authentication, we can use BP neural network of 3×3×1, use tangent Sigmoid function between input layer and hidden layer, and use linear function between hidden layer and output layer. 6 sets of data used as a sample for training network, set the maximum learning times is 1 000, learning rate is 0.01, the error sum of squares as learning objectives is 0.0001, input values are normalized to [0.1,0.9], Set the initial value of network connection weights in [-1,1].

Simulation calculation through Matlab, network convergence in 12 Epochs or so, and meet the expected distortion. Through section 7 sets of data, we can simulate that the residual of 11th year is 53.22. Then get its combined predictive value is 77.90.

Table 2. Comparison Table of 10th Actual and its Predicted Value

Year	Actual value	Grey Verhulst Model		Combination Model	
		Predicted Value	Relative Error(%)	Predicted Value	Relative Error(%)
10	79	24.68	68.76	77.90	1.39

From the table 2, it can be drawn that the accuracy of grey Verhulst neural network model is better than grey Verhulst model. Re-modeling by all raw data of table 1, then the predicted value of 11th year is 67.93.

6 Conclusion

Torpedo development cycle can not be too long, which determines that the sample points of annual development cost is less. Grey theory suitable for prediction of small sample, but its accuracy is lower. In the paper, we take full advantage of the complementarity of the two theories, mining information value of data as much as possible. Grey Verhulst neural network successfully predicted the development cost of torpedo next year. Experimental results show that the model has higher accuracy.

References

1. Chen, X.: Equipment Systems. National Defence Industry Press, Beijing (1995)
2. Jin, F., Cui, H., Zhang, Y., Li, Q.: Forecast and Emulation Research on the Cost of Development for Military Equipment Transactions of Beijing Institute of technology (2006)
3. Liang, Y., Li, B., Li, Y., Li, Z.: Forecast Development Cost Based on Improved BP Neural Network. Tactical Missile Technology (2006)
4. Li, D., Zhang, H.: Development Cost Estimation of Aircraft Frame Based on BP Neural Networks. Fire Control and Command Control (2006)
5. Deng, J.: Grey Predict and Grey Decision-making. Huazhong University of Science and Technology Publishing House, Wuhan (2002)
6. Wei, H.: Neural Network Structure Design Theory and Method. National Defense Industry Press, Beijing (2005)
7. Moravej, Z., Vishwakarma, D.N., Singh, S.P.: ANN Based Protection for Power Transformers. Electr.Mach, Power Sits (2000)
8. Liu, Y., Cao, J., Dai, Y.: Network Traffic Prediction Based on Compensated Grey Neural Network Model. Computer Applications (2007)
9. Zhang, X., Li, S.: Dynamic Data Sequence Forecasting Base on Combined Model of grey Neural Network. Electronic Measurement Technology (2009)
10. Zhao, Y.: Missile Weapons and Equipment of the Best Service in Life. College of the Air Force Missile (1998)

A New Automatic On-Situ Sampling and Monitoring Method for Water

Jingtao Liu, Jichao Sun, Jincui Wang, Xiaoping Xiang,
Guanxing Huang, Yuxi Zhang, and Xi Chen

The Institute of Hydrogeology and Environmental Geology, CAGS, China
aljingtao@foxmail.com, bgwwsun@263.net.cn, csinger18@sohu.com,
{d1053234108,fkobzhang}@qq.com, ehuangguanxing2004@126.com,
gchen501406@yahoo.com.cn

Abstract. In view of the fact that some difficulties exist in automatic and depth-set sampling and monitoring for water, a new sampling system is developed. Firstly, the future sampler design ingredients are analyzed. Then the sampling system structure and operating principles are introduced in detail. A model machine is designed and several experiments have been carried on. The results indicate that this sampling system can perform automatic sampling effectively and efficiently; it could sample water as deep as 30m, and the depth error is less than 0.5m; water parameters such as temperature, pH and electrical conductivity (EC) can be measured on-site and the accuracy is $0.1°C$, 0.01 and $1\mu S/cm$ respectively. Besides, the sampler can reduce volatile organic components loss effectively and is also suitable to collect water sample for inorganic component analysis.

Keywords: Water sampler, Automatic sampling, Depth-set sampling, On-situ monitoring, Experiment.

1 Introduction

There is an increasing requirement of water pollution investigations for more developed water sampling and on-situ monitoring techniques. Some advanced sampling techniques such as peristaltic pump, bailer sampler and direct pushing-forward on-site sampler are often implemented in water sampling. However, cross-contaminations may happen when using these samplers. Few water samplers have been reported to be both automatic and depth-set. Considering the ever developed automatic samplers, either the huge size or high cost limits their utilizations. In addition, function of the samplers is limited and water parameters cannot be measured on-site simultaneously. [1][2]. The on-situ monitoring techniques are developed to detect important physical and chemical indicators of water body, such as temperature, pH, electrical conductivity (EC) and so on, which are usually sensitive to environment[3].

Based on the investigation of sampling techniques, several ingredients that should be taken into account during future sampler development are analyzed at length in the following chapter. Then the design of an automatic on-situ sampling system is

Y. Yu, Z. Yu, and J. Zhao (Eds.): CSEEE 2011, Part II, CCIS 159, pp. 36–41, 2011.

proposed in detail. Several experiments have been carried on to test the sampling performance. Results show that this sampling system could work well and turn out to be a practical resolution for water sampling and monitoring.

2 Design Ingredients and Objectives

2.1 Design Ingredients

Through analysis, during design of future water samplers, ingredients that must be focused on are as follows:

1) Reduction of influences on sample components in sampling process.
2) Collection of the accurate target sample [4].
3) Integration of water sampling and on-situ monitoring techniques.
4) Automation of sampling and monitoring process.

2.2 Design Indicators

With development of microelectronic techniques and sensor technology, low cost, low energy consumption and high accuracy microprocessor provides better strategies to solve problems of automatic on-situ water sampling and monitoring. Our purpose is to design an automatic on-site sampler with microprocessor and sensors.

Depth of shallow water is usually less than 60m and the typical representative shallow water is well water. This sampling system is designed for static shallow water and design indicators are shown in Table 1.

Table 1. Design indicators of the sampling system

Depth/m	Sample volume/ml	Parameter measurement
0~30	200	Temperature, pH, Conductivity

3 Structure and Working Principles

Structure and working principles of the sampling system will be introduced in detail in this chapter.

3.1 Structure

This sampling system consists of the above-water control part which we call controller and the underwater sampling part which we call on-situ sampler (OS sampler). Both the controller and sampler are composed of mechanical and electric parts. The mechanical parts are shown in Fig. 1. Framework part in controller is used as a support of the sampler. It can be placed on the edge of the well with a small diameter.

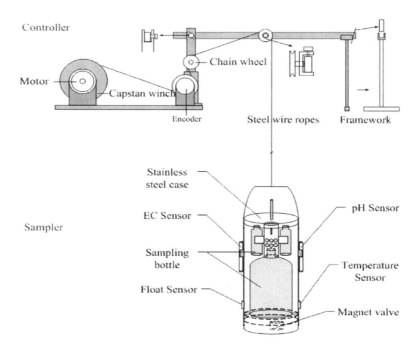

Fig. 1. Components of the sampling system

3.2 Principles

1) Controller principles: The controller will pass down the sampler to the set depth and pull it up when sampling has been completed.

According to parameters set by users, the controller sends drive command to motor. When the motor starts, it will rotate the capstan winch. With the gravity of sampler, steel wire will rotate the wheel connected to encoder. By capturing pulses feedback to MCU, motor rotating speed can be controlled. When the sampler get down to the water surface, a signal will be generated by float sensor and is transferred to controller MCU. Then sampler position begins to be calculated. Depth-set sampling will be realized.

2) Sampler principles: The sampler will sample water and guarantee the sample quality, and measure three water parameters such as temperature, pH and EC. Automatic sampling will be realized.

Real-time measurement function will be realized with sensors. Sensors are right beneath the sampling steel case. After sampler getting down to the set point, sensor detectors can execute the measurement. Water parameter signals are changed into electronic parameters and this will be finished by measurement circuit. These parameters will finally be stored in the sampler and displayed on LCD.

Automatic sampling will be carried on with the sampler. According to the set demands, magnet valve will be opened and closed automatically. When the valve opens, air in the sampling bottle will discharge through the valve port on top of the sampler and water will flow into the bottle through magnet valve. When the bottle is full of water sample, the magnet valve will be closed. Then sampling is completed.

For the sampler consumes little energy, rechargeable battery chosen as power supply. Power state is monitored by MCU and battery charging interface is left outside the sampler and water-proof measure has been taken into account.

3.3 Sampling Program

The sampling system can be operated either automatically or manually. When the condition is not appropriate for automatic sampling or the power is not enough, we can sample manually.

If water-sampling is to be conducted automatically, the working mode is chosen and then input the desired parameters such as water depth and measure times.

4 Experiments and Discussion

4.1 Design and Expeiments

One automatic sampling system is developed. Hardware of both the controller and sampler have already been designed. Several experiments have been taken.

1) Depth-set test

When the steel wire is driven moving by the motor, it rotates the wheel which is connected to photoelectric encoder. Diameter of the wheel is d=40mm. With every circle of the wheel, encoder emits n=200 pulses which are captured by MCU. Then depth (D) of sampler in the water is calculated by MCU, as in (1), where m represents the total number of pulses captured by MCU.

Set a certain depth to MCU and then MCU will drive motor to start and later to stop. By measuring the real length of the steel wire that has been passed down by motor, the set depth can be compared with the real length. Then the depth-set error can be achieved. With steel wire of 100m, it has been finally tested that the maximum error is limited within 0.5m, which is allowed in the depth-set water sampling. By analysis, we reckon that depth-set error mainly results from missing pluses captured by MCU and flexibility characteristics of steel wire.

$$D = \frac{\pi d}{2} \cdot \frac{m}{n} \tag{1}$$

2) Automatic sampling

When the sampler is down to water surface, a signal is sent by water level sensor to MCU of sampler. Then the signal is transmitted to MCU of controller. When sampler gets down to the set depth, it stops. One minute will elapse before the sampler MCU sends a signal, and then magnet valve opens. After a 2-minute delay, the magnet valve closes. During the 2-minute delay, on-site sampling has been already completed. The experiment proved that automatic sampling could work effectively and efficiently.

3) Parameter measurement

A thermal resistance is chosen as the temperature sensor. The bridge circuit is designed to achieve the accurate temperature. Measured temperature first turns out to be a voltage. This voltage is sampled by the 16 bit A/D converter and stored and transmitted. After calculated by MCU, the voltage finally changed to be a temperature and is displayed on LCD. The error is controlled within 0.1°C and it comes from sampling accuracy of A/D converter, noises of circuit and liner characteristics in thermal resistance itself.

PH and EC measurement circuits are also designed [5][6]. pH detectors and EC detectors are used. Signals detected will be changed in to voltage and then sampled by the 16 bit A/D converter. High accuracy of both the two parameters has been realized. pH accuracy is 0.01 and that of EC is 1μS/cm.

4) Componet experiments

Experiments are taken to test the performance of the sampler. It could sample water on-situ, which we call OS sampler.

Three water samples were collected parallel at the same time with three different methods and kept at 4°C. The sampling depth was 50cm below water surface. These samples were sent to Hong Kong ALS laboratory for organic components testing and Guangdong Province Material Testing Center for inorganic components test with 38 items. There were 57 items for volatile organic compounds test and 83 items for semi-volatiles organic compounds.

Results showed that the sampler could reduce volatile organic components loss effectively and was also suitable to collect water sample for inorganic component analysis.

4.2 Characteistics of the Sampling System

1) The system can sample water automatically.

2) Accurate depth-set and on-site sampling is realized and depth error is controlled within 0.5m. On-site sampling reduces the possibility of cross pollution of the water sample.

3) Parameters of the set point can be measured

4) Controller can be used with other water sampler or other similar substances' sampler.

5) Sampling way can be shifted between manual and automatic.

6) It is of small cubage and easy to operate.

7) The sampler can reduce volatile organic components loss effectively.

8) This sampler is also suitable to collect water sample for inorganic component analysis.

5   Conclusions and Future Work

The automatic on-situ sampling and monitoring system developed can effectively solve the automatic sampling problem of water. It works efficiently and helps to reduce labor. The water parameters measured on-site will provide much more accuracy information of water. However, in the flowing water, displacement would happen. Accordingly, the sampling system may not execute accurate depth-set sampling. We will solve this problem in the near future.

Acknowledgment

Financial support for this study, provided by Basic Scientific Research Professional Charge of Institute of Hydrogeology and Environmental Geology, Chinese Academy of Geological Sciences.

References

1. Roman, C., Camilli, R.: Design of a Gas Tight Water Sampler for AUV Operations. In: Oceans, Vancouver, pp. 1–6 (2007)
2. Conte, G., Scaradozzi, D., Vitaioli, G., et al.: Monitoring Groundwater Characteristics by Means of a Multi-parametric Probe and Sampling Device. In: 2009 IEEE Workshop on Environmental, Energy, and Structural Monitoring Systems, pp. 50–57. EESMS Press, Crema (2009)
3. Liu, J.T., Sun, J.C., Wang, J.C., Huang, G.X., Xiang, X.P., Jing, J.H., et al.: Development of the Shallow Groundwater Depth-setting Sampler. The Administration and Technique of Environmental Monitoring 20(3), 56–58 (2008) (in Chinese)
4. Jacobs, P.H.: A New Rechargeable Dialysis Pore Water Sampler for Monitoring Sub-aqueous In-situ Sediment Caps. Water Research 36(6), 3121–3129 (2002)
5. Zhao, Y.B., Zhang, J.M., Li, H.J.: The Design of an Intelligent PH Transducer in Industry. Experimental Technology and Management 23(4), 45–47 (2006)
6. Lin, B., Zhang, X.W., Jia, K.J., Liu, X.Y., Sun, Y.C.: Design of Control System for Water Conductance Instrument Based on Bi-directional Pulsed Voltage Technique. Chinese Journal of Electronic Devices 30(3), 921–925 (2007) (in Chinese)

Estimation of Walking Speed Using Accelerometer and Artificial Neural Networks

Zhenyu He[1,*] and Wei Zhang[2]

[1] Computer Center, Jinan University, Guangzhou, 510632, China
[2] Department of Electronic Engineering, Jinan University,
Guangzhou, 510632, China
hzy0753@126.com, tzhangw@jnu.edu.cn

Abstract. This paper proposes a practical method for estimating human walking speed using accelerometer data. A portable device based on accelerometer was developed to objectively record human walking signals. An Artificial Neural Network was developed to estimate the average speed of walking. We extracted six parameters, namely step number, subject's height, root mean square and difference between the maximum and minimum of vertical and frontal accelerometer signals as inputs for artificial neural network. To validate the performance of the proposed method, we tested accelerometer data collected from 35 subjects walking under free-living conditions. The experiments shows that the average accuracy of speed estimate is 96.96% which better than previous works. Based on the results of these experiments, is concluded that a tri-accelerometer is a promising tool for the accurate assessment of walking speed.

Keywords: Tri-axial Accelerometer Data, Walking Speed Estimation, Artificial Neural Networks.

1 Introduction

Step and speed estimation is an important issue in areas such as gait analysis, sport training or pedestrian localization. In recent years, accelerometry has become a promising technique for detecting the movement of the body and estimating the step, speed and energy expenditure because of some advantages such as small size, relatively low cost and measuring with minimal discomfort to the subjects [1-8]. For example, Aminian K. et al [1] developed an ambulatory monitoring system based on accelerometry for clinical applications. Significant parameters of body motion, namely, movement coordination, temporal parameters, speed, incline, covered distance, kind of physical activity and its duration, are extracted from the subject accelerations in its daily environment. By Moe-Nilssen R. [4], the study on a portable system to obtain cadence, step length, and measures of gait regularity and symmetry by trunk accelerometry was reported. Aminian K., et al. [3] present a neural network

---

* Corresponding author.

Y. Yu, Z. Yu, and J. Zhao (Eds.): CSEEE 2011, Part II, CCIS 159, pp. 42–47, 2011.
© Springer-Verlag Berlin Heidelberg 2011

based method to assess speed and incline from accelerometer signals. Kim J. W. et al. [5] proposed the step, stride and heading determination methods based on accelerometer, gyroscope and magnetic compass for the pedestrian navigation system.

Although in the literature there are already exist some approaches of using acceleration signals for walking step and speed estimation, little works have been done to validate the idea under real-world circumstances. Most results use data collected from a treadmill walking under laboratory conditions and very few subjects (often one and several subjects) [2, 3, 5]. The gait laboratory approach is accurate, however, it is costly, inefficient, and does not represent free-living activity. In general, existing methods for assessing human movement in free living have significant limitations [6] due to the complexities of human motion and the large variance of different subject.

Many parameters such as value of the vertical accelerations data, step length, forward velocity change in each step, and stride time are correlated to walking speed [1-8]. But these correlations can be varied according to subjects and the range of speed. It has been demonstrated that the use of artificial neural network techniques for decision making in certain gait analysis applications is more effective than biomechanical methods or conventional statistics [6]. In this paper, we design feedforward back-propagation neural networks to predict walking speed. We extract six parameters, namely step number, subject's height, root mean square (RMS) and difference between the maximum and minimum of vertical and frontal accelerometer signals as features for ANN. To validate the performance of the proposed methods, we test accelerometer data collected from 35 subjects walking under free-living conditions. The experiments show the very promising results.

The remainder of this paper is organized as follows. In section 2, we describe the sensor platform and data collection. Section 3 explains the detailed information about the walking speed estimation method and experiment results are given in section 4. Finally, conclusions are given in section 5.

2 Hardware Design and Data Collection

Figure 1 (a) shows the hardware architecture which consists of: the main CPU, sensors, power control, and wireless communication, LED and Keyboard. The main CPU is a ADuC7026 ARM7TDMI microcontroller. For the sensor we use a tri-axial accelerometer ADXL330 manufactured by Analog Devices, which is capable of sensing accelerations from −3.0g to +3.0g. The output signal of the accelerometer is sampled at 100 Hz. The data generated by the accelerometer is transmitted to a PDA wirelessly over Bluetooth. Figure 2(b) shows the acquisition platform and diagram of data collection is shown in Fig. 2 (a), a total of 35 subjects (28 males and 7 females) put the data collection apparatus in their clothes pocket and walked 50 meters long outdoor in free-living conditions. In order to test the accuracy of the speed estimation, we recorded walking time and walking step number of every subject. Thus we can obtain actual average speed of walking through calculating the waking distance divided by walking time. Fig. 2 (b) shows examples of raw accelerometer data.

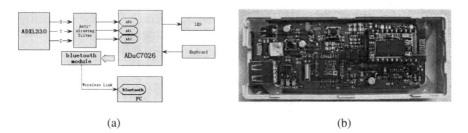

(a)   (b)

Fig. 1. (a) Hardware architecture and (b) acquisition platform

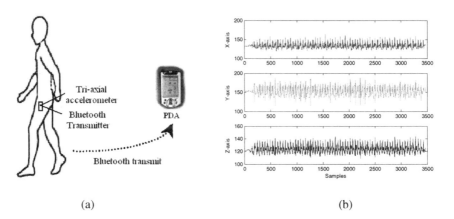

(a)   (b)

Fig. 2. (a) Diagram of data collection and (b) examples of raw accelerometer data

3 Speed Estimation Method

3.1 Parameters Extraction

As we known, walking speed are correlated to many parameters such as value of the vertical accelerations, step length, forward velocity change in each step, subject and stride time, subject's length of leg and weight etc. By Weinberg H. [7] walking distance was empirically determined using vertical movement. He established an approximation equation using a range of accelerometer signals measured in the vertical axis in a single stride. And Schutz Y. et al. [8] showed that the root mean square (RMS) of the vertical acceleration is strongly correlated with speed.

As this study test the subject a longer distance (50 m) under free-living condition outside the laboratory, it was difficult to measure the accurate instantaneous speed, thus we estimate the average speed during a period of time. In this work the following parameterization based both on physiological and statistical aspects of body accelerations is extracted. Firstly, the step number of every subject is used. Secondly, the RMS of vertical and frontal acceleration is computed. Thirdly, the difference of the maximum and the minimum of frontal accelerometer signals and vertical signals are extracted. Finally, each subject's height is used as subject-related factor. Therefore, six parameters from each subject are extracted to estimate walking speed.

3.2 Artificial Neural Network Design

Artificial neural network (ANN) [6] have been shown to be successful in finding complex relationships between patterns of different signals. In order to estimate the average speed, a feedforward back-propagation neural networks is designed here. The structure of the artificial neural network include three layers: one hidden layers of 13 units, one output layer with one node and 6 input units. Figure 3 show the structure of the ANN. Where N , H , R_y , R_z , D_y , D_z are respectably the step number, subject's height, RMS of vertical and frontal data, different between max and min of vertical and frontal data. The input data are transferred through one hidden layers to the output layer. Logistic sigmoid transfer function $f(x) = 1/(1+\exp(-x))$ is adopted in the hidden layers and linear transfer function $f(x) = x$ is applied to the output layer.

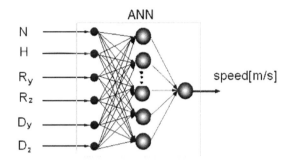

Fig. 3. The structure of the artificial neural network

The input to each node is the sum of the scalar products of the incoming vector components of all nodes of the previous layer with their respective weights. Thus, the input to a node h in layer j is given by

$$x_{j,h} = \sum_{m=1}^{k} W_{m,i} y_{m,i} \,. \tag{1}$$

Where $j = i+1$, $W_{m,i}$ is the weight connecting node m in layer i to node h in layer j , and $y_{m,i}$, is the output from node m in layer i .

Subsequently, this sum value is transformed by either the sigmoidal function or the linear function to generate the output of node h in layer j :

$$y_{j,h} = f(x_j, h) \,. \tag{2}$$

Which is then sented to all nodes in the following layer. This continues through all the layers of the network until the output layer is reached and the output vector is computed.

The mean square error (MSE) is minimized during the training process. The MSE equals the mean of the sum of the squares of the deviations from the target values, that is,

$$MSE = \frac{1}{n}\sum_{i=1}^{n}(O_l - T_i)^2 \; . \tag{3}$$

Where $O_l = l^{th}$ value of a group of n values of outputs, $T_l = l^{th}$ value of a group of n target values for the variable of interest [6].

4   Experiment Results

In the experiments, we validate the performance of our proposed speed estimation algorithm. We train the neural networks firstly by the 25 subject's data, and then test other 10 untrained subjects' data. Before training the networks, the input signals are normalized to values between 0 and 1. This range corresponds to the output ranges of the sigmoidal logistic transfer function that is used in the hidden layers of the network. The feedforward back-propagation algorithm is used to train the network. The relative error and average accuracy are defined as:

$$Relative\;err(\%) = \frac{|estimated - actual|}{actual} \times 100 \; . \tag{4}$$

$$Average\;accuracy(\%) = 100 - Average(Relative\;err(\%)) \tag{5}$$

Table 1. Estimated speed versus actual speed

Test NO.	Actual speed(m/s)	Estimated speed(m/s)	Relative Error (%)
1	1.58	1.56	1.27
2	1.58	1.61	1.90
3	1.31	1.40	6.87
4	1.30	1.32	1.54
5	1.60	1.64	2.50
6	1.67	1.58	5.39
7	1.40	1.38	1.43
8	1.54	1.46	5.19
9	1.64	1.57	4.27
10	1.42	1.42	0.00

The test results are shown in Table 1. A good agreement between actual and estimated value has been found. The average accuracy of speed estimate is 96.96%. The result we got is better than previous works. For example, Aminian K. [3] reports that the maximum of walking speed-predicted error was 16%. Song Y. [2] reports the speed relative error of running in outdoor is 9.57% and [6] report the speed error are less than 6%. Thus it can be shows that the proposed algorithm can be used to effectively estimate walking speed

5 Conclusions

This paper proposes a novel method to estimate walking speed using a single tri-axis accelerometer. A practical and reliable wearable platform based on a Tri-axial accelerometer was designed. In order to estimate average speed of walking, an artificial neural network (ANN) was developed. We extracted six parameters, namely step number, subject's height, RMS of vertical (Y-axis) and frontal (Z-axis) acceleration data, difference between of the maximum and the minimum of Y-axis and Z-axis data as ANN's input. Experiment based on acceleration data collected from 35 subjects walking under free-living conditions was conducted. The experiments shows that the average accuracy of speed estimate is 96.96% respectively which better than previous works. Therefore the tri-accelerometer can be provides a new approach for speed estimation.

Acknowledgement. This paper is partially supported by the following research fundings: Science and Technology Plan Foundation of Guangdong (No. 2010B050900016), the Fundamental Research Funds for the Central Universities (No. 21609606).

References

1. Aminian, K., De Andres, E., Rezakhanlou, K., Fritsch, C., Schutz, Y., Depairon, M., Leyvraz, P.-F., Robert, P.: Motion analysis in clinical practice using ambulatory accelerometry. In: Magnenat-Thalmann, N., Thalmann, D. (eds.) CAPTECH 1998. LNCS (LNAI), vol. 1537, pp. 1–11. Springer, Heidelberg (1998)
2. Song, Y., Shin, S., Kim, S., et al.: Speed Estimation From a Tri-axial Accelerometer Using Neural Networks. In: Proceedings of the 29th Annual International Conference of the IEEE EMBS, pp. 3224–3227. IEEE Press, France (2007)
3. Aminian, K., Robert, P., Jequier, E., Schutz, Y.: Estimation of Speed and Incline of Walking Using Neural Network. IEEE Trans. On Instrumentation and Measurement 44, 743–746 (1995)
4. Moe-Nilssen, R., Helbostad, J.L.: Estimation of Gait Cycle Characteristics by Trunk Accelerometry. Journal of Biomechanics 37, 121–126 (2004)
5. Kim, J.W., Jang, H.J., Hwang, D.-H., Park, C.: A Step, Stride and Heading Determination for the Pedestrian Navigation System. Journal of Global Positioning Systems 3, 273–279 (2004)
6. Zhanga, K., Sunc, M., Kevin Leste, D., et al.: Assessment of Human Locomotion by Using an Insole Measurement System and Artificial Neural Networks. Journal of Biomechanics 38, 2276–2287 (2005)
7. Weinberg, H.: Using the ADXL202 in Pedometer and Personal Navigation Applications. In: Application Note AN-602, Analog Devices, Inc. (2002)
8. Schutz, Y., Weinsier, S., Terrier, P., Durrer, D.: A New Accelerometric Method to Assess the Daily Walking Practice. International Journal of Obesity 26, 111–118 (2002)

Group Search Optimizer Algorithm for Constrained Optimization

Hai Shen[1,2,3], Yunlong Zhu[1], Wenping Zou[1,2], and Zhu Zhu[1,2]

[1] Key Laboratory of Industrial Informatics, Shenyang Institute of Automation,
Chinese Academy of Sciences, Shenyang 110016, China
[2] Graduate School of the Chinese Academy of Sciences, Beijing 100039, China
[3] College of Physics Science and Technology, Shenyang Normal University,
Shenyang 110034, China
{shenhai,ylzhu,zouwp,zhuzhu}@sia.cn

Abstract. In 2006, a novel Group Search Optimizer (GSO) inspired by animal behavioral ecology was proposed. On unconstrained optimization problems, GSO has shown its superior performance. In this paper, the performance of it in coping with constrained problems is investigated. Several experiments are performed on 13 well known and widely used benchmark problems. The obtained results are presented and compared with the best known solution obtained so far. The experimental results show that GSO can find the exact or close to global optimal solutions on most problems. GSO has an ability of solving constrained problem and is an alternative bio-inspired optimization algorithm.

Keywords: Bio-inspired Algorithm, Group Search Optimizer, Constrained Optimization, Penalty Function.

1 Introduction

Over the past few decades, bio-inspired algorithms have attracted more and more attention, and many algorithms have been proposed. Compared with traditional optimization method, bio-inspired algorithm utilizes stochastic search and direct "fitness" information instead of function derivatives or other related knowledge. Now, bio-inspired optimization algorithms have been successfully applied to a wide range of research and applications. Recently, a novel optimization algorithm, which was inspired by the social behavior of animals, called group search optimizer (GSO) algorithm has been proposed [1]. It adopts the scrounging strategies of house sparrows and employs especially animal scanning mechanism. S. He and Q. H. Wu test it on 23 unconstrained benchmark functions. In low and high dimensions benchmark problem, respectively, the GSO algorithm has a markedly superior performance to other EAs in terms of accuracy and convergence speed, especially on high-dimensional multi-modal problems.

GSO as a new state-of-the-art algorithm has gained some attention. GSO with early stopping scheme was applied to train artificial neural network (ANN) on two benchmark

Y. Yu, Z. Yu, and J. Zhao (Eds.): CSEEE 2011, Part II, CCIS 159, pp. 48–53, 2011.
© Springer-Verlag Berlin Heidelberg 2011

functions: Wisconsin breast cancer data set and Pima Indian diabetes data set [2]. S. K. Zeng presented two improved group search optimization algorithms: Quick GSO and Quick GSO with Passive Congregation [3]. They were applied to shape optimization design of truss structures with discrete variables. G.Qin successfully applied QGSO to optimal design of a large double layer grid shell structure [4]. H. Shen proposed an improved Group Search Optimizer (iGSO) for solving two mechanical design optimization problems [5].

Constrained optimization problem is a widespread but difficult in solving practical engineering problems, and therefore the research has very important theoretical and practical significance. In this field, one of the most difficult parts encountered is constraints handling. For solving it, many various methods were proposed [6]. In this paper, the performance of standard GSO is deeply investigated in solving constrained optimization problems. The results of experiments performed on well known 13 test problems are reported and discussed in comparison with best known solution obtained so far.

2 The Group Search Optimizer

The GSO algorithm was also inspired by animal searching behavior and is based on the Producer-Scrounger model [7], and also employs "rangers" role. There are three kinds of member in the group: (1)one producer, search for food; (2) scrounger, perform area copying behavior in order to keep searching for opportunities to join the resources found by the producer; (3) ranger, employ searching strategies of random walks for randomly distributed resources perform random walk motion. At each iteration, the member located the most promising resource is producer, a number of members except producer in the group are selected as scroungers, and the remaining members are rangers. In GSO, if a member is outside the search space, it will turn back to its previous position inside the search space.

Scanning is an important component of search orientation and it can be accomplished through physical contact. In GSO, vision as the basic scanning strategies introduced by white crappie is employed [8]. Every member has its current position $X_i^k \in R^n$, a head angle $\phi_i^k = (\phi_{i1}^k, \phi_{i2}^k, \hbar, \phi_{i(n-1)}^k) \in R^{n-1}$ and a head direction $D_i^k(\phi_i^k) = (d_{i1}^k, d_{i2}^k, \hbar, d_{in}^k) \in R^n$ which can be calculated from φ_i^k via a Polar to Cartesian coordinate's transformation:

$$d_{i1}^k = \prod_{p=1}^{n-1} \cos(\varphi_{ip}^k) \qquad (1)$$

$$d_{ij}^k = \sin(\varphi_{i(j-1)}^k) \cdot \prod_{p=i}^{n-1} \cos(\varphi_{ip}^k) \qquad (2)$$

$$d_{in}^k = \sin(\varphi_{i(n-1)}^k) \qquad (3)$$

(1) For producer: The producer scans at zero degree and then scans laterally by randomly sampling three points in the scanning field: one point at zero degree, one point in the right hand side hypercube and one point in the left hand side hypercube

$$X_z = X_g^k + r_1 l_{max} D_p^k(\varphi^k) \tag{4}$$

$$X_r = X_g^k + r_1 l_{max} D_g^k(\varphi^k + r_2\,\theta_{max}/2) \tag{5}$$

$$X_l = X_g^k + r_1 l_{max} D_g^k(\varphi^k - r_2\,\theta_{max}/2) \tag{6}$$

Where $r_1 \in R^1$ is a normally distributed random number with mean 0 and standard deviation 1 and $r_2 \in R^{n-1}$ is a random sequence in the range (0,1). If the producer finds the best point with the best resource from above three points than its current position, it will fly to this point, otherwise it will stay in its current position and turn its head to a new angle:

$$\varphi^{k+1} = \varphi^k + r_2 a_{max} \tag{7}$$

where, a_{max} is the maximum turning angle. If the producer cannot find a better area after a iterations, it will turn its head back to zero degree:

$$\varphi^{k+a} = \varphi^k \tag{8}$$

(2) For scroungers: The area copying behavior of scroungers can be modeled as a random walk towards the producer:

$$X_i^{k+1} = X_i^k + r_3(X_g^k - X_i^k) \tag{9}$$

where, $r_3 \in R^n$ is a uniform random sequence in the range (0, 1).

(3) For rangers: Rangers move to the new point through generate a random head angle and choose a random distance:

$$\varphi_i^{k+1} = \varphi_i^k + r_2 a_{max} \tag{10}$$

$$l_i = a \cdot r_1 l_{max} \tag{11}$$

$$X_i^{k+1} = X_i^k + l_i D_i^k(\varphi_i^{k+1}) \tag{12}$$

3   Experimental and Results

In the following experiments, 13 benchmark test problems have been used to test GSO algorithm performance [9]. These functions were tested widely in evolutionary computation domain to show the performance of optimization algorithm. For each problem, Table 1 shows some parameters: n is the number of variables; f is the type of objective function; ρ is the ratio between the feasible region and the whole search space; LI is the number of linear inequalities; NI is the number of nonlinear inequalities; NE is the number of nonlinear equations; a is the number of active constraints at the optimum. This paper adopts penalty function method to deal with constraint. There are problems with different features.

In GSO, each experiment was repeated 30 times. In every run, the max iterations T_{max}=3000. For all problems a population of 50 individuals is used. The initial population of GSO is generated uniformly at random in the search space. The initial head angle of each individual is set to be $\pi/4$. The constant a is given by $round(\sqrt{n+1})$. The parameter needed to be tuned is the percentage of rangers. The maximum pursuit angle Θ_{max} and maximum turning angle a is π/a^2 and $\pi/2a^2$, respectively. The L_i and U_i are the lower and upper bounds for the ith dimension, and then the maximum pursuit distance l_{max} is calculated by

$$l_{max} = |U_i - L_i| = \sqrt{\sum_{i=1}^{n}(U_i - L_i)^2} \tag{13}$$

In Table 2, the best results obtained in 30 independent runs are presented. It is very clear the problem where GSO was competitive was g08 and g11, which were located the exact optimum by GSO. GSO performed also well in highly constrained problems, which can be with various types of dimensionality, not only the low dimensionality problem g06 but also the moderated dimensionality problem g09, and the high dimensionality problems g01, g02, g03 and g07. In these highly constrained problems, GSO can find approximately global optimal solution. Furthermore, GSO is able to deal with large search spaces with a very small feasible region (g10). However, in functions g12, no feasible solutions were found in any single run. That due to the fact that g12 has even disjointed feasible regions. For problem g04 with moderately constraint, and g05 and g13, GSO can find the feasible solution of them, but which is far away from the global optimal solution.

In fact, the optimal solutions or feasible solutions of some problems were very difficulties to find. For example, the global optimum in some problems (g01, g02, g04, g06, g07, and g09) is lies on the boundaries of the feasible region. Problems g05 and g13 only have very small feasible regions. On those problems, the approximately optimal solution or feasible solution can been found by GSO. That's approved GSO has the powerful ability of handle constrained optimization problem, despite using the simply handling constraint method. The Fig.1 is the convergence curve on function g01 and g09, respectively. The values of each point in curves are the mean best values in 30 independent runs. From this figure, we can see that the convergence speed of GSO is fast and stable. For problem g01, along with the iteration, the feasible solution is also gradually close to the optimal solution. Especially for problem g09, GSO can find the optimal solution within in a short number of iterations.

Table 1. The characteristics of benchmark problems

No.	Characteristics						
	n	Type of f	ρ	LI	NI	NE	a
g01	13	quadratic	0.00000235%	9	0	0	6
g02	20	nonlinear	0.99996503%	1	1	0	1
g03	10	polynomial	0.00000000%	0	0	1	1
g04	5	quadratic	0.26962511%	0	6	0	2
g05	4	cubic	0.00000000%	2	0	3	3
g06	2	cubic	0.00006679%	0	2	0	2
g07	10	quadratic	0.00000103%	3	5	0	2
g08	2	nonlinear	0.00859082%	0	2	0	0
g09	7	polynomial	0.00524450%	0	4	0	2
g10	8	linear	0.00000522%	3	3	0	3
g11	2	quadratic	0.00000000%	0	0	1	1
g12	3	quadratic	0.04775265%	0	1	0	0
g13	5	nonlinear	0.00000000%	0	0	3	3

Table 2. The best value found by GSO in 30 runs.("---" means no feasible solutions was found).

No.	Optimal value	Best (GSO)
g01	-15	-14.558
g02	0.803619	0.78
g03	1	0.97
g04	-30665.539	-28679.8
g05	5126.4981	5281.1
g06	-6961.8139	-6871.4
g07	24.3062091	25.21
g08	0.095825	0.095825
g09	680.630057	680.93
g10	7049.25	7426.6
g11	0.75	0.75311
g12	1	---
g13	0.0539498	0.3046

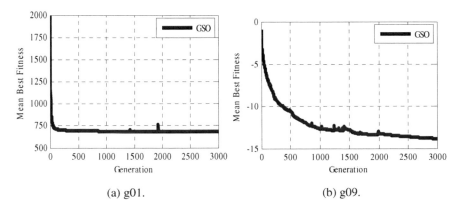

(a) g01.   (b) g09.

Fig. 1. Convergence curves of benchmark problems

4 Conclusions and Future Work

The capability of the GSO method to address constraint optimization problems was investigated through testing on 13 well-known and widely used benchmark problems. The results obtained imply that GSO is a good alternative for tackling this type of problems, particularly for highly constrained problems. In most cases GSO detected exact or approximately optimal solutions, in spite of using the default parameters of GSO. Moreover, GSO handles the constraints problem only using the simple penalty function method. That make the program is easily implementation on computer. The future research effort will focus on investigation of the GSO's performance in practical engineering problems, and improved the standard GSO though either on tuning parameters or development specialized operators for the aim of seeking better solutions. Also more experiments are required to determine why GSO fails on few problems and how to solve it.

Acknowledgment. This project is supported by the National 863 plans projects of China (Grant No. 2008AA04A105), and the Doctoral start fund of Liaoning province of China (Grant No. 09L3170301).

References

1. He, S., Wu, Q.H., Saunders, J.R.: A Novel Group Search Optimizer Inspired by Animal Behavioral Ecology. In: The Proceedings of the IEEE International Conference on Evolutionary Computation 2006, pp. 1272–1278. IEEE Computer Society, Washington (2006)
2. He, S., Wu, Q.H., Saunders, J.R.: A group search optimizer for neural network training. In: Gavrilova, M.L., Gervasi, O., Kumar, V., Tan, C.J.K., Taniar, D., Laganá, A., Mun, Y., Choo, H. (eds.) ICCSA 2006. LNCS, vol. 3982, pp. 934–943. Springer, Heidelberg (2006)
3. Qin, G., Liu, F., Li, L.J.: A Quick Group Search Optimizer with Passive Congregation and its Convergence Analysis. In: The Proceedings of the Computational Intelligence and Security, 2009, pp. 249–253. IEEE Computer Society, Washington (2009)
4. Qin, G., Liu, F., Li, L.J.: A Quick Group Search Optimizer and Its Application to the Optimal Design of Double Layer Grid Shells. In: The Proceedings of the 2nd International Symposium on Computational Mechanics. ADS, vol. 1233, pp. 718–723 (2010)
5. Shen, H., Zhu, Y.L., Niu, B., Wu, Q.H.: An Improved Group Search Optimizer for Mechanical Design Optimization Problems. Progress in Natural Science 19, 91–97 (2009)
6. Coello Coello, C.A.: Theoretical and Numerical Constraint-Handling Techniques Used with Evolutionary Algorithms: A Survey of the State of the Art. Computer Methods in Applied Mechanics and Engineering 191, 1245–1287 (2002)
7. Barnard, C.J., Sibly, R.M.: Producers and Scroungers: a General Model and its Application to Captive Flocks of House Aparrows. Animal Behaviour 29, 543–550 (1981)
8. O'Brien, W.J., Evans, B.I., Howick, G.L.: A New View of the Predation Cycle of a Planktivorous Fish, White Crappie (Pomoxis Annularis). Canadian Journal of Fisheries and Aquatic Sciences 43, 1894–1899 (1986)
9. Runarsson, T.P., Yao, X.: Stochastic Ranking for Constrained Evolutionary Optimization. IEEE Transactions on Evolutionary Computation 4, 284–294 (2000)

The Design and Realization of NAXI Pictograph Character Recognition Preprocessing System

Xing Li, Hai Guo, Guojie Suo, Zekai Zheng, and Zongwei Wei

Department of Computer Science and Engineering,
Dalian Nationalities University, DaLian 116600
li_xin_1987@163.com, dlguohai@gmail.com, 601519543@qq.com,
kiddkai@gmail.com, wei.zongwei@163.com

Abstract. NAXI character is a kind of pictograph which still in using archaic character at present. Our task group carried out the study of NAXI character recognition preprocessing, developed a system of NAXI character recognition preprocessing based on C#, accomplished binary processing, skew correction, document layout analysis, Character Segmentation and Normalization processing, Edge Detection and some other normal functions under the guidance of Guohai teacher, laid the foundation of the further development of recognition system.

Keywords: NAXI pictograph; Character recognition; Preprocessing; Image Segmentation; C#.

1 Introduction

NAXI character belongs to Tibeto-Burman languages, the NAXI language which branches of Yi, the unique spreading and in using archaic pictograph[1] character in the whole world. The study of NAXI character in the domestic and foreign dates from last century, including the United States, Japan and some museums of European, universities and research institutions have a lot of NAXI character materials and academics studying it all around the world. In 2003, the NAXI DONGBA ancient documents listed by UNESCO as world to remember its heritage, it is the only world heritage in Chinese minority language aspects. Because of ages ago, NAXI character were damaged greatly, and diffusing all over the world, in view of this NAXI DONGBA ancient documents protection, digital digitization work is imminent now. NAXI character belongs to pictograph, writing quite optionally. Early NAXI character monographs published mainly adopts hand-painted typesetting, existed problems of low efficiency, not normal and disunity of character pattern. In order to solve these problems, project team adopted "The NAXI pictographic standard sound writing dictionary" to wrote by LiLinCan as standard production NAXI character outline font, meanwhile developed NAXI pictographic character Latin pinyin input method and shape yards input methods information processing software [5-8] and so on. But an application input method typewriting manual input NAXI character

Y. Yu, Z. Yu, and J. Zhao (Eds.): CSEEE 2011, Part II, CCIS 159, pp. 54–59, 2011.
© Springer-Verlag Berlin Heidelberg 2011

literature has the following problems: Firstly, the users of NAXI character are not skilled at using computers. Secondly, ancient NAXI ancient literatures amount is larger, manual input efficiency is low. Thirdly, NAXI characters in promotion process needed to online handwritten identification system. At present, the graphics recognition research for NAXI character is still in a blank, NAXI character recognition research is imminent.

Chinese minority script recognition is a hot researching issue in the industry currently, such as Tibetan recognition which cooperatively studied by Northwest nationalities University and Tsinghua University, Dimension ha ko multilingual recognition of Xinjiang University, MengWen recognition research of Inner Mongolia University and Manchu script recognition study of Northeasstern University all gains great progress. But therewas no related reported on NAXI character recognition. NAXI character has its own features, such as no strokes, extremely casual writing, complex font, therefore cannot copy the other minorities and the han character in the recognition method.

Preprocessing is the foundation and key of recognition, this article mainly study the design and realization of NAXI character recognition preprocessing system. It mainly includes three parts, firstly brief introduced character recognition, secondly introduced NAXI character recognition preprocessing, thirdly the major function modules of preprocessing, fourthly realization and test, fifthly conclusion.

2 Character Recognition Introduction

Printing character recognition realization becomes people's dream early, in 1929, Tauscher obtains a patent on OCR in Germany. Occident in order to input increasingly and vast varieties of Newspapers, magazines, file information, documents statements, and writing material to computer to processing, started research of the technology of OCR(Optical Character Recognition) ,so as to replace artificial keyboard input.

The recognition of printing characters can be traced back to the early 1960s. In 1966, IBM Casey and Nagy published the first paper on the recognition of printing character; they used simple template matching method to identify 1,000 printed characters in this paper. Since the 1970s, Japan scholars did a lot of effort, including representative systems such as the system which can recognize 2000 characters printing character recognition developed by Toshiba comprehensive institute in 1977; In the early 1980s, Japan USASHI wild electric institute develops the system which can recognize 2,300 multi-body printing Chinese characters, represents the highest level of character recognition at that time. In addition, Japanese Sanyo, Panasonic, Ricoh and Fuji etc companies also have its development of printing character recognition system. These systems, mostly in method based on k-l digital transformation match scheme, used a lot of dedicated hardware, some equivalent to minicomputers mainframe in its equipment, the even price is extremely expensive, no widely used.

The research of printing character recognition in China dates from the late 1970s, early 1980s and divided into three phases roughly:

(1) First stage is from the late 1970s to the late 80s, mainly explore the algorithm and project.

(2) Second stage is from the early 1990, Chinese OCR went to the market from laboratory, using preliminarily.

(3) Third stage is the present, mainly improve the technology of printing Chinese character recognition and the property of the system, includes improve rate of identification and the Chinese and English synthesis and enhance stability.

Compared with foreign, national research of printing Chinese character recognition was established relatively late. But our government gives adequate attention and support to the research of Chinese character recognition and input automatically since 1980s, thanks to scientific research personnel 10 years of hard work, the development and application of printing Chinese character recognition technology have huge improvement. from simple individual word to various recognition of various scripts synchysis, from the recognition of Chinese printing materials to Chinese and English synchysis, effective quantitative analysis of simple layout, meanwhile Chinese character recognition rate has reached over 98%.

Original image is two-dimensional image signals obtained by Optical–Electronic Scanner, CCD device or electronic facsimile, perhaps Grayscale or Binary image. For simplicity, unless specifically mention, scanning input indicates the pattern of input image in this paper. So, printing Chinese character recognition technology mainly includes:

(1) Scanning and input text image.

(2) The preprocessing of image includes skew correction and such as filter disturbed noises.

(3)Document layout analysis and comprehend.

(4) Line segmentation and word segmentation of image.

(5) The feature selection and extraction based on the separate word images.

(6) Pattern classification based on the separate word images.

(7) Endow classified pattern recognition results.

(8) Editing, modify post-processing of recognition results.

Recognition preprocessing is the base and emphasis in the recognition, its function directly restricts the foundation of the recognition. We will have a simple introduce about it in the following.

3   The Basic Principle of NAXI Character Recognition Preprocessing

The pretreatment process including binary processing, skew correction document layout analysis, character segmentation and normalization processing, edge detection.[18]

Common binary processing method includes Niblack' Algorithm, Kittler's Algorithm and Otsu' Algorithm, the study in literature shown, Otsu' Algorithm fits NAXI character recognition preprocessing best .There are many methods of skew correction, NAXI character recognition skew correction mainly Hough transformation method, The test showed we can correct no more than 5 degrees tilt Angle documents effectively. Due to a few Charts and formula mixed arranging occurs in the presently printing document of naxi, we only study the publication for NAXI character layout analysis. There was no major breakthrough. NAXI character segmentation and normalization processing based on window projection method, and achieved good effect. There are many Edge Detection methods, we had several attempts to NAXI character, finally chose Canny process considering precision and the speed comprehensively.

4 NAXI Character Recognition Preprocessing System Composition

Major module consists of noise reduction, binary processing, Skew correction, layout analysis, character segmentation, normalization processing, extracting profile of character image. Customers can use the software to process NAXI character images, removed the interference information in the images, obtained images which beneficial to computers conducting information processing, and then conducted characters extracting and recognition. Specific functions are as following:

(1) The noise reduction of character image: selected noisy image randomly, respectively using such means as wavelet, median filter, Wiener filter to find the best solution.

(2) The binary processing of character image: the Division of pen-deficient couldn't happened in this process , otherwise repair it with expansion or smooth.

(3) Skew correction: take rapid correction algorithm based on principal direction and skew correction based on run-length smoothing processed document image in the experiment respectively.

(4) Layout analysis and character segmentation: the results show that taking the method of projection to do segmentation can achieve 100 percent accuracy, but the implementation of segmentation of section and character couldn't based on projection simply, we will find suitable algorithm to solve the problem by the experiment with relative references.

(5) Normalization processing: Separately using methods such as bilinear interpolation and three linear interpolation .

(6) Extracting profile. Separately using algorithm such as sobel, prewitt, canny to do experiment. Detailed function module shown in figure 1.

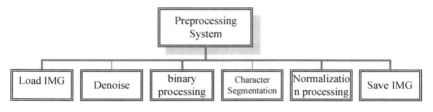

Fig. 1. Image recognition preprocessing system

5   The Test and Realization of NAXI Character Recognition System

Our NAXI character recognition system mainly uses the.net platform for development, due to .NET Framework contained two main core technology, respectively for Common Language Runtime(short for CLR) and .NET Framework categories library. .NET Framework shown in figure 2. C# is the core technology of .net, owning convenient, quick and high efficiency programming characteristics.

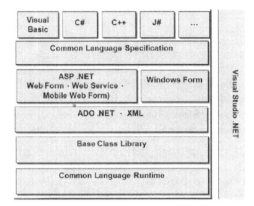

Fig. 2. NET Framework

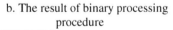

a. NAXI character recognition noise reduction

b. The result of binary processing procedure

c. Character Segmentation

Fig. 3. NAXI recognition system working sketch

Our system developed NAXI character recognition noise reduction system shown in figure 3.a, binary processing module shown in figure 3.b, character segmentation system shown in figure 3.based on C#.

6 Conlusion

Using C# to conduct NAXI character image of noise reduction, binary processing, character Segmentation, normalization effectively, laid the massy foundation of recognition in future, in the meantime , inspiring the other minorities character recognition operation and use for reference vitally.

Acknowledgment

This work is supported by National Natural Science Foundation of China (No. 60803096), Dalian Nationalities University independent research fund for the project, Dalian Nationalities University innovation studio project and Dalian Nationalities sunbird project. Meanwhile, thanks to Yang Maosheng, Zeng Fanpu, Wei Zongwei, Suo Guojie, Li Xing students programming work and so on. As well as Tao Ran, Yang Huanzhe students translation work for this article.

References

1. Lincan, L., Ku, Z.: The NAXI Pictographic Standard Sound Writing Dictionary. Yunnan National Publishing House, Kunming (2001)
2. Feizhou, Z.: About the NAXI DONGBA Character Information Processing Ideas. Academic Probe 2, 83–86 (2003)
3. Fuquan, Y.: NAXI Culture History Essay. Yunnan University Publishing House, Kunming (2006)
4. Rock, J.F.: Ancient Na-Khi Kingdom of southwest. Harvard University Press, China (1947)
5. Hai, G., Zhao, J., Liu, Y., Yu, H.: NaXi Pictographs Information Processing Based on Web Embedding Fonts Technology. Journal of Computational Information Systems (1), 495–501 (2009)
6. Hai, G., Jing-ying, Z.: NaXi Pictographs Input Method and WEFT. Journal of Computers 5(1), 117–124 (2010)
7. Hai, G., Zhao, J.-y., Da, M.-j., Li, X.-n.: NaXi Pictographs Edge Detection Using Lifting Wavelet Transform. JCIT: Journal of Convergence Information Technology 5(5), 203–210 (2010)
8. Hai, G., Zhao, J., Da. The Research, M.: Implementation of NaXi Pictographs Mobile Phone Dictionary Based on J2ME. Advanced Materials Research, 108-111, 1049–1054 (2010)

Study on the Safety Technology of Internet Protocol Television Based on Tri-Networks Integration

Weihai Yuan[1] and Yaping Li[2]

[1] School of Management Hefei University of Technology,
Hefei, China
yuanweihai@sina.com
[2] Anhui Economic Management Institute,
Hefei, China

Abstract. IPTV (Internet Protocol Television), which influenced by the country's related policy of merging tri-networks integration, business will face new opportunities and subjects under the background of tri-networks integration. Building a more reasonable safety system plays an important role for promoting the healthy development of IPTV business, and it will also be necessary and effective to introduce the information security related technologies into the IPTV business reasonably in the big trend of tri-networks integration.

Keywords: IPTV, Tri-Networks Integration, Information Security.

1 Introduction

The concept of tri-networks integration was first stated in the outline of Chinese 10th Five-Year plan in 2001. It is to promote the integration of telephone, TV and computer, but it hardly made substantial progress for tri-networks integration. On January 13, Chinese Premier Wen Jiabao chaired a State Council executive meeting to decide to promote the convergence of telecommunications, broadcasting networks and internet by issuing appropriate guidance and identifying integration timetable which is required telecommunications and broadcasting and television will step into the two-way pilot from 2010 to 2012 and to fully implement the development of tri-networks integration from 2013 to 2015[1] IPTV business has developed rapidly in foreign countries but slowly in China since Video Networks from the UK introduced it.

IPTV is defined as the manageable multimedia business by the transmission of IP networks including TV, video, text, graphics and data and providing QoS(Quality of Service) /QoE(Quality of Experience), security, interactivity and reliability by ITU (International Telecommunication Union). It is to state clearly IPTV should protect the security of its business operations and management of transmission in the definition. It will meet new development opportunities for IPTV in difficult situation in China with the continuous advance of tri-networks integration. In 2010 march, the market research institution iSuppli estimates that domestic IPTV users will be

Y. Yu, Z. Yu, and J. Zhao (Eds.): CSEEE 2011, Part II, CCIS 159, pp. 60–65, 2011.

doubled this year, because of the government's policies promulgation of promoting the tri-networks integration [2].

As far as business is concerned, the establishment of effective and unified safe system will also become increasingly urgent needs.

Therefore, in the big trend of domestic tri-networks integration, it is very necessary and effective to build perfect security mechanism for the futural IPTV healthy development as soon as possible.

2 Security Requirements and Security Situation of Domestic IPTV

2.1 Security Requirements

When it comes to IPTV security needs, domestic research is aimed at the security of contents and network. This is mainly due to IPTV services is in the form of "IP" and "TV" business integration, Which also makes the two securities distribute to telecommunications operators and broadcasting department. This paper says that the IPTV business comprehensiveness from security requirements and specificity under tri-networks integration will be ignored if only to grasp the IPTV security requirements from the content security and network security.

Therefore, IPTV security requirements should be from the perspective of information security to establish the whole security system from the information security requirements of content providers, network operators, users, regulators and etc [3], The analysis of IPTV security requirements of business as shown in Figure 1, this paper mainly divides IPTV system security requirements into:

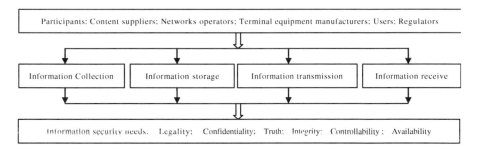

Fig. 1. The analysis of IPTV security requirements of business

1) Legitimacy: the program into the user terminal information is reviewed and legitimate;

2) Confidentiality: sensitive information of participants in interactive services, is confidential;

3) Authenticity: the identity of participants in the system is verifiable and true;

4) Integrity: program information in the transmission process is not altered;

5) Controllability: the protection of copyright interests, program information can not be illegally copied and transmitted;

6) Availability: the protection of user access to information timeliness.

2.2 Security Situation

For current domestic IPTV security situation, it is to sum up in two ways: one is that implementation of security means is far insufficient; the other is that accidents are uncommon. This is made by the particularity of domestic IPTV business. Domestic IPTV business is in the primary stage of accumulation; business and technical solutions are not ripe, and the positivity of operators for safe investment is not high; in addition, the market is scattered, smaller, single service and etc, I, which makes IPTV lack of enough attraction for the corresponding attackers. But with the advancement of tri-networks integration, IPTV business will enter a period of rapid development. More importantly, IPTV bearer network operations will gradually shift from the private network into public network. So a wide range of security attacks will be inevitable.

3 Security Technology Program

The comprehensive consideration of IPTV business security issues and providing appropriate solutions, we start to study the security technology of IPTV business of specific choice from the needs of information security of IPTV business system according to the above analysis of IPTV security needs.

3.1 Based on the Confidentiality of Information Encryption Technology

Whether for the different services of television programs or the confidentiality requirements for interactive services, modern encryption is the most effective and necessary. Symmetric encryption in efficiency will be far more than asymmetric encryption but on the whole weaker than the asymmetric encryption on confidentiality of information and not be able to achieve the non-repudiation of information. It is unrealistic to encrypt large amounts of data with asymmetric encryption in efficiency.

Considering the differences between TV and the interactive services in the number of information, encryption and asymmetric encryption should be cooperated and balanced to hedge against the use of the security and efficiency. A information encrypted scheme of viable programme is as in figure 2.

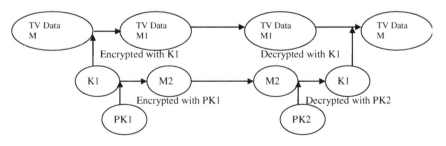

Fig. 2. A information encrypted scheme of viable programme

1) Using symmetric key K1 automatically generated to sent TV data to the terminal;

2) Use the terminal's public key PK1 encrypt K1 and sent to the terminal; only receiver can get K1 according to the theory of asymmetric encryption;

3) The receiver decry to get K1 using their private key PK2;

4) Recipient use k1 to decrypt the TV program.

The plan can guarantee that efficiency of the programme 's data encryption, and asymmetric encryption is the way to solve the symmetric key for the safety of transmission.

3.2 Filtering Identifies Based on Content

The legality of program content into IPTV terminal has been an obstacle to IPTV development. On April 12th, SARFT(The State Administration of Radio, Film and Television) issued to the provincial Radio and TV "41 Article"[4]. It is required in this article for regions to acquire the approval of SARFT to undergo IPTV business or they will be dealt with to stop the violations IPTV business in accordance with "Internet audio-visual program service regulations", which is based on the content of the requirements of legality.

The technical key to ensure the legitimacy of content lies in legal check the content of access to users terminal. It is necessary to increase unchanged and unduplicated legal identity data for program content data access to IPTV business system. Filtering identifies based on content are shown in Figure 3.

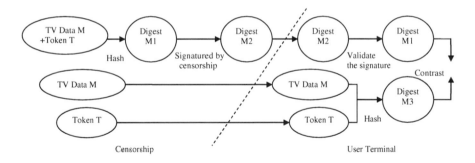

Fig. 3. Flitering identities of the legality of content

1) Show's data M is added legitimacy identification information T, use the hash function to generate the digest M1;

2) For digest M1 program content, the Censorship uses private key to generate the signature data M2;

3) Operators will send the programme of data, identity and signature data to the user terminal ;

4) Use the examiners' public key to validate the signature at user terminal and decrypt the signature M2 to the digest M1;

5) User terminal will receive the program data M and identification information T, use the same hash function to generating M3;

6) Comparing M1 with M3, if it is same, it illustrates the data is legal and valid, if different, it is invalid.

Meanwhile, in the above identity filter review, it also uses the corresponding information integrity mechanism to ensure the receiving program data to be checked in the process of transmission

3.3   Authentication Services Based on DRM

DRM (Digital Rights Management) or digital rights management, its basic principle is to build digital program authorized center[5] and   verify the copyright of digital program information of access to the user terminal to prevent   digital information from unauthorized copying and dissemination and provide corresponding solutions for the controllability of information transmission. DRM-based authentication services program are shown in Figure 4.

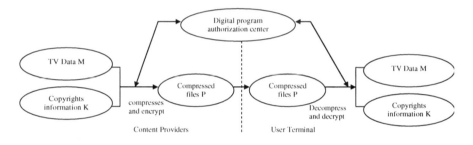

Fig. 4. DRM-based authentication services program

1) Content providers provide digital content and corresponding copyrights information, and ask for the authority from Digital program authorization center ;

2) The Center compresses and encrypt the digital information provided by content providers and send the information to users through network operators;

3) User terminal needs to show identification (digital certificate) before obtaining program content and asks for authorization;

4) Authority center checks the legitimacy of the identity of the user terminal to provide users with the authorization.

It needs clarifying that a user terminal copyright protection is crucial to the implementation of digital information from an illegal copy and spread, all the private key operation and user terminal hardware needs binding.

3.4   Network Security Technology Responding to DoS (Denial of Service) Attacks

To ensure users to get information service that is   the availability of information effectively, we also need to consider the network information service attacks on the basis of information security. IPTV business is information exchange service system mainly based on TCP / IP protocol technology. Its distinctive feature is to distribute large quantities of digital information from center comparing with networks. Therefore, the biggest security threat is DoS attacks and DDoS (Distributed Denial of Service) attacks in the layer of networks[6]. The basic principle of two attacks is to forge the legal address to send service requests to the server repeatedly, which caused server return information and wait for some time. This made resources depleted.

In response to a denial of service attacks, mostly in the server, it uses static and dynamic strategies. On one hand it is to configure the firewall reasonably to block specific attack from denial service flow; the other is to configure intrusion detection system to enhance anomaly detection. From the perspective of Internet security, network attacks and security protection are the process of dynamic equilibrium. It should be a subset of attack technology from network for network technology attack of IPTV business. It is necessary and feasible to dynamically apply the Internet network security protection technologies under IPTV service platform of tri-networks integration in a subset.

4 Conclusion

With tri-networks integration deepening gradually, it will become more urgent for security requirements with the further promoting IPTV business. Under the increasingly standardized environment of technology, market and regulatory mechanism. It is very important to build unified, efficient and dynamic sesecurity technology system for the specificity of the domestic IPTV. But only limitation to the security of technical aspects is not far enough. Improving security system, the supervision system, laws and regulations and to build a more complete IPTV security system is an important topic in front of the government, operators and users as well as researchers.

References

1. The Premier Presides Over the Executive Meetings to Decide to Promote Tri-Networks Integration, http://news.xinhuanet.com
2. This Year the Users of Iptv in China will Double, http://comm.ccidnet.com
3. Weng, S.H.: Awareness of Information Security from E-commerce. Hua nan Financial Computer 17, 68–70 (2009)
4. SARFT Ordered to Investigate the Difficult Situation of IPTV Tri-Networks integration, http://tech.163.com
5. Tian, H., Ji, W.: Components and Design of DRM System. TV Technology 33, 86–88 (2009)
6. Yu, Q.Y.: IPTV System Security Analysis. Information Security and Communication Security 9, 113–114 (2005)

Ontology-Assisted Deep Web Source Selection

Ying Wang[1,2], Wanli Zuo[1,2,*], Fengling He[1,2], Xin Wang[3],
and Aiqi Zhang[1,2]

[1] College of Computer Science and Technology, Jilin University, Changchun 130012, China
[2] Key Laboratory of Computation and Knowledge Engineering, Ministry of Education
[3] College of Software, Changchun Institute of Technology, Changchun 130012, China
wangying2010@jlu.edu.cn, wanli@jlu.edu.cn,
hefl@jlu.edu.cn, wangxccs@126.com

Abstract. Deep Web contains a significant amount of visited information, in order to effectively guide users to the appropriate searchable web databases, we need to organize it according to different domain. Ontology plays an important role in locating Deep Web content, therefore, this paper proposes a new Deep Web database selection framework based on ontology. Firstly, constructing domain ontology content model (DOCM), and then, designing the ontology-assisted similarity algorithm, which adds semantic information to form eigenvectors, lastly, selecting the mapping relational databases as domain-specific databases. Experiment shows that the method can effectively select Deep Web databases.

Keywords: Ontology, Deep Web, web database selection, DOCM.

1 Introduction

The web has been rapidly "deepened" by the prevalence of databases online, in order to effectively guide users to the appropriate searchable web databases, we need to organize it according to different domain. There are myriad useful databases, which may frequently undergo changes, artificial selection is a laborious and time-consuming task, so it is imperative to accelerate research on automatic selection of Deep Web databases.

In order to assist users accessing the information in Deep Web, recent efforts have focused on two kinds of approaches to select Deep Web databases automatically: Pre-Query and Post-Query approaches[1]. Pre-Query identifies the web databases by analyzing the features of HTML forms[2]. Post-Query approach identifies the web databases by submitting probing queries to the HTML[3]. However, when the database records have more attributes, it is difficult for Post-Query approach to obtain a better classification effects, what is more, it is some of wasting network and server resources by submitting a large number of queries only for the purpose of classification. Thus, the method of Pre-Query which depends on visual features of query interfaces, namely, attribute labels and other available resources, are usually used to determine the domain databases.

---

* Corresponding author.

Y. Yu, Z. Yu, and J. Zhao (Eds.): CSEEE 2011, Part II, CCIS 159, pp. 66–71, 2011.
© Springer-Verlag Berlin Heidelberg 2011

Domain ontology describes concepts and concept relationships in the application domain, which facilitates the semantic markups on the domain-specific aspects of web services. Making full use of ontology can improve decision accuracy by combing the clear definition and domain concept to achieve the Deep Web information. Therefore, this paper proposes a method of ontology-assisted Deep Web source selection.

2   A Framework of Web Database Selection

Finding the relevant Deep Web sites and accessing, retrieving and indexing the huge amounts of Deep Web data raises challenging research. The semantics of data can often be understood by viewing the data in the context of the user interface. Therefore, in the paper, we adopt Pre-Query method to select Deep Web databases based on semantic characteristic of query interface to obtain a higher degree of semantic relationship and lower query costs. Ontology plays an important role in recognizing Deep Web entry forms, which not only supports users in their web search efforts, but also increases the relevant web entry forms. Therefore, we utilize the ontology to map the attributes of query form, and then semantically select the Deep Web sources[4]. The framework of ontology-assisted Deep Web source selection is shown in Fig.1:

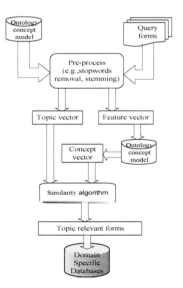

Fig. 1. The framework of web database selection

Step 1 Form information extraction: Extracting the attributes of each query form and converting each query form into eigenvector $T_d = \{t_{1,d}, t_{2,d}, ..., t_{m,d}\}$, in which $t_{i,d}$ denotes the i th feature $(1 \leq i \leq m)$ in searchable form.

Step 2 Converting DOCM into topic eigenvector $q = \{q_1, q_2, ..., q_n\}$, where q_j is a topic concept chosen from DOCM, n is the number of chosen topic concepts.

Step 3 It is possible that the same semantic concept may be represented quite differently over different forms. Therefore, for each form eigenvector $T_d = \{t_{1,d}, t_{2,d}, ..., t_{m,d}\}$, converting into form concept eigenvector $C_d = \{c_{1,d}, c_{2,d}, ..., c_{m,d}\}$, if feature $t_{i,d}$ can be searched in DOCM, then $c_{i,d}$ is the main class (concept) of $t_{i,d}$ in DOCM; if $t_{i,d}$ can not be searched in DOCM, then $t_{i,d}$ is the non-entry word, $c_{i,d} = t_{i,d}$. These features which are not in DOCM called non-entry words.

Step 4 Designing ontology-assisted similarity algorithm, and determining the relevance between form concept eigenvector and topic eigenvector by ontology-assisted similarity algorithm.

Step 5 Setting the similarity threshold σ, if matching degree is more than threshold σ, then it is topic relevant form, namely, the mapping relational database is a Domain-Specific database.

3 Web Database Selection

Definition 1. Domain Ontology Concept Model (DOCM): DOCM is a data model that describes a set of concepts and relationships that may appear in a specific domain. It should be understandable by machine so that it can be used to reason about the objects within that domain. DOCM has a good organizational structure, which represents high-level background knowledge with concepts and relationships[5][6].

In this paper, the concepts and relationships of DOCM are extracted from query interfaces and result pages, and the ontology is implemented by Protégé API and represented in the Web Ontology Language(OWL)[7]. To operate DOCM is equivalent to operate the OWL file.

Definition 2. Same-Branch: In DOCM, if concept A is the grandparent of concept B, then concept A and concept B are the Same-Branch concepts. For example, concept A and concept D are Same-Branch in Figure3. The concepts of Same-Branch(except root node) are semantically similar.

Definition 3. Different-Branch: In DOCM, if concept A is not the grandparent of concept B, at the same time, concept B is not the grandparent of concept A, then, concept A and concept B are the Different-Branch concepts. For example, concept B and concept C is Different-Branch in Figure3.

Definition 4. Equivalent: In DOCM, if concept A and concept B are the same, then concept A and concept B are Equivalent concepts.

There are only three relationships for different concepts in DOCM: Same-Branch, Different-Branch and Equivalent.

Definition 5. Concept Depth: In DOCM, concept depth is the hierarchical depth from root node to concept node. If $dep(A)$ is the hierarchical depth of node A, $dep(B)$ is the hierarchical depth of node B , then the hierarchical distance of node A and node B can be denoted in $dep(A,B)$, and $dep(A,B) = |dep(A) - dep(B)|$. For example, $dep(A,D) = 2$ means the hierarchical depth between node A and node D in Figure3.

Definition 6. Semantic Overlap: In DOCM, semantic overlap is the number of the hypernym concepts, which reflects the same degree between concepts. If the hierarchical depth of concept A is $dep(A)$, the hierarchical depth of concept B is $dep(B)$, then the semantic overlap of concept A and concept B is denoted in $overlap(A,B)$, and $1 \leq overlap(A,B) \leq Min[|dep(A)|,|dep(B)|]$.

Definition 7. Ontology Concept Similarity: Taking into account the above factors, suppose two concepts $t_{i,d}$ $(1 \leq i \leq m)$ and q_j $(1 \leq j \leq n)$, if $t_{i,d}$ is from query form d , and it can also be obtained in DOCM, q_j is from DOCM, then there are two cases for the matching degree between concept $t_{i,d}$ and q_j :

a) If $overlap(t_{i,d},q_j) \geq 2$, then set $Sim_1(t_{i,d},q_j) = 1$.

b) If $t_{i,d}$ and q_j are Same-Branch, then set $Sim_1(t_{i,d},q_j) = 1$.

c) If $t_{i,d}$ and q_j are Different-Branch, and $overlap(t_{i,d},q_j) = 1$, namely, there is no hypernym concept (except root) between $t_{i,d}$ and q_j , then set $Sim_1(t_{i,d},q_j) = 0$, that is to say, $t_{i,d}$ and q_j in DOCM are dissimilar.

Definition 8. Non-entry Words: the features from query forms can be divided into two parts: the features in DOCM and the features not in DOCM. These features which are not in DOCM called non-entry words.

Definition 9. Non-entry Words Similarity: suppose two concepts $t_{i,d}$ $(1 \leq i \leq m)$ and q_j $(1 \leq j \leq n)$, if $t_{i,d}$ is from query form d , and it can not be obtained in DOCM, namely, $t_{i,d}$ is the non-entry words, q_j is from DOCM, then the matching degree between concept $t_{i,d}$ and q_j is shown in formula (1):

$$Sim_2(t_{i,d},q_j) = \frac{2 * overlap(t_{i,d},q_j)}{depth(t_{i,d}) + depth(q_j)} \tag{1}$$

Definition 10. Ontology-assisted Similarity: Suppose that the eigenvector of query form is $T_d = \{t_{1,d}, t_{2,d}, ..., t_{m,d}\}$, the topic eigenvector for specific-domain is $q = \{q_1, q_2, ..., q_n\}$. Firstly, convert form eigenvector into concept eigenvector $C_d = \{c_{1,d}, c_{2,d}, ..., c_{m,d}\}$, and then compute the similarity between each feature $c_{i,d}$ in form and each feature q_j in DOCM. If feature $c_{i,d}$ is in DOCM, then call the method of ontology concept similarity to get the greatest similarity $\underset{q_j \in O}{Max}(Sim_1(c_{i,d},q_j))$, if feature $c_{i,d}$ is not in DOCM, then call the method of the non-entry words similarity to get the greatest similarity $\underset{q_j \in O}{Max}(Sim_2(c_{i,d}, q_j))$, in this way, we can get a greatest similarity eigenvector [8]. Lastly, the matching degree between form eigenvector and topic eigenvector is shown in formula (2):

$$Sim(T_d, q) = \frac{\alpha \sum_{i=1}^{a} (Max \sum_{j=1}^{n} (Sim_1(c_{i,d}, q_j))) + (1-\alpha) \sum_{i=1}^{b} (Max \sum_{j=1}^{n} (Sim_2(c_{i,d}, q_j)))}{m} \qquad (2)$$

Where $\alpha \in (0,1)$ is a factor to control the interaction force between ontology concept similarity(OCS) and non-entry word similarity(NEWS), m is the number of features for T_d, a is the number of concepts which appear in DOCM, b is the number of non-entry words. The method of ontology-assisted similarity is not only useful for query forms, but also for terms.

4 Experiment

We select 160 forms, and the evaluation metric for correctness is called precision, recall and F-measure. The precision is the percentage of the correctly identified domain-specific databases over all the identified domain-specific databases by the algorithm. The recall is the percentage of the correctly identified domain-specific databases over all the domain-specific databases. The F-measure is defined as a harmonic mean of precision and recall. The results are shown in Table1.

Table 1. The results of ontology-assisted deep web source selection

Threshold	$C_{precision}(p)$	$C_{recall}(p)$	$C_F(p)$
0.4	0.7533	1	0.8593
0.5	0.7902	1	0.8828
0.6	0.8235	0.9912	0.8996
0.7	0.8880	0.9823	0.9328
0.8	0.9821	0.9735	0.9778
0.9	1	0.7257	0.8411

We can see from Table1 that when the similarity threshold is set low, the results contain most relevant pages, and mistake a lot of irrelevant pages relevant, so the precision is low and the recall is high. When the similarity threshold is set high, it will ignore most relevant pages, so the precision is high and the recall is low. When $\sigma = 0.8$, there is a higher accuracy for recall, precision and F-measure, therefore, it is more reasonable for $\sigma = 0.8$. From the above results, the method of ontology-assisted Deep Web source selection can effectively obtain the classification accuracy of Deep Web databases, so it is feasible for this method to classify Deep Web databases.

5 Conclusions

We have presented a new approach to the Deep Web database selection based on ontology, which adds semantic information to form eigenvectors. Experiment reveals that the method can effectively select Deep Web databases.

Acknowledgement

This work is supported by the National Natural Science Foundation of China under Grant No.60973040; the National Natural Science Foundation of China under Grant No.60903098; the Science and Technology Development Program of Jilin Province of China under Grant No. 20070533; the Specialized Research Foundation for the Doctoral Program of Higher Education of China under Grant No.200801830021; the basic scientific research foundation for the interdisciplinary research and innovation project of Jilin University under Grant No.200810025.

References

1. Ru, Y., Horowitz, E.: Indexing the Invisible Web: A Survey. Online Information Review 29(3), 249–265 (2005)
2. Barbosa, L., Freire, J.: Combining Classifiers to Identify Online Databases. In: Proceedings of the World Wide Web Conference (www), pp. 431–440 (2007)
3. Ipeirotis, P.G., Gravano, L.: Classification-Aware Hidden-Web Text Database Selection. ACM Transactions on Information Systems (2008)
4. Lau, A., Tsui, E., Lee, W.B.: An Ontology-based Similarity Measurement for Problem-based Case Reasoning. Expert Systems with Applications 36(3), 6574–6579 (2009)
5. Su, W., Wang, J., Lochovsky, F.H.: ODE: Ontology-assisted Data Extraction. ACM Transactions on Database Systems 34(2), 1–35 (2009)
6. Zhang, W., Yoshida, T., Tang, X.: Using Ontology to Improve Precision of Terminology Extraction from Documents. Expert System with Applications, 9333–9339 (2009)
7. Horridge, M., Parsia, B., Sattler, U.: Explanation of OWL Entailments in Protege4. In: Proceedings of International Semantic Web Conference (2008)
8 Hong, J., He, Z., Bell, D.A.: An Evidential Approach to Query Interface Matching on the Deep Web. Journal of Information System 35(2), 140–148 (2010)

Key-Level Control in Hierarchical Wireless Sensor Network System

Zhifei Mao, Guofang Nan[*], and Minqiang Li

Institute of Systems Engineering, Tianjin University,
Tianjin, 300072, China
zhifei.mao@gmail.com,
{gfnan,mqli}@tju.edu.cn

Abstract. In this paper, key-level control method is proposed in hierarchical wireless sensor networks (HWSNs), and the key-level is composed of sensor nodes that only one hop distance to the sink node (OHSs). To guarantee energy efficiency and link performance for OHSs, power and admission control are both employed to compose the key-level control, which is modeled as a non-cooperative game afterward, by applying a double-pricing scheme for the game, solution with network properties can be derived.

Keywords: Hierarchical network, key-level control, game theory, pricing scheme.

1 Introduction

In HWSNs, sensor nodes are hierarchized into several levels in terms of hop number of transmissions from sensor to sink node [1], and those nodes belong to different levels play different roles in HWSNs [2], nodes from lower level are used to sense information from environment, while higher level nodes forward sensed data from low level nodes to other nodes or sink node. Since large proportion of energy consumption is caused by communications, nodes in higher level encounter significant challenge in terns of energy efficiency, especially in the key level, which is composed of OHSs representing the nodes that located only one hop to the sink node.

Note that more OHSs than one may transmit information to the sink node in a time period, and one-by-one strategy for data transmissions of these OHSs is not effective, this is because when one node is transmitting data, others have to wait, which leads to a worse delay performance with the increase of waiting nodes. Thus, the scenario that OHSs rather than only one OHS are admitted to transmit simultaneously should be adopted, which leads to interferences in transmitting OHSs however. With the purpose of addressing such interferences, key-level control is introduced in this paper.

Key-level control in HWSNs consists of two parts, i.e., power control and admission control of OHSs in the key level. 1) To avoid interference and improve energy efficiency, power control is used to dynamically tune the transmission power

---

[*] Corresponding author.

Y. Yu, Z. Yu, and J. Zhao (Eds.): CSEEE 2011, Part II, CCIS 159, pp. 72–77, 2011.
© Springer-Verlag Berlin Heidelberg 2011

of OHSs from the physical-layer perspective [3]. Besides, a balance criterion is applied to power control in order that each OHS get fair link quality while transmitting data. 2) It should be aware that, however, energy efficiency deteriorates and interference rises with the increasing number of OHSs, as in response to this problem, admission control [4] is employed, under which some OHSs with poor link gain should be abandoned (not allowed to transmit data to the sink node) so that each admitted (allowed to transmit data to the sink node) OHS could obtain desirable energy efficiency and link reliability guarantee.

Game theoretic approach is adopted to model the key-level control problem in HWSNs as a non-cooperative game, and it has been extensively used in wireless sensor networks [5]. OHSs which conflict with each other while transmitting simultaneously are considered as the players in this game. For considering these conflictions in the game model, a double-pricing scheme is introduced in utility function of players. Furthermore, by properly setting the pricing factors, the existence of a unique and Pareto efficient Nash Equilibrium is proved.

2 Network Model

A HWSN with one sink node denoted by S is considered. For concise formulation of key-level control, the HWSN is divided into two levels simply, i.e., the first level (key level) composed by OHSs and the second level composed by remanent nodes. We divide time into time intervals and assume there are n active OHSs denoted by $N = \{1, 2, \cdots, n\}$ which are intend to transmit data to S in time interval τ. Assume transmission powers of OHSs are tunable supported by sensor hardware devices, let t_i be the transmission power of OHS i which is the ith element of set N and t_{max} be the identical maximum transmission power of OHSs belong to N derived from the assumption that all OHSs are homogeneous, so for all $i \in N$, $0 \le t_i \le t_{max}$. Each OHS has a received power at S denoted by r_i for OHS i, and $r_i = g_i t_i$ where g_i is the link gain between OHS i and S. Given t_i is bounded, and so as to r_i that $0 \le r_i \le \overline{r_i} = g_i t_{max}$. There is no losing of generality to suppose that $g_1 < g_2 < \cdots < g_n$ due to the differences of OHSs in distance to S and environment condition around [6]. Since $r_i = g_i t_i$, it implies that the maximum received powers of OHSs at S are sorted as well, i.e., $\overline{r_1} < \overline{r_2} < \cdots < \overline{r_n}$. At last, AWGN (addictive white Gaussian noise) is adopted as our network channel model which is a basic noise and interference model. In AWGN based networks, SINR of OHS at S denoted by λ_i is defined as

$$\lambda_i = G \cdot \frac{r_i}{\sum_{j \neq i} r_j + N} = G \cdot \frac{g_i t_i}{\sum_{j \neq i} g_j t_j + N}, \text{ for all } i \in N . \tag{1}$$

where G is the processing gain and N is the received power of background noise at S.

3 Key-Level Control

Power Control. Observe the variation of λ_i caused by $\vec{t}$ from Eq. 1: λ_i increases with an increased t_i and on the contrary, for all $j \neq i$; λ_i decreases with an increased t_j . A vicious cycle comes out that one OHS will increase its transmission power for the purpose of obtaining a higher SINR which leads to a lower SINR for the other OHSs conversely, and the other OHSs will increase their transmission powers for the same purpose in turn. For avoiding the appearance of vicious cycle, power control is employed. Moreover, r -balance criterion is adopted in power control for fair resources allocation, i.e., the received power of all active OHSs at S should be identical, which is expressed as $r_1 = r_2 = \cdots = r_n$. Given $r_i = g_i t_i$, $i = 1, 2, \cdots, n$, it can be obtained that

$$t_i = \frac{g_1}{g_i} t_1, \ i = 2, 3, \cdots, n \ . \tag{2}$$

By taking Eq. 1, it's figured out that

$$\lambda_1 = \lambda_2 = \cdots = \lambda_n = G \cdot \frac{r_1}{(n-1)r_1 + \mathrm{N}} \ . \tag{3}$$

which indicates λ -balance criterion is satisfied as well, in other words, all OHSs possess identical SINR.

Admission Control. Review Eq. 3, the identical SINR decreases with the increasing number of OHSs so that the minimum SINR requirement for data transmitting denoted by λ_{th} will be dissatisfied, moreover, OHS with poor link gain needs a higher transmission power to achieve the identical SINR which implies that its energy consumption will be faster. Admission control is used here to handle this problem by admitting more OHSs as possible with the purpose that all admitted OHSs could obtain the link reliability no smaller than λ_{th} .

We define Λ_i as the identical SINR [7] under the condition that OHSs $1 \sim i-1$ are abandoned and the identical received power of admitted OHSs $i \sim n$ is $\bar{r}_i$ which can be formally expressed as

$$\Lambda_i = G \cdot \frac{\bar{r}_i}{(n-i)\bar{r}_i + \mathrm{N}}, \ i = 1, 2, \cdots, n \ . \tag{4}$$

And inequality $\Lambda_1 < \Lambda_2 < \cdots < \Lambda_n$ holds, from which it's concluded that OHSs will be abandoned in decreasing order of their link gain if needed. Furthermore, by considering the value of λ_{th} , it's derived that:

1) If $\lambda_{th} \leq \Lambda_1$, then the entire OHSs are admitted to transmit data to S;
2) If $\Lambda_l < \lambda_{th} \leq \Lambda_{l+1}$, $1 \leq l \leq n-1$, then OHSs $l+1 \sim n$ are admitted with OHSs $1 \sim l$ abandoned;
3) If $\lambda_{th} > \Lambda_n$, then all OHSs will shut off.

4 Non-cooperative Key-Level Control Game

In this section, a non-cooperative game is formulated for key-level control with the main aim to obtain a transmission power vector which satisfied goals proposed in key-level control in a distributed manner. In this game, n active OHSs in time interval τ are regarded as the players and the strategy space of each player is transmission power range. In addition, strategy profile is a transmission power vector in nature. Each player chooses a transmission power according to maximize its utility function. To obtain Nash equilibrium with property that Pareto efficiency and goal of key-level control are achieved, an appropriate utility function for each player is essential. Our utility function takes both link reliability in terms of SINR and interferences by using pricing scheme into consideration, which is expressed as

$$u_i(t_i, \vec{t}_{-i}) = a \log_2(1 + \lambda_i) - p_i, \ i \in N \ . \tag{5}$$

where a is a constant, $a \log_2(1 + \lambda_i)$ is a function proportional to link capacity of OHS i [8] which is an increasing function of SINR and p_i is a linear pricing function which can implicitly induce cooperation [9].

Under admission control, several OHSs are not allowed to transmit data to S. Let A denote the set of abandoned OHSs and R denote the set of admitted OHSs in a time interval τ, note that $A \cup R = N$. For simplicity and effectivity, linear pricing scheme is adopted in this paper, and it is a linear function of SINR different from [8] where a linear function of transmission power was used. Let $p_i = \alpha_i \lambda_i$ denote the pricing function of player i where α_i is a pricing factor, two types of values are assigned to α_i for all $i \in N$, i.e., α_A for OHSs from A, and α_R for OHSs from R, which is a useful method in [7]. Consequently, a specific version of utility function can be defined as

$$u_i(t_i, \vec{t}_{-i}) = \begin{cases} a \log_2(1 + \lambda_i) - \alpha_A \lambda_i & i \in A \\ a \log_2(1 + \lambda_i) - \alpha_R \lambda_i & i \in R \end{cases} . \tag{6}$$

Accordingly, the non-cooperative key-level control game called NKCG can be formulated as the following optimization problem:

$$\max_{t_i \in T_i} u_i(t_i, \vec{t}_{-i}) = \begin{cases} a \log_2(1 + \lambda_i) - \alpha_1 \lambda_i & i \in A \\ a \log_2(1 + \lambda_i) - \alpha_2 \lambda_i & i \in R \end{cases} . \tag{7}$$

Proposition. If $\alpha_1 = a/\ln 2$ and $\alpha_2 = a/\ln 2(1 + \lambda_{th})$, NKCG has a unique and Pareto efficient Nash Equilibrium which satisfies the goals of key-level control.

Proof. Assume that $1 \sim k$ OHSs should be abandoned and $k + 1 \sim n$ admitted accordingly in time interval τ under key-level control, $k = 1, 2, \cdots, n - 1$. Thus, $A = \{1, 2, \cdots, k\}$, $R = \{k + 1, k + 2, \cdots, n\}$. In addition, $\beta_i(\vec{t}_{-i}) = \arg \max_{t_i \in T_i} u_i$ is defined as the best reply of OHS i, for all $i \in N$ [10].

For $i \in A$, $\partial u_i / \partial t_i \leq 0$ which implies that u_i is a decreasing function of t_i. Since $0 \leq t_i \leq t_{max}$, u_i achieves maximum value when $t_i = 0$, i.e., 0 is the best reply to $\vec{t}_{-i}$, thus, $\beta_i(\vec{t}_{-i}) = 0$.

For $i \in R$,

$$\frac{\partial u_i}{\partial t_i} = \frac{a}{\ln 2} \cdot \frac{G g_i}{\sum_{j \neq i} g_j t_j + N} \cdot (\frac{1}{1 + \lambda_i} - \frac{1}{1 + \lambda_{th}}) . \tag{8}$$

First, we assume that $\lambda_i > \lambda_{th}$, then $\partial u_i / \partial t_i < 0$, which implies that u_i is a rigid monotone decreasing function of t_i. Since $0 \leq t_i \leq t_{max}$, u_i achieves maximum value when $t_i = 0$, i.e., 0 is the best reply to $\vec{t}_{-i}$, thus $\beta_i(\vec{t}_{-i}) = 0$. By computing λ_i according to Eq. 1, it's obtained that $\lambda_i = 0$ which is in contradiction with $\lambda_i > \lambda_{th} > 0$. Hence, the assumption $\lambda_i > \lambda_{th}$ is untenable, i.e., $\lambda_i \leq \lambda_{th}$. Accordingly, $\partial u_i / \partial \lambda_i \geq 0$ is valid, from which it's deduced that $\arg \max_{t_i \in T_i} u_i$ and $\arg \max_{t_i \in T_i} \lambda_i$ are equivalent. Since $\lambda_i \leq \lambda_{th}$, it means that $\max_{t_i \in T_i} \lambda_i = \lambda_{th}$, by taking the following equation (proved in [6])

$$p_i = \frac{\lambda_i}{h_i(\lambda_i + g)} \cdot \frac{N}{1 - \sum_{j=k+1}^{n} \frac{\lambda_j}{\lambda_j + g}} . \tag{9}$$

it's derived that

$$\beta_i(\vec{t}_{-i}) = \arg \max_{t_i \in T_i} u_i = \arg \max_{t_i \in T_i} \lambda_i = \frac{\lambda_{th} N}{g_i [G - (n - k - 1)\lambda_{th}]} . \tag{10}$$

Therefore, for all $i \in N$,

$$B(\vec{t}) = [\underbrace{0, \cdots, 0}_{k}, \frac{\lambda_{th} N}{g_{k+1}[G - (n - k - 1)\lambda_{th}]}, \cdots, \frac{\lambda_{th} N}{g_n[G - (n - k - 1)\lambda_{th}]}]^T \tag{11}$$

where $B(\vec{t})$ is defined as the best reply strategy profile, it's concluded that $B(\vec{t})$ is a Pareto efficient Nash Equilibrium [7].

Suppose another Nash Equilibrium exists besides $B(\vec{t})$ which denoted by $\vec{t}' = [t'_1, t'_2, \cdots, t'_n]^T$, by recalling the definition of Nash Equilibrium [10], it's obtained that, for all $i \in N$, $u_i(t'_i, \vec{t}'_{-i}) \geq u_i(\beta_i(\vec{t}'_{-i}), \vec{t}'_{-i})$. For $i \in A$, since u_i is a decreasing function of t_i, it can be deduced that $t'_i \leq \beta_i(\vec{t}'_{-i}) = 0$ which conflicts with $0 \leq t'_i$; for $i \in R$, since u_i is a increasing function of t_i, it can be deduced that $t'_i \geq \beta_i(\vec{t}'_{-i})$ which conflicts with $\beta_i(\vec{t}'_{-i}) = \arg \max_{t_i \in T_i} u_i$. That means another Nash Equilibrium different with $B(\vec{t})$ does not exist. Therefore, Nash Equilibrium $B(\vec{t})$ is unique in NKCG.

To figure out the solution for scenario that all OHSs should be abandoned/admitted under key-level control, only making $R = \phi / A = \phi$ particularly in the proof above is needed.

5 Summary

We presented key-level control for a hierarchical wireless sensor network in this paper, the control process was modeled as a non-cooperative game, and a double-pricing scheme was employed while defining the utility function. By each OHS choosing a transmission power for data transmission according to optimizing one's utility in a time interval, this game converged to a unique and Pareto efficient Nash Equilibrium, under which it's proved that energy efficiency and link reliability of the key level are both developed.

Acknowledgments. This paper is partially supported by a research grant from the National Science Foundation of China under Grant (no.70701025 and no.71071105), the Doctoral Foundation for young scholars of Education Ministry of China under Grant no. 20070056002, the Program for New Century Excellent Talents in Universities of China under Grant no.NCET-08-0396, and a National Science Fund for Distinguished Young Scholars of China under Grant no. 70925005.

References

1. Sekine, M., Sezaki, K.: Hierarchical Aggression of Distributed Data for Sensor Networks. In: TENCON 2004, vol. 2, pp. 545–548 (2004)
2. Al-Karaki, J.N., Kamal, A.E.: Routing Techniques in Wireless Sensor Networks: A Survey. IEEE Wireless Communication 11, 6–28 (2004)
3. Li, Y., Ephremides, A.: A Joint Scheduling, Power Control and Routing Algorithm for Ad Hoc Wireless Networks. In: Proceedings of the 38th Annual Hawaii International Conference on System Sciences (2005)
4. Chen, S.C., Bambos, N., Pottie, G.J.: Admission Control Schemes for Wireless Communication Networks with Adjustable Transmitter Powers. In: Proceedings of IEEE INFOCOM, Toronto, Canada, vol. 1, pp. 21–28 (1994)
5. Machado, R., Tekinay, S.: A Survey of Game-Theoretic Approaches in Wireless Sensor Networks. Computer Networks 52, 3047–3061 (2008)
6. Berggren, F., Kim, S.L.: Energy-Efficient Control of Rate and Power in DS-CDMA Systems. IEEE Trans. Wireless Commun. 3, 725–733 (2004)
7. Rasti, M., Sharafat, A.R., Seyfe, B.: Pareto-Efficient and Goal-Driven Power Control in Wireless Networks: A Game-Theoretic Approach with a Novel Pricing Scheme. IEEE/ACM Transactions on Networking 17, 556–569 (2009)
8. Fattahi, A.R., Paganini, F.: New Economic Perspectives for Resource Allocation in Wireless Networks. In: Proceedings of the American Control Conference, Portland, vol. 5, pp. 3690–3695 (2005)
9. Saraydar, C.U., Mandayam, N.B., Goodman, D.J.: Efficient Power Control Via Pricing in Wireless Data Networks. IEEE Transactions on Wireless Communications 50, 291–303 (2002)
10. Mackenzie, A.B., Dasilva, L.A.: Game Theory for Wireless Engineers. A Publication in the Morgan&Claypool Publishers' Series (2006)

An Exact Inference Method for Credal Network Based on Extended Relational Data Model

Ying Qu[1,2] and Jingru Wu[1]

[1] School of Economics and management, HeBei University of Science and Technology,
Shijiazhuang, 050081, China
[2] School of Management and Economics, Beijing Institute of Technology,
Beijing 100081, China
aquying1973@126.com,
bwujingru.teacher@163.com

Abstract. The extended relational data model was extended with the two operators of extended product join and extended Marginalization defined. A method to make credal network exact inference was proposed by computing the joint probability distribution and posterior probability of variables through the above operators. Furthermore its arithmetic was developed by SQL statement. Its validity and efficiency could be showed in the application case. The relational data structure of credal network was constructed, thus the query optimization mechanism of relational database could be fully used and the efficiency of credal network inference could be improved greatly.

Keywords: Credal network; extended relational data model; extended product join; extended marginalization.

1 Introduction

Credal network is extended by bayesian network which uncertainty is introduced on parameter probability value by probability distribution set or convex distribution theory and it also shows imperfection and fuzziness of parameter probability set which coincides with objective rules of specialist knowing problem, so under such parameter model obscure uncertain information offers academic support to the study of decision-making. Nature of credal network is to broaden the strict requirement of Bayesian network to parameter probability which can not specify unique probability distribution of its parent-node under configuration for every node but use different probability measurement method such as parameter distribution set, parameter distribution interval and inequality restriction to express uncertainty of parameter probability value formed by specialist or data information [1, 2].

Inference of credal network is based on Bayesian network and calculates max/mini parameter (up and down scope of interval) of classification value of some variables. Under strong extension type searching posterior probability calculation of variables can be regarded as global optimization problem of Bayesian network inference under credal network set restriction [1]. The difficulty to resolve this optimization is that

Y. Yu, Z. Yu, and J. Zhao (Eds.): CSEEE 2011, Part II, CCIS 159, pp. 78–83, 2011.

credal network set of every marginalization or condition credal network set brings a large quantity of vertexes combination to credal network strong extension even if it is easy to bring vertexes combination exploration problem to small type network.

In 1996 S.K.M put forward to using extended relational model to express Bayesian network [3, 4], which made Bayesian network combine with relational data base and made good use of the advantage of traditional relational model and made realize Bayesian network inference more easy. The article has the further extension of the concept of extended relational model and marginalization operation and product join in references [4, 5] and puts forward to the method of realizing credal network inference by further extended relational model and specific realizing progress.

2 Further Extended Relational Data Model

In the traditional relational model a piece of relation R includes n attributes $\{A_1, A_2, \cdots, A_n\}$. In extended relational model a piece of extended relation form Φ_R includes not only n attributes but also a series of f_{ϕ_R}, f_{ϕ_R}, which is regarded as a mapping of tuple t to non-negative real number field, $\Phi_R : \{t | t \in \Phi_R\} \to [0, +\infty]$ [3, 4]. That is to say that on the basis of traditional relational data model extended relational data model adds an attribute f_{ϕ_R}, value of which is decided by function ϕ_R. If $\phi_R(t_i) = 1$, the value of $t[f_{\phi_R}]$ of every tuple weight is 1, which the column can be omitted at this time such model is traditional relational data model and so traditional relational data model can be regarded as the special example of extended relational data model.

Credal network is the extended of Bayesian network, in which parameter of every node can be the different probability measurement method including probability distribution set, probability distribution interval and inequality restriction and here we just consider parameter as the format of probability interval. In order to adopt the format of interval parameter f_{ϕ_R} in extended relational model is factorized $f_{\phi_R}^L$ and $f_{\phi_R}^U$ and in it $t[f_{\phi_R}^L] \leq t[f_{\phi_R}^U]$, as the following Table 1, it expresses up and under parameter of every node of credal network and this forms the further extended format Φ_R^e of extended relational model.

Table 1. The basic form of further extended relational data model Φ_R^e

A_1	A_2	$\cdots$	A_m	$f_{\phi_R}^L$	$f_{\phi_R}^U$
t_{11}	t_{12}	$\cdots$	t_{1m}	$\phi_R^L(t_1)$	$\phi_R^U(t_1)$
t_{21}	t_{22}	$\cdots$	t_{2m}	$\phi_R^L(t_2)$	$\phi_R^U(t_2)$
$\vdots$	$\vdots$	$\vdots$	$\vdots$	$\vdots$	$\vdots$
t_{s1}	t_{s2}	$\cdots$	t_{sm}	$\phi_R^U(t_m)$	$\phi_R^U(t_m)$

Extended relational data model supports basic data operation of traditional relational data model, such as selection, projection, join and so on and adding two new operations *marginalization* and *product join*[4,5]. Here we expand the above two operations.

Definition 1. (Extended marginalization) If X is the sub-set of attribute set U, $\Phi_R^e(\downarrow X)$ shows the marginalization to attribute X under Φ_R^e and the result is one relation   including attributes $X \cup f_{\phi R}$ and constructing $\Phi_R^e(\downarrow X)$ in Φ_R^e is as the following steps:

If projecting Φ_R^e on $X \cup f_{\phi R}^L \cup f_{\phi R}^U$ and not deleting repeated tuple, we can get $\Phi_R^e\left[X \cup f_{\phi R}^L \cup f_{\phi R}^U\right]$

If t is one tuple in $\Phi_R^e\left[X \cup f_{\phi R}^L \cup f_{\phi R}^U\right]$ we combine all the same tuple in t[X] as one tuple   t', it can make   every configuration X_c to   X , $t'[X]=t[X]=X_c$,
$$\Phi_R^e(\downarrow X)\phi_R^L(t)(t') = \sum_{t[x]=x}\phi_R^L(t) \qquad \Phi_R^e(\downarrow X)\phi_R^U(t)(t') = \sum_{t[x]=x_c}\phi_R^U(t)$$
,

In Table 2, marginalizing Φ_R^e to A1, A2 can get extended marginalization relational $\Phi_R^e(\downarrow A_1 A_2)$

Table 2. An example of extended marginalization

	A_1	A_2	A_3	A_4	$f_{\phi R}^L$	$f_{\phi R}^U$		A_1	A_2	$f_{\phi R}^L$	$f_{\phi R}^U$
	0	0	0	1	d_{11}	d_{12}		0	0	$d_{11}+d_{21}$	$d_{12}+d_{22}$
Φ_R^e	0	0	1	1	d_{21}	d_{22}	$\Phi_R^e(\downarrow A_1 A_2)$				
	0	1	0	0	d_{31}	d_{32}		0	1	$d_{31}+d_{31}$	$d_{32}+d_{32}$
	0	1	1	0	d_{31}	d_{32}					
	1	0	0	1	d_{41}	d_{42}		1	0	d_{41}	d_{42}

Definition 2. (Extended product join) If Φ_X^e and Ψ_Y^e are extended relational model of two extendeds, its relation schema is $\Phi_X^e(X + f_{\phi X}^L + f_{\phi X}^U)$ and $\Psi_Y^e(Y + f_{\phi Y}^L + f_{\phi Y}^U)$ respectively and extended product join $\Phi_X^e \times \Psi_Y^e$ of Φ_X^e and Ψ_Y^e can be constructed as the following three steps:

Calculating the natural join $\Re_{\Phi\Psi}$, $\Re_{\Phi\Psi} = \Phi_X^e \infty \Psi_Y^e$.

Adding new row $f_{\phi X \phi Y}^L$ and $f_{\phi X \phi Y}^U$ to $\Re_{\Phi\Psi}$, $f_{\phi X \phi Y}^L$ can be defined as $\phi_X^L(t[x]) \bullet \varphi_Y^L(t[Y])$ and $f_{\phi X \phi Y}^U$ can be defined as $\phi_X^U(t[x]) \bullet \varphi_Y^U(t[Y])$ and t is one tuple in $\Re_{\Phi\Psi}$.

If projecting $\Re_{\Phi\Psi}$ to attribute set $X \cup Y \cup f_{\phi X \phi Y}^L$, we can get $\Phi_X^e \times \Psi_Y^e$.

In Table 3, after having expanding product join operation of Φ_X^e and Ψ_Y^e we can get relation $\Phi_X^e \times \Psi_Y^e$.

Table 3. An example of extended product join

	A_1	A_2	$f^L_{\varphi X}$	$f^U_{\varphi X}$		A_1	A_2	A_3	$f^L_{\varphi X \varphi Y}$	$f^U_{\varphi X \varphi Y}$
Φ^e_X	0	0	a_{11}	a_{12}		0	0	0	$a_{11}.b_{11}$	$a_{12}.b_{12}$
	0	1	a_{21}	a_{22}		0	0	1	$a_{11}.b_{21}$	$a_{12}.b_{22}$
	1	0	a_{31}	a_{32}		0	1	0	$a_{21}.b_{31}$	$a_{22}.b_{32}$
	1	1	a_{41}	a_{42}	$\Phi^e_X \times \Psi^e_Y$	0	1	1	$a_{21}.b_{41}$	$a_{22}.b_{42}$
	A_2	A_3	$f^L_{\varphi Y}$	$f^U_{\varphi Y}$		1	0	0	$a_{31}.b_{11}$	$a_{32}.b_{12}$
Ψ^e_Y	0	0	b_{11}	b_{12}		1	0	1	$a_{31}.b_{21}$	$a_{32}.b_{22}$
	0	1	b_{21}	b_{22}		1	1	0	$a_{41}.b_{31}$	$a_{42}.b_{32}$
	1	0	b_{31}	b_{32}		1	1	1	$a_{41}.b_{41}$	$a_{42}.b_{42}$
	1	1	b_{41}	b_{42}						

3   Exact Inference Principle of Credal Network by Using Extended Relational Model

Inference is on the condition of given evidence $X_E = e$ to infer parameter of query variable X_q under different situations and that is to solve $K(X_q | X_E = e)$. Its calculation method is shown as the following Eq. 1.

$$K(X_q | X_E = e) = \frac{K(X_q, X_E = e)}{K(X_E = e)} \tag{1}$$

If we can get joint probability distribution of variables *marginalization* can be used to get condition probabilities. The above extended relational data model can be used to store parameter distribution table of every node and further using every relational operation solves the joint probability distribution of all variables. The specific operation is as the following 5 steps:

To transfer variable parameter credal network set to vertex of credal network set;

To use extended relational table Φ^e_X to store condition probability table on every node of credal network and the n pieces of extended relational tables to n nodes, i.e. every CPT in the network corresponds with an extended relational table, and in the table $f^i_{\varphi R}$ is the parameter of the ith configuration of node;

If having extended product join to all tables we can get extended relational table of all variables. In it $f^{ij...}_{\varphi X \varphi Y...}$ is joint probability distribution of all variables;

Using the joint probability distribution to have *extended marginalization* operation can get $P_{ij...}(X_q, X_E = e)$, $P_{ij...}(X_E = e)$ and then get $P_{ij...}(X_q | X_E = e)$.

After calculating max$\{ P_{ij...}(X_q | X_E = e) \}$ and min$\{ P_{ij...}(X_q | X_E = e) \}$ we can get $K(X_q | X_E = e)$.

The key of using extended relational data model to having inference to credal network is to use product join operation among tables and now relational data system is more mature and it has query optimization system and under the system product join operation can speed up and therefore it can improve inference efficiency.

4   Application and Methods Comparison

The article uses the above method to have inference to the network (as the Fig. 1) in the references [7]. If $C = c_1$, solve the posterior probability of $K(D|c_1)$.

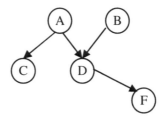

Fig. 1. An example of credal network

At first we should transfer section parameter of every node to vertexes of credal network set and we can get 4608 kinds of configurations. The operation to every allocation is as the following: parameter distribution table of 5 nodes transfers to 5 extended relational tables and we have extended product join to 5 tables and then get variable joint probability distribution table and have *marginalization* and get *extended marginalization* relational table of *CD* and *C*. Using $K(d_I|c_I) = K(c_Id_I)/K(c_I)$ gets parameter. The result is $K(d_I|c_I)$=[0.2047, 0.4041].

To the above example, by traditional enumeration method in the reference [8], using BNT tool box in Matlab to have different inference we can get the same result. The results of two calculating methods and time comparison are as the following Table 4.

Table 4. The comparison of the two methods

Result Methods	Application case (Fig.1) (vertex combinations 4608)	
	calculation result	operation time [second]
The method in the references [8]	[.2477, .4041]	2470
The method in the article	[.2477, .4041]	5

From Table 4 we can see clearly that the exact inference method which the article put forward and the method in the reference [8] can get the exact result but the efficiency of the method in the article is higher than the later. This is because all operations to adopting extended relational model having inference can be realized by

SQL which is simple and convenience and at the same time join operation and projection operation among tables can use strong query optimization system of relational data base to finfish rapidly and to improve inference efficiency. Thus the method in the article relieves vertex combination exploration problem of credal network set to some extent and offer new thought to study of credal network inference method.

Acknowledgments. This paper is supported by the project from national natural Foundation of China (71071049). It also gets help from the Foundation of science and technology plan of Hebei Province (09213509D) and the one from department of education of Hebei Province (2007205).

References

1. Fagiuoli, E., Zaffalon, M.: 2U: An Exact Interval Propagation Algorithm for Polytrees with Binary Variables. Artif. Intel. 106(1), 77–107 (1998)
2. Li, Y.L., Wu, Q.Z., Zheng, H.: The Application of Credal Network in Harbor Production Safety Evaluation. Comp. Eng. Appl. 43(29), 214–216 (2007)
3. Wong, S.K.M.: An Extended Relational Model for Probabilistic Reasoning. J. Intel. In. 9(1), 181–202 (1997)
4. Wong, S.K.M., Butz, C.J., Xiang, Y.: A Method for Implementing a Probabilistic Model as A Relational Database. In: 11th Conference on Uncertainty in Artificial Intelligence (UAI 1995), pp. 556–564. Morgan Kaufmann, San Francisco (1995)
5. He, Y.G., Liu, W.Y.: SQL Realizing Uncertainty Inference of Bayesian. Journal of Yunnan University (Natural Science Edition) 23(2), 100–103 (2001)
6. Wang, M.W.: Bayesian Decision Method Based on Probability Interval. Systems Engineering Theory and Practice 11(15), 79–82 (1997)
7. Xia, Y.L., Zhang, Z.X., Li, W.H., Liu, W.Y.: The Similar Inference of Bayesian Network with Section Parameter. Journal of Yunnan University (Natural Science Edition) 30(2), 135–141 (2007)
8. Li, Y.L.: The Problem and Application Study on Credal Network. Dissertation in Beijing Institute of Technology, Beijing (2008)

WSN Connected Coverage Analysis Based on Clifford Algebra

Lijuan Pu[1,2], Wenming Cao[2], and XaoJiang Liang[3]

[1] School of Electronic Engineering, Xidian University,
Xian, Shanxi 710071, China
[2] Intelligent Information Institute, Shenzhen University,
Shenzhen, Guangdong 518060, China
[3] Shen zhen Chao Yue Wu Xian Technology Co.,Ltd. Shenzhen,
Guangdong 518060, China
13823682521@139.com, doctor1105@126.com, liangxj28@163.com

Abstract. This paper built a 3-D Clifford sensor network model on the basis of ref [1], and analyzed the metric relationship of the model. The connection graph of Clifford sensor network is independent of coordinate and is consistent with different targets in different dimensional spaces. We proposed an algorithm---CSNCCCA, which builds a connecting coverage graph under the principle of nearest direction distance. At the same time the deployment of the sensor network achieves full coverage and connectivity. Finally, we tested and verified the rationality of the model and the algorithm.

Keywords: Clifford Algebra, Connecting Coverage, Homogeneous Model.

1 Introduction

Sensors in sensor network may be distributed on a variety of moving platform in space. Practice shows that, so far there is not a single kind of sensor can replace all other sensors to process targets in different dimensional space. It is necessary to establish a unified model for hybrid sensor network, which consists of different types of sensor. So Xie Weixin, etc. [1] proposed a method, which is based on Clifford algebra, for analyzing space sensor network coverage theory. They set up a consistent model for different targets in different dimensional space. This model does not depend on a specific coordinate system. It is a unified theoretical model for the performance analysis of distributed sensor network in complex environment. Based on this work, we built a 3-D Clifford sensor network model.

In recent years, literatures [2-10] studied the connecting coverage of sensor network. Ref [3] only considered the connectivity of discrete point targets. A heuristic algorithm was proposed to divide the sensors into several subsets. Each subset maintains target coverage and connectivity between working nodes and agents. Maximum network lifetime was achieved. Ref [4] proved that when the transmission radius is at least twice of the perception radius, the complete coverage of convex region and the connectivity of work nodes is equivalent. But this may not work well in discrete point target

Y. Yu, Z. Yu, and J. Zhao (Eds.): CSEEE 2011, Part II, CCIS 159, pp. 84–89, 2011.

coverage. An example in ref [5] offers the certification with the hexagonal grid perception model. In ref [2] Himanshu Gupta, etc. designed a centralized approximation algorithm that constructs a topology involving a near-optimal connecting sensor cover. The size of the constructed topology is within an O(logn) factor of the optimal size, where n is the size of the network. But, the perception model is limited to a rectangular area. Many researchers [6-10] studied connecting coverage of regions, but didn't involve connecting sensor coverage of both point targets and region targets.

On the basis of ref [1], which proposed a method for analyzing space sensor network coverage with Clifford algebra, this paper proposed a model, which didn't depend on a specific coordinate and were consistent for different targets in different dimensional space. We studied the properties of this model and verified its rationality. Experiments demonstrate that this model has practical value in applications.

2 WSN Clifford Algebra Connection Graph Theory and Properties

2.1 Three-Dimensional Model of Clifford Sensor Network

Commonly referred subspaces off the origin are points, lines, and planes, etc. In Clifford algebra, they are expressed as follows: to represent n-dimensional space with off-set subspaces, we embed it in an $(n+1)$-dimensional space, we express these off-set subspaces of G_n with pure blade by extending the space with an 'extra vector' e perpendicular to the n-space. A point P in 3-space, at location p, is represented by the vector: $P=(e+p)\wedge1$. A line in 3-space, with direction U and some given point P on it is represented as the 2-blade $P\wedge U=(e+p)\wedge U$. etc.

There are point, line, and plane of targets and sensor nodes in hybrid sensor network. These sensor nodes and targets are expressed as blade subspace in Clifford algebra, and the sensor network they form is called a Clifford sensor network. So, we treat sensor nodes and targets as blades in the following description.

2.2 Metric Relations of Clifford Sensor Network

Definition 1. the rout geometric entity A passes to have a non-trivial intersection with B is called track, denoted by (AB). The minimum length of track (AB) is called the distance of A,B denoted by $d(AB)$.

Theorem 1. in the homogeneous model of Clifford sensor network, the distance of A,B is $\quad d(A,B)=\left|rej(B,T)/(e,B)-rej(A,T)/(e,A)\right|$,
$T=e\wedge(join(e\cdot A,e\cdot B)/meet(e\cdot A,e\cdot B))$, $\quad rej(B,T)=B-(B\cdot T)/T$
$|d_{AB}|=\sqrt{d_{AB}d_{AB}}, d_{AB}=(-1)K_1(K_1-1), K_1=grade(d_{AB})$ [11].

2.3 WSN Clifford Algebra Connection Graph

$\mathcal{G}_n$ space is a linear combination of multidimensional geometric entities. So a mixed sensor network can be denoted as a multiple vector in $\mathcal{G}_n$ space. G is a multi-dimensional graph in $\mathcal{G}_n$ space, the vertex of G is a pure blade $A = (e + p) \wedge V$ in $\mathcal{G}_n$ space, the edge of G is the join of the corresponding vertexes, the length of the edge is the distance of the corresponding vertexes. It is a non-negative function over the edge set $S(G)$.

Definition 2. According to definition (1), if the distance between two sensor nodes is not greater than their maximum communication radius r_c, the corresponding edge is connected.

If arbitrary edge of graph is G connected, and the corresponding vertexes don't repeat, then graph G is called WSN Clifford algebra connection graph --CCG, denoted as $G(V, S)$. The path from one vertex to another is called connection path.

3   Algorithm Based on Clifford Sensor Network Connecting Coverage Theory

A key feature of wireless sensor network is the limited battery power. The energy efficiency in wireless sensor network operation is extremely important. This section proposed a method to find work nodes in redundant Clifford sensor network. The algorithm is called Clifford Sensor Network connecting - Coverage Concentrate Algorithm — CSNCCCA.

The probability of target B^j detected by node V^k is p_j^k. Each node has the same maximum communication radius r_c and the same maximum sensing radius r_s. When $d(B^j, V^k) \leq r_s$, $p_j^k = 1$, otherwise, $p_j^k = 0$.

Definition 3. Clifford sensor network connecting coverage
If the sub graph $H(\Gamma, E)$ of CCG $G(V, S)$ can cover all the targets in set B under the metric relationship of Clifford sensor network connecting coverage model, $H(\Gamma, E)$ is called Clifford sensor network connecting coverage.

In Clifford algebra connection graph $H(\Gamma, E)$, the path from source node V^0 to node V^i is denoted as $\Gamma_i : V_i^0 \to V_i^1 \cdots V_i^{k^i} \to V_i^{k^i+1}, d(V_i^j, V_i^{j+1}) \leq r_c (0 \leq j \leq k^i)$ $(V_i^0 = V^i, V_i^{k^i+1} = V^0)$. The length of Γ_i is $length(\Gamma_i) = \sum_{k=0}^{k^i} d(V_i^k, V_i^{k+1})$. k^i is the number of relay nodes in Γ_i. V_i^{j-1} is the predecessor node of V_i^j, V_i^{j+1} is the successor node of V_i^j. We build one connected path for each target, so, there are m connected paths in $H(\Gamma, E)$.

The problem of connecting coverage is to find a sensor subset $\Gamma \in V$ which covers the target set B with the probability of p_{th}, and the sensing data is sent to the central node through the connection path of Γ. That is $p_j^k = 1, \forall B^j \in B$, $V^k \in \Gamma_k \subseteq \Gamma$.

This algorithm can spontaneously activate the dormant state nodes, search the shortest connection path under the principle of nearest direction, and update the WSN Clifford algebra connection graph.

1. Establish neighbor node set, $\forall V^k \in V, N(V^k) = \{V^j \mid d(V^j, V^k) \le r_c, 1 \le j \le n\}$; 2. Establish sensing set V_j: $\forall V^k \in V_j, d(B^j, V^k) \le r_s, 1 \le j \le m$; 3. Select the nearest node V^k in V_j as the monitoring node of B^j and add V^k to path Γ_j, then delete V^k from V_j; 4. Compute the distance between V^k and central node V^0. If the distance is not greater than r_c, a connecting coverage path Γ_j is built from V^k to V^0. Otherwise, build a connecting coverage path including relay node as follow: 4.1. Set V^k as the starting node and search the next edge and relay node V^{k+1} from $N(V^k)$, according to the nearest direction principle. 4.2. If the distance between V^{k+1} and V^0 is not greater than r_c, the connecting coverage path is built. Otherwise, turn to 4.1 until the connecting coverage path between V^k and V^0 is built. 5. Turn to 1 to build the connecting coverage for the next target; 6. The connecting coverage paths of all targets form the connecting coverage graph $H(\Gamma, E)$, $\Gamma = \bigcup_{i=1}^{m} \Gamma_i$. Output the connection graph $H(\Gamma, E)$. 7. When node V' in $H(\Gamma, E)$ fails, if V' is a source node, select a nearest alternative node from the sensing node set. If V' is a relay node, its predecessor node select a node from the neighbor set according to the nearest direction principle and replace V' with the selected node, and turn to 4 to build a new path. The successor of V' is set into sleep mode. Finally, a new connected graph $H'(\Gamma', E')$ is formed.

4 Experiment and Analysis

Parameter are as follows: $n = 135, r_c = 6km, r_s = 8km$. The central node locates on the ground. Targets randomly generated in the space 2km above sea level. The probability of node failing is 50%. We compared CSNCCCA with SDCCA(Shortest-Distance connecting -Coverage Algorithm). Figure 1 and figure 2 shows the computing results. The target number $m = 34, 68, 102$ and 136.

Figure 2 shows the relation between the number of work node and the number of targets after each update of the Clifford algebra connection graph. The five-point star and point represent the number of targets covered by CSNCCCA and SDCCA, respectively. The triangle and open circle represent the number of work nodes in CSNCCCA and SDCCA, respectively. From the experiment results we can see that with the same amount of work nodes CSNCCCA can cover equal or more targets than SDCCA. And CSNCCCA can keep the complete coverage in more update times. In addition, CSNCCCA can completely cover the targets with less work nodes.

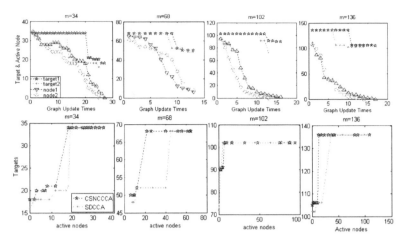

Fig. 1. connecting coverage of different number of targets

CSNCCCA and SDCCA have the same length of connecting coverage path in the initial WSN Clifford algebra connection graph. After updating, CSNCCCA get the paths whose lengths are all far more less than that of SDCCA. It's about one order of magnitude less for both the sensing distance and communication distance. With the increase of the number of targets, the length of the SDCCA coverage paths multiplied (from $185602.2km$ to $1427718km$), while the length of CSNCCCA connecting coverage path maintains within 10^4 (the variance of the length is from $7332.273km$ to $22186.40km$ in the experiment). The shortest path of the connecting coverage is only 3.95% of that of SDCCA. The longest path of the connecting coverage is 11.95% of that of SDCCA. So CSNCCCA saves the energy of sensing and communication and prolongs the network lifetime.

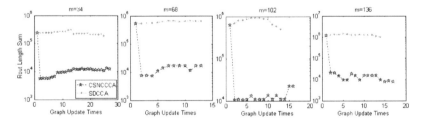

Fig. 2. Energy efficient compare of CSNCCCA with SDRSA

5 Conclusion and Outlook

We built the Clifford sensor network model in 3D space on the basis of ref [1]. This model does not depend on any specific coordinate system and is consistent with different targets in different space. Then we analyzed the metric relation in this model.

We farther study the capacity theorem of WSN Clifford algebra connection graph. At last, we verify the rationality of our model and algorithm.

Clifford sensor network aims at monitoring and tracking moving targets in practice. So, the connecting coverage of Clifford sensor network in dynamic is the main objective of further study.

Acknowledgments. This work is supported by the Natural science Funds of China No.60872126.

References

1. Xie, W., Cao, W., Meng, S.: Coverage Analysis for Sensor Networks Based On Clifford Algebra. Science in China Series F: Information Sciences 51, 460–475 (2006)
2. Gupta, H., Zhou, Z., Samir, R., Das, Q.: GU: Connected Sensor Cover:Self-Organization of Sensor Network for Efficiency Query Execution. IEEE/ACM Transactions on Networking 14, 55–67 (2006)
3. Zhao, Q., Gurusamy, M.: Lifetime Maximization for Connected Target Coverage in Wireless Sensor Networks. IEEE/ACM Transactions on Networking 16, 1378–1391 (2008)
4. Zhang, H., Hou, J.C.: Maintaining Sensing Coverage and Connectivity in Large Sensor Networks. Ad Hoc & Sensor Wireless Networks 1, 89–124 (2005)
5. Lu, M., Wu, J., Cardei, M., Li, M.: Energy-Efficient Connected Coverage of Discrete Targets in Wireless Sensor Networks. In: International Conference on Computer Networks and Mobile Computing, pp. 181–184. IEEE Press, New York (2005)
6. Wang, Y., Hu, C., Tseng, Y.: Efficient Placement and Dispatch of Sensors in a Wireless Sensor Network. IEEE Transactions on Mobile Computing 7, 262–274 (2008)
7. Vivek, P., Mhatre, Rosenberg, C., Kofman, D., Mazumdar, R., Shroff, N.: A Minimum Cost Heterogeneous Sensor Network with a Lifetime Constraint. IEEE Transactions on Mobile Computing 4, 4–15 (2005)
8. Wang, Q., Xu, K., Takahara, G., Hassanein, H.: Device Placement for Heterogeneous Wireless Sensor Networks: Minimum Cost With Lifetime Constraints. IEEE Transactions on Wireless Communications 6, 2444–2453 (2007)
9. Zou, Y., Chakrabarty, K.: Distributed Mobility Management for Target Tracking. Mobile Sensor Networks. IEEE Transactions on Mobile Computing 6, 872–887 (2007)
10. Wan, P., Yi, C.: Coverage by Randomly Deployed Wireless Sensor Networks. IEEE Transactions on Information Theory 52, 2658–2669 (2006)
11. Dorst, L., Mann, S., Bouma, T.: GABLE: A Matlab Tutorial for Geometric Algebra (December 2002), http://www.cgl.uwaterloo.ca/~smann/GABLE/

Research on Self-adaptive Wireless Sensor Network Communication Architecture Based on Ptolemy II

Jia Ke[1], Xiao-jun Chen[2], Man-rong Wang[3], and Ge Deng[4]

[1] School of Business Administration, Jiangsu University, Zhenjiang, Jiangsu, China
[2] Affiliated Hospital of Jiangsu University Zhenjiang, Jiangsu, China
[3] Institute of Computer&Communication Engineering, Jiangsu University, China
[4] University of Alabama, 1108 14th AVE Apt 308 Tuscaloosa, Alabama, USA
{kejia,cxj}@ujs.edu.cn, zjdxwmr@163.com, gdeng@crimson.ua.edu

Abstract. For the existing heterogeneous wireless sensor network architectures, different data link layers lack a common structure. To solve the problem mentioned above, this paper has proposed the concept of attribute assembly layer, and based on the hierarchical heterogeneous modeling of Ptolemy II, an adaptive architecture for wireless sensor network has been put forward. Experimental results show that this architecture has a low memory occupation and time cost. It unifies the data link layers for heterogeneous networks, and it is well compatible with the existing platforms, communication protocols and network mechanisms. The proposed architecture has good adaptive capacity, and it can apply to potential communication protocols and mechanisms.

Keywords: WSN, Network Architecture, Ptolemy II, Self-adaptive, Attribute Assembly Layer.

1 Introduction

With the rapid development of wireless communication technology and wireless sensor networks, there emerge lots of typical wireless communication technologies, mainly including 3G [1] based wireless WAN, IEEE 802.16 [2] based wireless MAN, IEEE 802.11 [3] based wireless LAN , and wireless PAN based on IEEE 802.15[4]. For the diversity of existing wireless communication technology, applications and protocols of the wireless network would run over different link layers, and data transferring also can't depend on the same link-layer mechanism. Therefore, we need to consider that, over the existing communication protocol architecture for wireless sensor network, if there is a mechanism independent of link layers to develop applications and protocols.

Communication protocol architecture with common operability has a good code reuse, and it can meet the needs of compatibility, interoperability and scalability for the link layer. Thus, such architecture has a great advantage in the development for wireless sensor network. Many organizations associated with sensor network have put forward various communication protocol architectures for heterogeneous networks. For instance, J. Sachs.A has proposed the concept of general link layer (GLL) [5].

Y. Yu, Z. Yu, and J. Zhao (Eds.): CSEEE 2011, Part II, CCIS 159, pp. 90–95, 2011.

IETF has designed the architecture of 6LowPAN [6], shown as Fig 1.1 (b). This architecture can be well extended on the application and protocol. However, it hasn't fundamentally solved the problem of automatic matching for protocols, but just adds a 6LowPAN adaptation layer to perform the protocol format conversion from IPv6 to IEEE802.15.4.

In order to ensure the adaptive capacity of communication architectures, this paper has proposed an adaptive general structure for sensor networks, shown as Fig 1.1(c), which is based on traditional network architecture, shown as Fig 1.1(a), and inspired by 6LowPAN. The proposed architecture has reassembled the network structure and brought in attribute assembly layer. The attribute assembly layer consists of two parts, which are the attribute factory and the assembly factory. The proposed architecture has been designed to adapt to different types of potential communication protocols, and such ability is revealed by the compatibility with the existing different link layers. The simulation experiments show that the adaptive network architecture can be well mapped to traditional wireless sensor network protocols, including packet distribution, packet collection, mesh routing, etc.

2 Adaptive Communication Architecture

2.1 Attribute Factory

The protocol stack of attribute factory has defined a serial of communication prototypes. Applications and protocols above the attribute factory can call any one or more communication prototypes. The protocol stack has a hierarchical structure, and a complicated protocol is composed of relatively simple protocols. The hierarchical structure for the protocol stack of attribute factory is shown as Fig 1.3.

The attribute factory supports the transmission prototypes of single-hop and multi-hop. The multi-hop prototype doesn't specify how the packets through the network perform routing, and the routing work is assumed by the communication protocol above. The communication protocol chooses an appropriate next-hop address according to the header field of a packet. The separation of multi-hop model from routing protocol contributes to implementing any one or more types of routing protocol. When multiple types of routing protocol are needed, we only need to add them, and the specific transmission of multi-hop is done by communication prototype. In a certain sense, the communication model has provided a high abstraction of data transmission, and it has separated the transmission model from the protocol logic.

The attribute factory substitutes the header of a packet with some packet attributes, which contain the same information required in communication as the traditional packet header. Therefore, packet attributes are abstract representation for the data of header field. The attribute factory has predefined some packet attributes, including id of a packet, type of a packet, addresses of the sender and receiver, hop count, lifetime, number of retransmission, reliability, link conditions, etc. The application and protocol can also define additional attributes.

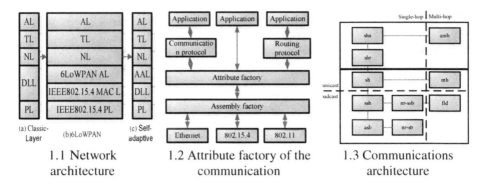

| 1.1 Network architecture | 1.2 Attribute factory of the communication | 1.3 Communications architecture |

Fig. 1. Network and communications Architectures

2.2 Assembly Factory

The assembly factory can not only generate the packet header encapsulated in the way of bit, it can also generate the packet header for a specific link layer. The application and protocol pass data of the application layer to the attribute factory, which is response for adding packet attributes, then the attribute factory passes the application data and packet attributes to the assembly factory. According to the packet attributes and the specific link layer, the packet header format conversion module of the assembly factory will generate a corresponding header and then deliver it to the sending module. Finally, the MAC layer of the sending module will decide how to transmit a packet according to the meaning of the packet header field.

Packet headers are generated by independent third party header conversion modules. These modules encapsulate corresponding data and packet attributes, and consequently generate packets with typical formats. All the work mentioned above is done by the various assembly factories of the adaptive adaptation layer. Different assembly factories make different standards for the packet header, and generate corresponding packets. For example, through the adaptive adaptation layer, data of the application layer can be easily encapsulated into packets with 802.15.4 MAC layer headers and those with UDP/IP headers.

Compared with the traditional protocol architecture, the advantage of adaptive adaptation layer is that applications running over it have no need to consider whether to be compatible with the underlying communication protocols or potential ones. This is because the assembly factory can generate packet headers required according to the communication protocol of link layer.

3   Ptolemy II Based Adaptive Communication Architecture for WSN

Based on the WSN prototype proposed in the literature, this paper has extended the processing module of the heterogeneous architecture for WSN. The main UML class structure is shown as Fig 2.

The logic layer structures for attribute factory and assembly factory based on wireless channel are shown as the red boxes in Fig 4. The attribute factory encapsulates behaviors of the upper applications, communication protocols and routing protocols, while the assembly factory is response for producing packet headers and distributing the packets to corresponding networks.

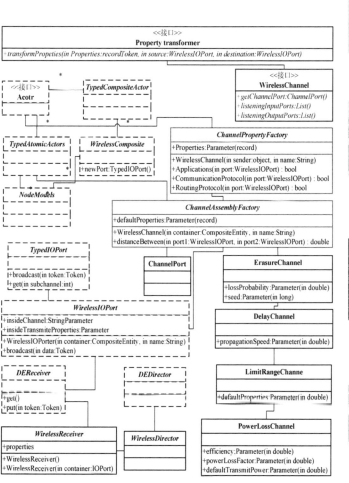

(a) All nodes

(b) Topological structure

(c) Operation

(d) Topological structure

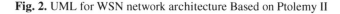

Fig. 2. UML for WSN network architecture Based on Ptolemy II

Fig. 3. Experimental simulation

4 System Test

4.1 Analysis of Network Communication

As shown in Fig 3a, 200 sensor nodes are randomly deployed in the test area. The simulation diagrams, shown as Fig 3b, Fig 3c, Fig 3d, show that the context of the virtual topology is very clear and there is good connectivity among the nodes. In Fig 3b and Fig 3d, the black line represents the communication link based on 802.15.4, the blue one represents that based on 802.11 and the red one represents that based on Ethernet. Whenever it is in the initial condition or after the attenuation of the communication capacity, effective network communication architecture can be well built.

4.2 Transmission Rate and Throughput

To compare the performance of the adaptive architecture protocol stack with that of IEEE802.15.4, 802.11 and Ethernet, their channel transmission capacity and throughput have been measured respectively, and we have made a comparison for them. As shown in Fig 4, in the initial stage, channel transmission rates of the four architectures have similar linear changes. With the increase of the node load, the transmission rates of the architectures have slight changes but still maintain at a high level. Owing to the channel adaptation of protocols, the channel utilization of the adaptive architecture can reach more than 98%, which is higher than that of the other two architectures.

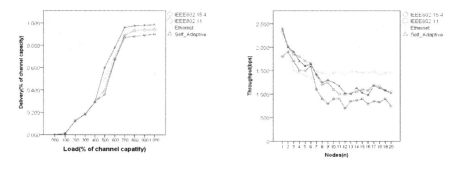

Fig. 4. Compared with the transmission   **Fig. 5.** Compared with the Throughput

As shown in Fig 5, the node throughput capacity will decrease with the increase of node number, which is because of the channel congestion caused by the increase of node number. As the protocol stack of IEEE802.15.4 is simple and it needs much less buffers, the throughput of IEEE802.15.4 is the highest among the three and it remains relatively stable. Both the adaptive structure and 802.11 add the adaptation layer, and for the implementation of congestion control, the protocol complexity is increased. In addition, the common operability and matching support for different channels need to be considered in certain situations. For the above reasons, the node throughput of

802.11 and the adaptive structure would reduce accordingly. However, their node throughputs tend to be unchanged with the increase of node number.

5 Conclusion

In view of current development situation of the wireless sensor networks and the existing various architectures, this paper has proposed novel adaptive communication architecture. The architecture with adaptive adaptation mechanism can well solve the problem of cross-layer communication. For the protocol stack with hierarchy, it can ensure the application to mask the underlying platform, and it can also provide unified attribute interface services for the upper layer and there is no need to break the existing protocol rules. Moreover, using the lightweight adaptive communication architecture can reduce the developing complexity, which is caused by the need of changing the code frequently to implement different communication protocols. The architecture proposed in this paper has good scalability and operability. Not only can it be well compatible with the MAC layer of 802.11, Ethernet and IEEE802.15.4, it is also independent of the link layer and can interconnect with any network seamlessly. As proved by the simulation tests, the memory occupation and time cost of the proposed architecture are relatively low. What's more, it has a high channel utilization and moderate throughput. In a word, the proposed architecture can adapt to all kinds of wireless sensor network applications

Aknowledgement. This research has partially been supported by Natural Science Foundation of ShanXi Province under Grant No. 20100321066, Natural Science Foundation of Jiangsu Province under Grant No. BK2009199, College Graduate Research & Innovation Plan of Jiangsu Province under Grant No. CX08B_097Z, Key Technology R&D Program of ZhenJiang City under Grant No. SH2009002 and Education Reform Project of JingJiang College of JiangSu University under Grant No. J08C014.

References

1. 3GPP TS 23.234 V6.1.0. 3G Security Architecture (2004)
2. IEEE Std 802.16-2004. IEEE Standard for Local and Metropolitan Area Networks Part 16: Air Interface for Fixed Broadband Wireless Access Systems. IEEE Std 802.16c-2002. xii+78 (2004)
3. IEEE 802.11s Workgroup. Joint SEE-Mesh/Wi-Mesh Proposal to 802.11 TGs. IEEE Document, Number: IEEE 802 11-06/328 (2005)
4. Part 15.4: Wireless MAC and PHY Layer Specifications for Low-Rate Wireless Personal Area Networks. IEEE Std 802.15.4d-2009. ix+27 (2006)
5. Sachs, J.: A generic Link Layer for Future Generation Wireless Networking. In: IEEE International Conference on Communications, vol. 2, pp. 834–838 (2003)
6. Bormann, C.: Draft-bormann-6lowpan-cbhc-00.txt. Context-based Header Compression for 6lowpan. The 6LoWPAN Working Group (2008)

$ALCI_{R+}$ Reasoning for Semantic Web Services Composition

Junyan Qian[1,2], Rongliang Gao[1], and Guoyong Cai[1]

[1] Department of Computer Science and Engineering,
Guilin University of Electronic Technology, Guilin 541004, China
[2] National Laboratory for Parallel and Distributed Processing,
National University of Defense Technology, Changsha 410073, China
qjy2000@gmail.com, {grl,ccgycai}@guet.edu.cn

Abstract. Description logic (DL) $ALCI_{R+}$ is a very expressive knowledge representation and reasoning language formed by adding transitive roles and inverse roles to the basic DL ALC. Transitive roles in $ALCI_{R+}$ are appropriate for describing accessibility relationship between services, while the relationship is encoded as a ServiceProcess in DAML-S as an external understanding. The transitive closure of the special transitive role which represents accessibility relationship between the first and final services can be used to find an execution sequence of the composite service. We have presented a procedure for computing it. Existing standards DAML-S and convenient transformation from ServiceProcess in DAML-S to transitive roles in $ALCI_{R+}$ helps to make our method simple and viable.

Keywords: Semantic Web, Services Composition, Transitive Closure.

1 Introduction

The Semantic Web and Web Services are now two increasingly promising technologies in the field of Internet Information Technology. Their combination results in the Semantic Web Services, a much more ambitious research ,which attends to allow the automated annotation, advertising, discovery, selection, composition, and execution of inter-organization business logic, making the Internet become a global common platform where organizations and individuals communicate among each other to carry out various commercial activities and to provide value-added services [3]. The intelligent combination of Web services and the semantic Web can start off a technological revolution with the development of semantic Web processes [4]. These technological advances can ultimately lead to a new breed of Web-based distributed applications with dynamic, adaptability and reconfiguration features.

The discovery of a single service and often the composition of several indispensable services that meet the client's business requirements play very significant role in the general lifecycle of Semantic Web processes. Most of the Semantic Web Services Composition approaches focus on AI Planning methods and DL reasoning methods. But the reasoning of automatic composition under different DL frameworks is still a great challenge.

Y. Yu, Z. Yu, and J. Zhao (Eds.): CSEEE 2011, Part II, CCIS 159, pp. 96–101, 2011.
© Springer-Verlag Berlin Heidelberg 2011

DLs are a family of knowledge representation languages that can be used to represent the knowledge of an application domain in a structured and formally well-understood way. As the basis of DAML-S (DARPA Agent Markup Language for Services) and OWL-S (Web Ontology Language for Services), it can be easily used for tackling this problem.

This paper attempts to tackle the Web services composition problem through logic reasoning in $ALCI_{R+}$, a special variant of DL which enables transitive roles and inverse roles [2]. We have shown that the service composition sequence can be derived by computing the transitive closure of the accessibility relationship between the first and the final services. In next section we described the specification of Web service using DAML-S which is appropriate for definition of the execution sequence of services. Section 3 simply introduced $ALCI_{R+}$ and emphasized service composition problem in $ALCI_{R+}$ reasoning. Finally, in section 4 we compared our approach with others through DL reasoning and concluded this paper with a look into the future.

2 Describing Services with DAML-S

In order to make all the software agents knowing the Web services, a general language is needed to specify the services. Some initiatives have been taken for this specification language. DAML-S [5] is such a rich representation of Web services that it can help the software agents to automatically discovering, selecting, executing, and even monitoring services.

In DAML-S framework, a Service was annotated by three Classes: **ServiceProfile**, **ServiceModel** and **ServiceGrounging**. **ServiceProfile** and **ServiceModel** provide information for services discovery and composition, while **ServiceGrounding** is useful to services execution.

ServiceProfile represents two aspects of functionality of the service: the information transformation and the state change produced by the execution of the service. It contains four kinds of properties or parameters of a service: **Input**, **Output**, **Precondition** and **Postcondition**. These four parameters are very important for services matching and composition in all existing services composition methods.

The DAML-S **ServiceModel** models service as a process giving information about invoking orders. The subclass **ServiceProcess** of **ServiceModel** identifies a service into: **AtomicProcess**, **SimpleProcess** and **CompositeProcess**. For example, a single service S_A can be viewed as an **AtomicProcess**; a sequential execution of services S_A and S_B (simply written in $<S_A, S_B>$) will be seen as a **SimpleProcess** where S_A and S_B are both **AtomicProcess**. Otherwise $<S_A, S_B>$ will be seen as a **CompositeProcess** which has several structures such as **Sequence**, **Split**, **Unordered**, **Choice**, **If-Then-Else**, **Iterate**, and **Repeat-Until**. When reasoning about the composite services in DL frameworks it often only takes the two basic structures **Sequence** and **Split**. The others can be transformed into them.

Actually the process information in DAML-S identified with the accessibility relationship between services. In our method ServiceProcess is very useful when reasoning about composite services in Description Logic $ALCI_{R+}$. The processes of services are transformed into transitive roles in $ALCI_{R+}$.

3 Reasoning Web Services Composition in $ALCI_{R+}$

3.1 Description Logic $ALCI_{R+}$

$ALCI_{R+}$ is an expressive DL which allows for rather expressive roles, such as transitive roles, and inverse roles. Transitive roles are introduced for aggregated concepts in DL originally. In this paper we used it to describe the Sequence process of services. For a simply introduction of $ALCI_{R+}$, here we borrowed the definition in article [15] and introduced a transitive closure definition of the $ALCI_{R+}$ transitive roles.

Definition 1 ($ALCI_{R+}$-roles formulas). Let R be a set of role names, which is partitioned into a set R_+ of transitive roles and a set R_P of normal roles. The set of all $ALCI_{R+}$-roles is $R \cup \{r^- | r \in R\}$, where r^- is called the inverse of the role r. A role inclusion axiom is of the form $r \sqsubseteq s$; where r, s are $ALCI_{R+}$-roles.

Suppose that there exists a role $r \in R_+$ and an interpretation I.
If $<x, y> \in r^I$ and $<y, z> \in r^I$ hold, then $<x, z> \in r^I$ can be derived.

Definition 2 ($ALCI_{R+}$-concepts formulas). Let N_C be a set of concept names. The set of $ALCI_{R+}$-concepts is the smallest set such that:

- every concept name $A \in N_C$ is an $ALCI_{R+}$-concept.
- if C and D are $ALCI_{R+}$-concepts and r is a $ALCI_{R+}$-role, then $C \sqcup D, C \sqcap D \neg C, \exists r.C$ and $\forall r.C$ are $ALCI_{R+}$-concepts.

Assertion Axioms in $ALCI_{R+}$ ABox take the form of $C (a)$ or $r (a, b)$. Terminology Axioms in $ALCI_{R+}$ TBox have forms: $C \sqsubseteq D$, $C \equiv D$ and $r \sqsubseteq s$, where C, D are concept names and r, s role names.

Definition 3 (transitive closure of $ALCI_{R+}$-roles) Let r, s, t, be $ALCI_{R+}$ roles, I be an interpretation. If the three conditions below are all satisfied: (i) s^I is transitive; (ii) $r^I \sqsubseteq s^I$; (3) for any other roles t, if t^I is transitive, and $r^I \subseteq t^I$ implies $s^I \subseteq t^I$

Then we say that s^I is a transitive closure of r^I, written in $(r^I)^+$.

So the transitive closure of a role r is the smallest transitive relationship including r under certain interpretation I. For this reason, it can be used to get a shortest path of services execution when coding accessibility relationship between Web services with $ALCI_{R+}$ roles. It just achieves the target that we want to find a shortest execution sequence of a composite service, i.e., the service composition problem.

The transitive closure of r^I can be computed in a usual manner and in $ALCI_{R+}$ separately:

$$(r^I)^+ = (r^I) \cup (r^I)^2 \cup (r^I)^3 \cup ...$$
$$\forall r^+.C = \forall r.C \cup \forall r. \forall r.C \cup \forall r. \forall r. \forall r.C \cup ...$$

It denotes concept whose individuals are directly accessible or transitive-enabled accessible to concept C.

Unfortunately, the complexity of this computation in DL is ExpTime-complete. However, according to the current state of the art, some highly optimized DL systems, such as FACT++, practiced well with transitive roles.

3.2 From DAML-S Specification to $ALCI_{R+}$ Formulas

DAML-S is suitable for describing Web services externally, but cannot be directly used for reasoning in. $ALCI_{R+}$ is a very well candidate for classification reasoning, which is not only critical to matching of Web service's **Input, Output, Precondition** and **Postcondition**, but also to describing relationships between services using transitive roles. So in our method DAML-S specification was transformed into $ALCI_{R+}$ formula for reasoning. We introduced several concepts and roles to describe DAML-S service information:

- Services names, Types of **Input** and **Output** consist of a finite set of concepts N_C in TBox,
- **Preconditions** and **Postconditions** consist of a finite set of ABox assertions,
- Relations between services and Properties of services consist of a finite set of role assertions in ABox.

For example, a service S_A with an Input of type C and an Output of type D can be described in $ALCI_{R+}$ as:

$$S_A \sqsubseteq \exists HasInput.C \sqcap \exists HasOutput.D.$$

Here *HasInput* and *HasOutput* are all properties of services. If we still know that S_A has a successor S_B, i.e., services S_A and S_B have a **Sequence** relationship in DAML-S, we can formulated it in $ALCI_{R+}$ as:

$$S_A \sqsubseteq \exists HasSuccessor.\ \{\ S_B\ \}.$$
$$HasSuccessor\ (S_A, S_B)$$
$$HasForward\ (S_B,\ S_A).$$

We defined role *HasForward* as an inverse of *HasSuccessor* $\in \mathbf{R_+}$. It is easy to find that these two roles are all transitive roles. For example, if there are *HasSuccessor* (S_A, S_B) and *HasSuccessor* (S_B, S_C), then *HasSuccessor* (S_A, S_C) holds. But a directly sequential relationship is different from a transitive-enabled sequential relationship, because services in the first and second roles can be invoked directly while services in the third one can't be. In order to distinguish them we introduced two kinds of roles: *IsdirectSequence* and *IsTransitiveSequence*, corresponding to the DAML-S *SimpleProcess* and *CompositeProcess*.

In the example preceding, if *HasSuccessor* (S_A, S_B) and *HasSuccessor* (S_B, S_C) are included in *IsdirectSequence*, then *HasSuccessor* (S_A, S_C) is ascribed to *IsTransitiveSequence*. This kind of intended distinguish is not only important for final execution but also critical for composition reasoning in our method.

Except for the sequence relationship between services, there is a parallel relationship between services which corresponds to **Split** relationship in DAML-S; we introduced the role *HasParallel* to represent it. Web services are often deployed and executed in a typical distributed environment. They can actually be executed at the same time. For example, in a multi-agent-based web services framework, services are seen as actions that can be invoked by different agents at the same time. So role *HasParallel* is a very important structure in service composition. It also can be seen as a transitive role.

$$HasParallel\ (S_A, S_B),\ HasParallel\ (S_B, S_c) \models HasParallel\ (S_A, S_C)$$

While there is also situations much freedom than transitive roles like:

$$HasParallel\ (S_A, S_B),\ HasParallel\ (S_A, S_c) \models HasParallel\ (S_B, S_C)$$
$$HasParallel\ (S_A, S_B),\ HasParallel\ (S_C, S_B) \models HasParallel\ (S_A, S_C)$$

There are also some relationships between services in DAML-S CompositeProcess, but they can either be transferred into basic relationships **Sequence** or **Split**. So we don't introduce anymore roles for ease to talk.

3.3 Service Composition in $ALCI_{R+}$ Reasoning

A composite service is a finite sequence of services S_i ($i=1 \ldots n$), S_i be an atomic service or composite service. The service composition problem is to find such a sequence where the service requirement can be fulfilled after the execution of every appropriate service.

Suppose we have extracted an input type **X** and an output type **Y** from a service requirement. Let S_{first} be the first service in the composite service that matches the Input and Precondition, and S_{final} the last service that matches the Output and Postcondition. So the type of S_{first}'s Input must be a type of **X** and the type of S_{final}'s Output must be a type of **Y**. We are going to find a service composition sequence from S_{first} to S_{final}.

Thus the composition is translated into whether formula:

$$I \models S_{first} \subseteq \forall HasSuccessor^+.S_{final}.$$

stands or not in $ALCI_{R+}$ reasoning. It is obviously that there is a service composition sequence started from service S_{first} and ends in service S_{final} that can accomplish the requirement when this formula is satisfied. So in our method, the choice of a final service is very necessary, the service composition sequence can be computed after some times of iteration.

The computation procedure can be stated as follows:

- First step, let set $Q = \forall HasSuccessor. \{S_{final}\}$ and check whether S_{first} is included in **Q**. If it is true, the procedure ends and the simple process $< S_{first}, S_{final} >$ is just the required composition service, else let $Z = \{< S_a\ S_{final} >| S_a \in Q$ **and** $< S_a\ S_{final} > \in HasSuccessor^{1}\}$ and go to the second step.
- Second step, compute $\forall HasSuccessor.Q$, and if **Q** equals to $\forall HasSuccessor.Q$, procedure ends, else let $Q = \forall HasSuccessor.Q$, check whether $S_{first} \subseteq Q$. If this formula is true, procedure ends too, else let $Z = \{< S_a, S_b >| S_b \in Q$ and $< S_a\ S_b > \in HasSuccessor^{1}\}$ and repeat second step until procedure ends.

In this procedure result set **Z** is used during the computation to keep the transitive closure of the accessibility relation $< S_{first}, S_{final} >$ from the first service to the last service.

Although the transitive closure has been computed, it contains all the possibilities that may satisfy the requirement. A short and better execution path must be chosen among these possibilities. This work can be done according to the non-functional properties described in the DAML-S service profile, such as the service price, execution time and so on.

4   Conclusion and Related Works

We are aim at tackling the automation of Semantic Web services matching and composition under the DL $ALCI_{R+}$ framework. There are some related works that focused on different DL frameworks for services composition problem such as article [6], but they did not give any approaches for automatic composition. Our method for semantic Web services composition is based on DL $ALCI_{R+}$ reasoning. The computation of the transitive closure of transitive roles that represent accessibility relations between the first and final services is critical. Compared with other DL reasoning methods depending on proving from the preconditions to postconditions, we directly used preconditions and postconditions as well as Input and Output types. It makes the best use of the information extracted from service requirement. Our method of computing transitive closure in $ALCI_{R+}$ is simple and viable although the complexity is ExpTime-complete. We can use the state of the art DL reasoners such as FACT++ that are proved to be practical for DL with transitive roles. The next work we want to do conclude: first, to optimize the computation of the transitive closure; second, to explore the distributed transitive closure which is introduced in article [7], it gives us some due for composition in a multi-agent-based services environment.

Acknowledgment

This work is supported by Natural Science Foundation of China (No. 60663005, No. 60803033, No. 61063002), China Postdoctoral Science Foundation (No. 20090450211), State Key Laboratory of Software Engineering (No.SKLSE20080710), Guangxi Natural Science Foundation of China (No. 0728093), Guangxi Graduate Innovation Foundation (No. 2009105950812M22).

References

1. Baader, F., Calvanese, D., McGuinness, D., et al.: The Description Logics Handbook: Theory, Implementations, and Applications. Cambridge University Press, Cambridge (2003)
2. Baader, F., Horrocks, I.: A Description Logic with Transitive and Inverse Roles and Role Hierarchies. Journal of Logic and Computation 9(3), 385–410 (1999)
3. Lee, T.B., Hendler, J., Lassila, O.: The Semantic Web. Scientific American 284(5), 28–37 (2001)
4. Cardoso, J., Sheth, A.P.: Semantic Web Processes: Semantics Enabled Annotation, Discovery, Composition and Orchestration of Web Scale Processes. In: 4th International Conference on Web Information Systems Engineering (WISE 2003), Roma, Italy, pp. 375–391 (2003)
5. The DAML Services Coalition: DAML-S: Semantic Markup for Web Services (December 2001)
6. Baader, F., Lutz, C., Milicic, M., Sattler, U., Wolter, F.: A Description Logic Based Approach to Reasoning about Web Services. In: Proceedings of the WWW 2005 Workshop on Web Service Semantics: Towards Dynamic Business Integration, pp. 199–210 (2005)
7. Houstma, M.A.W., Apers, P.M.G., Ceri, S.: Distributed Transitive Closure Computation: The Disconnection Set Approach. In: 16th International Conference on Very Large Data Bases, Brisbane, pp. 335–346 (1990)

Index-Based Search Scheme in Peer-to-Peer Networks

Jin Bo and Juping Zhao

Science and Technology Department, Shunde Polytechnic
Foshan 528333, China
wlmingyue@163.com

Abstract. Peer-to-Peer (P2P) systems are proposed to provide a decentralized infrastructure for resource sharing. In particular, file sharing was the initial motivation behind many of the pioneering P2P systems. As P2P systems evolve to support sharing of a large amount of data, resource locating issues must be addressed. In this survey, we propose reference architecture for P2P systems that focuses on the data indexing technology required to support resource locating. Based on this reference architecture, we present and classify technologies that have been proposed to solve the data indexing issues particular to resource locating in P2P systems.

Keywords: Peer-to-Peer, Indexing; Search, Distributed Hash Table.

1 Introduction

Peer-to-Peer (P2P) systems are massively distributed and highly volatile systems for sharing large amounts of resources. Nodes communicate through a self-organizing and fault tolerant network topology which runs as an overlay on top of the physical network. When we construct such P2P network, one important issue is query routing.

P2P routing algorithms can be classified as "structured" or "unstructured". Early instantiations of Gnutella were unstructured-keyword queries were flooded widely [12]. Napster [13] had decentralized content and a centralized index, so only partially satisfies the distributed control criteria for P2P systems. Early structured algorithms included Pastry [1], Chord [3], CAN [4] and Tapestry [8]. Researchers in [14] classified P2P systems by the presence or absence of structure in routing tables and network topology.

Some researchers in [1], [8] have regarded unstructured and structured algorithms as competing alternatives. Unstructured approaches are called "first generation", which are implicitly inferior to the "second generation" structured algorithms. When generic key lookups are required, these structured, key-based routing methods can guarantee location of a target within a bounded number of hops [6]. However, the broadcasting unstructured approaches may have much routing traffic, or fail to find the target content [5].

Recently, the "structured versus unstructured routing" taxonomy has been becoming not very useful, and there are two reasons: Firstly, most "unstructured" proposals have evolved and incorporated structure. Secondly, there are emerging schema-based P2P designs [9], with super-node hierarchies and structure within documents. All these above are quite distinct from the structured DHT proposals.

Y. Yu, Z. Yu, and J. Zhao (Eds.): CSEEE 2011, Part II, CCIS 159, pp. 102–106, 2011.
© Springer-Verlag Berlin Heidelberg 2011

Based on the analysis above, a more instructive taxonomy should be proposed to describe the structure of P2P systems. In this survey, we propose new reference architecture for P2P systems that focuses on the data indexing technology required to support resource locating. Based on this reference architecture, we present and classify technologies that have been proposed to solve the data indexing issues particular to resource locating in P2P systems.

The rest of this paper is organized as follows: Section 2 describes the related work. Then, we introduce the indexing strategies, and also propose the reference architecture for P2P systems that focuses on the data indexing in Section 3. We conclude this paper in Section 4.

2 Related Work

Some researchers have argued the various issues of P2P networks. The authors in [21] gives a mini-literature survey on the recent work of the P2P media streaming systems, attempting to overview and summary the ideas and techniques handling above challenges. The survey also covers five recent P2P media streaming systems and discusses their approaches.

In [22], researchers propose a framework for analyzing p2p content distribution technologies. The proposed approach focuses on nonfunctional characteristics such as security, scalability, performance, fairness, and resource management potential, and examines the way in which these characteristics are reflected in-and affected by-the architectural design decisions adopted by current p2p systems. Authors also study current p2p systems and infrastructure technologies in terms of their distributed object location and routing mechanisms, their approach to content replication, caching and migration, their support for encryption, access control, authentication and identity, anonymity, deniability, accountability and reputation, and their use of resource trading and management schemes.

Researchers in [23] provide taxonomy for resource discovery systems by defining their design aspects. This allows comparison of the designs of the deployed discovery services and is intended as an aid to system designers when selecting an appropriate mechanism. The surveyed systems are divided into four classes that are separately described. Finally, they identify a hiatus in the design space and point out genuinely distributed resource discovery systems that support dynamic and mobile resources and use attribute-based naming as a main direction for future research in this area.

3 Indexing Strategies

A P2P index can be local, centralized or distributed. With local index-based search scheme, a node can only keep the references to its own data, and does not receive references for data at other nodes. In central index-based search scheme, a single server just keeps references to data on many nodes. With distributed index-based search scheme, pointers towards the target reside at several nodes, and distributed indexes are used in most P2P designs nowadays.

3.1 Local Index-Based Search Scheme

P2Ps with a purely local data index have become rare. In such designs, nodes flood queries widely and only index their own content. They enable rich queries the search is not limited to a simple key lookup. However, they also generate a large volume of query traffic with no guarantee that a match will be found, even if it is shared by some nodes in the network. For example, to find potential nodes on the early version of Gnutella, 'ping' messages were broadcast over the P2P network and the 'pong' responses were used to build the node index. Then small 'query' messages, each with a list of keywords, are broadcast to nodes which respond with matching filenames [4].

There have been numerous attempts to improve the scalability of local-index P2P systems. Gnutella uses the fixed time-to-live (TTL) rings, where the query's TTL is set less than 7-10 hops [2]. Small TTLs reduce the network traffic and the load on nodes, but also reduce the query success rate. Researchers in [10] reported that the fixed "TTL-based mechanism does not work". To address the TTL selection problem, they proposed an expanding ring, known elsewhere as iterative deepening [7]. It uses successively larger TTL counters until there is a match. The flooding, ring and expanding ring methods all increase network load with duplicated query messages. A random walk, whereby an unduplicated query wanders about the network, does indeed reduce the network load but massively increases the search latency. One solution is to replicate the query k times at each node. Called random k-walkers, this technique can be coupled with TTL limits, or periodic checks with the query originator, to cap the query load [10].

3.2 Central Index-Based Search Scheme

Centralized index-based search schemes like Napster [13] are significant because they were the first to demonstrate the P2P scalability that comes from separating the data index from the data itself. Ultimately its service was shut down because of the massive copyright violations of music and film media as well as other intellectual property [15].

There has been little research on P2P systems with central data indexes. Such systems have also been called 'hybrid' since the index is centralized but the data is distributed. Authors in [16] devised a four-way classification of hybrid systems: (1) unchained servers, where users whose index is on one server do not see other servers' indexes; (2) chained servers, where the server that receives a query forwards it to a list of servers if it does not own the index itself; (3) full replication, where all centralized servers keep a complete index of all available metadata; (4) hashing, where keywords are hashed to the server where the associated inverted list is kept. The unchained architecture was used by Napster, but it has the disadvantage that users do not see all indexed data in the system. Strictly speaking, the other three options illustrate the distributed data index, not the central index.

3.3 Distributed Index-Based Search Scheme

An important early P2P proposal for a distributed index was Freenet [17]. While its primary emphasis was the anonymity of nodes, it did introduce a novel indexing scheme. Files are identified by low-level "content-hash" keys and by "secure signed-subspace" keys which ensure that only a file owner can write to a file while anyone can read from it. To search for a file, the requesting node first checks its local table for the node with keys closest to the target. When that node receives the query, it too checks for either a match or another node with keys close to the target. Finally, the query either finds the target or exceeds TTL limits. The query response traverses the successful query path in reverse, depositing a new routing table entry (the requested key and the data holder) at each node. The insert message similarly steps towards the target node, updating routing table entries as it goes, and finally stores the file there. Whereas early versions of Gnutella used breadth-first flooding, Freenet utilizes a more economic depth-first search [18].

There have been some extensions of Freenet work. To avoid the slow startup in Freenet, Mache *et al.* proposed to amend the overlay after failed requests and to place additional index entries on successful requests they claim almost an order of magnitude reduction in average query path length [18]. Clarke also highlighted the lack of locality or bandwidth information available for efficient query routing decisions in Freenet [19]. He proposed that each node gather response times, connection times and proportion of successful requests for each entry in the query routing table. When searching for a key that is not in its own routing table, it was proposed to estimate response times from the routing metrics for the nearest known keys and consequently choose the node that can retrieve the data fastest. The response time heuristic assumed that nodes close in the key space have similar response times. This assumption stemmed from early deployment observations that Freenet nodes seemed to specialize in parts of the key space it has not been justified analytically. Kronfol pay attention to Freenet's inability to do keyword searches [20]. He suggested that nodes cache lists of weighted keywords in order to route queries to documents, using Term Frequency Inverse Document Frequency measures and inverted indexes.

4 Conclusion

Resource locating is an important issue in P2P network field, and data indexing has great impact on the performance of the system. In this paper, we fist discusses various data indexing techniques in P2P networks. Next, the evolvements of P2P systems in each category are presented. The above analysis is very useful to design resource locating scheme when we build new P2P network in the future.

References

1. Rowstron, P.D.: Pastry: Scalable, Distributed Object Location and Routing for Large-Scale Peer-To-Peer Systems. In: IFIP/ACM International Conference on Distributed Systems Platforms (Middleware), pp. 329–350. IEEE Press, New York (2001)
2. Klingberg, T., Manfredi, R.: Gnutella 0.6,
   http://www8.cs.umu.se/~bergner/thesis/html/thesis.html

3. Stoica, R., Morris, D., Liben-Nowell, D., Karger, M., Kaashoek, F., Dabek, H., Balakrishnan, C.: A Scalable Peer-To-Peer Lookup Service for Internet Applications. In: ACM Annual Conference of the Special Interest Group on Data Communication, pp. 149–160. ACM Press, San Diego (2001)
4. Ratnasamy, S., Francis, P., Handley, M., Karp, R., Shenker, S.: a Scalable Content-Addressable Network. In: ACM Annual Conference of the Special Interest Group on Data Communication, pp. 161–172. ACM Press, San Diego (2001)
5. Huebsch, R., Hellerstein, J.M., Lanham, N., Loo, B.T., Shenker, S., Stoica, I.: Querying the Internet with PIER. In: 29th International Conference on Very Large Database, pp. 149–160. ACM Press, New York (2003)
6. Hellerstein, J.M.: Toward Network Data Independence. ACM SIGMOD Record 32, 34–40 (2003)
7. Yang, H.G.-M.: Efficient Search In Peer-To-Peer Networks. In: 22nd International Conference on Distributed Computing Systems, pp. 127–151. IEEE Press, Vienna (2002)
8. Zhao, B., Huang, L., Stribling, J., Rhea, S., Joseph, A., Kubiatowicz, J.: Tapestry: A Resilient Global-Scale Overlay for Service Deployment. IEEE Journal on Selected Areas in Communications 22, 41–53 (2004)
9. Nejdl, W., Siberski, W., Sintek, M.: Design Issues and Challenges for RDF and Schema-Based Peer-To-Peer Systems. ACM SIGMOD Record 32, 41–46 (2003)
10. Lv, Q., Cao, P., Cohen, E., Li, K., Shenker, S.: Search and Replication in Unstructured Peer-To-Peer Networks. In: 16th International Conference on Supercomputing, pp. 84–95. ACM Press, Baltimore (2002)
11. Plank, J., Atchley, S., Ding, Y., Beck, M.: Algorithms for High Performance, Wide-Area Distributed File Downloads. Parallel Processing Letters 13, 207–223 (2003)
12. Clip2, The Gnutella Protocol Specification, http://www.clip2.com
13. Napster, http://www.napster.com
14. Mishchke, J., Stiller, B.: A Methodology for the Design of Distributed Search in P2P Middleware. IEEE Network 18, 30–37 (2004)
15. Loo, A.W.: The Future or Peer-To-Peer Computing. Communications of the ACM 46, 56–61 (2003)
16. Yang, H.G.-M.: Comparing Hybrid Peer-To-Peer Systems. Technical Report, Stanford University (2001)
17. Clarke, S., Miller, T., Hong, O., Sandberg, Wiley, B.: Protecting Free Expression online with Freenet. In: IEEE Internet Computing, vol. 6, pp. 40–49 (2002)
18. Mache, J., Gilbert, M., Guchereau, J., Lesh, J., Ramli, F., Wilkinson, M.: Request Algorithms in Freenet-Style Peer-To-Peer Systems. In: 2nd IEEE International Conference on Peer-To-Peer Computing, pp. 1–8. IEEE Press, Sweden (2002)
19. Clarke: Freenet's Next Generation Routing Protocol,
    http://freenetproject.org/index.php?page=ngrouting
20. Kronfol,A.Z.: FASD: A Fault-Tolerant, Adaptive Scalable Distributed Search Engine. Master's Thesis, http://www.cs.princeton.edu/~akronfol/fasd/
21. Liu, X.: A Survey of Peer-To-Peer Media Streaming Systems. Course Project Report,
    http://people.cs.ubc.ca/~liu/course/cpsc538/
    asurveyofp2pmedia//streamingsystems.pdf
22. Androutsellis-Theotokis, S., Spinellis, D.: A Survey of Peer-To-Peer Content Distribution Technologies. ACM Computing Surveys 36, 335–371 (2004)
23. Vanthournout, K., Deconinck, G., Belmans, R.: A taxonomy for resource discovery. In: Müller-Schloer, C., Ungerer, T., Bauer, B. (eds.) ARCS 2004. LNCS, vol. 2981, pp. 78–91. Springer, Heidelberg (2004)

Field Theory Based Anti-pollution Strategy in P2P Networks

Xian-fu Meng and Juan Tan

School of Electronic and Information Engineering, Dalian University of Technology,
Dalian 116024, China
xfmeng@dlut.edu.cn, tanjuan865@gmail.com

Abstract. To deal with the file pollution problem in unstructured P2P file sharing systems, an anti-pollution strategy based on field theory is proposed. Based on the management of pollution degree, a pollution field is constructed, which is composed by pollution slicks. In order to reduce the impact on user experience while curbing the propagation of polluted files, classification management is achieved by applying different punitive measures to peer in different pollution slick. The experimental results show that this strategy can curb the propagation of polluted files quickly.

Keywords: Peer-To-Peer, File Pollution, Pollution Field, File Popularity, Classification Management.

1 Introduction

Peer-to-Peer (P2P) file-sharing systems have achieved a significant success until now. The normal operation of the P2P system is often affected by pollution attack because of its autonomy and openness. Researches show that pollution is pervasive in P2P file-sharing system definitely, especially for the popular files. In KaZaA there are more than 50% of the copies of popular songs have been polluted [1], ZHANG Min et al found that more than 59% servers in the eDonkey network are polluted servers [2].

To address the pollution problem, some reputation systems have been proposed, which can be categorized into three classes, i.e., peer-based reputation systems, object-based reputation systems and hybrid reputation system. The P-Grid [3] is a peer-based reputation system, a peer will collects the evaluations of other peers. But a high communication overhead is cost in this system. Credence system [4] uses object-based reputation, one collects evaluation to an object file during gossip procedure, however, but with a slow convergence rate. XRep [5]and X2Rep [6] are proposed based on hybrid reputation system which both evaluate the peers and the file, but it couldn't conducive to the development of P2P networks.

Based on the modern physics theory, this paper builds a pollution field in order to implement classification management of pollution peers effectively. And an anti-pollution strategy is proposed.

Y. Yu, Z. Yu, and J. Zhao (Eds.): CSEEE 2011, Part II, CCIS 159, pp. 107–111, 2011.

2 Related Definitions and Modeling

2.1 Related Definitions

According to the modern physics theory, if there are some kind of physical quantity distributed in space or one part of the space, we can say that a field is formed [7], such as density field, gravitational field, and so on. The impact of pollution peers has the basic elements of field: a specific space (file-sharing system) is available, a physical quantity (pollution degree) distributed in the space, and a specific distribution of physical quantity. In order to facilitate the specification of the pollution field, three definitions are given as follows:

Definition 1 pollution degree: Also known as pollution intensity. It refers to the strength of a peer that spread pollution files.

Definition 2 pollution slick: Refers to a set of peers whose pollution degree is located in the specified range.

Definition 3 pollution field: Refers to a continuous regional composed by a number of pollution slicks, peers in the pollution field will be called pollution peer.

2.2 The Calculation of Pollution Degree

Pollution degree of a peer is under the charge of its neighbors. Each peer has an evaluation matrix which is stored by its neighbors. Entries in the matrix are the evaluation of files shared by the peer. According to the quality of files provided by the peer, assessment is divided into two levels $\{-1, 1\}$, 1 means that the file was polluted, -1 means that the file wasn't polluted. When a query message is forwarded to the peer i, its neighbors will calculate the pollution degree of the peer i, Let V_i denote the matrix of peer i.

$$V_i = \begin{bmatrix} v_{11} & v_{12} & \cdots & v_{1m} \\ v_{21} & v_{22} & \cdots & v_{2m} \\ \cdots & \cdots & \cdots & \cdots \\ v_{n1} & v_{n2} & \cdots & v_{nm} \end{bmatrix}$$

Definition of pollution degree shows as Equation (1):

$$p_i = \begin{cases} (1-\delta)\sum v / N_v + \delta p_i' - \delta / N_{vb}\theta & \sum v > 0 \\ (1-\delta)\sum v / N_v + \delta p_i' - \delta / N_{vg}\theta & \sum v \le 0 \end{cases} \qquad (1)$$

where p_i denote the history pollution degree of peer i, and δ is the weight of it. v is the evaluation, N_v is the number of v, N_{vb} is the number of positive evaluation, N_{vg} is the number of negative evaluation. θ is the excitation factor, $\theta = N_{feedback} / N_{down}$, $N_{feedback}$ denotes the number of evaluation the peer send, and N_{down} denote the number of files the peer downloaded. From equation (1) we can see that $p_i \in (-1,1)$.

2.3 Independent Construction of Pollution Field

Pollution field is built up with the process of file retrieval, and it is composed of several pollution slicks. Let $P_{set} = \{P_k | 0 \leq k \leq d\}(P_0 < P_1 < ... < P_k < 1)$, $P_0 = 0$, $P_d = P_{max} = 1$, $P_k = P_{k-1} + P_{max}/k + 1$ $(k = d-1, d-2 ... 1)$, then any peer who's pollution degree between $[P_k, P_{k+1}]$ $(k = 0, 1, ... d)$ is located in the k-level field slick, denoted by symbol PS_k. Let φ_k denoted the average pollution degree of PS_k., definition of φ_k is shown as Equation (2):

$$\phi_k = P_k + (P_{k+1} - P_k)/2 \tag{2}$$

Through the process of query message, the neighbors will determine the pollution sick peer i located in, a pollution field was constructed which is illustrated in Figure 1.

3 Anti-pollution Strategy Based on Pollution Field

3.1 File Popularity

Polluters more likely choose the popular file as a carrier [1], so we introduce the file popularity factor into the Anti-pollution strategy. Because of the feature of unstructured P2P network, the local popularity will be calculated. Let T denotes the cycle of calculating file popularity, the definition of file popularity is given below.

Definition 4 single point popularity: c_{ij} is the access probability of file F_j of peer i. It is called the single point popularity of F_j, $c_{ij} = R_j/R$, where R_j denotes the access number of F_j, and R denote the access number of all files of peer i. If peer i didn't share F_j, then $c_{ij} = 0$, according to the definition, we know that $c_{ij} \in [0,1]$.

Definition 5 local popularity: H_{ij} is the estimate popularity value of F_j across the P2P network. It is called the local popularity of F_j,

$$H_{ij} = ac_{ij} + (1-a)\sum\nolimits_{peer_k \in path(i)} c_{kj} / N_k \tag{3}$$

where a is a constant and $1 \leq a \leq 0$, $path(i)$ denotes a set of peers which the query message passed before it arrived at the neighbor of peer i, N_k refers the number that the value of c_{kj} is not zero in $path(i)$.

3.2 Process of Query Message and Response Message

Repulsing query messages by the pollution field to control the pollution peers is the important part of our strategy. Now giving a description of the process of query message, Let peer i want to query the file F_j, it will send a Query message to its neighbors, the neighbor peers will process the message as follows after received it:

Step 1. Peer s will calculate p_i the pollution degree of peer i, then determine PS_x the pollution sick peer i located in, and the x will be filled into the Query message.

Step 2. Peer s will select a peer k from its neighbors, at first calculate p_k the pollution degree of peer k, if $p_k > 0$ transfer to Step 3, else transfer to Step4.

Step 3. Peer s will get the pollution degree of peer i and the file popularity of F_j from the Query message, then calculate H_{sj} the local popularity of F_j, $M(Q,i,F_j)$ the request intensity of message, its definition shows as Equation (4):

$$M(Q,i,F_j)=\begin{cases}(1-H_{mj})\mid p_i\mid & p_i\leq 0 \\ (1-H_{mj})(1-p_i) & p_i>0\end{cases} \qquad (4)$$

Then peer s will determine PS_n the pollution slick peer k located in first. The action intensity to the Query message from peer k isφ_n. At last, determine the value of Query State of the Query message in accordance with rules as follows:

(1) if $M(Q,i,F_j)$-$\varphi_n\geq|p_k-p_i|$, then Query State=0,which means the message will be forwarded to peer k, and it could return the response message.
(2) if $0\leq M(Q,i,F_j)$-$\varphi_n<|p_k-p_i|$, then Query State=1,which means the message will be forwarded to peer k, but it couldn't return the response message.
(3) if $M(Q,i,F_j)$-$\varphi_n<0$, then Query State=2,which means the message won't be forward to peer k.

If Query State=0 or 1, then transfer to Step 5; else transfer to Step 5.

Step 4. Peer s will forward the Query message directly to peer k.

Step 5. Peer s won't forward the Query message to peer k, and will select another peer, repeat Step 2 to Step 3 until forward the Query message to another peer.

Step 6. Peer k will continue to process the Query message as above.

Now giving the process of Response message, Let peer k revive a Query message forwarded by peer s, and Query State=0, it will process the message as follows:

Step 1. Peer k will search the file in the local shared directory, if hit it, then transfer to Step2, else forward the Query message using the algorithm in the previous section.

Step 2. Peer k calculate the popularity of file, and get x the level of pollution sick of the peer i, if $H_{kj}<1/(d-x-1)$, then send the Response messages to its neighbor peer s in accordance with the path the Query message carried, and transfer to Step3;else peer k won't send the Response message.

Step 3. Forward the message in accordance with the path Query message carried.

4   Evaluations

To evaluate the performance of our strategy, we simulate the P2P file sharing system in the P2P simulator PeerSim. There are 1000 peers and 10,000 initial files in the P2P network. The pollution degree of each peer for 0 initial, and d=3, δ=0.3, a= 0.5. There are two types of peers, i.e., honest peers and pollution peers. Pm present the fraction of pollution peers.

We take the case which did not add any anti-pollution strategy for the baseline. We present the comparison results by the daily fraction of unpolluted downloads by changing the value of Pm to verify the effectiveness of our strategy. As shown in Figure 2, on the 9th day, our strategy achieves a high ratio of success, maintaining roughly 95% correct classification. Even with the value of Pb is 0.6, our strategy also has got a good performance.

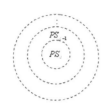

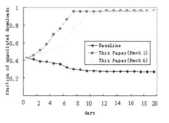

Fig. 1. Pollution Field   **Fig. 2.** Performance of strategy

5   Conclusions and Future Work

To address the pollution attack of P2P network, this paper presents an anti-pollution strategy based on pollution field. When curbing the pollution file we also implement the classification punish of pollution peers. For our future work, we plan to improve the ability of our strategy to distinguish the fake evaluation.

References

1. Jian, L., Rakesh, K., Yongjian, X., Keith, W.R.: Pollution in P2P File Sharing Systems. In: IEEE INFOCOM, pp. 1174–11859. IEEE Press, Miami (2005)
2. Min, Z., Changjia, C., Jinkang, J.: Fake Servers in Edonkey Networks. In: International ICST Qshine, Hong Kong (2008)
3. Karl, A., Zoran, D.: Managing Trust in a Peer-To-Peer Information System. In: 10th CIKM, pp. 310–317. ACM Press, Atlanta, Georgia (2001)
4. Kevin, W., Emin, S.: Fighting Peer-To-Peer SPAM and Decoys with Object Reputation. In: P2PECON Workshop, pp. 138–143. ACM Press, New York (2005)
5. Ernesto, D., Sabrina, D.C., Stefano, P., Pierangela, S., Fabio, V.: A Reputation-Based Approach for Choosing Reliable Resources in P2P Networks. In: ACM CCS, Washington, pp. 207–216 (2002)
6. Nathan, C., Rei, S.N., Willy, S.: X2Rep: Enhanced Trust Semantics for the Xrep Protocol. In: Proc. of Applied Cryptography and Network Security, Yellow Mountain, pp. 205–219 (2004)
7. Dongsheng, M.: System Theory. China Renmin University, M. Beijing (1999)

Semantic Web Service Composition: From OWL-S to Answer Set Programming[*]

Junyan Qian[1,2], Guowang Huang[1], and Lingzhong Zhao[1]

[1] Department of Computer Science and Engineering, Guilin University of Electronic Technology, Guilin 541004, China
[2] State Key Laboratory of Software Engineering, Wuhan University, Wuhan 430072, China
qjy2000@gmail.com, hv_32@yahoo.com.cn, zhaolingzhong@guet.edu.cn

Abstract. This paper presents a method that adapting planning description to bring the semantic information into play for service composition through action language C. It shows how service descriptions can be expressed by preconditions and effects and the action language C provides a richer syntax and semantic for complex service descriptions. We also presents the algorithm of translating semantic web service described by OWL-S to action language C. Thanks to the structured description and the powerful expression of C, we only consider the initial Situation and the desired goal ignoring details of transition and planning. At last we use satisfiability planning to solve the planning problem by translating the action language into disjunctive logic program.

Keywords: Service composition, Action language, Answer set programming.

1 Introduction

Semantic Web services are created and updated on the fly at present. There are many initiatives that use to describe web services, such as WSDL, BPEL4WS and OWL-S, which are focused on representing services composition. We are only interested in OWL-S which bases on describe logic whose process ontology provides a vocabulary for describing the composition of Web Services. Several methods for services composition have been proposed, in particular, most researches conducted fall in the realm of AI planning. The idea of satisfiability planning [1] is to deal with the problem of plan generation by reducing it to the problem of finding a satisfying interpretation for a set of proposition. Answer set programming method uses logic program that is easier than the previous, this is also the preponderance of this approach simultaneously. In this paper our aim is to shows how composition procedures are described in the action language to produce specific plans for the desired goals. Our main contribution is the translation of OWL-S to action language and how planning with complex actions is done within our algorithm.

[*] This work is supported by Natural Science Foundation of China (No.60663005, No.60803033, (No.61063002), China Postdoctoral Science Foundation (No.20090450211), State Key Laboratory of Software Engineering (No.SKLSE20080710), Guangxi Natural Science Foundation of China (No.0728093).

Y. Yu, Z. Yu, and J. Zhao (Eds.): CSEEE 2011, Part II, CCIS 159, pp. 112–117, 2011.
© Springer-Verlag Berlin Heidelberg 2011

2 Preliminaries

A program called extended disjunctive logic programs [2] is only consider here, which not only allows for two kinds of negation, the classical negation ¬ and negation as failure not ,but also the ideal of rules with disjunctive heads. An extended disjunctive logic programs P is a finite set of rules r of the form:

$$h_1|...|h_k \leftarrow l_1,...,l_m ,not\ l_{m+1},...,not\ l_n$$

Where $1 \leq m \leq n$,each l_i is a literal (a literal is an atom p or its negation $\neg p$. An answer set of a program can be obtained as follow:

First let P be an extended disjunctive program without negation as failure. Lit stands for the set of literal in the language of P. An answer set of P is any subset S of Lit such that

(I) for each rule $h_1|...|h_k \leftarrow l_1,...,l_m$ of P, if $l_1,...,l_m \in S$, then some $h_i \in S, 1 < i$

(II) if S contains a pair of complementary literals, then $S=Lit$

Secondly, we consider the program P with not. The reduct $\prod^s$ of P is obtained using the procedure as follow, that is, all rule with unsatisfiable negative literals in the body are dropped as well as all remaining negative literals. If S is an answer set of $\prod^s$, then it also is the answer set of P.

An action language planning problem can be described by a triple $<D, I, G>$, Where: (D, I) is an action theory G is fluent formula which represents goal state. An action sequence $a_1,...,a_m$ is a plan of G if there exists a trajectory $s_0a_1s_1,...,s_{m-1}a_ms_m$ such that s_0 and s_m satisfies I and G, respectively [4]. We use high-level action language C [5] to represent action theories, which is a very expressive language, for it is able to deal with the problem of nondeterministic actions and the concurrent execution of actions .

3 Translate OWL-S into Action Language

In this paper we consider the Web service composition problem as an action language planning problem. One of the reasons why we adopt the planning is that both the planning problem and the composition problem search for a sequence of operations that could get the desired goal starting from the initial state. The other, the Web services operations have input and output data, preconditions and effects as well as the AI planning domain actions, which make them are appropriate for planning approach [3].

To solve a planning problem in a given planning domain, the answer set planner needs to be given the knowledge about that domain in the action language C formalism. Now we should convert the Web service description to action language C, so as to apply the answer set planning. In the first step, Web service description and user goal are converted into action problem description, in the second step, we translate language C into answer set programming specification.

3.1   Atomic Processes to C Formalism Translation

An atomic processes ontology has several properties including a set of input and output parameters, precondition and effect. Preconditions provide state pre-required to invoke the service, Effect represent what is changed after the execution of a service. Input and Output correspond to knowledge preconditions and effects. Our method of translating OWL-S atomic processes into equivalent action language is illustrated as the following table 1.

Table 1. Translating OWL-S atomic processes into equivalent action language

OWL-S	Action language C
Atomic process	Action name
Precondition	Physical precondition
Input	Knowledge precondition
Effect	Physical precondition
Output	Knowledge effect

The precondition of an atomic process is the necessary condition to execute the service, which is the same to an action corresponding to action language C. We consider both the precondition and the input as precondition in the action language C. To avoid confusion, we write the precondition which comes from the precondition of atomic process as knowledge precondition, the others as physical precondition.

A physical precondition is encoded in C as:

$$caused \text{ executable}(a) \text{ } if \text{ holds}(p_1) \wedge ... \wedge \text{holds}(p_n)$$

where p_i is precondition and a is the name of the service. While knowledge precondition provides the value of the input parameters for an agent to execute the service operation, which is encoded as follow:

$$caused \text{ executable}(a) \text{ } if \text{ knows}(i_1) \wedge ... \wedge \text{knows}(i_n)$$

The necessary precondition of an action is the mergence of all the preconditions:

$$caused \text{ executable}(a) \text{ } if \text{ holds}(p_1) \wedge ... \wedge \text{ holds}(p_n) \wedge \text{ knows}(i_1) \wedge ... \wedge \text{knows}(i_n)$$

Analogously, while dealing with the Effects and Output of an atomic process, we sign them as $\text{holds}(e_1),..., \text{holds}(e_n)$ and $\text{knows}(o_1),...\text{knows}(o_n)$ respectively which are both considered as the effect of an action in C. We represent an algorithm for this translation procedure as below:

```
Definition AP<A,I,P,E,O> where:
    A: atomic process name
    I: set of input:{i₁,..,iₙ}
    P: Precondition: conjunction of literals: (p₁,…, pₙ)
    E: Effect: conjunction of literals (e₁,…, eₙ)
    O: set of output:{o₁,.., oₙ}

program APtoC
    Input: an atomic process AP<A,I,P,E,O>
    Output: a law of action language C
```

```
Begin
   If  pᵢ∈ P
      preᵢ=holds(pᵢ);
      Pre = pre₁∧...∧ preₙ;
   If  iᵢ∈ I
      inᵢ= knows(iᵢ);
      In = in₁∧...∧inₙ
   If  eᵢ∈ E;
      effᵢ=eᵢ;
      Eff =eff₁∧...∧effₙ;
   If  oᵢ∈O;
      outᵢ =oᵢ;
      Out=out₁∧...∧outₙ;
   Return caused Eff∧Out if Pre∧In after AP;
End
```

3.2 Compound Processes to C Formalism Translation

Composite processes are constructed by several subprocesses through OWL-S control constructs, which represent the relation of execution between the different subprocesses. In order to define how composite processes are mapped to the corresponding action language C, we should first define the constants, actions and fluent. The similar problem has been studied earlier in different contexts [7]. Each service in the composite processes is represented by a constant name (say, A_i). The action executed(a) denotes the execution of an action which is corresponding to a service, while executable(a) represents the a can execute or not.

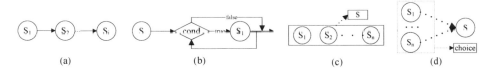

Fig. 1. OWL-S control constructs. (a) Sequence, (b) Split, (c) Any-order, (d) Choice

Sequence

The Sequence construct contains the set of all component processes to be executed in order. The sequential transition is given in Fig.1(a).

This transition can be described in C as follows:

$$S_1 \text{ caused executed } (S_1) \text{ if } PreS_1 \wedge InS_1.$$
$$S_i \text{ causes executed } (S_i) \text{ if executed } (S_{i-1}), PreS_i \wedge InS_i.$$
$$\text{nonexecutable ser if executed (ser).}$$

Loops (Repeat-while and Repeat-until)

The loop condition controls the execution of service in the loop, the value of which can be changed by the execution of service in the loop. The service in the loop executes tautologically until the loop condition is false. The most common situation is that the number of execution of the loop for a given integer φ.

S_1 *caused* count(J) *if* count(I) $\wedge$sum(J, I, 1) $\wedge I<\varphi\wedge$PreS$_1\wedge$InS$_1$.
        caused ¬count (I) *if* count (J) $\wedge$¬equal (I, J).

where sum(i, j, k) means the calculates $i = j + k$. The second rule is to state that only one of the count(number) fluent can be true any time.

Any-order

The subprocesses $S_1...S_n$ can execute in arbitrary order, but none of them executed in the same time. In other words they can not execute concurrently. After all of the subprocesses have executed, the service S is executed. This transition can be described in C as follows:

$S_i(T_i)$ *caused* executed(S_i, T_i) *if* PreS$_i\wedge$InS$_i$.
        nonexecutable S$_j(T_i)$ *if* S$_i(T_i)$ $\wedge$¬equal (I, J).
    caused executed(S) *if* executed(S_1, T_1) $\wedge...\wedge$executed(S_n, T_n)

   In order to describe the unconcurrency relation, we import the auxiliary variable time T. The predicate $S_i(T_i)$ means S_i executes at time T. We ignore the time of executing, the activity starts at time T and accomplishs at the same moment. In the first two rules, we state that there one and only one of the services can execute in the same moment. With the third rule we get that, after all of the services are executed, the process S is executed.

Choice

Different from the any-order structure, one of the subprocesses $S_1...S_n$ is executed, the Choice is completed. For convenience, we add S which is the successor process of choice in fig.1(d). This transition is represented by rules of the form:

S_i *caused* executed(S_i) *if* PreS$_i\wedge$InS$_i$
        caused executable(S) *if* executed(S_i).

where 1$\leq$M$\leq$N

4   Translate C into ASP

Answer set planning refers to the approach to planning in which a problem is translated into a logic program whose answer sets correspond to plans. Lifschitz and Turner showed a well known approach of translating from C to disjunctive logic program [8]. Given an action language planning problem $<D, I, G>$, Let notation $\prod(D, I, G)$ be the disjunctive logic program correspond to the problem. For every positive integer T called the time stamp of the literal. A logic program $\prod_t(D)$ whose answer sets correspond to paths of length T in the transition system described by D. The following result is adapted from Lifschitz's translation. Program $\prod_t(D, I, G)$ will be given in several steps:

   Step One: Translation of the initial state I :
$$\prod_I (I) = \{\text{holds}(f, 0). f\in I\}$$
   Step Two: Translation of the final state G:
        For every $f_i\in G$,
            goal(T) $\leftarrow$ holds(f_1, T),...,holds(f_n,T) $\leftarrow$ not goal(T)
   Step Three: Translation of the action description D:

(i) for every static law
$$caused \ F(t) \ if \ l_1(t) \wedge ... \wedge l_m(t)$$
in D, the rules
$$F(t) \leftarrow not \ \neg holds(l_1, t), \ ..., not \ \neg holds(l_m, t)$$
(ii) for every dynamic law
$$caused \ F(t) \ if \ l_1(t) \wedge ... \wedge l_m(t) \ after \ l_{m+1}(t) \wedge ... \wedge l_n(t)$$
in D, the rules
$$F(t) \leftarrow not \ \neg l_1(t), \ ... \ , not \ \neg l_m(t), l_{m+1}(t), ..., l_n(t)$$
(iii) for very fluent name or action name, we add the following rules
$$\neg B(t) \leftarrow not \ B(t) \quad B(t) \leftarrow not \ \neg B(t)$$
where B is a fluent atom end with the time stamp 0 or an action atom. for all $t = 0,...,$
$T-1$,

5 Conclusions and Future Work

In this paper we presented an approach that convert the semantic web services composition problem to an answer set planning problem which is powerful in representing planning problems with incomplete information about the initial situation and non-deterministic actions. We are currently focused on semantic service descriptions using action language C and the algorithm of translation service composition problem to answer set panning. The answer set panning seems to be reasonably efficient in searching for comparatively short, highly parallel plans, in complex domains. We believe that new answer set solving algorithms and systems will substantially expand its applicability. We should make a tool to realize our algorithm of translation Atomic Processes and composition Processes into action language C automatically and achieve converting the action language planning problem $<D, I, G>$ to the corresponding answer set planning $\prod(D, I, G)$. We hope the method that semantic web service composition using answer set planning showed in this paper is promising and deserves further application.

References

1. Kautz, H., Selman, B.: Planning as Satisfiability. In: Proc. ECAI-WS 1992, pp. 359–363 (1992)
2. Gelfond, M., Lifschitz, V.: Classical Negation in Logic Programs and Disjunctive Databases. New Generation Computing, pp.365–385 (1991)
3. Subrahmanian, V.S., Zaniolo, C.: Relting Stable Models an AI Planning Domains. In: Proc. ICLP-1995, pp. 233–247. MIT Press, Cambridge (1995)
4. Lifschitz, V., Turner, H.: Representing Transition Systems by Logic Programs. In: Gelfond, M., Leone, N., Pfeifer, G. (eds.) LPNMR 1999. LNCS (LNAI), vol. 1730, pp. 92–106. Springer, Heidelberg (1999)
5. Lifschitz, V.: Action Languages, Answer Sets and Planning. The Logic Programming Paradigm: a 25–Year Perspective, 353–373 (1999)
6. Lifschitz, V., Turner, H.: Representing Transition Systems by Logic Programs. In: Gelfond, M., Leone, N., Pfeifer, G. (eds.) LPNMR 1999. LNCS (LNAI), vol. 1730, pp. 92–106. Springer, Heidelberg (1999)
7. Nihan, P.K., Koksal, P., Cicekli, N.K., Toroslu, I.H.: Specification of Workflow Processes Using the Action Description Language. In: Answer Set Programming. Spring 2001 Symposium Series, pp. 103–109 (2001)

Improving K-Nearest Neighbor Rule with Dual Weighted Voting for Pattern Classification

Jianping Gou[*], Mingying Luo, and Taisong Xiong

School of Computer Science and Engineering, University of Electronic Science and
Technology of China, Chengdu, 610054, P. R. China
{cherish.gjp,mingyingluo,xiongtaisong}@gmail.com

Abstract. In this paper, we propose a dual distance-weighted voting for KNN,
which can solve the oversmoothing of increasing the neighborhood size k. The
proposed classifier is compared with the other methods on ten UCI data sets.
Experimental results suggest that the proposed classifier is a promising
algorithm due to its satisfactory classification performance and robustness over
a large value of k.

Keywords: The k-nearest neighbor rule, Weighted voting, The distance-
weighted k-nearest neighbor rule.

1 Introduction

In pattern classification, the k-nearest neighbor (KNN) rule is one of the oldest and
simplest nonparametric techniques, due to its simplicity and effectiveness. This rule
assigns an unclassified pattern by the majority label among its k nearest neighbors in
the training set. It has a good asymptotic performance: (i) its error rate approximates
the optimal Bayesian error rate if the number of training objects approaches infinity,
and (ii) if k=1, its asymptotic error rate is bounded above by twice the optimal
Bayesian error rate [1], [2]. Dudani [3] firstly proposed a weighted voting method,
called the distance-weighted KNN rule (WKNN), which weights more heavily close
neighbors based on their distances to the query. It is an improvement for KNN.
However, the good asymptotic performance of the KNN usually can not be obtained
in the presence of small sample size, and oversmoothing of increasing neighborhood
size k degrades the classification performance. In addition, KNN gives an identical
weight to each neighbor of the query, and results in unnecessary classification error.

In order to overcome these problems above, we propose a dual weighted voting
based on WKNN, to improve the classification performance of KNN. In this new rule,
we employ dual weights of k nearest neighbors to determine the class of the query by
the majority weighted voting. The experimental results show the effectiveness of our
proposed classifier in many practical situations.

---

[*] Corresponding author.

Y. Yu, Z. Yu, and J. Zhao (Eds.): CSEEE 2011, Part II, CCIS 159, pp. 118–123, 2011.
© Springer-Verlag Berlin Heidelberg 2011

2   The Proposed Classifier

In this section, we firstly review the related work briefly, and then give our proposed classifier.

2.1   The KNN Classifier

The k-nearest neighbor rule (KNN), introduced by Cover and Hart [1], is one of the well-known classifiers. In this rule, a query is assigned to the class, represented by a majority of its k nearest neighbors in the training set. NN is the simplest form of the KNN rule, when $k=1$. According to the KNN rule, we give a high-level summary of the nearest neighbor classifier. Let $T = \{x_i, y_i\}_1^N$ be the training class-labeled set, with m-dimensional feature vectors $x_i \in R^m$ and the corresponding class labels y_i. Given a query (x', y'), its unknown class y' is determined in the following:

1) select the set $T' = \{x_1^{NN}, ..., x_k^{NN}\}$ of k nearest objects to x', arranged in an ascending order, based on the distance measure between x_i^{NN} and x', $d(x', x_i^{NN})$. $d(x', x_i^{NN})$ is popularly computed by Euclidean distance metric:

$$d(x', x_i^{NN}) = \left\| x' - x_i^{NN} \right\|_{L_2} . \tag{1}$$

2) assign the class label to the query object, based on the majority voting class of its k nearest neighbors:

$$y' = \arg\max_{y} \sum_{(x_i^{NN}, y_i^{NN}) \in T'} I(y = y_i^{NN}) . \tag{2}$$

where y is a class label, y_i^{NN} is the class label for the i-th nearest neighbor among its k nearest neighbors. $I(y = y_i^{NN})$, an indicator function, takes a value of one if the class y_i^{NN} of the neighbor x_i^{NN} is the same as y, and zero otherwise.

2.2   The WKNN Classifier

As stated above, the KNN rule implicitly assumes that k-nearest neighbors have an identical weight in making decision, and neglects that closer nearest neighbor should contribute more to the classification. It is intuitively appealing to define a weighted voting to give k nearest neighbors different weights. To weigh the closer neighbors more heavily than the farther ones, Dudani proposed the distance-weighted k-nearest neighbor rule (WKNN), in which the votes of the different members of the k nearest neighbors set are computed by a function of their distances to the query [3].

In this scheme, the i-th weight of the corresponding nearest neighbor is defined as below:

$$w_i = \begin{cases} \dfrac{d_k^{NN} - d_i^{NN}}{d_k^{NN} - d_1^{NN}}, & d_k^{NN} \neq d_1^{NN}, \\ 1 & d_k^{NN} = d_1^{NN}. \end{cases} \qquad (3)$$

where d_i^{NN} is the distance to the query of the i-th nearest neighbor, d_1^{NN} is the distance of the nearest neighbor and d_k^{NN} is the distance of the k-furthest neighbor.

Then, the query is assigned the majority weighted voting class label $y_{j_{max}}$ using the following rule:

$$y_{j_{max}} = \arg\max_{y_j} \sum_{(x_i^{NN}, y_i^{NN}) \in T'} w_i \times I(y = y_i^{NN}) \cdot \qquad (4)$$

3   The Proposed Dual Weighted Voting for KNN Rule

As for the KNN classifier, the decision rule for pattern classification is determined by the estimate of the conditional probabilities from objects in a local region of the data space. The local region contains k nearest neighbors of the query, and its radius depends on the distance of the k-furthest neighbor. So k determines conditional class probabilities. If k is very small, the local estimate tends to be very poor due to the data sparseness and the noisy points. In order to smooth the estimate, we can increase k and take into account a large region around the query. However, a large value of k easily makes the smoothing of the estimate become oversmoothing, because the dramatic degradation of the classification performance results from the introduction of more outliers from other classes [4]. In addition, the supposition of each neighbor with an identical weight also has a negative influence on the estimate.

Motivated by these issues, we propose a dual distance-weighted voting for k-nearest neighbor rule (DWKNN), based on WKNN. The new classifier addresses the problems and improves the classification performance of KNN. Let $x_1^{NN}, ..., x_k^{NN}$ denote the k-nearest neighbors to the query x', and $d_1^{NN}, ..., d_k^{NN}$ be the corresponding distances, arranged in an increasing order in terms of the dissimilarity between x_i^{NN} and x'. A dual distance-weighted voting scheme is defined as follow:

1) select the set $T' = \{x_1^{NN}, ..., x_k^{NN}\}$, the same as the representation of KNN.

2) assign a dual weight w_i to the i-th nearest neighbor, x_i^{NN}, using dual distance-weighted function.

$$w_i = \begin{cases} \dfrac{d_k^{NN} - d_i^{NN}}{d_k^{NN} - d_1^{NN}} \times \dfrac{d_k^{NN} - d_i^{NN}}{d_k^{NN} - d_1^{NN}}, & d_k^{NN} \neq d_1^{NN}, \\ 1 & d_k^{NN} = d_1^{NN}. \end{cases} \qquad (5)$$

3) predict the class of the query object by the majority weighted voting, the same as Eq. 4.

It is obvious that the dual weight of each nearest neighbor is the square of the weight in the WKNN rule and each weight also depends on the corresponding distance. As can be seen from Eq. 5, it is clear that the dual weight w_i is less than that of WKNN, except the weights of the first and k-further nearest neighbors. Hence, the corresponding neighbor x_i^{NN} has less influence on the classification of the query. In the new rule, the nearest neighbor gets weight of 1, the furthest k-th neighbor gets a weight of 0, and the other weights are scaled linearly between 0 and 1.

4   Experimental Results

In this section, we evaluate the classification performance of the proposed classifier. In order to obtain the reliable classification performance, we conduct experiments via error or accuracy rate on ten real data sets respectively. The real data sets are selected from the UCI machine learning repository [5] and the Euclidean distance is used as the distance metric. To verify the classification performance to be reliable, we adopt the Leave-One-Out (LOO) method.

4.1   Experimental Data Sets

We briefly describe the real data sets in Table 1. The number of instances varies from 214 (Glass) to 20000 (Letter) while the number of attributes varies from 10 (Glass) to 90 (Libras Movement). Amongst the ten real data sets, eight data sets are multi-class classification tasks, and two data sets are two-class classification tasks.

Table 1. Some characteristics of the UCI data sets

Data set	Attributes	Instances	Classes
Pendigits	16	10992	10
Optdigits	64	5620	10
Ionosphere	34	351	2
Glass	10	214	7
Sonar	60	208	2
Landsat Satellite	36	6435	7
Libras Movement	90	360	15
Wine Quality-White	11	4898	11
Letter	16	20000	26
Image Segmentation	19	2310	7

4.2   Experimental Comparison

We conduct experiments to investigate the classification performance of the proposed classifier on the real data sets, compared with KNN and WKNN.

Table 2. The lowest error of each method with the corresponding k on all data sets

Data set	KNN	WKNN	DWKNN
Ionosphere	13.39(1)	13.39(1)	*12.54(11)*
Landsat Satellite	8.44(4)	8.45(7)	*8.30(14)*
Libras Movement	12.78(1)	12.78(1)	*12.22(5)*
Glass	26.64(1)	26.64(1)	*25.70(4)*
Wine Quality-White	38.38(1)	38.36(4)	*37.93(9)*
Sonar	16.83(1)	16.35(4)	*15.38(5)*
Pendigits	0.58(3)	0.53(6)	*0.52(10)*
Optdigits	1.00(4)	*0.96(7)*	*0.96(9)*
Image Segmentation	3.33(1)	*3.12(6)*	3.20(8)
Letter	3.64(4)	3.29(8)	*3.12(19)*
Average Error	12.50	12.39	*11.99*

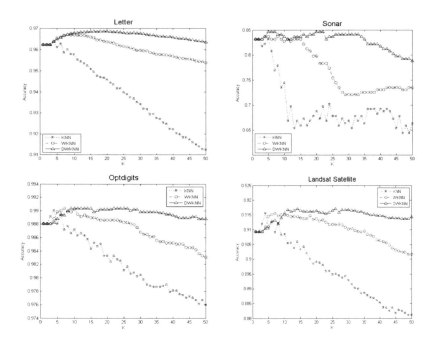

Fig. 1. The classification accuracies via the neighborhood size k

The optimal error rates (%) of each classifier with the corresponding optimal k in the parenthesis on each data set are shown in Table 2. Note that the neighborhood size k varies from 1 to 50 in our experiments. The italic error rates are the smallest error rates among them on each data set. As shown in Table 2 at the best cases, the average error rate of DWKNN on all the real data sets is lower than other methods. We can also see that the classification performance of DWKNN is superior to KNN on whole data sets, and WKNN in 8 out 10 data sets. Furthermore, the most importantly experimental result is such that the optimal neighborhood size k on many data sets is one as shown in Table 2. It has been found that the error rate of NN is bounded above by twice the optimal Bayes error [1], [2]. Only if the distribution of samples around a query is dense enough [6], a larger value of k guarantees a lower error rate. However, in the case of the data sets with small sample size in our experiments, a large value of k may not guarantee the good classification performance due to outliers from the other classes [4]. Consequently, NN ($k=1$) results in the best performance for many data sets in Table 2. On the other hand, with a small value of k, the classification result can be unreliable due to data sparseness or noise objects [7], [8]. It is more appealing that, unlike the other two methods, DWKNN has the large optimal values of k on all data sets, and the optimal values of k are always equal or greater than 4 and are larger than that of KNN and WKNN. In addition, Fig. 1 only depicts the classification accuracies on four data sets via k, because the experimental results of other data sets result in the similar pattern as the four ones. As shown in Fig. 1, the accuracies of DWKNN decrease slowly as k increases, however, KNN and WKNN decrease rapidly. It follows that DWKNN can deal with the negative effects of oversmoothing with increasing the value of k.

As a consequence, we can draw a conclusion that the proposed classifier has three benefits: (a) it can employ more nearest neighbors to keep the estimate of the majority voting smooth. (b) it can be robust to large values of the neighborhood size k. (c) the satisfactory classification performance results from a larger value of k than that of other methods. So the proposed classifier is an effective algorithm in many practical situations.

5 Conclusion

In this paper, we proposed dual distance-weighted voting method for KNN in pattern classification. The experimental results show that this proposed classifier always outperforms the other classifiers with a large value of k. Therefore, the dual weighted voting method for KNN is a promising algorithm.

References

1. Cover, T.M., Hart, P.E.: Nearest Neighbor Pattern Classification. IEEE Transactions on Information Theory 13, 21–27 (1967)
2. Hastie, T., Tibshirani, R., Friedman, J.: The Elements of Statical Learning. Springer, New York (2001)
3. Dudani, S.A.: The Distance-weighted K-Nearest Neighbor Rule. IEEE Transactions on System Man and Cybernetics 6, 325–327 (1976)
4. Fukunaga, K.: Introduction to Statistical Pattern Recognition. Academic Press, London (1990)
5. Frank, A., Asuncion, A.: UCI Machine Learning Repository. University of California, School of Information and Computer Science, Irvine, CA (2010),
   http://www.archive.ics.uci.edu
6. Duda, R.O., Hart, P.E., Stork, D.G.: Pattern Classification. New York, NY, USA (2001)
7. Wu, X.D., Kumar, V., et al.: Top 10 Algorithms in Data Mining. Knowledge Information System 14, 1–37 (2008)
8. Zavrel: An Empirical Re-examination of Weighted Voting for K-NN. In: Proceedings of the 7th Belgian-Dutch Conference on Machine Learning, Tilburg, pp. 139–148 (1997)

Building a WBC Software Testbed for the Ubiquitous Consumer Wireless World

Zhanlin Ji[1], Ivan Ganchev[2], and Máirtín O'Droma[2]

[1] CAC Dept, Hebei United University, TangShan, 064300, P.R. China
[2] ECE Dept, University of Limerick, Limerick, Ireland
Zhanlin.ji@ieee.org, Ivan.Ganchev@ul.ie, Mairtin.ODroma@ul.ie

Abstract. The wireless billboard channels (WBCs) are used by the service providers (xSPs) for broadcasting advertisements of their services to the mobile terminals (MTs) in the ubiquitous consumer wireless world (UCWW). In this paper, a 3-layer model and a software testbed for WBCs based on the digital video broadcasting - handheld standard (DVB-H) are set out.

Keywords: Ubiquitous Consumer Wireless World (UCWW); Wireless Billboard Channel (WBC); Digital Video Broadcasting - Handheld (DVB-H), Simulink.

1 Introduction

The wireless billboard channels (WBCs) [1, 2] used for service advertisement, discovery and association (ADA) are fundamental to the consumer-centric business model (CBM), which is integral to the ubiquitous consumer wireless world (UCWW) evolution [3, 4]. The WBCs are broadcasting channels used to push service advertisements simultaneously to a large number of mobile terminals (MTs).

A set of technologies, both terrestrial and satellite, such as digital radio mondiale (DRM), digital audio broadcasting (DAB), digital video broadcasting - handheld (DVB-H), digital multimedia broadcasting (DMB), multimedia broadcast multicast service (MBMS) etc, can act as a WBC carrier. Among these, the DVB-H standard deserves particular attention. This is a new digital broadcasting standard for IP-datacasting (IPDC) to portable and battery limited MTs. Several novel features are included in the DVB-H standard [5].

- *Transmission parameter signaling* - used to enhance and speed up the service discovery;
- *4K mode* - offers an additional trade-off between the single-frequency network cell size and mobile reception performance;
- *In-depth symbol interleaving* - increases the flexibility of the symbol interleaving thus improving the robustness in mobile environments and impulse noise conditions.

Building a WBC software testbed based on the DVB-H is the subject of this paper.

Y. Yu, Z. Yu, and J. Zhao (Eds.): CSEEE 2011, Part II, CCIS 159, pp. 124–129, 2011.

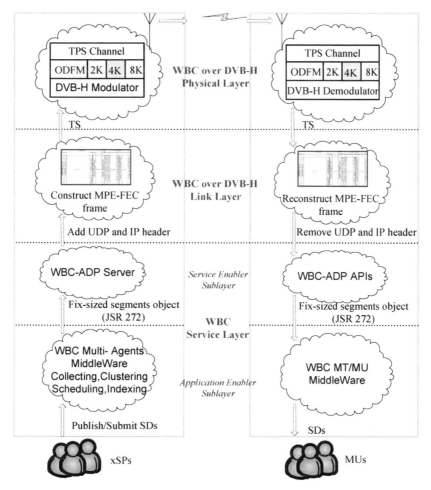

Fig. 1. The 'WBC over DVB-H' layered architecture

2 'WBC over DVB-H' Architecture

The WBCs are being developed along three layers: a service layer, a link layer, and a physical layer. Figure 1 shows the 'WBC over DVB-H' layered architecture. At the service layer of the WBC-SP, the service descriptions (SDs) submitted by the corresponding service providers (xSPs) are first collected, clustered, scheduled and indexed by the content server, and the output is captured by a WBC advertisement delivery protocol (ADP) [2] for subsequent UDP/IP packets generation. At the 'WBC over DVB-H' link layer and physical layer, the IP packets are encapsulated into a transport stream (TS) and broadcast via a DVB-H modulator. On the mobile user (MU) side, the SDs are processed in a reversed order.

3 WBC Layered Functional Model

3.1 WBC Service Layer

There are two sub-layers in the WBC service layer. A WBC application enabler sub-layer is used for SDs collecting, clustering, scheduling, and indexing on the WBC-SP side, and for service discovery and association on the MU side. A goal-oriented, flexible multi-agent system (MAS) [6] is used for facilitating the proper coordination of all involved algorithms (Figure 2). The other sub-layer is the service enabler sub-layer which is used for wireless services' IPDC. A reliable and scalable ADP protocol was designed at this sub-layer to facilitate the encapsulation of WBC segments into IP packets.

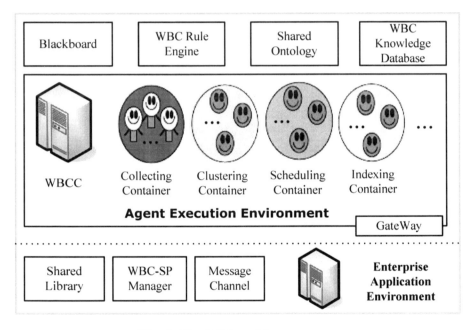

Fig. 2. The WBC multi-agent system

3.2 'WBC over DVB-H' Link Layer and Physical Layer

The WBC service layer is a software layer with a common structure for all WBC nodes. On the contrary, the link layer and physical layer are hardware-dependent layers and thus may have different structures depending on the carrier technology used. The 'WBC over DVB-H' physical layer follows the DVB standard specified in [5]. The 'WBC over DVB-H' link layer adds the following two features:

• The size of the IP packet is set to 1024B so as each MPE-FEC column cannot install more than one IP packet;

• A smart 8-byte segment table (ST) is inserted at the end of the MPE-FEC frame to help decode it in an efficient way. way.

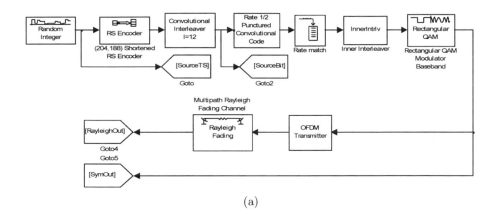

(a)

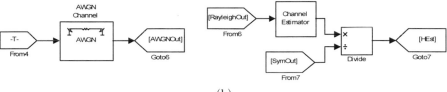

(b)

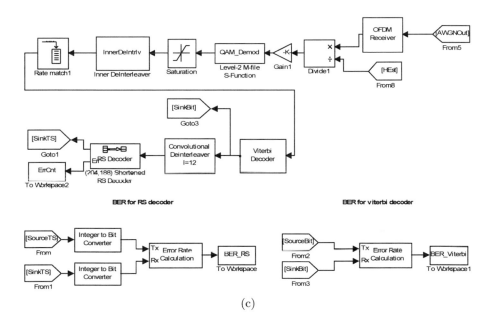

(c)

Fig. 3. The 'WBC over DVB-H' physical layer simulation testbed: (a) Transmitter part; (b) Channel part; (c) Receiver part

4   'WBC over DVB-H' Software Testbed

4.1   Service Layer Implementing

A set of intelligent components are involved in building a flexible and extensible WBC content server. Object Oriented (OO) software process was chosen for implementing the application enabler sub-layer of the service layer, and the corresponding 3-tier architecture has been introduced in [7].

The WBC-ADP protocol running at the service enabler sub-layer of the service layer is implemented in C++ on a Linux platform. To enable ADP working in an efficient way, a multi-thread scheme running under a Finite State Machine (FSM) was designed for optimizing the business logic of sending and receiving processes [2].

At the MT service layer, a lightweight 3-tier software architecture has been designed which follows the Model-View-Controller (MVC) design pattern.

4.2   Link Layer and Physical Layer Implementation

To make the 'WBC over DVB-H' testbed flexible and extensible in a lab testing environment, the link layer was implemented in C++ and the physical layer was implemented with Simulink (Figure 3).

The link layer includes five sections: an IP packet cache; a MPE-FEC encoder/decoder; a MPE encapsulator/decapsulator; a TS encoder/decoder; and a time slicer.

The physical layer includes six sections: an outer coding/decoding; an outer interleaving/de-interleaving; an inner coding/decoding; an inner interleaving/de-interleaving; a mapping/de-mapping and modulation/de-modulation; and a channel section.

5   Conclusions

A 3-layer software testbed for operational probing and testing of wireless billboard channels (WBCs) established over the digital video broadcasting - handheld standard (DVB-H) has been presented in this paper.

Acknowledgments. This publication has been supported by the Irish Research Council for Science, Engineering and Technology (IRCSET) and the Telecommunications Research Centre, University of Limerick, Ireland.

References

1. Flynn, P., Ganchev, I., O'Droma, M.: Wireless Billboard Channels: Vehicle and Infrastructural Support for Advertisement, Discovery, and Association of UCWW Services. Annual Review of Communications 59, 493–504 (2006)

2. Zh, J., Ganchev, I., O'Droma, M.: Reliable and Efficient Advertisements Delivery Protocol for Use on Wireless Billboard Channels. In: 12th IEEE International Symposium on Consumer Electronics, pp. 116–120. IEEE Press, New York (2008)
3. O'Droma, M., Ganchev, I.: Toward a ubiquitous consumer wireless world. IEEE Wireless Communications 14, 52–63 (2007)
4. O'Droma, M., Ganchev, I.: The Creation of a Ubiquitous Consumer Wireless World through Strategic ITU-T Standardization. IEEE Communications Magazine 48, 158–165 (2010)
5. ETSI: Digital Video Broadcasting (DVB); DVB specification for data broadcasting (2004)
6. Bellifemine, F., Poggi, A., Rimassa, G.: JADE: A FIPA2000 Compliant Agent Development Environment. In: AGENTS 2001. IEEE Press, New York (2001)
7. Zh, J., Ganchev, I., Ganchev, I., O'Droma, M.: On WBC Service for UCWW. In: 9th IFIP/IEEE International Conference on Mobile and Wireless Communications Networks, pp. 106–110. IEEE Press, New York (2007)

Optimization of Oil Field Injection Pipe Network Model Based on Electromagnetism-Like Mechanism Algorithm

Hanbing Qi, Qiushi Wang, Guozhong Wu, and Dong Li

The College of Architecture and Civil Engineering, Northeast Petroleum University,
Hei Longjiang, Daqing 163318, China
qihanbing@sina.com,
{qiushi19870418,WGZDQ,lidonglvyan}@126.com

Abstract. The accuracy of water injection pipe network hydraulic model is directly related to the fitting degree of water injection pipe network actual running state. In order to improve the accuracy, using optimization algorithm of electromagnetic-like mechanism to correct the pipe friction coefficient was introduced. To meet constraint conditions, such as the pipe network hydraulic balance and range of parameters, establishing a new oil field injection water distribution network optimization model, the experience estimate of the pipe friction coefficient was introduced to the electromagnetic mechanism optimization algorithms in order to achieve the pipe friction correction. The results show that the node flow calculated value obtained from using the electromagnetism-like mechanism optimization algorithm calibrate water injection pipe network hydraulic model is in good agreement with node actual flow.

Keywords: Water injection Pipe Network, Pipe Friction, Electromagnetism-like Mechanism Algorithm, Optimization.

1 Introduction

The accuracy of water injection pipe network hydraulic model is directly related to the fitting degree of water injection pipe network actual running state, and accurate parameters can make the hydraulic model more meet the actual situation, simulate the behavior of water supply network more better, analyze the running water system accurately, design water distribution system more rationally[1]. With the long-term use of the oil injection pipe network system, the pipe friction have changed, resulting in the pipe network hydraulic model and the actual don't match, so the pipe friction has become an important factor of affecting the accuracy of the hydraulic model, it is necessary to correct pipe friction[2], making the water supply network model to simulate the operation consistent with the actual situation as possible, so as to provide a scientific basis for study the actual pipe network design, operation management, and save energy values.

When calibrating water injection network pipe friction, usually through the use of the node pressure and node flow, under the constraint of meeting the water injection

Y. Yu, Z. Yu, and J. Zhao (Eds.): CSEEE 2011, Part II, CCIS 159, pp. 130–135, 2011.
© Springer-Verlag Berlin Heidelberg 2011

system hydraulic balance and the range of parameters, the optimization calculation to determine the optimal values of the pipe friction. Now the method of multiple working conditions[3] was used often, collecting large amounts of data by the testing fire, then reverse calculates correction friction via genetic algorithm[4], but there is still the error correction. the electromagnetic-like mechanism[5] algorithm was initially applied to the field of function optimization, and now has begun to be applied to other optimization field, for example, Peitsang Wu and others applied electromagnetic-like mechanism algorithm to neural network training, and solved the problems in the operation of textile retailing; Dieter Debels[6] and others combined electromagnetism-like mechanism algorithm with scattered search method and evolution search method based on population, and successfully resolving project scheduling problem; P.Kaelo and M.M.Ali[7] introduced electromagnetism-like mechanism algorithm as a variation to the differential evolution strategy algorithm, to solve function optimization problems, and achieved good results.

2 Mathematical Model

According to hydraulic theory, establish the energy conservation equation of pipe, referred to as the pipe energy equation. All the pipe energy equations simultaneous comprised of pipe energy equations.

$$H^{F_i} - H^{T_i} = hi \quad i = 1, 2, 3, ..., M. \tag{1}$$

Where, Fi is the number of pipe upper endpoint; Ti is the number of pipe lower endpoint; HFi is the head of pipe upper endpoint; HTi is the head of pipe lower endpoint; hi is the pressure drop of pipe i; M is the total pipe number of the network model. According to the head loss formula,'

$$hi = s_i q_i^n \quad i = 1, 2, 3, ..., M. \tag{2}$$

Where, hi is the pressure drop of pipe, m; si is pipe friction; qi is pipe flow, m3/s.

According to hydraulic theory, establish the flow continuity equation, referred to as nodal flow equation. All the nodal flow equations simultaneous comprised of nodal flow equations.

$$\sum_{j \in S_j} \pm q_i + Q_i = 0 \quad j = 1, 2, 3, ..., N. \tag{3}$$

Where, qi is the flow of pipe i; Qi is the flow of node i; N is the total node number of network model; $\sum_{j \in S_j} \pm$ representatives summing the flow of pipes in the associated set of node j, when the pipe direction pointing the node, take a minus, or take a positive, that is when flow out the node is positive, flow into the node is negative.

The idea of optimization model is to meet the pipe network constraints such as the hydraulic balance and range of parameters, to find the optimal values of the parameters to be checked so that the objective function value that composed of the square of the difference between pipe measured and calculated value is minimal. In this paper, an optimization model for the parameter calibration, the mathematical expression of the objective function is the follow form:

$$f(X) = \sum_{j=1}^{N} |Q_{jm} - Q_{j0}|$$ (4)

Where, f is the objective function to be optimized; X is the parameter vector to be optimized; Q_{jm} is the calculated flow of the node j; Q_{j0} is the measured flow of the node j; N is the number of pressure nodes. According to the experience value of the Hazen-Williams coefficient, the constraint conditions of the objective function is :

$$\min C_i < C_i < \max C_i$$ (5)

While constraints in full, the node calculated flow as state variable should also meet implicitly the condition of the hydraulic balance of the pipe network (flow continuity equation and energy balance equations). Such problems are complex constrained nonlinear programming problem, its complexity is proportional to the size of pipe network.

3 Principles of Electromagnetism-Like Mechanism and Its Realization

Electromagnetism-like mechanism algorithm is proposed as a new population-based random intelligent optimization algorithm by Birbil and Fang, which inspired the attraction-rejection mechanism between charged particles in electromagnetic field in 2003. Optimization of the electromagnetism-like mechanism is to generate a set of randomly initial solution from feasible region as the initial particles that constitute the initial population, then the objective function value of each solution to determine the domain of attraction, simulation the mechanism of attraction and repulsion between charged particles in electromagnetic field, each solution will be compared to a charged particle, and then makes the search particles move toward the global best particle according to certain criteria.

The main steps of the electromagnetism-like mechanism algorithm are as follows:

    1: Initialize( )
    2: iteration ← 1
    3: **While** iteration < T **do**
    4: Local search(L, δ)
    5: F←Calculation( )
    6: Move(F)
    7: iteration←iteration+1
    8: **end while**

Parameter setting (N, T, L, δ): N is population size, T is the maximum number of iteration, L is iteration maximum number of each local search, $\delta \in (0,1)$ is the local search parameter.

Algorithm consists of four stages , particles initialization, calculation forces on each particle, moving the particles along together forces vector , and the local search.

4 Correcting Pipe Friction

One of the characteristics of water injection pipe network is node pressure and node flow of the pipe ends is known. The pipe friction coefficient is a function of the Hazen-Williams coefficient, while the pipe Hazen-Williams coefficient is unknown, so the pipe flow can not be directly obtained. In the pipe end nodes pressure fixed case, the change in pipe friction can change the pipe flow, which can change the relevant node flow, so you can change the pipe Hazen-Williams coefficient to calculate the different node flow, until the flow close to the known node flow, then the Hazen Williams coefficient is what we seek.

In this paper, we applied the optimization idea of electromagnetism-like mechanism to find the optimal value of Hazen-Williams coefficient, making the objective function value is minimal. The steps of applying Electromagnetism-like mechanism algorithm to calibrate pipe friction are as follows:

(1) Input the known information of pipe network, the number of pipes and nodes, pipe length and diameter, node pressure, node measured flow;

(2) Generate M particles randomly; the coordinate of each particle consist of the friction of each pipe and the friction refers to the Hazen-Williams coefficient within the range(75-150) of experience, using the pipe equation method to calculate the node flow;

(3) Node flow calculated and measured values constitute the objective function, each particle corresponds to an objective function value;

(4) According to the objective function value of particle, calculate charge particle represented and total force particle suffered, the particles move according to total force vector to generate new particles;

(5) Repeat process of steps (2), (3), (4), end when the maximum number of iterations.

According to the mathematical model and the electromagnetism mechanism algorithm, The solving process of friction coefficient can be mapped to optimization problem, use software matmab realize correction process of pipe friction. Select the network shown in Fig.1. to optimize.

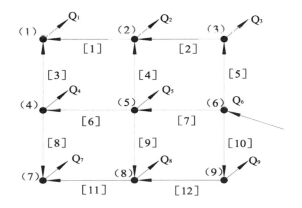

Fig. 1. Ring Pipe Network Diagram

Setting the running parameters of electromagnetism-like mechanism algorithm: the number of particles m=10, the dimension n=12, the maximum number of iterations of local search LSITER=5n, the maximum number of iterations MAXTER=1000n, the pipe correction results are shown in Table 1.

Table 1. Results of Calibration

number of pipe	1	2	3	4	5	6
accurate value	103	103	103	103	103	103
calculated value	103. 00	103. 67	103. 96	119. 48	103. 45	103. 31
error	0. 50%	0. 65%	0. 93%	16. 00%	0. 44%	0. 30%
number of pipe	7	8	9	10	11	12
accurate value	103	103	103	103	103	103
calculated value	102. 47	102. 91	110. 04	103. 24	103. 09	103. 53
error	0. 51%	0. 09%	6. 83%	0. 23%	0. 09%	0. 51%

As can be seen from Table 1, most of calibration results of the friction coefficient are consistent with the measured values.

5 Conclusion

The coefficient calibration of pipe friction of water injection pipe network where node pressures were known was studied. According to information value of experience friction, the continuity equation and the energy balance equation were satisfied by the friction of pipeline which were generated randomly, the pipe friction coefficient of water injection pipe network was optimized and corrected using the electromagnetism-like mechanism which is a new optimization algorithm, the calculation example shows that the algorithm provides availability for the network optimization study.

References

1. Wang, Y., Wang, Y.: Friction Coefficient of The Injection Pipe Network Calibration Based on Particle Algorithm. Oil-Gas Field Surface Engineering 27, 27–28 (2008) (in Chinese)
2. Xin, K., Liu, S.: The Accuracy Factors of Hydraulic Model For Urban Water Distribution Network. China Water & Wastewater 4, 52–55 (2003) (in Chinese)
3. Zhang, T., Xu, G., Lv, M., Zhuo, M.: The Research for Correction of Water Supply Network Pipe Friction. Journal of Zhejiang University (Engineering Science) 40, 1201–1205 (2006) (in Chinese)

4. Xu, G., Zhang, T., Lv, M., Hong, B.: Multiple Working Conditions Genetic Algorithms For Pipe Friction Factor Correction. China Water & Wastewater 20, 50–53 (2004) (in Chinese)
5. Birbil, S.I., Fang, S.C.: An Electromagnetism-Like Mechanism for Global Optimization. Journal of Global Optimization 25, 263–282 (2003)
6. Debels, D., Dereyck, B., Leus, R., et al.: A Hybrid Scatter Search/Electromagnetism Meta-Heuristic for Project Scheduling European. J. Operational Research 169, 638–653 (2005)
7. Kaelo, P., Ali, M.M.: Differential Evolution Algorithms Using Hybrid Mutation. Computer Optimization Applications 37, 231–246 (2007)

Cost Aggregation Strategy with Bilateral Filter Based on Multi-scale Nonlinear Structure Tensor

Li Li[1,2] and Hua Yan[1]

[1] School of Computer Science and Technology
Shandong Provincial Key Laboratory of Digital Media Technology,
Shandong Economic University
[2] Shandong University, 250014, Jinan, China
lily_jn@sina.com, yhzhjg@sdu.edu.cn

Abstract. This paper proposed a novel cost aggregation method with modified bilateral filter. By smoothing each element of the structure tensor considering both the spatial and gradient distances of neighboring pixels, the nonlinear structure tensor for each pixel is constructed. We adopt the Log-Euclidean calculus as tensor dissimilarity function to compute the structure tensor distance of two considering pixels. Then the multi-scale value is computed by summing of the tensor distances in each scale. So a new weight basing on multi-scale structure tensor distance is set up and included in bilateral filter for cost aggregation. By adding the new weight in cost aggregation, more pixels similar with central pixel could be aggregated in a support window and the final disparity map could be more accurate. Experimental results have confirmed the effectiveness of our proposed method.

Keywords: Cost aggregation, Multi-Scale nonlinear structure tensor, Log-Euclidean distance, Bilateral filter.

1 Introduction

Stereo matching is a key problem in computer vision. According to authors [1] cost aggregation is mandatory for local stereo matching algorithms to improve signal noise rate (SNR) and often adopted by global ones. The cost aggregation step is to aggregate initial matching costs in a support window. An ideal support window should be adjusted according to image content to include only the pixels with the same disparity. Many cost aggregation methods have been presented while this behavior is far from ideal. This paper proposed a novel cost aggregation strategy with modified bilateral filter based on multi-scale nonlinear structure tensors.

Many adaptive window methods have been proposed to include more pixels having the same disparity values with the central pixel by varying the size, the shape and the position of the support window [2]. Different from the adaptive window methods, the adaptive weight method (AW) [3] adopts a fixed window and assigns a weight to each pixel in the support window according to the spatial and color similarity with the central pixel and gains a high performance. The AW method is based on bilateral

Y. Yu, Z. Yu, and J. Zhao (Eds.): CSEEE 2011, Part II, CCIS 159, pp. 136–141, 2011.
© Springer-Verlag Berlin Heidelberg 2011

filter which is a non-iterative feature-preserving image smoothing technique. In AW method, the weight of a pixel within the support window is obtained by applying two independent bilateral filters in the neighborhood of potential correspondence. To further improve disparity accuracy and decrease the computational time, many modified approaches against the AW method have been presented in recent years [4-6]. Danny Barash [7] pointed out that the nature of bilateral filter resembles that of anisotropic diffusion. So many stereo matching methods with PDE models have also been presented recently. Usually these methods have a solution of disparity maps by constructing partial differential equations (PDE) which all base on structure tensors of images. Because the structure tensor can be used as a local indicator to analyze the geometric structure of image and widely used in image segmentation, corner detection and object tracking areas and so on [8-9]. However they have not been used in cost aggregation method directly until now.

To fill the void in existing stereo approaches, this paper added a new weight in bilateral filter based on structure tensor proximity of corresponding pixels for cost aggregation. By using structure tensor information, the final weight can take into account the local geometric structure of image and then can more accurately detect the similarity of two pixels which results in the obtained disparity map after cost aggregation step more accurately. The rest of paper is organized as follows. Section 2 presents our proposed cost aggregation algorithm in detail. Experimental results are given in section 3. Section 4 gives conclusions and an outlook to possible future work.

2   Our Proposed Method

Given a reference image I_L and a matching image I_R, the matching cost indicates similarity of the two images with a disparity map. Many stereo matching algorithms adopt a criteria basing on the color difference between two corresponding pixels. To improve the robustness to noise or distortions, we adopt a truncated L1 norm as initial matching cost function given by

$$C(p_l, p_r, d)^2 = \| I_L(p_l) - I_R(p_r) \|^2 + \lambda_M \| \nabla I_L(p_l) - \nabla I_R(p_r) \|^2$$

$$C_0(p_l, p_r, d) = -\log[\delta_M + (1 - \delta_M) \exp(-C(p_l, p_r, d) / \sigma_M)] \qquad (1)$$

where $p_l(x, y)$, $p_r(x-d, y)$ are corresponding two pixels in the reference image I_L and the matching image I_R respectively assuming the disparity is d, $\nabla = (\partial x \quad \partial y)^T$ is the gradient operator, $\lambda_M, \delta_M, \sigma_M$ are predefined parameters and $C_0(p_l, p_r, d)$ is the initial matching cost which will be aggregated using the bilateral filter method.

A pixel in the support window will be assigned a weight based on the spatial and color distances with the central pixel. To more accurately detect the similarity of considering two pixels, a new weight is added based on the proximity of structure tensors. We will explain our cost aggregation step in detail as below.

The classic differential geometry theory [10] provides a method to analyze the local geometric structure of an image. Let us consider a multi-valued reference image: $I_L(p): \Omega \to R^n$ defined on a domain $\Omega \in R^2$ where $n \in N^+$ is the number of the image channel, $p(x, y)$ is a pixel in the domain. The local variations of the vector norm $\| dI_L \|$ can be given by

$$\| dI_L \|^2 = dX^T G dX, G = \sum_{i=1}^{n} \nabla I_i \nabla I_i^T, dX = \begin{pmatrix} dx \\ dy \end{pmatrix}$$

(2)

where G is 2×2 symmetric as well as semi-positive-definite matrix.

We call G a structure tensor because it indicates the local geometry of the image. In fact, the eigenvalues λ_+, λ_- of G are the maximum and minimum of $\| dI_L \|^2$ while the orthogonal eigenvectors θ_+, θ_- of G are corresponding variation directions. Structure tensor has been used in many image applications to present the local geometric feature of the image. However, the computing derivatives is sensitive to noise, it needs to smooth the derivatives for noise reduction. Usually an isotropic Gaussian kernel is used to smooth each of the four elements in the 2×2 structure tensor in a local window. As we all know, such a smoothing operation will smooth out some weak features and the results will not be accurate. So we construct a nonlinear structure tensor using bilateral filter same as [9]. During the smoothing the structure tensor, we consider both the spatial distance $D_S(p_l, q_l)$ and gradient distance $D_G(p_l, q_l)$ which both are Euclidean in the averaging weight assignment. Then a bilateral weighting function of smoothing the structure tensor for each pixel p_l in the reference image I_L is defined as

$$\hat{g}_{i,j}(p_l) = \frac{\sum_{q_l \in S(p_l)} W_G(D_G(p_l, q_l)) W_S(D_S(p_l, q_l)) g_{i,j}(q_l)}{\sum_{q_l \in S(p_l)} W_G(D_G(p_l, q_l)) W_S(D_S(p_l, q_l))}$$

(3)

where $g_{i,j}$ is the element of structure tensor for each pixel in the reference image I_L, the range of indices is $i, j = 1, 2$, $\hat{g}_{i,j}$ is the filtered element value of structure tensor, $W_G = \exp(-D_G / 2\sigma_G)$ and $W_S = \exp(-D_S / 2\sigma_S)$ both are Gaussian based on the spatial distance D_G and the gradient distance D_S respectively. Compared with the Gaussian kernel, the weighting function is anisotropic and adaptive to the image local structure by using bilateral filter.

One key factor in the tensor space analysis is a proper choice of the tensor distance norm to measure the similarity between tensors. For simplicity, we use the recently proposed Log-Euclidean calculus [11] as the measure function of tensor similarities. The Log-Euclidean distance between two tensors G_{p_l} and G_{q_l} for pixels p_l and q_l respectively is given by

$$D_T(p_l, q_l) = \sqrt{tr((\log m(\hat{G}_{p_l}) - \log m(\hat{G}_{q_l}))^2)}$$

(4)

where $tr(.)$ is the trace operator of a matrix, $\log m(.)$ is the logarithm operator of a matrix, the "hat" denotes that the structure tensor has been bilateral filtered as described before.

Multi-scale structure tensor was firstly defined in [12]. By considering the scale difference of the multi-valued image, we construct a multi-scale structure tensor using tensor information at each scale. The tensor distance for multi-scale structure tensor can be defined as the square root of the sum of the Log-Euclidean distances for all scales and can be rewritten as

$$\tilde{D}_T(p_l, q_l) = \sqrt{\sum_{m=0}^{L-1} tr((\log m(\hat{G}_{p_l}^m) - \log m(\hat{G}_{q_l}^m))^2)}$$

(5)

where L is the total number of the scales. Then the new weight $W_T = \exp(-\tilde{D}_T / 2\sigma_T)$ between two considering pixels p_l and q_l is computed based on the similarity of their tensors.

$$\tilde{C}(p_l, p_r, d) = \frac{\sum_{q_l \in S(p_l)} W_C(D_C(p_l, q_l))W_S(D_S(p_l, q_l))W_T(\tilde{D}_T(p_l, q_l))C_0(q_l, q_r, d)}{\sum_{q_l \in S(p_l)} W_C(D_C(p_l, q_l))W_S(D_S(p_l, q_l))W_T(\tilde{D}_T(p_l, q_l))}$$

(6)

By the above procedures, the final cost aggregation equation can be expressed by where w_s and w_c are spatial distance and range distance which both are Gaussian same as the function used in the AW method. Compared with the AW method, a new weight is included which reflects the similarities of two considering pixels based on their structure tensors in our approach. A neighboring pixel is assigned a high weight to the central pixel not only if their spatial and range distances are small but also if they have similar local geometric structures. The final weight could more accurately distinguish the similarity of two relevant pixels. It is worth noting that, in the above equation, we simply execute the filter on the reference image only different from the adaptive weight method.

Finally the disparity map is obtained in a WTA framework as below

$$d(p_l) = \arg\min_{d \in D}(\tilde{C}(p_l, p_r, d))$$

(7)

where D represents the set of all allowed disparities.

3 Experimental Results

In this section, we aim at assessing the performance of our proposed cost aggregation approach based on the modified bilateral filter with nonlinear multi-scale structure tensor. We used the Middlebury test bed provided by authors of [1] to evaluate our approach performance compared with the original bilateral filter and the other state-of-art cost aggregation methods.

Firstly we compared the results of our proposed method with that of the method using the original bilateral filter (OBF). Both methods adopt asymmetric strategy which executes filter on the reference image only for simplicity. For comparison two methods both have the L1 norm as initial cost function which is different from the AW method. A constant set of parameters are run for all test images. The corresponding disparity maps are plotted in Fig. 1. From the figures, we can see that our proposed method has better results than the method using the original bilateral filter. This manifests that including new weight in cost aggregation function can actually improve the disparity accuracy, especially in discontinuity area, because of structure tensor reflecting the local geometric feature of the image.

Then we compare our proposed method with the state-of-art cost aggregation strategies. For our method, we adopt the optimal parameters minimizing the Vis.+Dis. error on the whole test images. The results of the other four top methods used here were reported in [5]. It is worth noting that these results reported by [5] are obtained using the original cost function proposed by the authors of the each paper and the results for AW and SS available on the Middlebury evaluation sites including the post processing steps are not used. We have reported quantitative comparative results in Table 1. From the table, we can see that our proposed method have results comparable to the best performing cost aggregation strategies. The most similar method with our approach is the AW method basing on the bilateral filter. The AW method outperforming our results is mainly due to its symmetric strategy while our approach adopted the asymmetric strategy and the run time can reduce by half.

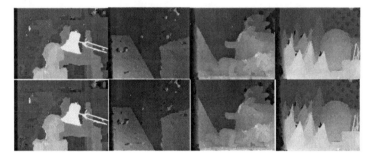

Fig. 1. The disparity maps of two methods. The first row is the results of the OBF method. The second row is our results.

Table 1. Quantitative comparative results of our method with the other four top algorithms

	Tsukuba		Venus		Teddy		Cones	
	Vis.	Dis.	Vis.	Dis.	Vis.	Dis.	Vis.	Dis.
SS[4]	2.19	7.22	1.38	6.27	10.50	21.20	5.83	11.80
AW[3]	3.33	8.87	2.02	9.32	10.52	20.84	3.72	9.37
FBS[6]	2.95	8.69	1.29	7.62	10.71	20.82	5.23	11.34
VW[2]	3.12	12.40	2.42	13.30	17.70	25.50	21.20	27.30
Our method	2.70	10.72	4.93	11.83	10.86	19.70	6.41	12.77

4   Conclusions

In this paper, we proposed a cost aggregation algorithm based on modified bilateral filter with multi-scale nonlinear structure tensor for local stereo matching. A new weight is included in cost aggregation equation based on the structure tensor distance. The proposed new algorithm not only considers the spatial and range distances of two pixels same as the original bilateral filter, but also the local geometric feature distance of them. Our new weight can more accurately reflect the similarity of two pixels which is important for cost aggregation approach. The experimental results confirm the effectiveness of our approach compared with the OBF method and the other state-of-art strategies.

In the future, we plan to devise new cost aggregation methods based on the structure tensor to further improve the accuracy of disparity maps and decrease the computational time. We are also interested in using the diffusion equation constructed by structure tensor in the variational framework to devise the global stereo matching method.

Acknowledgments. The authors would like to thank financial supports from National Natural Science Foundation of China under Grant Nos. 60970048, Natural Science Foundation of Shandong Province Grant Nos. 2009ZRB019SF.

References

1. Scharstein, D., Szeliski, R.: A Taxonomy and Evaluation of Dense Two-frame Stereo Correspondence Algorithms. International J. Computer Vision, 7–42 (2002)
2. Veksler, O.: Fast Variable Window for Stereo Correspondence Using Integral Images. In: Proc. of CVPR, pp. 556–561 (2003)
3. Yoon, K.J., Kweon, I.S.: Adaptive Support-weight Approach for Correspondence Search. PAMI 28(4), 650–656 (2005)
4. Tombari, F., Mattoccia, S., Di Stefano, L.: Segmentation-based adaptive support for accurate stereo correspondence. In: Mery, D., Rueda, L. (eds.) PSIVT 2007. LNCS, vol. 4872, pp. 427–438. Springer, Heidelberg (2007)
5. Gong, M., Yang, R.G., Liang, W., Gong, M.W.: A performance Study on Different Cost Aggregation Approaches Used in Real-time Stereo Matching. International J. Computer Vision 75(2), 283–296 (2007)
6. Mattoccia, S., Giardino, S., Gambini, A.: Accurate and Efficient Cost Aggregation Strategy for Stereo Correspondence Based on Approximated Joint Bilateral Filtering. In: Proc. of ACCV (2009)
7. Barash, D.: A Fundamental Relationship between Bilateral Filtering Adaptive Smoothing and the Nonlinear Diffusion Equation. IEEE TPAMI 1(24) (June 2002)
8. Han, S., Tao, W., Wang, D., Tai, X.C., Wu, X.L.: Image Segmentation Based on GrabCut Framework Integrating Multi-scale Nonlinear Structure Tensor. IEEE Transactions on Image Processing 18(10), 289–302 (2009)
9. Zhang, L., Zhang, L., Zhang, D.: A Multi-Scale Bilateral Structure Tensor Based Corner Detector. In: Proc. of ACCV 2009 (2009)
10. Carmo, M.D.: Differential Geometry of Curves and Surfaces. Prentice Hall, Englewood Cliffs (1976)
11. Arsigny, V., Fillard, P., Pennec, X., Ayache, N.: Log-Euclidean Metrics for Fast and Simple Calculus on Diffusion Tensors. Magnetic Resonance in Medicine 56(2), 411–421 (2006)
12. Weickert, J., Romeny, B., Viergever, M.A.: Efficient and Reliable Schemes for Nonlinear Diffusion Filtering. IEEE Transactions on Image Processing 7, 398–410 (1998)

Active Set Strategy for the Obstacle Problem with a T-Monotone Operator

Shuilian Xie and Hongru Xu

School of Mathematics, Jiaying University
Meizhou, Guangdong, 514015, China
shuilian6319@163.com,
hrxu001@163.com

Abstract. In this paper, we consider the numerical solution of finite-dimensional variational inequalities of obstacle type associated with some free boundary problem with T-monotone operator. Algorithm based on active set strategy is proposed for the problem. Each iteration consists of two steps. In the first step, the index set is decomposed into active and inactive parts, based on a certain criterion. In the second step, a reduced nonlinear system associated with the inactive set is solved. Convergence theorem of the algorithm is established.

Keywords: Active Set Strategy, Obstacle Problem, T-Monotone Operator, Convergence.

1 Introduction

In this paper, we consider the numerical solution of finite-dimensional variational inequalities of obstacle type associated with some free boundary problem with T-monotone operator. This kind of obstacle problems has many applications, such as the diffusion problem involving Michaelis-Menten or second-order irreversible reactions, see for example [1] and the references therein for details.

Active set strategy is an efficient way to solve discrete obstacle problems, see for example [2-5] and the references therein. In [2], based on some kind of active set strategy, two iterative schemes are presented for the numerical solution of unilateral variational inequality. Both schemes requires the solution of an algebraic system of equations at each iteration. In [4], an active set strategy is presented for obstacle problems described by variational inequality. At each step of the iteration the method leads to a reduced linear algebraic system which is solved by a multigrid algorithm. In this paper, the active set strategy is extended for obstacle problem with a T-monotone operator. In our algorithm, each interation consists of two steps. In the first step, the index set is decomposed into active and inactive parts, based on a certain criterion. In the second step, a reduced nonlinear system associated with the inactive set is solved.

The rest of paper is organized as follows: In section 2, the model problem and some preliminaries are presented. In section 3, active set strategy is presented for obstacle

Y. Yu, Z. Yu, and J. Zhao (Eds.): CSEEE 2011, Part II, CCIS 159, pp. 142–147, 2011.
© Springer-Verlag Berlin Heidelberg 2011

problem with a T-monotone operator, and the convergence theorem of the algorithm is established.

2 Model Problem and Some Preliminaries

In this section, we present the model problem we considered and give some preliminaries which are useful in the following sections.

First, we introduce some notations. Let K be a subset of R^n and F an operator from K to R^n. Let every element $v \in R^n$ be expressed by $v = v^+ + v^-$ with $v^+ = \max\{v, 0\}$ and $v^- = \min\{v, 0\}$. Let $N = \{1, 2, \cdots, n\}$ be the index set. We introduce the definition of T-monotone as follows:

Definition 1 [6] Operator F is called T-monotone over K, if $\langle F(v) - F(w), (v - w)^+ \rangle \geq 0$, $\forall v, w \in K$, where $\langle, \rangle$ denotes Euclidean inner product. Moreover, if for all v and $w \in K$, $\langle F(v) - F(w), (v - w)^+ \rangle = 0$ is equivalent to $(v - w)^+ = 0$, F is called strictly T-monotone over K.

In this paper, we consider the following model problem:

$$\begin{aligned} &\text{find} && u \in K, \\ &\text{such that} && \langle F(u), v - u \rangle \geq 0, \quad \forall v \in K, \end{aligned} \tag{1}$$

where $K = \{v \in R^n : v \leq \psi\}$, F is a continuous, coercive and strictly T-monotone operator.

We present some properties of T-monotonicity, which are very important and useful for out active set strategy.

Lemma 1. Let F be a continuous T-monotone operator over K, I and J be the subsets of N satisfying $J = N \backslash I$. For any vectors $y, z \in K$, if $y_I \leq z_I$ and $y_J \geq z_J$, then $F_I(y) \leq F_I(z)$.

Proof. For any $y, z \in K$ satisfying $y_I \leq z_I$ and $y_J \geq z_J$, let $\hat{I} = \{j \in I : F_j(y) > F_j(z)\}$ and $\hat{J} = N \backslash \hat{I}$. Suppose that $\hat{I}$ is not empty. Let $w_2 = y$ and w_1 be defined by $(w_1)_j = \begin{cases} z_j + \delta, & j \in \hat{I}, \\ z_j, & j \in \hat{J}, \end{cases}$ where δ is a positive constant. Since F is continuous, we let δ be small enough, such that $F_j(w_1) < F_j(w_2)$ for all $j \in \hat{I}$.

Then $(w_1 - w_2)_j^+ = \begin{cases} \delta + (z_j - y_j)^+, & j \in \hat{I}, \\ 0, & j \in \hat{J}, \end{cases}$ and therefore

$$0 \leq \langle F(w_1) - F(w_2), (w_1 - w_2)^+ \rangle = \sum_{j \in \hat{I}} (\delta + (z_j - y_j)^+)(F_j(w_1) - F_j(w_2)) < 0,$$

which is a contradiction. Hence $\hat{I} = \varnothing$, which means that $F_I(y) \leq F_I(z)$ and the proof is completed.

Lemma 2. Let F be a continuous strictly T-monotone operator over K, I and J be the subsets of N satisfying $J = N\backslash I$. For any vectors $y, z \in K$, if $y_I \leq z_I$ and $F_J(y) \leq F_J(z)$, $y \leq z$.

Proof. Let $\hat{I} = \{i \in N : y_i \leq z_i\}$ and $\hat{J} = N\backslash\hat{I}$. Without loss of generality, we assume that $\hat{I} = \{1, 2, \cdots, k\}$ and $\hat{J} = N\backslash\hat{I}$. If $\hat{J}$ is not empty, then by the definition of $\hat{I}$, we have that $I \subset \hat{I}$ and $\hat{J} \subset J$. Moreover, we have

$$F_j(y) \leq F_j(z), \quad y_j > z_j, \quad y_i \leq z_i. \tag{2}$$

What's more, let $w_1 = (z_j, y_j)$ and $w_2 = z$. It is easy to verify that

$$F_j(w_1) = F_j(z_j, y_j) \leq F_j(y_j, y_j) = F_j(y) \leq F_j(z) = F_j(w_2), \tag{3}$$

where the first inequality comes from (2) and Lemma.1.

The strictly T-monotonicity of F combining with (2) and (3) implies $0 \leq \langle F(w_1) - F(w_2), (w_1 - w_2)^+ \rangle = \langle F_j(w_1) - F_j(w_2), (y_j - z_j)^+ \rangle \leq 0$. Hence, $\langle F(w_1) - F(w_2), (w_1 - w_2)^+ \rangle = 0$, which means $(w_1 - w_2)^+ = 0$ and thereby $y_j \leq z_j$. The contradiction implies that $\hat{J}$ is empty and hence $y \leq z$. The proof is then completed.

3 Active Set Strategy

In this section, we extend the active set strategy proposed in [3, 5] for unilateral linear obstacle problem to our problem and establish its convergence theorem. We let

$$\lambda^* = -F(u^*). \tag{4}$$

It is easy to verify the solution of (1) is equivalent to the following problem: find $u^\square$, such that

$$\begin{cases} \lambda_j^* > 0 \quad \text{and} \quad u_j^* = \psi_j, \quad j \in J^*, \\ \lambda_j^* = 0 \quad \text{and} \quad u_j^* \leq \psi_j, \quad j \in I^*, \end{cases} \tag{5}$$

where $N = I^* \cup J^*$. Now we can present our active set strategy suggested by (5). Algorithm 1 (Active Set Strategy for Unilateral Obstacle Problem with Upper Obstacle)

Step.1. Initialize $u^0 \in K$ and $\lambda^0 \geq 0$. Set $k := 0$.

Step.2. Compute

$$\hat{\lambda}^k = \max\{0, \lambda^k + u^k - \psi\}. \tag{6}$$

Determine the active and inactive sets by

$$\begin{cases} J^k = \{j \in N : \hat{\lambda}_j^k > 0\}, & \text{(active)}, \\ I^k = \{j \in N : \hat{\lambda}_j^k = 0\}, & \text{(inactive)}. \end{cases} \tag{7}$$

Step.3. If $k \geq 1$ and $J^k = J^{k-1}$, then stop; the solution is u^k.

Step.4. Let u and λ be the solution of the nonlinear system

$$\begin{cases} u_j = \psi_j, & j \in J^k, \\ \lambda_j = 0, & j \in I^k, \\ F(u) + \lambda = 0. \end{cases} \tag{8}$$

Set

$$u^{k+1} = u, \quad \lambda^{k+1} = \max\{0, \lambda\}, \tag{9}$$

$k := k + 1$, and go to Step 2.

We can easily see that the only laborious step in Algorithm 1 is the realization of the following nonlinear system:

$$\begin{cases} u_{J^k} = \psi_{J^k}, \\ F_{I^k}(u) = 0, \\ \lambda = -F(u). \end{cases} \tag{10}$$

It is easy to verity the following lemma.

Lemma 3. If Algorithm 1 stops at Step 3, the current iterate (u^k, λ^k) satisfies condition (5).

Lemma 4. Let $u^* \leq u^0 \leq \psi$ be given. Define $\lambda^0 = \max\{0, -F(u^0)\}$ and assume that we have a disjoint decomposition $N = I^{-1} \cup J^{-1}$, with $I^{-1} = \{j \in N : \lambda_j^0 = 0\}$ and $J^{-1} = \{j \in N : \lambda_j^0 > 0 \text{ and } u_j^0 = \psi_j\}$. Then for all $k \geq 0$,

$$u^* \leq u^{k+1} \leq u^k \quad \text{and} \quad \lambda^* \leq \lambda^{k+1} \leq \lambda^k. \tag{11}$$

Proof. Let J^* be defined as in (5). Since $u^* \leq u^0 \leq \psi$, we have

$$J^* \subset C^* = \{j \in N : u_j^* = \psi_j\} \subset C^0 = \{j \in N : u_j^0 = \psi_j\}. \tag{12}$$

By (12), we get

$$
\begin{aligned}
\lambda_{j^*}^0 &= \max\{0, -F_{j^*}(u^0)\} \\
&\geq -F_{j^*}(u^0) \\
&\geq -F_{j^*}(u^*) \\
&= \lambda_{j^*}^*,
\end{aligned}
\tag{13}
$$

where the second inequality comes from $u_{j^*}^0 = u_{j^*}^*$, $u_{I^*}^0 \geq u_{I^*}^*$ and Lemma 1. This together with $\lambda_{I^*}^0 \geq 0$ and $\lambda_{I^*}^* = 0$, we have $\lambda^0 \geq \lambda^*$.

Let J^0 and I^0 be determined by (7). Since $I^{-1} \cap J^{-1} = \varnothing$ and $u^0 \leq \psi$, we have actually $\hat{\lambda}^0 = \lambda^0$ and $J^0 = J^{-1}$ and $I^0 = I^{-1}$. Hence $I^0 \subset I^*$ since $\lambda^0 \geq \lambda^*$. Noting that (u^1, λ) satisfies the following nonlinear system

$$
\begin{cases}
u_j^1 = \psi_j, & j \in J^0, \\
\lambda_j = 0, & j \in I^0, \\
F(u^1) + \lambda = 0,
\end{cases}
\tag{14}
$$

we have for all $j \in I^0$ that

$$
\begin{aligned}
F_j(u^1) - F_j(u^0) &= -F_j(u^0) \\
&\leq \max\{0, -F_j(u^0)\} \\
&= \lambda_j^0 = 0.
\end{aligned}
\tag{15}
$$

This together with $u_{j^0}^1 = \psi_{j^0} = u_{j^0}^0$, we have $u^1 \leq u^0$ by Lemma.2 Furthermore,

$$
\begin{aligned}
\lambda_{j^0} &= -F_{j^0}(u^1) \\
&\leq -F_{j^0}(u^0) \\
&\leq \lambda_{j^0}^0,
\end{aligned}
\tag{16}
$$

where the first inequality comes from $u_{j^0}^1 = u_{j^0}^0$, $u_{I^0}^1 \leq u_{I^0}^0$ and Lemma.1 Hence,

$$
\lambda^1 = \max\{0, \lambda\} \leq \lambda^0,
\tag{17}
$$

$$
I^1 = I^0 \cup J_2^0 \quad \text{and} \quad J^1 = J_1^0.
\tag{18}
$$

Since $I^0 \subset I^*$, we have $F_{I^0}(u^1) = F_{I^0}(u^*) = 0$ by (4), (5) and (8). This together with $u_{j^0}^1 = u_{j^0}^0 \geq u_{j^0}^*$, we have $u^1 \geq u^*$ by Lemma 2. By $I^0 \subset I^*$, we have $J^* \subset J^0$ and then

$$
\begin{aligned}
\lambda_{j^*}^1 &= \max\{0, -F_{j^*}(u^1)\} \\
&\geq -F_{j^*}(u^1) \\
&\geq -F_{j^*}(u^*) \\
&= \lambda_{j^*}^*,
\end{aligned}
\tag{19}
$$

where the second inequality comes from $u_{j^*}^* = u_{j^*}^1$, $u_{I^*}^* \leq u_{I^*}^1$ and Lemma 1. Noting that $\lambda_{I^*}^* = 0$ by (5), we have $\lambda^1 \geq \lambda^*$. This together with (17) concludes that $\lambda^* \leq \lambda^1 \leq \lambda^0$ and then $I^0 \subset I^1 \subset I^*$. By the principle of induction, we can conclude (11) and complete the proof.

Now, we can establish the convergence theorem of Algorithm 1. Theorem 5 Algorithm 1 converges in finite steps.

Proof. By (6) and (11), we have $\hat{\lambda}^{k+1} \leq \hat{\lambda}^k$. This together with (7), we have $J^{k+1} \subseteq J^k \subseteq N$. Since N is finite index set, there exists $k \geq 0$, such that $J^k = J^{k-1}$.

References

1. Meyer, G.H.: Free Boundary Problems with Nonlinear Source Terms. Numer. Math. 43, 463–482 (1984)
2. Hoppe, R.: Two-Sided Approximations for Unilateral Variational Inequalities by Multigrid Methods. Optim. 18, 867–881 (1987)
3. Hoppe, R., Kornhuber, R.: Adaptive Multilevel Methods for Obstacle Problems. SIAM J. Numer. Anal. 31, 301–323 (1994)
4. Hoppe, R.: Multigrid Algorithms for Variational Inequalities. SIAM J. Numer. Anal. 24, 1046–1065 (1987)
5. Kärkkäinen, T., Kunisch, K., Tarvainen, P.: Augmented Lagrangian Active Set Methods for Obstacle Problems. J. Optim. Theory Appl. 119, 499–533 (2003)
6. Zeng, J.P., Zhou, S.Z.: A Domain Decomposition Method for a Kind of Optimization Problems. J. Comput. Appl. Math. 146, 127–139 (2002)

A New Approach of Capturing System Call Context for Software Behavior Automaton Model

Zhen Li and Junfeng Tian

Mathematics and Computer College, Hebei University, Baoding, China
lizhen_hbu@yahoo.com.cn, jftian@hbu.cn

Abstract. According to the problems of traditional methods of capturing calling context, a new approach of capturing system call context is proposed and applied to the software behavior automaton model based on system call. The approach represents system call context by context value computed, which can capture system call context accurately with low time overhead. The experimental results show that our approach is better than traditional methods of capturing calling context in the aspect of accuracy or time overhead for software behavior monitoring.

Keywords: Software Behavior, Automaton, System Call, Context.

1 Introduction

With the development of computer and network technology, software plays an important role in the information society and software security has become a current research focus. Because system call is the interface provided by the operating system to access system resources, it can reflect the features of software behavior to a great extent. A lot of software behavior models based on system call have appeared [1], such as Var-gram model [2], finite state automaton (FSA) model and pushdown automaton (PDA) model [3], HPDA model [4], HFA model [5], anomaly detection model of program behavior combining system call with Markov chain [6]. For software behavior models, a natural approach for capturing properties of system call arguments is using sets. This approach combines information about the system call that appears in any trace. Such a combination can lead to significant losses in accuracy. A good method is to partition the argument value set of each system call into subsets according to the calling context, and capture properties of each subset separately.

Calling context can provide a rich representation of program location. Some of the above software behavior models capture calling context by walking the stack which is the simplest method for capturing calling context. While, walking the stack more than infrequently is too expensive for monitoring software. An alternative to walking the stack is to build a calling context tree (CCT) dynamically [7]. Unfortunately, tracking the program's current position in a CCT adds a factor of 2 to 4 to program runtimes.

Y. Yu, Z. Yu, and J. Zhao (Eds.): CSEEE 2011, Part II, CCIS 159, pp. 148–153, 2011.
© Springer-Verlag Berlin Heidelberg 2011

These overheads are unacceptable for most deployed systems. Bond et al. [8] proposed an approach called probabilistic calling context (PCC) for object-oriented programs. PCC computes the context value by evaluating a function at each call site. It uses the following function to compute the next PCC value from the current PCC value: $f(V,cs):=3\times V+cs$, where V is the current value of the calling context, cs is used to identify the current call site and can be computed statically for a call site with a hash of the method and line number. PCC is effective for object-oriented programs, while, PCC cannot represent calling context accurately when it is applied to intrusion detection based on system call. System call number can identify a system call, so it can be used as cs. The context value of current system call depends on the value of V, that is, last PCC value, because the value of cs is fixed. When two system calls with the same calling context are in different positions of program, the PCC values of these two system calls computed are different unless hash collision occurs because the PCC value is accumulated generating.

According to the above problems, we propose a new approach for presentation of system call context by context value. Our approach can well solve the above problems.

2   The Automaton Model Based on System Call Context

The software behavior automaton model improves HPDA model [4] capturing system call context by walking the stack, introduces a new approach for presentation of system call context by context value, and provides system call argument policies based on context. Fig. 1 shows the source code for an example program and the corresponding automaton model based on system call context. The automaton model is a 5-tuple: (S, $\sum$, T, s, A). S is a finite set of states. $\sum$ is a finite set of input symbols. T is the set of transition functions: $(S\times\sum)\rightarrow S$. s is the start state: $s\in S$. A is the set of accept states: $A\subseteq S$. In our model, the set of input symbols is defined as

$$\sum = (X\times Addr)\cup(Y\times ContVal)\cup\varepsilon . \qquad (1)$$

X={Entry, Exit}, Y is the set of system calls, Addr is the set of all possible addresses defined by the program executable, ContVal is the set of context values of system calls. The types of inputs are as follows: (1) Entry and Exit are two new added system calls. Entry is added before each function call and Exit is added after each function call. Entry and Exit get the return address before and after the function call respectively. (2) Each system call in Y is the invocation point of a system call, where the associated context value represents the context of current system call. (3) ε is the empty string.

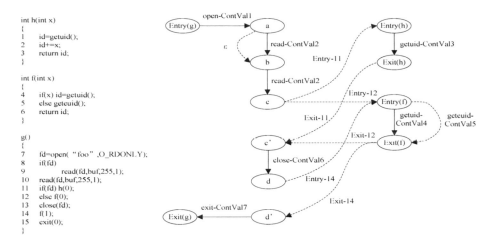

```
int h(int x)
{
1      id=getuid();
2      id+=x;
3      return id;
}

int f(int x)
{
4      if(x) id=getuid();
5      else geteuid();
6      return id;
}

g()
{
7      fd=open( "foo" ,O_RDONLY);
8      if(fd)
9            read(fd,buf,255,1);
10     read(fd,buf,255,1);
11     if(fd) h(0);
12     else f(0);
13     close(fd);
14     f(1);
15     exit(0);
}
```

Fig. 1. Program code and its corresponding automaton model based on system call context

3   Capturing System Call Context

3.1   System Call Context

System call context is represented by call stack of system call. If the call stacks of two system calls are the same, the system call contexts of two system calls are the same; otherwise, the system call contexts of two system calls are different. For current call stack, if the return address of function main is a_0, the return addresses of internal functions are $a_1,\ldots,a_{m-1}$ and the return address of last system call is a_m, then the system call context can be expressed as $(a_0, a_1, \ldots, a_{m-1}, a_m)$.

Different system call contexts can be expressed by different context values, and the same context value expresses the same system call context. Context value should have the following properties: (1) Context values must be basically distributed randomly so that the number of value conflicts is close to the ideal; (2) The context value must be deterministic, i.e., a given calling context always computes the same value.

Equation (2) is used to determine the number of expected conflicts given population size n and 32- or 64-bit values [9]:

$$E[\text{conflicts}] := n - m + m\left(\frac{m-1}{m}\right)^n . \tag{2}$$

m is the size of the value range (e.g., $m=2^{32}$ for 32-bit values). Bond et al. [8] analyzed the feasibility of calling context representation by context value based on equation (2). If a program executes 10 million distinct calling contexts, we expect to miss contexts at a rate of just over 0.1%, which is likely good enough for many clients. For programs with many more distinct calling contexts, or for clients that need greater probability guarantees, 64-bit values should suffice. For example, one can expect only a handful of conflicts for as many as 10 billion distinct calling contexts.

3.2 Context Value Calculation

The process of gaining system call argument policies based on system call context value is as follows:

(1) Form the system call context string.

For system call sc_i, concatenate function names of multilevel functions which call sc_i and system call sc_i with proper separators between two function names or function name and system call, and form a string $SysContStr_i$ which can represent the current system call context uniquely. For example, if function f_1 calls function f_2, function f_2 calls function f_3, ..., function f_{k-1} calls function f_k, function f_k calls system call sc_i, then the context string $SysContStr_i$ can be expressed as follows:

$$SysContStr_i = strcat(strcat(\cdots strcat(strcat(f_1, f_2), f_3), \cdots, f_k), sc_i)$$

Because full context provides too much sensitivity and results in many false positives on real program, especially recursive programs, we limit context to the top h function calls on the stack, that is, k≤h.

(2) Compute the hash value $SysContVal_i$ of the context string $SysContStr_i$.

(3) Determine the position p_i of system call context in hash table. If the size of hash table is Hashsize, then $p_i = SysContVal_i \% Hashsize$, where % is the modulus operator.

(4) Gain the argument policies of system call sc_i in context $SysContStr_i$ from the position p_i of hash table.

It is possible that multiple context strings are mapped into one context value, that is, hash table collision occurs. We use three hashes to resolve hash table collision problem. One for the hash table offset, two for verification. These two verification hashes are used in place of the actual context string. Of course, this leaves the possibility that two different context strings would hash to the same three hashes, but the chances of this happening are safe enough for just about anyone. Instead of using a linked list for each entry, when a collision occurs, the entry will be shifted to the next slot, and the process repeats until a free space is found.

4 Experiments and Analysis

We have implemented our model in Linux. Each system call is intercepted by loadable kernel module (LKM) and modified to capture the context and arguments of the system call. We made experiments on a PC with AMD Phenom (tm) 8400 Triple-Core Processor 2.10GHz and 2GB of main memory running Linux kernel 2.4.20. According to four common programs (gzip, cat, find and tar) in Linux, we compare three methods of capturing system call context that is call stack, PCC and our method. Among them, call stack method adopts __builtin_return_address() function to gain the return address of function in system call stack; PCC method adopts system call number of current system call as cs; Our method limits context to the top five function calls on the stack, that is, h=5 and the size of hash table in PCC and our method is 2^{20}. The test programs and statistical results are shown in Table 1 and Table 2. When two or more different calling contexts map to the same value, conflict occurs. When there are k different calling contexts mapping to the same value, the conflict number is k-1.

Table 1. The test programs

Program	Workload	Total number of system calls	Number of different system calls
gzip	Compress a 4.2MB file	309	17
cat	Display the content of a 160KB file	184	11
find	Find a file in directory including 1980 files	1537	17
tar	Pack 200 files to a 13.6MB tar file	4088	19

Table 2. Statistical results of system call context

Program	Number of system call context			Conflict number	
	Call stack	PCC	Our method	PCC	Our method
gzip	44	307	44	2	0
cat	33	183	33	1	0
find	64	1533	64	4	0
tar	83	4078	83	10	0

We test the time overhead of these three methods and take open system call in four common programs (gzip, cat, find and tar) for example. Table 3 shows the average time of capturing open system call context by call stack, PCC and our method. The time of capturing system call context by call stack is the longest, while the time by PCC is the shortest. The time of former is about 2.5~10 times longer than that of latter. The time of capturing system call context is different for different number of open system calls and realized functions in different programs. The time capturing system call context by our method is between call stack and PCC, and is closer to PCC. Although the time of PCC method is the shortest, capturing system call context by PCC is inaccuracy. Our method can capture the system call context accurately with low time overhead.

Table 3. Average time of capturing open system call context

Program	Time(µs)		
	Call stack	PCC	Our method
gzip	37.6	3.4	5.7
cat	43.2	4.0	9.3
find	8.1	1.0	2.3
tar	4.5	1.3	2.7

5 Conclusion

The paper proposes a new approach for presentation of system call context by context value and builds an automaton model based on system call context for software behavior monitoring. The model can capture system call context accurately with low time overhead. Our approach of capturing context uses storage for time, so it is suitable for enough memory that can be provided. It can also be applied to any anomaly detection system to capture system call context.

Acknowledgement. Supported by the National Natural Science Foundation of China (Grant No. 60873203), the Outstanding Youth Foundation of Hebei Province (Grant No. F2010000317), the Natural Science Foundation of Hebei Province (Grant No. F2008000646), and the Science Foundation of Hebei University (Grant No. 2008Q09).

References

1. Tao, F., Yin, Z.Y., Fu, J.M.: Software Behavior Model Based on System Calls. Computer Science 37(10), 151–157 (2010)
2. Wespi, A., Dacier, M., Debar, H.: Intrusion detection using variable-length audit trail patterns. In: Debar, H., Mé, L., Wu, S.F. (eds.) RAID 2000. LNCS, vol. 1907, pp. 110–129. Springer, Heidelberg (2000)
3. Wagner, D., Dean, D.: Intrusion Detection via Static Analysis. In: IEEE Symposium on Security and Privacy, pp. 156–169. IEEE Computer Society, Oakland (2001)
4. Liu, Z., Bridges, S.M., Vaughn, R.B.: Combining Static Analysis and Dynamic Learning to Build Accurate Intrusion Detection Models. In: 3rd IEEE Int'l Workshop on Information Assurance, pp. 164–177. IEEE Computer Society, College Park (2005)
5. Li, W., Dai, Y.X., Lian, Y.F., Feng, P.H.: Context Sensitive Host-Based IDS Using Hybrid Automaton. Journal of Software 20(1), 138–151 (2009)
6. Frossi, A., Maggi, F., Rizzo, G.L., Zanero, S.: Selecting and improving system call models for anomaly detection. In: Flegel, U., Bruschi, D. (eds.) DIMVA 2009. LNCS, vol. 5587, pp. 206–223. Springer, Heidelberg (2009)
7. Spivey, J.M.: Fast, Accurate Call Graph Profiling. Software-Practice and Experience 34(3), 249–264 (2004)
8. Bond, M.D., McKinley, K.S.: Probabilistic Calling Context. In: 2007 Object-Oriented Programming Systems, Languages, and Applications, pp. 97–112. ACM, Montreal (2007)
9. Mitzenmacher, M., Upfal, E.: Probability and Computing: Randomized Algorithms and Probabilistic Analysis. Cambridge University Press, New York (2005)

A Software Trustworthiness Evaluation Model Based on Level Mode

Tie Bao[1], Shufen Liu[1], Xiaoyan Wang[1], Yunchuan Sun, and Guoxing Zhao[2]

[1] College of Computer Science and Technology, Jilin University,
Changchun 130012, China
[2] College of Information Science and Technology, Beijing Normal University,
Beijing 100875, China
apche@126.com, {liusf,wangxy}@jlu.edu.cn,
yunch@bnu.edu.cn, zgxlogic@gmail.com

Abstract. This paper, aiming at issues of software trustworthiness research, proposed a trustworthiness level mode evaluation model based on actual evidence, analyzed the collection and measure method for evidence of software practical application, built trustworthiness evidence model and level model based on actual evidence, and finally described the main part of trustworthiness level matching algorithm in software trustworthiness evaluation. Practical application proved that the above-mentioned evaluating model was provided with sound operability and practical applicability.

Keywords: Trusted software, Trustworthiness evaluation Model, Actual Evidence.

1 Introduction

Modern information society relies on computing system, to be large extent, relies on software in fact. Software is the spirit of information foundation. Software system becomes more huge and hard to control as the function requirements increasing. It is difficult to avoid the fault and leak, and the system is much weaker. More often, the system works not as people's intention, and errors always happened which leads to the lost of customers. For instance, on August 14th, 2007, errors of the computer system of the Los Angeles airport happened, and sixty scheduled flights and twenty thousand passengers could not enter. They could enter until the next day morning. After investigation, people realized that this accident was caused by the disability of customs computer system. In consequence, software is not always reliable, and this is the subject "software trustworthiness" that we talk about.

Software trustworthiness refers the computer system could satisfy the requirements of customers, and it is reliable [1]. ISO/IEC specification [2] defines it as follows: "A trusted component, operation or process is one whose behavior is predictable under almost any operating condition". From the aspect of user experience, Bill Gates of Microsoft regards dependable computing as a dependable and safe computing which could be obtained at any time [3]. Developing high quality and high reliable software is always one research objective of software engineering discipline, and related theory

Y. Yu, Z. Yu, and J. Zhao (Eds.): CSEEE 2011, Part II, CCIS 159, pp. 154–159, 2011.
© Springer-Verlag Berlin Heidelberg 2011

and technology has enhanced software credibility [4], directly or indirectly. Yet traditional software engineering discipline did not apply software credibility as the primary research objective. Due to the emergence of trustworthiness issues, software trustworthiness research has become a hot topic aroused wide concern. China has enrolled trusted software in important research programs including Natural Science Foundation and 863 Program. Trustworthy software tools and integration environment (Trustie) platform and resource pool were opened to public in Oct, 2008, and came into practice [5]. All these above mentioned researching work has driven trusted software research to a deeper and more comprehensive direction, rapidly.

Above mentioned researching work [6-8] concentrated on definition and attributes of software trustworthiness, as well as guarantee of trustworthiness in terms of certain software. Yet, former researches didn't provide specific solution for defining evaluation model in practical application as well as map building of software and corresponding trustworthiness level. For this reason, this paper is going to research into trustworthiness evaluation model, and put forward a trustworthiness evaluation model based on actual evidence which can be used to evaluate trustworthiness level of software in practical application.

2 Trustworthiness Evidence Model

At present, trustworthy evidences in trustworthiness evaluation of common softwares are classified according to software life circle; collected evidences are used to evaluate trustworthiness according to attributes of trustworthiness. However, without considering practical application of software, these evidences might be difficult to collect during application processes. To tackle the issues in trustworthiness evidence research, the trustworthiness evidence in this paper will be based on actual evidences which are collected or disintegrated by common method in software developing and applying process, including software function requirement, software test data, user exchanging file, project acceptance report and user feedbacks, etc. Actual trustworthy evidences in this paper will be collected by common method, and thus ensures operability and handleability of trustworthiness evidence model building, as showed in figure 1.

With regard to one actual evidence, which can be described as a tetrad-<Ed_ID, Ed_Desc, Ed_Data, Ed_Evalue >, of which Ed_ID refers to unique identifier of the mentioned evidence, Ed_Desc refers to explaination, Ed_Data refers to evidence data which may be the form of text or number and Ed_Evalue refers to numerical value generated in evidence measure process. One evidence item of model for trustworthiness evidence can be described as a two-tuples < Ed_ID, $ed\_measure(\ )$ >, of which $ed\_measure(\ )$ refers to the specific measure method for actual evidence Ed_ID. Because the evidences are collected in software practical application, we can reference the research on evidence measure methods in specific application fields. With regard to subjective evidences which are difficult to quantify, such as application scale, users feedback, etc., an integrated evaluation method which combines mathematical evaluation method and domain experts can be adopted.

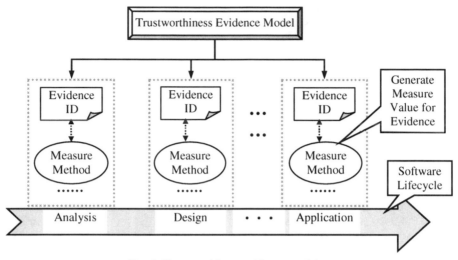

Fig. 1. Trustworthiness evidence model

3 Trustworthiness Level Model

Mapping between software and trustworthiness level should be established for software trustworthiness analysis, for the purpose of conducting an integrated measurement and comparison over trustworthiness of different software. Therefore, trustworthiness level should be defined firstly; this is too a difficult point of trustworthiness evaluation. In that overall and accurate leveling system is difficult to achieve due to variation of application domain and requirement, and, at the same time, level-mapping method with favorable operability is also difficult to work out, operable level evaluation method, aiming at practical application, are absent at present. Trustworthiness leveling system should be defined according to characters of application domain, considering the handleability at the same time. This paper proposes a trustworthiness leveling method for software based on actual evidence; this method can effectively tackle the existing problem, and is provided with favorable operability and practical applicability, as showed in figure 2.

Figure 2 indicates that trustworthy level model is a finite ordered set. One level *TL* of model for trustworthiness level (*TLM*) can be described as a triple-< *LevelID*, *NecLib*, *OptLib* >. *LevelID* refers to unique identifier of trustworthiness level. *NecLib*, which includes *NecSet* and *NecTotalMin*, refers to necessary library; *NecSet*, finite set of < *Ed_ID*, *Ed_Con* > pair, refers to necessary set of evidence in this level; *Ed_Con* refers to limiting condition on measure value. *OptLib*, which includes *OptSet* and *OptTotalMin*, refers to selectable library. *OptSet*, a finite set consisting of *Ed_ID*, refers to selectable set of this level, while there is no limit for measure value on single selectable evidence. *NecTotalMin* refers to the minimum value which measure value of necessary set must reach, and is equal to summation of all measure value in necessary set of evidence; *OptTotalMin* refers to the minimum value which measure value of selectable set must reach, and is equal to summation of all measure value in selectable set of evidence.While model for trustworthiness level (*TLM*) is a finite ordered set consisting of *TL*, which can be described as {TL_0, TL_1, ... , TL_n}, fulfilling $TL_0 < TL_1 < ...... < TL_n$; here the condition $TL_i < TL_j$ can be described as:

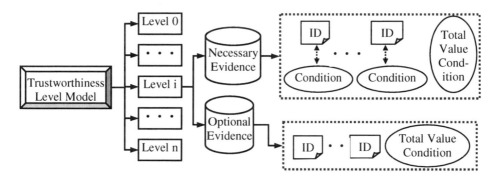

Fig. 2. Trustworthiness level model

$$TL_i.OptLib.OptSet \subseteq TL_j.OptLib.OptSet \wedge TL_i.NecLib.NecTotalMin <$$
$$TL_j.NecLib.NecTotalMin \wedge TL_i.OptLib.OptTotalMin \leq TL_j.OptLib.OptTotalMin$$
$$\wedge \left((\forall em_i \in TL_i.NecLib.NecSet)(\exists em_j \in TL_j.NecLib.NecSet)em_i.Ed\_ID = em_j.Ed\_ID \right)$$

4 Application of Evaluation Model

In order to build the mapping between software and trustworthiness level, we will describe the main part of trustworthiness level matching algorithm.

Algorithm 1. trustworthiness level matching algorithm
   Input: Actual evidence set *ActualEdSet*, Evidence Model *TEM*, Target level *TL_i*
   Output: Boolean variable //show whether software match the level
   Initialize *necsum=0, optsum=0*;
   For each *ed* in *ActualEdSet* //generate measure value for every actual evidence
   Begin
     Extract *med* which belongs to *TEM* and *med.Ed_ID* equals to *ed.Ed_ID*;
     ed.Ed_Evalue = med.ed_measure (ed.Ed_Data);
   End
   For each *em* in *TL_i.NecLib.NecSet* //judge whether actual evidence set satisfy
limiting conditions of *TL_i*
   Begin
     Extract *ed* which belongs to *ActualEdSet* and *ed.Ed_ID* equals to *em.Ed_ID*;
     If *ed == null*
        return *false*;
     Else
       { If !*satisfied(ed.Ed_Evalue, em.Ed_Con)*
       return *false*;

$necsum = necsum + ed.Ed\_Evalue;\ \}$
End
For each *oeid* in $TL_i.OptLib.OptSet$
Begin
    Extract *ed* which belongs to *ActualEdSet* and *ed.Ed_ID* equals to *oeid*;
    If *ed!=null*
       $optsum = optsum + ed.Ed\_Evalue;$
End
If $(necsum<TL_i.NecLib.NecTotalMin)||(optsum<TL_i.OptLib.OptTotalMin)$
  return *false*;
  return *true*

5   Conclusions and Further Work

Along with the continuously getting deeper of computer and software application in human society, software becomes more and more important in key application of domain. This highlights the importance of software trustworthiness research as people urgently hoping to guarantee reliability, safe and stable operation of software by means of measurement and evaluation on software trustworthiness. In the research on software trustworthiness, the most important question is how to provide a software trustworthiness evaluation method with favorable operability and handleability.

For tackling the above-mentioned issues, this paper has conducted a study on trustworthiness evaluation model, put forward trustworthiness evidence and level model based on actual evidence. The model in this paper based on actual evidences, and is provided with favorable operability, and, when adopting in practical application of software, is easy for collecting related trustworthiness evidences.

This paper provides effective model for trustworthiness evaluation of software; but measure of actual evidence and analysis still needs the participation of existed methods and tools in software application. Therefore, effective measure method for evidence should be provided in further work. In addition, how to bring formal description and certification for trustworthiness into our evaluation model should be one of our future jobs.

Acknowledgment. This work is supported by the National High Technology Research and Development Program of China (Grant No. 2009AA010314).

References

1. TCG Specification Architecture Overview V1.2,
   https://www.trustedcomputinggroup.org
2. Information Technology, Security Techniques Evaluation Criteria for IT Security,
   http://standards.iso.org
3. Trustworthy Computing-Wired News, http://www.wired.com
4. Chen, H.W., Wang, J., Dong, W.: High Confidence Software Engineering Technologies. Chinese Journal of Electronics 31, 1933–1938 (2003)

5. A Trustworthy Software Production Environment for Large Scale Software Resource Sharing and Cooperative Development, http://www.trustie.net
6. Ryu, S.H., Casati, F., Skogsrud, H., Benatallah, B., Saint-Paul, R.: Supporting the Dynamic Evolution of Web Service Protocols in Service-Oriented Architectures. ACM Trans. 2, 1–46 (2008)
7. Software Trustworthiness Classification Specification (TRUSTIE-STC V 2.0), http://www.trustie.net
8. Zeng, J., Sun, H.L., Liu, X.D., Deng, T., Huai, J.P.: Dynamic Evolution Mechanism for Trustworthy Software Based on Service Composition. Journal of Software 21, 261–276 (2010)

Acoustic Source Localization Based on Multichannel Compressed Sensing and Auditory Bionics in Hearing Aids

Ruiyu Liang[1,2], Li Zhao[2], Ji Xi[1], and Xuewu Zhang[1]

[1] College of Computer and Information, Hohai Univ.,Nanjing 210098,China
[2] School of Information Science and Engineering, Southeast Univ., Nanjing 210096,China
liangry@hhuc.edu.cn, zhaoli@seu.edu.cn,
xijie952611@gmail.com, zhangxw@hhuc.edu.cn

Abstract. The microphone array is an effective means in acoustic location. However, it has become very difficult to acquire adequate positional parameters when the space between the microphone pairs is too small. To this problem, the paper proposed a signal model which contains source information constructed by differential microphone. Moreover, to decrease computation complexity, we obtain acoustic source data by multichannel compressed sensing(CS). Finally, acoustic source location is obtained by the estimation of energy in reconstructed signal. Theoretical analysis and simulation results conclude that the proposed approach exhibits a number of advantages over other source localization techniques.

Keywords: Microphone Array, Compressed Sensing, Auditory Bionics, Digital Hearing Aid.

1 Introduction

The acoustic source localization based on microphone array has been used widely [1, 2]. In the instance of enough space and array nodes, the signals received by different microphone node are very different. And it is easier to get the localization parameters (e.g. TDOA). In practice, it has restriction in the size of microphone array, e.g., intelligent hearing equipment [3].For instance, resolving ITD cues at 8 bits resolution using microphones placed at a distance of 1mm would require 100MHz sampling frequency. The higher sampling frequency means to deal with more data. The higher frequency sets, the higher power consumed, and the easier the circuit noises affect. In order to overcome these defects, the paper [4] proposed the acoustic source localization technology with Gradient Flow based on zoology prototype of Ormia ochracea. It evaluates the acoustic source location by minimum mean square error technology with differential microphone recorded data. Ref [5] proposed to integrate the acoustic sound localization arithmetic into the $\sum\Delta$ modulated modules, and its precision can reach to 2. But the researches described above are mainly a realization of auditory bionics localization structure. It is not based on the traditional microphone array.

Y. Yu, Z. Yu, and J. Zhao (Eds.): CSEEE 2011, Part II, CCIS 159, pp. 160–164, 2011.

In the far field localization model, the acoustic source can be considered as a dot source. Furthermore, its location is sparse comparing with localization space. So some scholars proposed an acoustic source localization method based on the reconstruction of sparse signals [6]. The CS theoretic is a new technology combined sampling and compressing [7]. Ref.[8] used a style of one reference node and several compressed ones. The reference node samples with nyquist sampling theorem, and used as base node. So its localization performance is fully depended on that of the reference node.

To the problem of small node space and large sampling data, this paper simulate the amplitude difference of Orimia ochracea hearing system to construct signal model which is based on the differential microphone array. And it samples data with multi-channels CS with no reference node, and takes the average of signal frames as the sample input. After constructed model, it will reconstruct the acoustic source signal with subgradient projection method. Then it evaluates the location of maximum energy value to obtain the multi-acoustic source.

2 Methodology

2.1 Signal Model

Suppose N far field narrowband signals $s_1(t),...,s_N(t)$ radiate to the M array respectively with $(\theta_1,...,\theta_N)$. The radius of array is d. If the signal frequency is f_c, the signal received by the node m is $f_m(t)$:

$$f_m(t) = \sum_{n=1}^{N} g_{mn} s_n(t - \tau_{mn}) + n_m(t) \tag{1}$$

Here, g_{mn} is the plus of the node (m) to the signal (n); τ_{mn} is the relative time-delay of the signal (n) arriving at the node (m); and $n_m(t)$ is the noises received by the node (m) at time (t), that is Gauss white noises. Taking no account of noises and in a far model, $g_{mn} \approx 1$, then

$$\Delta f_m(t) - \frac{1}{M}\sum f_m(t) = -\sum_{n=1}^{N} \tau_{mn} \dot{s}_n(t) + n_m(t) \tag{2}$$

Here, $\Delta f_m(t)$ is the differential signal of every microphone[5]. Because it evaluates current angel's space spectrum, $\dot{s}_n(t)$ equals to $s_n(t)$.

Combined every microphone differential signal, it can obtain

$$\Delta F(t) = \begin{bmatrix} f_1(t) \\ f_2(t) \\ \vdots \\ f_m(t) \end{bmatrix} = -\begin{bmatrix} \tau_{11} & \tau_{12} & \cdots & \tau_{1N} \\ \tau_{21} & \tau_{22} & \cdots & \tau_{2N} \\ \cdots & \cdots & \cdots & \cdots \\ \tau_{m1} & \tau_{m2} & \cdots & \tau_{mN} \end{bmatrix}\begin{bmatrix} s_1(t) \\ s_2(t) \\ \vdots \\ s_n(t) \end{bmatrix} + \begin{bmatrix} n_1(t) \\ n_2(t) \\ \vdots \\ n_m(t) \end{bmatrix} = -\Gamma S + N \tag{3}$$

Using the model like Eq.3, it can avoid the complex number operation. It calculates the differential signal $\Delta F(t)$ according to the received signals, then evaluates the vector S's spectrum energy. And the location of the maximum energy value is just that of the acoustic source θ_n.

2.2 Structure of Acoustic Source Localization

If the sampling frequency is higher and the nodes of the microphone array is much more, the system calculation is still big with the model like Eq.3. And it will be bigger if the channel number increases. So this paper proposes a method of multichannel CS sampling to reduce workload. Setting $\Psi = -\Gamma$, Eq.3 changes to:

$$Y = \Phi\Delta F(t) = \Phi\Psi S + \Phi N = \Phi\Psi S + N_\Phi \tag{4}$$

At this time, acoustic source location evaluation often changes to the convex optimization problem. This paper uses the subgradient projection method. It comes from convex set projection method which has good robustness to noises [9].It can modify the parameter adaptively which improve the performance of convergence and reconstruction precision.

Change the Eq.4 to the standard limited hypo-programming style [10]:

$$\min_z \ F(z) = c^T z + \frac{1}{2} z^T Bz \quad s.t. \quad z \geq 0 \tag{5}$$

According to the formula (10), it defines convex set $\{C_k, k = 1,...,m\}$ to meet the following restriction:

$$C_k(\rho) = \left\{ \hat{z}_k \in \mathcal{H} : \left\| c^T \hat{z}_k + \frac{1}{2}\hat{z}_k^{\ T} B\hat{z}_k \right\|^2 \leq \rho \right\} \tag{6}$$

Define a convex function

$$g(\hat{z}) = \left\| c^T z + \frac{1}{2} z^T Bz \right\|^2 - \rho, \forall \hat{z} \in \mathcal{H} \tag{7}$$

Put them into the projection Eq.8 and $\hat{z}_k$ iteration updating Eq.9, it can obtain the optimized reconstruction signal z_{opt}:

$$P_{H^-(\hat{z}_k)}(\hat{z}_k) = \begin{cases} \hat{z}_k & \hat{z}_k \in H^-(\hat{z}_k) \\ \hat{z}_k + \dfrac{-g_k(\hat{z}_k)}{\left\| \nabla g_k(\hat{z}_k) \right\|^2} \nabla g_k(\hat{z}_k) & \hat{z}_k \notin H^-(\hat{z}_k) \end{cases} \tag{8}$$

$$\hat{z}_{k+1} = \hat{z}_k + \tau_k \left[P_{H^-(\hat{z}_k)}(\hat{z}_k) - \hat{z}_k \right], \forall k \in \mathbf{N} \tag{9}$$

Af ter getting the reconstruction signal $\hat{z}$, change it to the acoustic source signal $\hat{S}$, then obtain the source number and their location by evaluating the maximum energy value of $\hat{S}$.

3   Experimental Results and Analysis

In this session, it compares some experiment result to prove the method of acoustic source localization proposed in this paper. With two acoustic source localization method Capon and MUSIC which are based on space spectrum, it compares their ability of acoustic source localization under different conditions. The experiment data is set up based on the analog room pulse responding model [11]. It selects an 8-node microphone array whose radius is 3mm.There are 2 acoustic sources with sampling number of 10000. The localization space is $0°\sim360°$, and every $1°$ is one sector. It uses the reconstruction method of subgradient projection and initializes the parameters: $\lambda = 0.1\left\|\Upsilon^T Y\right\|_\infty$, τ , u , v and the ending iterative times $M = 150$, $\delta = 0.01$ .In the actual experiment, we uses the quadratic reconstruction. First, it divides the space into 18 sector. At this time, S_1 is a 18×1 vector, and Ψ_1 is a 16×18 matrix. It will judge which sector the acoustic source belongs to by reconstruction. Then according to the result of step 1, it divides the sector into 20 grids with space of $1°$. Now S_2 is a 20×1 vector and Ψ_2 is a 16×20 matrix.

If the SNR is 20dB, the comparisons of spectrum performance with different method see Fig 1. Only the method proposed by this paper has obvious spectrum performance. It gets two acoustic source locations, but the other two methods lost one.

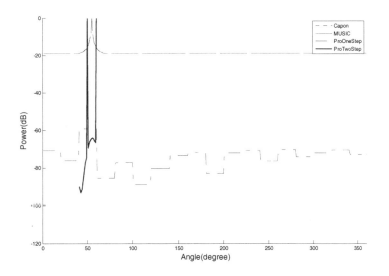

Fig. 1. Space spectrum comparison of two relevant acoustic source in $50°$ and $60°$ sector

4   Conclusions

This paper proposed an acoustic source localization method of multi-channel CS based on bionic. Firstly it constructed signal model with differential microphone array. It avoided complex number calculation, improved the efficiency and restrained the noises. Secondly, it reconstructed the acoustic source signal with adaptive sub-gradient method. It adjusted the expand coefficient adaptively in the iterative process. It used the advantage of the convergent speed in different stage and convergent performance in stable state, which improved the convergent speed and had Robustness to noises. Finally, it got the acoustic source location by evaluating the maximum energy value of reconstruction signals. The theoretical analysis and simulation results conclude that the proposed method has lower calculation complexity and quicker convergent speed. It also has Robustness to noises. Although it has some deviation in the instance of low SNR, it is still the direction for the further research.

Acknowledgments. This work was supported by the Fundamental Research Funds for the Central Universities: No. 2009B32614.

References

1. Sarradj, E.: A Fast Signal Subspace Approach for the Determination of Absolute Levels from Phased Microphone Array Measurements. J.Sound and Vibration 329, 1553–1569 (2010)
2. Weiping, C., Shikui, W., Zhenyanga, W.: Accelerated Steered Response Power Method for Sound Source Localization Using Orthogonal Linear Array. Applied Acoustics 71, 134–139 (2010)
3. Miles, R.N., Hoy, R.R.: The Development of a Biologically-Inspired Directional Microphone for Hearing Aids. Audiology and Neuro-Otology 11, 86–94 (2006)
4. Stanacevic, M., Cauwenberghs, G.: Micropower Gradient Flow Acoustic Localizer. IEEE Trans. on Circuits and Systems 52, 2148–2157 (2005)
5. Gore, A., Fazel, A., Chakrabartty, S.: Far-Field Acoustic Source Localization and Bearing Estimation Using $\sum \Delta$ Learners. IEEE Trans. on Circuit and Systems 57, 783–792 (2010)
6. Malioutov, D., Cetin, M., Willsky, A.S.: A Sparse Signal Reconstruction Perspective for Source Localization with Sensor Arrays. IEEE Trans. on Signal Processing 53, 3010–3022 (2005)
7. Guangming, S., Danhua, L., Dahua, G., et al.: Advances In Theory and Application of Compressed Sensing. Acta Electronic a Sonica 37, 100–1081 (2009)
8. Gurbuz, A.C., Mcclellan, J.H., Cevher, V.: A Compressive Beamforming Method. In: Acoustics, Speech and Signal Processing, Las Vegas, NV, pp. 2617–2620 (2008)
9. Yukawa, M., Slavakis, K., Yamada, I.: Adaptive Parallel Quadratic-Metric Projection Algorithms. IEEE Trans. On Audio, Speech and Language Processing 15, 1665–1680 (2007)
10. Figueiredo, M.A.T., Nowak, R.D., Wright, S.J.: Gradient Projection for Sparse Reconstruction: Application To Compressed Sensing and Other Inverse Problems. IEEE J. Selected Topics in Signal Processing 1, 586–598 (2007)
11. A Model for Room Acoustics, http://www.2pi.us/rir.html

Service Advertisements' Formatting for Wireless Billboard Channels

Zhanlin Ji[1], Ivan Ganchev[2], Paul Flynn[2], and Máirtín O'Droma[2]

[1] CAC Dept, Hebei United University, TangShan, 064300, P.R. China
`Zhanlin.ji@ieee.org`
[2] ECE Dept, University of Limerick, Limerick, Ireland
{`Ivan.Ganchev,Paul.Flynn,Mairtin.ODroma`}`@ul.ie`

Abstract. This paper addresses the formal description of services for advertisement on wireless billboard channels (WBCs). As using as little bandwidth as possible is one of the WBC's desired properties, the basic data structure -the service description (SD)- is formatted with the abstract syntax notation (ASN.1) and encoded/decoded with packed encoding rules (PER). A SD's template division and an efficient software encoding/decoding architecture are proposed.

Keywords: Ubiquitous Consumer Wireless World (UCWW); Wireless Billboard Channel (WBC); Service Descriptions (SDs); Abstract Syntax Notation (ASN.1); Packed Encoding Rules (PER).

1 Introduction

The wireless billboard channel (WBC) [1] is a fundamental element for wireless service advertisement, discovery and association (ADA) in the consumer-centric business model (CBM), which is integral to the ubiquitous consumer wireless world's (UCWW) evolution [2, 3]. The aim of CBM is to enable mobile users (MUs) to be always best connected and best served (ABC&S) [4], i.e. to use the best service anytime-anywhere-anyhow through the best available wireless access network (AN). To archive this, users must be aware of all services provided in a particular location/area. This could be done by a proactive continuous push-based advertisement of services over WBCs provided by WBC service providers (WBC-SPs).

The details of desired WBC characteristics are listed below and in [1]:
• Simplex and broadcast communication;
• Limited bandwidth;
• Maximum coverage area;
• Different versions for different areas, i.e. local, regional, national and international WBCs;
• Not operated by access network providers (ANPs).

The push-based WBC's service ADA model operates as follows:

Registration: Service providers (xSPs) register their service descriptions (SDs) with a WBC-SP's central registry using some external method, e.g. via a web portal.

Y. Yu, Z. Yu, and J. Zhao (Eds.): CSEEE 2011, Part II, CCIS 159, pp. 165–170, 2011.
© Springer-Verlag Berlin Heidelberg 2011

Service advertisement: WBC-SP broadcasts all SDs repeatedly on a WBC.

Service discovery: User's mobile terminal (MT) tunes to WBC and listens to broadcast to receive desired SDs.

Association: MT associates with chosen 'best' service provider to use the desired services.

As the SD is the basic element in WBC and because using as little bandwidth as possible is one of the WBC desired properties, the abstract syntax notation (ASN.1) and its packed encoding rules (PER) were selected as the notation language and encoding/decoding scheme, respectively [1]. This paper mainly focuses on the SD's formatting, encoding and decoding procedures.

2   WBC SD's Format

An appropriate SD format was defined for storing and exchanging the service's ADA information in WBCs. A SD consists of a set of fields, such as a service type, scope list, length, composite capability/preference profiles (CC/PP) [5], QoS, and attribute list, as shown below:

```
ServiceDescription ::= SEQUENCE
{
    serviceType        Service-Type,
    length             Length,
    scopeList          ScopeList,
    ccpp               CCPP,
    qos                QoS,
    attrList           ServiceTemplate.
}
```

The attrList field is the main field of a SD to carry the ADA information. Each service type has its own attribute template. The templates are defined by a global WBC authority.

Access Networks' Communication Services (ANCSs) and teleservices (TSs) are the two main groups of wireless services [1]. TS is the actual service available to MUs, whereas ANCS provided by the physical ANs is used to get access to the teleservice. In practice, some ANPs may only provide ANCSs, or both ANCS and TS. Figure 1 shows the relationship between MUs, teleservices, and ANCSs.

The attribute template of teleservices includes the name, cost, software, text, uniform resource locator (URL), and supported AN list, as shown below:

```
AttributeTemplate ::= SEQUENCE
{
    name        Name,
    cost        Cost,
    software    SoftwareInfo,
    text        Text,
    url         URL,
```

```
    feature             Features,
    divisionSpecific    CHOICE {
            requirements        Requirements,
            supportedANList     SEQUENCE OF  AccessNetID,
        }
}
```

The attributes of Internet connection services (ICSs) includes the name, connection parameters, cost, and supported AN list:

```
AttributeTemplate ::= SEQUENCE
{
    name                ICSName,
    connectionParams    ConnectionParameters,
    cost                ICSCost,
    supportedANList     SEQUENCE OF AccessNetID
}
```

The attribute of AN services includes the name, identifier, and access point information [1]:

```
AttributeTemplate ::= SEQUENCE
{
    name                ANName,
    identifier          AccessNetID,
    accessPointInfo     SEQUENCE OF AccessPoint
}
```

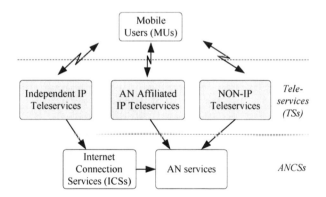

Fig. 1. The Layered Model of MUs, TSs and ANCSs

3   WBC SD's Encoding and Decoding

The ITU-T ASN.1 [6, 7] is well known as a reliable description language that uses compactable encoding rules for specifying data in telecommunications protocols. It is closely tied to the Java programming language [8].

To integrate the ASN.1-PER scheme into the WBC service layer, all SD templates were compiled into Java classes with an ASN.1 compiler (Figure 2).

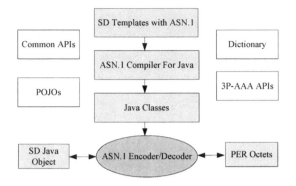

Fig. 2. The ASN.1-PER encoding/decoding of SDs

There are several important APIs in the ASN.1-PER encoder/decoder:

• *SD POJOs*: All the parameters of the SD are carried by the SD plain old Java objects (POJOs).

• *3P-AAA APIs*: These APIs help WBC-SP to obtain the SD's usage data from a third-party authentication, authorization and accounting server [2].

• *Dictionary*: This decodes values transmitted on the WBC to the names represented by those values. For instance, a service type with the value "0x5002004001" might represent a "StockTicker".

• *Common APIs*: To improve the software development efficiency, a number of basic WBC ASN.1 APIs were implemented, such as, *log, SDtoByte, Hexprint*, etc.

4   Software Architecture and Implementation

The WBC system includes a WBC-SP node and multiple MTs. The WBC-SP node runs an enterprise application which is developed under the Java platform enterprise edition (Java EE) and multiple agent system (MAS) environment. On the contrary, the MT runs a portable mobile application under the Java standard edition (Java SE) and lightweight MAS environment.

To develop the WBC SD's encoding/decoding module quickly and efficiently with low complexity and cost, a set of open-source/academic license IDEs/APIs /frameworks have been used as described in Table 1.

Figure 3 depicts the time-sequence UML diagram of a xSP submitting a SD to the WBC-SP enterprise application.

In Figure 3, the xSP submits a SD via the WBC-SP portal application. In the presentation web tier, a POJO is built in the controller, and then sent to the business tier. The SDManager is a service manager, which saves the POJO

Table 1. IDEs/APIs/Frameworks for SD's Encoding and Decoding

Name	Descriptions
Eclipse	An open-source Java IDE platform.
GlassFish	A Java EE Application Container.
JADE	The Java agent development environment (JADE) is an efficient MAS released by the Telecom Italia lab with FIPA specifications standard.
OSS	The Nokalva OSS ASN.1 tool for Java API generates a PER octets stream from the ASN.1 BNF file.
ANT	A pure Java building tool, developed by Apache.
DB4O	A high-performance, embeddable, open-source and one-line-of-code object-oriented (OO) database.
Log4j	A pure Java logging framework.

to an OO database, creates a new blackboard object, and then sends it to the personal assistance agent (PAA). The PAA acts as a gateway agent in the WBC-SP enterprise application. It receives the blackboard object, formats an agent communication language (ACL) message, and then sends the message to the SD agent for encoding. The latter requests creating a home object, which then creates the enterprise Java bean (EJB) object. With the reference of EJB object, the SD agent invokes the SD encoding process to generate the byte stream.

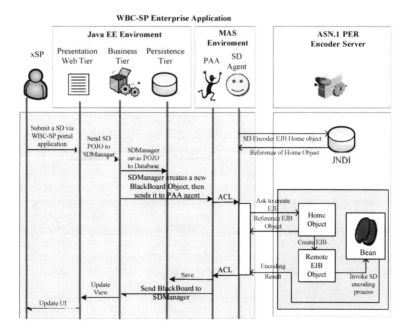

Fig. 3. The SD's Encoding Time-Sequence UML diagram

5   Conclusions

A WBC service description's (SD) structure, encoding and decoding design and implementation has been described in this paper. The ASN.1 notation is used as a flexible solution to describe the wireless service advertisements. PER is used as a space efficient method to encode SDs to a byte stream. Comparing with the other encoding rules, PER uses only the very minimum of bits to encode fields. The main details of the SD structure, encoder and decoder have been presented. In order to integrate the SD encoder into the WBC-SP enterprise application and integrate the SD decoder into the WBC MU portable application, a novel software architecture built on a number of open-source/academic IDEs/APIs /frameworks has been set out.

Acknowledgments. This publication has been supported by the Irish Research Council for Science, Engineering and Technology (IRCSET) and the Telecommunications Research Centre, University of Limerick, Ireland.

References

1. Flynn, P., Ganchev, I., O'Droma, M.: Wireless Billboard Channels: Vehicle and Infrastructural Support for Advertisement, Discovery, and Association of UCWW Services. Annual Review of Communications 59, 493–504 (2006)
2. O'Droma, M., Ganchev, I.: The Creation of a Ubiquitous Consumer Wireless World through Strategic ITU-T Standardization. IEEE Communications Magazine 48, 158–165 (2010)
3. O'Droma, M., Ganchev, I.: Toward a ubiquitous consumer wireless world. IEEE Wireless Communications 14, 52–63 (2007)
4. Passas, N., Paskalis, S., Kaloxylos, A.: Enabling technologies for the 'Always Best Connected' concept. Wireless Communications & Mobile Computing 6, 523–540 (2006)
5. Kiss, C.: Composite capability/preference profiles (CC/PP): Structure and vocabularies 2.0. W3C Working Draft 8 (2006)
6. ITU-T: Rec. X.680 Information technology - Abstract Syntax Notation One (ASN.1): Specification of basic notation, http://www.itu.int/ITUT
7. ITU-T: Rec. X.691 - Information technology - ASN.1 encoding rules: Specification of Packed Encoding Rules (PER), http://www.itu.int/ITUT/studygroups
8. Larmouth, J.: ASN.1 complete. Morgan Kaufmann Publication, San Francisco (2000)

Outage Probability of Selection Cooperation with Fixed-Gain Relays in Nakagami-*m* Fading Channels

Haixia Cui

School of Physics and Electronic Engineering,
Guangzhou University, 510006, China
cuicuihang0715@gmail.com

Abstract. In this paper, we analyze the outage performance of decode-and-forward selection cooperation systems, under a fixed-gain relay setting, in independent but not identically distributed Nakagami-*m* fading channels at arbitrary signal-to-noise ratios (SNRs) and number of available relays. In particular, an exact closed-form expression for the outage performance is derived employing partial relay selection in the first-hop transmission. Our analysis extends previous results pertaining to identically distributed fading channels under an adaptive relay setting. Furthermore, simulation and numerical results are presented to show the validity of the analytical results.

Keywords: Nakagami fading, selection cooperation, outage probability, performance analysis.

1 Introduction

Relay selection techniques [1], have been shown to be promising candidates to implement cooperative diversity in future wireless networks. The key ideas are to increase the connectivity, capacity and potential diversity of networks and also to reduce complexity of multi-relay cooperative communication systems, without the need of high power levels at the transmitters. More specifically, several neighboring nodes listen to the source, and only a single relay out of the set of nodes encodes and retransmits the received signal to the destination, which leads to copies of independent fading signal paths at the destination and thus brings diversity. Within the research of relay selection techniques, one of the most commonly used method is selection cooperation [1], in which the best relay is selected before the source transmission by taking into account the channel quality of the first hop.

The performance analysis for the selection cooperation with adaptive relays has been extensively evaluated in terms of outage probability[2]-[4]. In [5], the authors derived exact closed-form solutions for the fixed-gain opportunistic relays. However, to the best of the authors' knowledge, for a fixed-gain scenario with partial relay selection in the first-hop transmission and independent but not identically distributed Nakagami-*m* fading channels, there is few work in the literature pertaining to the selection cooperation case. Motivated by the this, we extend previous results [5] pertaining to the adaptive opportunistic relays case. In this paper, an exact closed-form

Y. Yu, Z. Yu, and J. Zhao (Eds.): CSEEE 2011, Part II, CCIS 159, pp. 171–176, 2011.

expression in term of outage probability for the decode-and-forward selection cooperation systems is derived at arbitrary signal-to-noise ratios (SNRs) and number of available relays. In addition, we verify the obtained closed-form results using numerical methods.

2 System and Channel Models

Consider a dual-hop wireless cooperative communication system, where a source node S communicates with a destination node D with K available relay nodes R_l, $l = 1,...,K$ and partial channel state information (CSI), i.e., the first hop channel condition information. All the channel links are mutually independent and orthogonal. Further, an selection cooperation strategy [1] is adopted in a half-duplex dual-hop time division scenario, and there is no a direct link. Therefore, each data packet is assigned two time slots. Let $\gamma_{1,l}$ and $\gamma_{2,l}$ be the instantaneous SNRs pertaining to the first- and second-hop transmissions through the l-th relay, respectively. In the first hop transmission , partial relay selection [6] is employed. That is to say, from the set of K available relays, only one of them may forward the received source information to the destination in the second-hop transmission by a distributed fashion, and the selected relay has the best source-relay link quality for the information tranmission. In fixed-gain relaying scenarios, the dual hop source-destination SNR can be expressed as [1]

$$\gamma_l = \frac{\gamma_{1,l}\gamma_{2,l}}{C_l + \gamma_{2,l}} \tag{1}$$

where C_l is a constant parameter [1] from which the relay gain can be obtained. Assuming that the i-th relay is selceted as the best relay. It can be easily get that the probability density functions (pdfs) and cumulative distribution function (cdfs) of $\gamma_{1,i}$ and $\gamma_{2,i}$ re, respectively, written by

$$f_{\gamma_{1,i}}(\gamma) = \sum_{k=1}^{K} f_{\gamma_{1,k}}(\gamma) \prod_{\substack{l=1 \\ l \neq k}}^{K} F_{\gamma_{1,l}}(\gamma), \gamma \geq 0 \tag{2}$$

$$F_{\gamma_{1,i}}(\gamma) = \prod_{l=1}^{K} F_{\gamma_{1,l}}(\gamma), \gamma \geq 0 \tag{3}$$

$$f_{\gamma_{2,i}}(\gamma) = \sum_{i=1}^{K} f_{\gamma_{2,i}}(\gamma) \int_{0}^{\infty} f_{\gamma_{1,i}}(x) \prod_{\substack{l=1 \\ l \neq i}}^{K} F_{\gamma_{1,l}}(x)dx, \gamma \geq 0 \tag{4}$$

$$F_{\gamma_{2,i}}(\gamma) = \sum_{i=1}^{K} F_{\gamma_{2,i}}(\gamma) \Pr\{i = \arg\max_{l \in 1,...,K} \{\gamma_{1,l}\}\} = \sum_{i=1}^{K} F_{\gamma_{2,i}}(\gamma) \int_{0}^{\infty} f_{\gamma_{1,i}}(x) \prod_{\substack{l=1 \\ l \neq i}}^{K} F_{\gamma_{1,l}}(x)dx, \gamma \geq 0 \tag{5}$$

where $\int_0^\infty f_{\gamma_{1,i}}(x) \prod_{\substack{l \in DS(p) \\ l \neq i}} F_{\gamma_{1,l}}(x)dx$ in (4) is the probability of selecting the best

relay R_i; $f_{\gamma_{1,l}}(\gamma)$, $F_{\gamma_{1,l}}(\gamma)$, $f_{\gamma_{2,l}}(\gamma)$, and $F_{\gamma_{2,l}}(\gamma)$ denote the pdfs and cdfs of $\gamma_{1,l}$ and $\gamma_{2,l}$, respectively. Assuming that $m_{1,l}$, $m_{2,l}$ and $\Omega_{1,l}$, $\Omega_{2,l}$ denote the Nakagami fading figure and `the variance of the first and second hop links, respectively; $\Gamma(\cdot)$ and $\Upsilon(\cdot,\cdot)$ denote gamma function and incomplete gamma function, respectively; γ_0 denotes the transmit SNR per hop; $\chi_{1,l} = m_{1,l}/(\gamma_0\Omega_{1,l})$, and $\chi_{2,l} = m_{2,l}/(\gamma_0\Omega_{2,l})$. The pdfs and cdfs of $\gamma_{1,l}$ and $\gamma_{2,l}$ can be obtained easily as [6]

$$f_{\gamma_{1,l}}(\gamma) = \frac{\chi_{1,l}^{m_{1,l}} \gamma^{m_{1,l}-1}}{\Gamma(m_{1,l})} \exp(-\chi_{1,l}\gamma), \gamma \geq 0 \tag{6}$$

$$F_{\gamma_{1,l}}(\gamma) = \frac{\Upsilon(m_{1,l}, \chi_{1,l}\gamma)}{\Gamma(m_{1,l})}, \gamma \geq 0 \tag{7}$$

$$f_{\gamma_{2,l}}(\gamma) = \frac{\chi_{2,l}^{m_{2,l}} \gamma^{m_{2,l}-1}}{\Gamma(m_{2,l})} \exp(-\chi_{2,l}\gamma), \gamma \geq 0 \tag{8}$$

$$F_{\gamma_{2,l}}(\gamma) = \frac{\Upsilon(m_{2,l}, \chi_{2,l}\gamma)}{\Gamma(m_{2,l})}, \gamma \geq 0 \tag{9}$$

3 Performance Analysis

Giving the target rate of r, below which the signal undergoes outage, the source-destination outage probability can be expressed as

$$P_{out} = \Pr\{\frac{1}{2}\log_2(1+\gamma_i) < r\} \tag{10}$$

where $\Pr\{\cdot\}$ denotes probability. Le $q = 2^{2r} - 1$, the above equality (10) can be rewritten as

$$P_{out} = \Pr\{\gamma_i < q\} = \int_0^\infty \Pr\{\frac{\gamma_{1,i}x}{C_i + x} < q\} f_{\gamma_{2,i}}(x)dx \tag{11}$$

$$= \int_0^\infty \Pr\{\gamma_{1,i} < \frac{q(C_i + x)}{x}\} f_{\gamma_{2,i}}(x)dx = \int_0^\infty F_{\gamma_{1,i}}(\frac{q(C_i + x)}{x}) f_{\gamma_{2,i}}(x)dx$$

Therefore, substituting (3)-(9) into (11) yields

$$P_{out} = \int_0^\infty \prod_{l=1}^{K} F_{\gamma_{1,l}}\left(\frac{q(C_l+x)}{x}\right)\sum_{i=1}^{K} f_{\gamma_{2,i}}(x)\int_0^\infty f_{\gamma_{1,i}}(y)\prod_{\substack{l=1\\l\neq i}}^{K} F_{\gamma_{1,l}}(y)\,dy\,dx \tag{12}$$

$$= \sum_{i=1}^{K}\int_0^\infty f_{\gamma_{1,i}}(y)\prod_{\substack{l=1\\l\neq i}}^{K} F_{\gamma_{1,l}}(y)\,dy\int_0^\infty \prod_{l=1}^{K} F_{\gamma_{1,l}}\left(\frac{q(C_l+x)}{x}\right)f_{\gamma_{2,i}}(x)\,dx$$

where [6]

$$\int_0^\infty f_{\gamma_{1,i}}(y)\prod_{\substack{l=1\\l\neq i}}^{K} F_{\gamma_{1,l}}(y)\,dy = 1 + \frac{\chi_{1,i}^{m_{1,i}}}{\Gamma(m_{1,i})}\sum_{u=1}^{p-1}\frac{(-1)^u}{u!}\underbrace{\sum_{n_1=1,\ldots,K}\cdots\sum_{n_j=1,\ldots,K}}_{n_1\neq n_2\neq\ldots\neq n_j\neq i}\times\sum_{k_1=0}^{m_{1,n_1}-1}\cdots\sum_{k_j=0}^{m_{1,n_j}-1}\left(\prod_{t=1}^{u}\frac{(\chi_{1,n_t})^{k_t}}{k_t!}\right)\frac{\Gamma(m_{1,i}+\sum_t^u)}{(\chi_{1,i}+\sum_{t=1}^u\chi_{1,n_t})^n} \tag{13}$$

$$\int_0^\infty \prod_{l=1}^{K} F_{\gamma_{1,l}}\left(\frac{q(C_l+x)}{x}\right)f_{\gamma_{2,i}}(x)\,dx \tag{14}$$

$$= \int_0^\infty \prod_{l=1}^{K}\frac{\Upsilon\left(m_{1,l},\chi_{1,l}\frac{q(C_l+x)}{x}\right)}{\Gamma(m_{1,l})}\frac{\chi_{2,i}^{m_{2,i}}x^{m_{2,i}-1}}{\Gamma(m_{2,i})}\exp(-\chi_{2,i}x)\,dx$$

$$= \frac{\chi_{2,i}^{m_{2,i}}}{\Gamma(m_{2,i})}\int_0^\infty \prod_{l=1}^{K}\left[1-\exp\left(-\chi_{1,l}\frac{q(C_l+x)}{x}\right)\sum_{k=0}^{m_{1,l}-1}\frac{[\chi_{1,l}q\frac{(C_l+x)}{x}]^k}{k!}\right]x^{m_{2,i}-1}\exp(-\chi_{2,i})$$

$$= \frac{\chi_{2,i}^{m_{2,i}}}{\Gamma(m_{2,i})}\int_0^\infty \prod_{l=1}^{K}[1-\xi_l]x^{m_{2,i}-1}\exp(-\chi_{2,i}x)\,dx$$

$$= \frac{\chi_{2,i}^{m_{2,i}}}{\Gamma(m_{2,i})}\int_0^\infty \sum_{l=0}^{K}\frac{(-1)^l}{l!}\underbrace{\sum_{n_1=1}^{K}\cdots\sum_{n_l=1}^{K}}_{n_1\neq\ldots\neq n_l}\prod_{t=1}^{l}\xi_{n_t}x^{m_{2,i}-1}\exp(-\chi_{2,i}x)\,dx$$

$$= \frac{\chi_{2,i}^{m_{2,i}}}{\Gamma(m_{2,i})}\sum_{l=0}^{K}\frac{(-1)^l}{l!}\underbrace{\sum_{n_1=1}^{K}\cdots\sum_{n_l=1}^{K}}_{n_1\neq\ldots\neq n_l}\prod_{t=1}^{l}\int_0^\infty \xi_{n_t}x^{m_{2,i}-1}\exp(-\chi_{2,i}x)\,dx$$

$$= \frac{\chi_{2,i}^{m_{2,i}}}{\Gamma(m_{2,i})}\sum_{l=0}^{K}\frac{(-1)^l}{l!}\underbrace{\sum_{n_1=1}^{K}\cdots\sum_{n_l=1}^{K}}_{n_1\neq\ldots\neq n_l}\prod_{t=1}^{l}\sum_{k=0}^{m_{1,n_t}-1}\frac{[\chi_{1,n_t}q]^k}{k!}\exp(-\chi_{1,n_t}q)$$

$$\times\int_0^\infty \exp\left(-\chi_{1,n_t}qC_{n_t}\frac{1}{x}-\chi_{2,i}x\right)\left(\frac{C_{n_t}}{x}+1\right)^k x^{m_{2,i}-1}\,dx$$

where

$$\xi_l = \exp\left(-\chi_{1,l}\frac{q(C_l+x)}{x}\right)\sum_{k=0}^{m_{1,l}-1}\frac{[\chi_{1,l}q\frac{(C_l+x)}{x}]^k}{k!} \tag{15}$$

$$\prod_{l=1}^{K}[1-\xi_l]=\sum_{l=0}^{K}\frac{(-1)^l}{l!}\underbrace{\sum_{n_1=1}^{K}\cdots\sum_{n_l=1}^{K}}_{n_1\neq\ldots\neq n_l}\prod_{t=1}^{l}\xi_{n_t} \qquad (16)$$

Note that the equality in (12), (13) is based on the operation of exchange integral and the following formula [7]:

$$\Upsilon(n,x)=(n-1)![1-\exp(-x)\sum_{k=0}^{n-1}\frac{x^k}{k!}] \qquad (17)$$

Finally, substituting (13)-(16) into (12), we obtain the exact closed-form expression for the outage probability of selection cooperation with fixed-gain relays in independent but not identically distributed Nakagami-m fading channels at arbitrary signal-to-noise ratios (SNRs) and number of available relays.

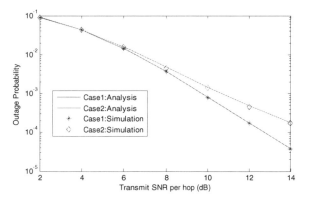

Fig. 1. The outage probability against the transmit SNR per hop

4   Results and Discussions

In this section, we consider a scenario with $K=3$ relays, and $r=1$Kb/s with M-PSK. For convenience of comparison, in case1, the channel gain parameters are set with $\{m_{1,l}\}_{l=1}^{K}=\{2,2,1\}$, $\{m_{2,l}\}_{l=1}^{K}=\{1,2,3\}$, $\{\Omega_{1,l}\}_{l=1}^{K}=\{0.9,0.8,0.7\}$, $\{\Omega_{2,l}\}_{l=1}^{K}=\{0.6,0.8,0.3\}$. And in case2, the channel gain parameters are set with $\{m_{1,l}\}_{l=1}^{K}=\{1,1,1\}$, $\{m_{2,l}\}_{l=1}^{K}=\{1,2,2\}$, $\{\Omega_{1,l}\}_{l=1}^{K}=\{0.9,0.8,0.7\}$, $\{\Omega_{2,l}\}_{l=1}^{K}=\{0.6,0.8,0.3\}$. All the channels are independent and not identically distributed. Fig. 1 plots the outage probability against the transmit SNR per hop. From this figure, it is seen that the analytical results very well match with the simulation results through numerical integrations, which validates our provided theoretical analysis. Finally, by comparing the two cases situation, we observe that the outage performance becomes better for high average SNR and large Nakagami figures.

5 Conclusion

In this paper, we have investigated the outage probability of selection cooperation with fixed-gain relays in independent but not identically distributed fading channels. An exact expression for the outage performance is derived. Furthermore, our analytical derivations have been verified by simulation resluts with Monte Carlo simulations. The analysis and conclusions provided herein are not only novel but also practical for the design of cooperatve relay networks.

References

1. Costa, D.B., Aissa, S.: Capacity Analysis of Cooperative Systems with Relay Selection in Nakagami-m Fading. IEEE Communications Letters 13, 637–639 (2009)
2. Costa, D.B., Aissa, S.: Dual-hop Decode-and-forward Relaying Systems with Relay Selection and Maximal-ratio Schemes. Electronics Letters 45, 1 (2009)
3. Xu, F., Lau, F., Zhou, Q., Yue, D.: Outage Performance of Cooperative Communication Systems Using Opportunistic Relaying and Selection Combining Receiver. IEEE Signal Processing Letter 16, 113 (2009)
4. Yan, K., Jiang, J., Wang, Y.G., Liu, H.T.: Outage probability of selection cooperation with MRC in Nakagami-m fading channels. IEEE Signal Processing Letter 16, 1031 (2009)
5. Costa, D.B., Aissa, S.: Performance Analysis of Relay Selection Techniques with Clustered Fixed-gain Relays. IEEE Signal Processing Letter 17, 201 (2010)
6. Cui, H.X., Wei, G., Wang, Y.D.: Effects of CSI on ASEP Based Opportunistic DF Relaying Systems. Submitted
7. Duong, T.Q., Bao, V.N.Q., Zepernick, H.J.: On the Performance of Selection Decode-and-forward Relay Networks over Nakagami-m Fading Channels. IEEE Communications Letters 13, 172 (2009)

A New Kind of Chaos Shift Keying Secure Communication System: FM-CDSK

Jianan Wang and Qun Ding

Hei Longjiang University, Electronic Engineering Key Laboratory of
Universitiesin Hei Longjiang Province Harbin, China
nannan0914@126.com, qunding@yahoo.cn

Abstract. Considering the shortcomings of DCSK (differential chaos shift keying) with bad secrecy and low data transfer rate, this paper designs and researches a new chaos shift keying digital modulation scheme—FM-CDSK(frequency-modulated correlation delay shift keying). The scheme has the both merits of CDSK and FM-DCSK. The simulation analysis of data transmission rate, confidentiality and BER verify that this new scheme has more superior performance than DCSK and CDSK, and the transmission rate is twice as much as original shift keying schemes. It can be clearly seen that this scheme is realizable in digital secure communication.

Keywords: Chaotic digital secure communication, Chaos shift keying, FM-CDSK.

1 Introduction

In recent years, using chaos system to achieve real-time secure communication is still main goal of the research. According to digital communication needs, this paper researches chaos shift keying communication system design proposal. In the existing chaos secure communication system, chaotic masking can not only mask analog signals but also conceal digital signals [1]. Chaotic parameter modulation and chaos shift keying mainly transmit digital signals. But there are many shortcomings in covering up digital signals in chaotic masking. When the digital signal amplitude is greater than 5V, the covered signals will reveal the useful signal characteristics, so chaotic masking can only transmit small energy signals. Because it is easy to get analyzers' attacks, it shows the poor confidentiality of chaotic masking [2].

Like chaotic masking, chaotic parameter modulation is also only suitable for slowly varying signals, and it can not handle quickly changing signals and time-varying signals very well. The capability of chaos shift keying resisting noise and parameter mismatch is better than the first two communication modes [3].

DCSK is a non-coherent chaos shift keying communication scheme. In the half of a symbol period, it doesn't transmit useful information, which limits the transmission rate. And in the transmission channel, there are two identical or opposite chaotic signals, which brings down the confidentiality of transmitting information. Because of non-periodic characteristic of chaotic signal, the modulation of DCSK makes unit bit

Y. Yu, Z. Yu, and J. Zhao (Eds.): CSEEE 2011, Part II, CCIS 159, pp. 177–182, 2011.

energy vary according to the change of time. Because the demodulation of FM-DCSK is the same as DCSK [4][5], which makes FM-DCSK have the same shortcomings as DCSK, such as lower data transmission rate and lower system confidentiality [6][7]. The modulation scheme of CDSK overcomes the deficiency of DCSK. The transmitter of CDSK can work continuously, and there aren't repeated chaotic signals in the communication channel, which improves the confidentiality of the system [8]. Now, we improve CDSK to combine FM with CDSK. FM-CDSK eliminates mutative variance of symbol energy. It can effectively reduce bit error rate and the transmission rate is twicc as much as DCSK and FM-DCSK, and its confidentiality of system is better than the others.

2   FM-CDSK Scheme

This paper proposes a new non-coherent detection system improvement based on the correlation delay shift keying. In this scheme, the per bit output energy of the FM modulator becomes constant, and the receiver uses simple correlation delay detection to modulate. Therefore, the receiver only need analyze and process the received waveform, and then we can simply estimate the sending data of the transmitter. At the receiver of FM-CDSK, it does not need to restore the chaotic carrier wave, which makes reception more convenient and quicker. The goal of FM-CDSK is to generate a wideband CDSK signal with constant bit energy E_b. Fig. 1 shows the transmitter and receiver structures of a FM-CDSK system, T_b is a code transmission time.

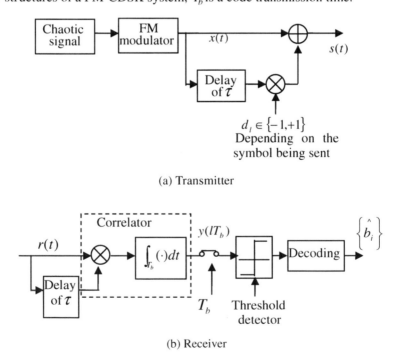

(a) Transmitter

(b) Receiver

Fig. 1. Block diagram of the FM-CDSK system

Chaotic signal $u(t)$ is modulated by FM frequency modulator, the output of which is $x(t)$:

$$x(t) = A\cos(2\pi[ft + m\int_0^t u(\tau)d\tau] + \theta) = A\cos(wt + \alpha(t) + \theta) = A\cos(\varphi) \qquad (1)$$

Where A is carrier amplitude, f is carrier frequency, m is modulation factor and θ is initial phase. Denote the lth transmitted symbol by $d_l \in \{-1,+1\}$. The transmitted signal $s(t)$, is the sum of the chaotic signal $x(t)$ and the delayed version of the signal $x(t-\tau)$, which is modulated by the symbol $d_l \in \{-1,+1\}$, where τ denotes the delay. Thus, the transmitted signal is given by

$$s(t) = \begin{cases} x(t) + x(t-\tau) & d_l = +1 \\ x(t) - x(t-\tau) & d_l = -1 \end{cases} \qquad (2)$$

As shown in Eq. 2, the first segment $x(t)$, is the reference signal of $\pm x(t-\tau)$, while the second $\pm x(t-\tau)$ denotes the information-bearing signals, and the transmitted signal $s(t)$ is never repeated. Since no individual reference signal is sent, the bandwidth efficiency is improved. Also, the transmitted signal of FM-CDSK is more homogeneous and less prone to interception than DCSK.

At the receiving side, the output of the correlator is $y(lT_b)$,

$$y(lT_b) = \int_{(l-1)T_b}^{lT_b} r(t)r(t-\tau)dt \qquad (3)$$

Where the $r(t)$ denotes the sending of signals transmitted through noisy channel mixed with noise signals $\xi(t)$.

$$y(lT_b) = \int_{(l-1)T_b}^{T_b}[s(t)+\xi(t)][s(t-\tau)+\xi(t-\tau)]dt$$

$$= \int_{(l-1)T_b}^{T_b} s(t)s(t-\tau)dt + \int_{(l-1)T_b}^{T_b} s(t)\xi(t-\tau)dt + \int_{(l-1)T_b}^{T_b}\xi(t)s(t-\tau)dt + \int_{(l-1)T_b}^{T_b}\xi(t)s(t-\tau)dt \qquad (4)$$

As shown in Eq. 4, when noise exists in the channel, the first item is useful signal, the second and third items are the related product of noise signal with useful signal, and the last item is the related product of noise signal itself. It is irrelevant between chaotic signal $s(t)$ and noise signal $\xi(t)$, as well as noise signal $\xi(t)$ and $\xi(t-\tau)$. Therefore, except the first useful signal, the rest of integral values are zero.

In non-noise environment, when carrier amplitude A is 1 and initial phase θ is 0, if the lth sending signal is "+1", $y(lT_b)$ is given by Eq. 1-3.

$$y(lT_b) = \int_{(l-1)T_b}^{T_b}[\cos^2(\varphi_{t-\tau}) + \cos(\varphi_t)\cos(\varphi_{t-\tau}) + \cos(\varphi_t)\cos(\varphi_{t-2\tau}) + \cos(\varphi_{t-\tau})\cos(\varphi_{t-2\tau})]dt \qquad (5)$$

Similarly, if the *lth* sending signal is "-1", we can get $y(lT_b)$.

$$y(lT_b) = \int_{(l-1)T_b}^{lT_b} [-\cos^2(\phi_{t-\tau}) + \cos(\phi_t)\cos(\phi_{t-\tau}) - \cos(\phi_t)\cos(\phi_{t-2\tau}) + \qquad (6)$$
$$\cos(\phi_{t-\tau})\cos(\phi_{t-2\tau})]dt$$

As shown in Eq. 5 and Eq. 6, the first item is useful signal, while the second, third and forth items are the related product of adjacent chaotic signals. The chaotic signals modulated by FM frequency modulator still have chaotic characteristic. According to these characteristics, we can judge that the last three items are zero. The threshold value is set to zero, if $y(lT_b) > 0$, thus, we can decide that the sending signal is "+1"; if $y(lT_b) < 0$, the sending signal is "-1".

The BER performance of CDSK had been analyzed and derived in the references [9][10]. Let us assume that the reference signal contains M chaotic samples in half a symbol period, while the length of the information-bearing part is M too. The sampling interval of the chaotic signal is σ, and f_{max} is the maximum frequency of FM signals. For the FM-CDSK receiver, we assume the received signal is correlated according to the sampling interval σ, and it should at least satisfy $1/\sigma \geq 2f_{max}$ [11]. Let R_t be the chip rate of the discrete chaotic signal, thus, a chaotic chip will be detected through $T_b/2\sigma$ sampling intervals when correlating. So, the M becomes

$$M = \frac{T_b/2\sigma}{1/R_t\sigma} = T_b R_t/2 \qquad (7)$$

E_b/N_0 is the ratio of average signal bit energy and noise power spectral density. In the AWGN channel, the bit error rate of FM-CDSK is

$$BER = \frac{1}{2}erfc(\sqrt{\frac{E_b}{8N_0}(1 + \frac{19}{20M}\frac{E_b}{N_0} + \frac{M}{4}\frac{N_0}{E_b})^{-1}})$$
$$= \frac{1}{2}erfc(\sqrt{\frac{E_b}{8N_0}(1 + \frac{19}{20}\cdot\frac{2}{T_b R_t}\cdot\frac{E_b}{N_0} + \frac{1}{4}\cdot\frac{T_b R_t}{2}\cdot\frac{N_0}{E_b})^{-1}}) \qquad (8)$$

3   Simulation Results and Discussion

Here we choose logistic chaotic mapping as transmitter system. Let us assume $T_b = 2\mu s$, subsequently, the output of the zero-order-hold module is frequency modulated by the FM modulator, that is

$$x(t) = A\cos(2\pi[ft + m\int_0^t u(\tau)d\tau]) \qquad (9)$$

The parameters of the FM modulator in this simulation are chosen as A = 1V, f = 36MHz, m=7.8MHz/V. We choose delay time $\tau = T_b/4$, and we can see that the transmission rate is twice as fast as DCSK and FM-DCSK.

In the transmission channel, comparing the modulated chaotic signal of COOK with CSK, DCSK and FM-CDSK, it is clearly that the confidentiality of transmission signal of FM-CDSK is better than the others. The modulated chaotic signal transmitted in channel of COOK, CSK and DCSK reveal more or less the traces of useful signals, as shown in Fig. 2- Fig. 5.

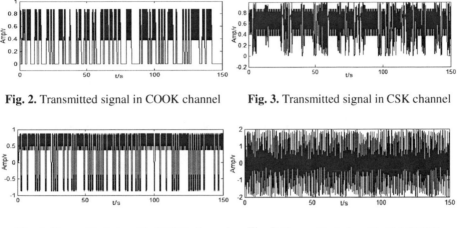

Fig. 2. Transmitted signal in COOK channel   **Fig. 3.** Transmitted signal in CSK channel

Fig. 4. Transmitted signal in DCSK channel   **Fig. 5.** Transmitted signal in FM-CDSK channel

In AWGN channel, we compare the BER performance of FM-CDSK with CDSK, and the result is shown in Figure 6. The BER curves of FM-CDSK and CDSK indicate that the system performance is limited by parameters M. Seen from Figure 6, BER performance of FM-CDSK is close to CDSK in AWGN channel when E_b/N_0 is relatively low. Under $E_b/N_0 > 12dB$ condition, the BER performance of FM-CDSK system is much better than CDSK system, and as E_b/N_0 increases, the BER performance of FM-CDSK will increase in relation to CDSK; when M=16, $BER = 5 \times 10^{-2}$, comparing with CDSK, SNR of FM-CDSK increases by 4.2 ~ 4.7dB.

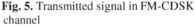

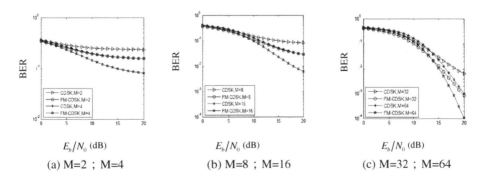

(a) M=2 ; M=4   (b) M=8 ; M=16   (c) M=32 ; M=64

Fig. 6. Comparison of BER performance of the improved FM-CDSK system and CDSK

4 Summary

This paper proposes a new chaos shift keying system—FM-CDSK, and verifies that the transmission rate, confidentiality and BER of FM-CDSK are better than DCSK and CDSK. Comparing with text information, image information is two-dimensional or three-dimensional, and its data quantities are much greater than general text data, meanwhile, image information has high redundancy. Especially image information isn't strict for the integrality of information transmission, while the demand for image signal transmission rate and confidentiality is higher in practice. Considering these characteristics of image data, the improved FM-CDSK is very suitable to be applied in the secrecy transmission of Video images in the practical chaotic digital secure communication. It can clearly be seen that the proposed improvement of FM-CDSK is meaningful and it is easy to realize.

Acknowledgments. The authors acknowledge the support of the National Natural Science Foundation of China (No.60672011) and (No.61072072).

References

1. Gao, J., Liao, N., Liang, Z.: A Hybrid Secure Communication Method Based on the Synchronization of Hyperchaos Systems. J. Circuits and Systems 57, 128–130 (2005)
2. Si-Yuan, H.: The Research of Synchronization Communication Technology Based on Chaotic Masking. In: IEEE International Conference on Intelligent Computing and Intelligent Systems, pp. 267–270. IEEE Press, New York (2009)
3. Yao, J., Lawrence, A.J.: Performance Analysis and Optimization of Multi-User Differential Chaos-Shift Keying Communication Systems. IEEE Transaction on Circuits and Systems-I: Regular Papers (2006)
4. Han, J., Zhu, Y.: A Kind of Method for Enhancing the Transmission Efficiency in FM-DCSK. Acta Electronic, 1032–1035 (2005)
5. Chen, S., Wang, L., Chen, G.: Data-Aided Timing Synchronization for FM-DCSK UWB Communication Systems. IEEE Transactions on Industrial Electronics, 1538–1545 (2010)
6. Li, H., Feng, S.F.: UWB DCSK System Design and Simulation. In: IEEE International Conference on Test and Measurement, pp. 311–314. IEEE Press, New York (2009)
7. Wren, T.J., Yang, T.C.: Orthogonal Chaotic Vector Shift Keying in Digital Communications. IET Communications, 739–753 (2010)
8. Arai, S., Nishio, Y.: No coherent Correlation-based Communication Systems Choosing Different Chaotic Maps. In: IEEE International Symposium on Circuits and Systems, pp. 1433–1436 (2007)
9. Galias, Z., Maggio, G.M.: Quadrature Chaos-Shift Keying: Theory and Performance Analysis. In: IEEE Transactions on Circuits and Systems I: Fundamental Theory and Applications, vol. 167, pp. 1510–1519. IEEE Press, New York (2001)
10. Lau, F.C.M., Ye, M., Tse, C.K., Hau, S.F.: Anti-jamming Performance of Chaotic Digital Communication Systems. IEEE Transactions on Circuits and Systems I: Fundamental Theory and Applications, 1486–1494 (2002)
11. Zhang, Y., Shen, X., Zhu, X.: An FM-QCSK Chaotic Communication System and It Performance Analysis. Journal of Xidian University, 259–263 (2007)

Automatic Deep Web Table Segmentation by Domain Ontology

Kerui Chen, Wanli Zuo*, Fengling He, and Yongheng Chen

College of Computer Science and Technology, Jilin University, Changchun 130012, China
Key Laboratory of Symbol Computation and Knowledge Engineering of the Ministry of
Education, Changchun, China
chenke0616@163.com, wanli@jlu.edu.cn, hfl@jlu.edu.cn,
cyh771@163.com

Abstract. For deep web, abundant data information will be presented to users in the form of tables through query forms. Because of the isomerism of data source, to understand and integrate these tables is a very challenging task. This paper proposed a domain ontology based technique for integrating tables. As a method that is completely independent from the structure of tables, it could efficiently solve the nested and conjoined problems existed in complex tables. Experimental results prove that method's effectiveness of improving the accuracy of integration.

Keywords: Deep Web, Table Segmentation, Domain Ontology, Table Extraction.

1 Introduction

World Wide Web provides abundant resources for each user, especially in deep web. Compared with surface web, deep web usually has a backstage database and only provides users special query interface. Once the user submits query qualifications to web, after the server response, deep web website will return information that meets our query qualifications in the form of web pages to the user. Based on the observation, it's clear that deep web's information will normally be shown to users through query forms in the form of tables. In addition, as different deep webs return page information with different layouts and contents, it's a challenging task to understand and integrate these tables.

Table contains rows and columns. Generally, a row represents a record, while a column represents data with the same type. Although the structure of table is pretty simple, the data relationships in the table are quite complex and many tables have nested and conjoined problems.

In this paper, we suggested a domain ontology based method for combining these tables. We use multiple heterogeneous tables as the input of system, and then a unified table will be outputted. This method is independent from the file structure of web pages, and able to solve nested and conjoined problems existed in tables appropriately.

---

* Corresponding author.

Y. Yu, Z. Yu, and J. Zhao (Eds.): CSEEE 2011, Part II, CCIS 159, pp. 183–188, 2011.

2 Related Work

Existing approaches to extracting table data from web page can be classified as manual approach, wrapper induction or automatic extraction.

Manual approach is to observe web pages and their source codes, and identity some modes by programmers. Based on those modes, a program will be complied to extract target data. Although this approach has a very high accurate rate of extracting, it's hard to suit the extraction of multiple data sources.

Wrapper induction is partially supervised learning. Wrapper induction approaches learn a model of data from a set of labeled training examples using hand-selected features.

Currently, there are many researchers study about the table segmentation, and gain some achievements. Lerman [1] describes using the structure of web sites to automatic extraction and segmentation of records. This approach refers to two algorithms: constraint satisfaction-based technique and probabilistic inference approach.

Deng Cai [2] propose VIPS (a Vision based page segmentation) algorithm, he presents an automatic top-down, tag-tree independent approach to detect web content structure. It is similar understanding the layout of web structure by user visual information. The VIPS algorithm first extracts all the suitable blocks from the web page, and finds the separators between these extracted blocks. Finally, the semantic structure is constructed based on these separators.

Cui Tao [3] offers a conceptual model solution for sibling pages. The sibling pages we consider are pages on the Deep web, commonly generated from underlying databases. It compares them to identify and connect non-varying components (category labels) and varying components (data values).

3 Overview of the Problem

3.1 Table Recognition and Canonicalization

We can define a canonical table as follows. A Schema for a canonical table is a set $L = \{ L_1, \ldots, L_n \}$ of label names or phrases. $\forall L_i \in L$, $1 \leq i \leq n$, the domain of L_i is a set D_i. Let $D = D_1 \cup D_2 \ldots \cup D_n$. A canonical table T with table schema S is a set of functions $T = \{t_1, \ldots, t_m\}$ from S to D with the restriction, $\forall t \in T$, $t(L_i) \in D_i$, $1 \leq i \leq n$. For a relational database, we often display tables in two dimensions: row and column. Each row in the table composes the domain values for the corresponding labels in the column headers. Like this, for example, we can display the canonical table $\{\{(A,1),(B,2)\}, \{(A,3),(B,4)\}\}$, A and B are column name, 1 and 3 are the value of column A.

3.2 Ontology Description

Ontology is a formal, explicit specification of a shared conceptualization in terms of concepts, attributes and relations. The logic structure of an Ontology could be formalize as [4]: two-tuple O=(C, A); 'C' represents a concept, described in the form of rectangle, for example, people, music, book and so on. 'A' described the attribute of the concept 'C'. A concept 'C' could have multiple attributes 'A'.

Generally, an attribute model 'A' could be described as a seven-tuple A= {N, V, FA, DT, R, cons, count}. For every attribute, N represents the name of the attribute 'A', it is unique. V stand for the value of attribute, also name as the instances of 'A', and it is obtains from query interfaces and query result pages

FA is the associate attribute of 'A', in other word; it means the synonyms for 'A'. Such as, attribute "Author" is the synonyms for attribute "first name" and "last name", if "Author" is the name of attribute 'A', then "first name" and "last name" will become the associate attribute.

DT described the data types of A. the common data types contain int, real, price, datetime, string.

R shows the relationship between concepts with individual attribute. Cons represents constraint condition, for example, the value of attribute "age" should never be less than zero.

Count indicates the frequency of 'A' appearance in training text and typically is used for detect the important of that attribute.

4   Table Extraction

After getting some result pages, system first parses the source code and locates all HTML components embedded by <table> and </table> tags. To compare and match tables, system transforms each HTML tables into a DOM tree. Usually, one document has many tables, but we assume that there is only one actual query result table. To detect which of the remaining data regions corresponds to the actual query result section, system uses the following some simple heuristics.

(1) From the visual point of view to observe a query result page, the query result tables usually occupies a large space and located at the center of the query result pages [5]. For each remaining data region DR, an area weight is calculated as DR's area divided by the largest area region, a center distance weight is calculated as the smallest center distance divided by DR's center distance.

(2) Usually, the query result tables contain more raw data strings than the record in other sections. For each remaining data region DR, a value weight is calculated as the average number of raw data strings in a record divided by the largest average number of data values.

5   Table Segmentation

The extracted data needs further handling, which is mainly divided into two parts: data value alignment and label alignment. In the handling of that problem, system adequately utilizes some characteristics of table layouts, such as the position information of label and data value, and the left-right and up-down relation; label and data value usually is close to each other, which means the gap is relatively small; there are some special separators existing between label and data value.

The next step is table segmentation. The system uses multiple tables as input, and also uses the ontology of that domain as a global variable, to finally output a unified table by performing integration through the mapping relation of tables and ontology. The mapping relation between tables and ontology mainly falls into two categories:

(1) Label Level matching:

The label Level similarity is calculated based on the edit distance between labels and attribute model. The label matching similarity is defined as Eq. (1), where $EditDist(l_i, A_i)$ is levenshtein distance between label l_i and the N_i, FA_i of attribute A_i, $L(l_i)$ and $L(A_i)$ are the string length of l_i and A_i.

$$labelSim(l_i, A_i) = 1 - \frac{EditDist(l_i, A_i)}{\max(L(l_i), L(A_i))} \qquad (1)$$

(2) Data value matching:

We used cosine similarity to calculate the mapping between data value DV_i and the V_i of A_i in a vector space model using Eq(2), where V_{l_i} and V_{A_i} are two vectors, n is the dimension of the vectors, $V_k^{DV_i}$ and $V_k^{A_i}$ are k th value in the vectors V_{l_i} and V_{A_i}, $|V_{DV_i}|$ and $|V_{A_i}|$ are the lengths of the two vectors, respectively,

$$DataValueSim(DV_i, A_i) = \frac{\overrightarrow{V_{DV_i}} \cdot \overrightarrow{V_{A_i}}}{|V_{DV_i}| \, |V_{A_i}|} = \frac{\sum_{k=1}^{n} (V_k^{DV_i} * V_k^{A_i})}{\sqrt{\sum (V_k^{DV_i})^2} \sqrt{\sum (V_k^{A_i})^2}} \qquad (2)$$

6 Results

We now present results of automatic segmentation of records using domain Ontology described in the sections above.

Table 1. Experimental results

URL	Cor	InCor	FN	FP
www.foyles.co.uk	65	8	1	3
www.ecampus.com	212	30	23	12
www.abc.nl	11	2	0	1
www.abebooks.com	266	29	14	33
www.thebookbeat.com	23	2	0	1
www.amazon.com	322	37	62	36
bookshop.blackwell.co.uk	128	12	5	12
www.thebookery.com	28	1	1	3
www.netlibrary.com	19	2	0	1
www.ippbooks.com	46	3	1	4
Precision	82.84%			
Recall	91.28%			
F-measure	0.87			

6.1 Experimental Setup

We choose UIUC web integration repository [6] as the query interface of deep web. This dataset collects the original query interfaces of 447 deep Web sources from 8 representative domains, in the Travel group: Airfares, Hotels, and Car Rentals; in the Entertainment group: Books, Movies, and Music Records; in the Living group: Jobs and Automobiles. We have chosen ten sites about Book fields. The query result pages are returning from the query server based on the randomly inputted data by query interface.

6.2 Evaluation

There are three main indicators of the evaluation: Precision, recall and F measure.

$$Precision = Cor / (Cor + InCor + FP) \tag{3}$$

$$\mathrm{Re}\,call = Cor / (Cor + FN) \tag{4}$$

$$F - measure = 2 Precision \times \mathrm{Re}\,call / (Precision + \mathrm{Re}\,call) \tag{5}$$

Cor means the number of correctly segmented, InCor stand for the number of incorrectly segmented record. About remaining two characters, the FN is unsegmented records and FP represents non-records.

Our experimental results are show in table 1. In the first column, it gives the list of site URL. We input randomly some keyword into the Query interface of each web page. About remaining four columns described the different data. Using Eq.(3,4,5), respectively, were computed the value of Precision, Recall, F-measure. There are two main factors for table segmentation error: one is that attributes with different names have same meanings, and the other one is that attributes with same names have different meanings. Especially for the second one, there is still no perfect solution. Experimental results prove that method's effectiveness of improving the accuracy of integration.

7 Conclusion

In this paper, we analyzed existing extraction and integration techniques, and suggested a strategy of combining tables based on domain ontology. The system use multiple heterogeneous tables as its input, after calculating the similarity between terms in the table with the concepts and attributes of ontology, the system will finally output a unified table. This approach is independent from the file structure of web pages, and able to solve nested and conjoined problems existed in tables appropriately.

References

1. Lerman, K., Getoor, L., Minton, S., Knoblock, C.: Using the Structure of Web Sites for Automatic Segmentation of Tables. In: Proceedings of the 2004 ACM SIGMOD International Conference on Management of Data, pp. 119–130 (2004)

2. Cai, D., Yu, S., Wen, J.-R., Ma, W.-Y.: VIPS: A Vision-based Page Segmentation Algorithm. Microsoft Technical Report. MSR-TR-2003-79 (2003)
3. Tao, C., Embley, D.W.: Automatic Hidden-web Table Interpretation by Sibling Page Comparison. In: Proceedings of the 26th International Conference on Conceptual Modeling, pp. 566–581 (2007)
4. Chen, K., Zuo, W., Zhang, F., He, F., Peng, T.: Automatic Generation of Domain-specific Ontology from Deep Web. Journal of Information and Computational Science 7(2), 519–525 (2010)
5. Zhao, H., Meng, W., Wu, Z., Raghavan, V., Yu, C.: Fully Automatic Wrapper Generation for Search Engines. In: Proceedings of the 14th international conference on World Wide Web, pp. 66–75 (2005)
6. Kushmerick, N.: Wrapper Induction: Efficiency and Expressiveness. Artificial Intelligence 118(3), 171–181 (2000)

A Position-Based Chain Cluster Routing Protocol for Strip Wireless Sensor Network

Gangzhu Qiao and Jianchao Zeng

Complex System and Computational Intelligence Laboratory,
Taiyuan University of Science and Technology, 030024 Taiyuan, China
qiaogangzhu@sohu.com, zengjianchao@263.net

Abstract. In some Wireless sensor network application the nodes have to be deployed in a narrow strip area and the traditional routing algorithms can not be directly applied in such strip WSN. In the strip WSN the node near the base station has to receive and forward massive data, and are of no use after energy exhausting very quickly which will reduce survival time of the network and aggravate the "hot spot" question of the network. In order to improving communication efficiency of the strip network, a Position-based Chain Cluster routing (PCCR) protocol proposed is proposed in this paper. In PCCR protocol the strip WSN is divided into many belt-shaped region cluster which elected the cluster head through node' position and dump energy and the cluster head chain is set up as the backbones of the strip WSN to forward the data of nodes. The simulation experiment result indicated that the PCC protocol is more suitable for the strip WSN than LEACH and can prolong the life cycle of the strip WSN effectively.

Keywords: Wireless Sensor Network, Routing Protocol, Leach, Cluster, Chain, Strip Area.

1 Introduction

Energy consumption is one of the key research questions of Wireless Sensor Networks (WSN) [1, 2]. Due to the large amount of nodes, limited energy, and energy supplement inconvenience, extending the network life cycle and efficient using node energy is the primary goal of WSN design. The WSN Routing protocol [3, 4] plays a decisive role in improving communication efficiency, saving communication energy and balanced using the energy the network. The WSN route protocol is closely related to the WSN application, therefore it's very necessary to research and develop energy efficient route protocol in accordance with the characteristics of wireless application, which plays a decisive role in improving the quality of monitoring and survival time of WSN.

With WSN widely used mines, bridges, tunnels, canyons and so on in recent years, a new research questions is proposed for wireless sensor network routing algorithm. In aforesaid application, the length of the monitored objects is typically longer and the width is often very narrow. Meanwhile, the WSN has to be deployed into a narrow

Y. Yu, Z. Yu, and J. Zhao (Eds.): CSEEE 2011, Part II, CCIS 159, pp. 189–194, 2011.

strip structure and the base station would normally be deployed in on side of the area. Due to the narrow and strip network structure, the traditional routing algorithms can not be directly applied in such strip WSN. An energy efficient routing protocols suit for the narrow and strip network structure is proposed in this article, which can improve the network performance and meet the monitoring needs of a variety of WSN application scenarios by combined the LEACH [2, 5, 6] protocol and PEGASIS [7] protocol.

2 Analysis on Requirement of Routing Protocol

According to the Characteristics of the WSN application environment, the strip WSN' network model is as follows.

1) The monitor area is a rectangular region, whose length is L and width is N and $L \gg N$.

2) The WSN consists of 1 base station and M common nodes, the base station has plenty of energy, memory and computing resources, and is often far away from the network and can not move.

3) The nodes is deployed in the monitoring area uniformly, each node in the network have the same initial energy, computing and communication capabilities and plays the same role in the network. The position of node can not change having been deployed.

4) Each node can obtained its own position information through location mechanism, and can estimate the distance to the sending node through the received signal strength.

5) Each node has the capacity of adjusting its wireless transmission power, and can adjust the size of transmission power according to the distance to the destination node. Each node in the network can calculate its current sump energy.

In the strip WSN, the node limited by the energy usually needs forward the data to the base station through multi-hop, which will cause the node energy consumption unbalance。 In the network the data far away from the base station flows together to the base station, the node near the base station has to receive and forward massive data, which will exhaust the node' energy very quickly and lead to the node failure. This will reduce survival time of the network and aggravate the "hot spot" question of the network.

In view of the deployment characteristic and the communication requirement of the strip WSN, a Position-based Chain Cluster routing (PCCR) protocol proposed is proposed in this paper, combining the position information, cluster routing and chain routing on the basis of Leach protocol and PEGASISI protocol. In the PCCR protocol a cluster head chain which connects all the adjacent cluster heads is established to form the backbones of strip WSN to reduce the delay of the data collection effectively. The protocol is simple and effective and the less cost of route can upgrade the route information immediately and guarantee the connectivity of the network as far as possible when the topology of the strip WSN changes.

3   The PCCR Protocol

The PCCR protocol is compose of the Cluster Divide phase, The Cluster Head Elect phase, the Cluster Head Chain Establishment phase and Steady-state phase.

3.1   The Cluster Divide Phase

Having deployed the WSN in strip area, the base station divides the strip area into several belt-shaped regions of the same size and next to each other with no overlap. The length of the region should be less than or equal to half of the communication distance in order to ensure that every node of current region can communicate directly with the node in adjacent area. The nodes in a belt-shaped region form a cluster, after that the formation of the cluster and the elect of cluster-heads is limited to the node in the current belt-shaped regions, shown in Fig 1.

3.2   The Cluster Head Elect Phase

In each round of the protocol, at first, the cluster head of every cluster will be elected based on the topology of the strip WSN. The Leach protocol does not take the position and the energy level of the cluster head into account during the period of the cluster head elect, which will lead to uneven energy dissipations of the network [8]. In order to avoid this kind of situation, the PCCR protocol has set up the energy threshold value of the cluster head, only the node whose dump energy is bigger than the threshold value can be elected as cluster head.

At Cluster Formation phase, every member node whose dump energy is bigger than the threshold value calculates the distance between it itself and base station periodically, the node with the nearest distance will be elected as the cluster head at that round. After the cluster head has been elected, it will broadcast a packet including the id and the position of the cluster head to the network. After the member nodes of the cluster have received the packet, they can send their data to the cluster head, shown in Fig 1. The cluster head can reduce the amount of the upload information through data fusion technology after the data from the member nodes have been received. If the dump energy level of every node in the cluster is smaller than the energy threshold, the energy threshold value can be reduced according to the actual need.

3.3   Cluster Head Chain Establishment Phase

If the adjacent Cluster heads can connect each other and form the Cluster Head Chain, it will greatly reduce the energy consumption and delay of data forwarding. As the length of region is less than or equal to half of the communication distance, the cluster head in the adjacent region can communicate with each other, which makes the establishment of cluster head chain possible. Shown in Fig 2, the cluster head of the region C will send a Cluster Head Information packet (CHIP) to region B after the cluster head of region C has been elected. Having received the CHIP, the cluster head of region B will connect the cluster head of region C, then the cluster head chain between region B and C is set up. The cluster head of other regions can be interconnected with the same method to establish the cluster head chain, which forms the backbone of the strip WSN to transmission data between clusters. Having established the cluster head chain, the protocol enters the Steady-state phase.

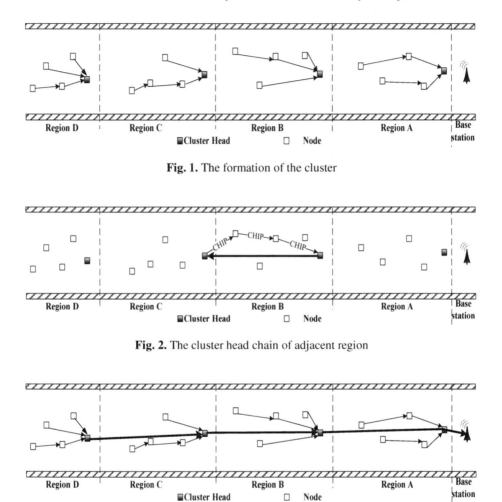

Fig. 1. The formation of the cluster

Fig. 2. The cluster head chain of adjacent region

Fig. 3. The cluster head chain of the whole area

3.4 Steady-Stat Phase

The last phase of PCCR protocol is Steady communication phase, in this phase the member nodes, the cluster head and the base station forward the data each other. The data of the member nodes are gathered together and forward to the cluster head. Having fused the data from cluster member and other clusters sent by the member node and adjacent cluster node, the cluster head forward the data to the adjacent cluster node nearer to the base station. One by one, the data will reach the base station eventually through the cluster head chain of the Strip WSN, shown in Fig 3. The data forward method of PCCR is similar with the PEGASIS protocol.

4   Simulation and Results

In order to assess the performance of the PCCR protocol, the Matlab is used to simulate the strip WSN. The application scenarios of simulation are as follow, 100 blind nodes are random deployed in a narrow and strip area of 500 meter long and 20 meter wide, the position of base station is (0,10) , the strip area is 10m far away the base station , shown as Fig 2. The initial node energy is 2J, the length of broadcast packet is 50 bytes and the length of the data packet is 2000 bytes and the parameters of simulation experiment are as follows: the distance threshold d0 is 75m, the energy consumption parameter of wireless transmission ε_{fs} is 13 pJ/b/m^2, the energy consumption parameter of amplifier E_{tr} is 0.0013 pJ/b/m^4, the energy consumption parameter to launch the transceiver circuit E_{elec} is 10 nJ/bit, and the energy consumption parameter of data aggregation E_{fusion} is 5 nJ/bit. To compare with LEACH protocol, a model has also been built for LEACH with the same parameter.

The round of the first node failure during the course of routing protocol process is a important parameters to measure the lifetime of the WSN [9], here the round of the first node failure is defined as the lifetime of the WSN. The result of simulation is shown in Fig 4. It can be seen that the lifetime of Leach is about 63 rounds and the lifetime of PCCR is about 1997 rounds, about 30 times higher than Leach protocol. The result shows that the proposed protocol can ensure equal usage of the energy of node and, with an obvious advantage in strip WSN, is more energy-efficient than Leach protocol. Compared with the PEGASIS protocol, the time delay of the PCCR protocol is also much smaller, for example, in Figure 3, every nodes in the bar area A can transmit data to the base station only with 2 hops, the hops can be 3 in the bar area B to transfer the data to the base station, however the hops of the same node to transfer the data to the base station will increase a lot if the PEGASIS protocol is used in strip WSN.

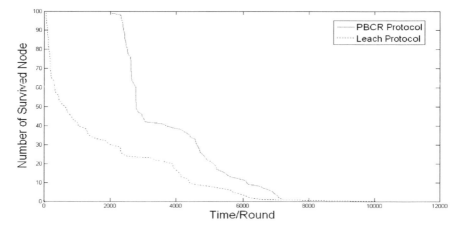

Fig. 4. Numbers of survived nodes with both protocols

5 Summary

Through the analysis on the routing requirement of strip WSN and the limitation of the LEACH protocol and the PEGASIS protocol in this strip WSN, the PCCR protocol is proposed. In PCCR protocol the strip WSN is divided into many belt-shaped region cluster which elected the cluster head through the position and dump energy and the cluster head chain is set up as the backbones of the strip WSN to forward the data of nodes. The PCCR protocol can reduce the network energy consumption and the network latency effectively and balance the node load. The simulation experiment result indicated that The PCC protocol is more suitable for the strip WSN than LEACH and can prolong the life cycle of the strip WSN effectively.

Acknowledgments. Financial support for this work, provided by the National Natural Science Foundation of China (No: U0970124) is gratefully acknowledged.

References

1. Akyildiz, I.F., Wang, X., Wang, W.: Wireless Mesh Networks: A Survey. Computer Networks Journal 47(4), 445–487 (2005)
2. Heinzelman, W.R., Chandrakasan, A.P., Balakrishnan, H.: Energy Efficient Communication Protocol for Wireless Micro-Sensor Networks. In: Proceedings of the 33rd Annual Hawaii International Conference on System Sciences, pp. 181–184 (2000)
3. Akkaya, K.: A Survey on Routing Protocols for Wireless Sensor Network. Ad Hoc Network 3(3), 325–349 (2005)
4. Sohrabi, K., Gao, J., Ailawadhi, V., Pottie, G.J.: Protocols for Self Organization of a Wireless Sensor Network. IEEE Personal Communications 7(5), 16–27 (2000)
5. Heinzelman, W., Chandrakasan, A., Balakrishnan, H.: An Application-specific Protocol Architecture for Wireless Microsensor Networks. IEEE Transactions on Wireless Communications 1(4), 660–670 (2002)
6. Younis, O., Fahmy, S.: HEED: A Hybrid Energy Efficient Distributed Clustering Approach for Adhoc Sensor Networks. IEEE Transactions on Mobile Computing 3(4), 366–379 (2004)
7. Lindsey, S., Raghavendra, C.S.: PEGASIS: Power-Efficient Gathering in Sensor Information Systems. In: Proceeding of the IEEE Aerospace Conf., vol. 3, pp. 1125–1130. IEEE Computer Society, Montana (2002)
8. Choi, W., Shah, P., Das, S.K.: A Framework for Energy Saving Data Gathering Using Two Phase Clustering in Wireless Sensor Networks. In: Proceedings of the International Conference on Mobile and Ubiquitous Systems: Networking and Services, pp. 203–212 (2004)

SNS Communication Model Applying in Network Education System

Lina Lan, Xuerong Gou, and JianXiao Xi

School of Network Education, Beijing University of Posts and Telecommunications, China
lindalan2002@sina.com, xrgou@126.com, xijianxiao@sina.com

Abstract. Social Network Service (SNS) is an emerging Internet communication model. It has unique advantages in communication and interaction. This article analyzes the existing communication models in network education platform and discusses how to use SNS communication model. It presents a design of roles and activities of SNS communication model in network education platform. This design improves the problems of lack communication and alone learning among the remote learning students.

Keywords: Network Education System, SNS (Social Network Service), Communication Model, Interactive, Role, Activity.

1 Introduction

In distance education environment, teachers and students are separated in time and space. The interactions among them are achieved through network education platform. Teachers know the study progress, learning method, learning effect of the students through communication, so the teachers can provide personalized education to different students. Students also can share their ideas and learning experience through network education platform, which can promote their study. Employment of effective communication model in network education platform can improve quality of network education significantly.

Currently, the main forms of communication in network education system are HTML web page, BBS forum, Chat Room (Online Text), E-mail, network meeting etc [1]. These communication models can be categorized as un-real-time model and real-time model. Different communication models have different characteristics and advantages. All of them are important in the distance education. However, these traditional Internet-based communication models between students and teachers have many disadvantages such as less content, small effectiveness, and so on. Therefore, it is significant to do research on the network communication model in network education system to improve the quality of distance education.

SNS (Social Networking Service) is a kind of service network based on 'six degrees of separation theory'. It helps people to build social network [2]. In recent years, many SNS web sites had been published such as Facebook, Myspace in foreign country and Renren, Kaixin in China. They grow fast because they effectively support

Y. Yu, Z. Yu, and J. Zhao (Eds.): CSEEE 2011, Part II, CCIS 159, pp. 195–200, 2011.

the social interaction of people in internet. SNS is an emerging Internet social interaction model. It's a new idea to apply the SNS communication model into distance education to improve the interaction of students and teachers. This paper analyzes the SNS communication model and presents a design of SNS functions in the network education platform.

2   Traditional Communication Models in Network Education

The traditional communication model in network education system can be divided into real-time communication and un-real-time communication. Real-time communication includes virtual classroom, net-meeting, instant message etc. Un-real-time communication includes E-mail, web message boards, BBS forums etc. Comparison of traditional Internet communication model is shown in Table 1. The advantage of real-time communication model is real-time and directly. A variety of teaching activities including teaching, home-work layout, and problem sets and exams analysis can be done directly. But the shortcomings of traditional classroom teaching still exist, that is the pressure of the teachers and students has not decreased. The problems of learning interests and learning methods have not been fundamentally solved. Traditional teaching can not accommodate more people to a session or a topic in discussion. This is inconsistent with the cooperative learning theory and individualized education in modern distance education. So it is difficult to improve the teaching quality [3].

Table 1. Comparisons of the Traditional Communication Modes in Internet

Communication mode Comparison item	Real time		Un-real-time	
	Network meeting	Instance communication	BBS	E-mail
Mode and content	Online at same time, Traditional classroom mode, Multi-media communication	Text chat online, Video or audio communication online	Post, Reply	Send Email, Reply Email
Advantage	Good real time, Face to face in network	Good real time, One-on-one communication	Any time, Any issue	Un-real-time, Any time
Disadvantage	Teaching indirection from teachers to students, No communication among students	Difficult to support discussion among many people	Single form, Free content, Easy to wander	Single form, Difficult to support discussion among many people

In un-real-time communication model, the response is not immediately and the feeling communication is lack. But un-real-time communication like BBS can largely alleviate the pressure of teachers and students in real classroom teaching. Knowledge and learning experience can be shared more effectively in this model. But the inherent flaws in BBS like dull pages, nonsense posts, irrelative with learning, fewer active people in forum, prone to deviate from the theme of interaction and so on are unhelpful for learning. It's difficult to ensure the contents and objectives of the communication [4].

3   Analysis on SNS Communication Model

SNS (social network service) platform is successful which depends on the employed communication model. Chinese social networking site appeared in 2005. Many SNS site has emerged so far, such as 'Kaixin net', 'Renren network' and other successful SNS sites. In order to promote interaction between users, social networking sites introduce some novel methods and models.

SNS communication model generally consists of the following parts:

1. Friend Space. Friends are the fixed 'social circle'. They are the user's most frequently-seen friends in the SNS. Users can visit the personal home page, leave messages between each other.

2. Fresh news. Users can see the latest news of their friends, such as friend's speech, update logs, etc. Users can comment on friend's status and logs, communicate with other friends on interested topics.

3. Share. Users can share favorite photos, videos, music and articles. Their interests will be recommended to friends. Sharing can deepen understanding between friends, make user know each other's interest and promote exchange.

4. Public topics. Some hot topics, celebrities, theme and super star are listed as a public topic. Users with relevant interests can discuss under the same topic.

5. Online games. In order to promote interactions, SNS websites provide games embedded in web pages, such as 'Happy Farm', 'Friends sale', 'grab spaces'. Participants in the game shall be friends. This design enhances the participation of the game. Playing games improved the interpersonal relationship between friends in the virtual community.

6. Instant messaging. SNS adds real-time web chat, similar to MSN, QQ and other IM. IM ensure the users can communicate online anytime.

Gene Smith defines cellular model diagram of social software. It summed up the elements of social network or software. It is shown in Fig. 1[4].

In Fig. 1, the user has a unique ID in the SNS. User can belong to one or more group. The relationship between users can be friends or other relationship. The users have a variety of status. The status can be shared with friends. There can be conversation between groups. SNS users can build their reputation in virtual community. User is the center in SNS model. All interactions are based on the user social network. This is similar to social network in the real world. No matter what kind of communication model SNS adopts, its core purpose is to gather users in the interaction as more as possible and create a virtual network society to enhance user participation sense [5]. SNS communication model meets the human characteristics and meet people's social emotional needs, so it is quite suitable for distance education. SNS communication model in distance education would inspire teachers and students' sense of participation and meet the teachers and students' the emotional needs. Eventually it can promote students' learning [6].

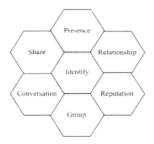

Fig. 1. Gene Smith's beehive model of SNS

4   SNS Model Design in Network Education System

4.1   Relationship between SNS and Network Education System

SNS Communication model is designed as Fig.2 according to above principles.

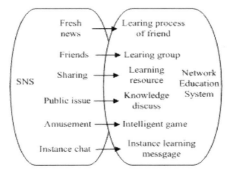

Fig. 2. Relationship between SNS and network education system

The SNS modules in network education platform include study partners, study groups, sharing of learning resources, study and discussion, puzzle games, real-time learning communication and so on. These modules are corresponding to friends, friends circle, sharing, public topics, entertainment communication, instant messaging in SNS website. With the purpose for promoting learning partnerships was built between students.

4.2 Roles and Activities Design in SNS Model

The roles and activities of SNS Communication model in network education system is defined in following Fig. 3.

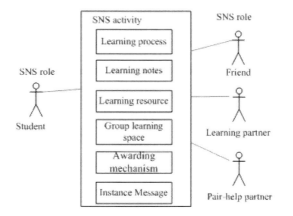

Fig. 3. Roles and activities design in SNS

The roles of SNS module in network education platform are set as the following types:

Friend. Students can add other students into their friends list. The friend number is unlimited. They can become friend relationship if they require or accept as friend by each other.

Learning partner. Students can choose their learning partners. The system also can recommend learning partners to students based on their study habits, behavior, hobbies and so on. In order to make students choose learning partners carefully, the number of learning partners is limited (eg 10). A study group is formed with several learning partners.

Pair-help partner. The student can specify a pair-help partner from his friends and learning partners. Pair-help partner is the most important friend in learning process. If two students become pair-help partners, they could not dismiss the relationship in a short term until time expired.

The activities of SNS module in network education platform are designed as the following:

Status update. Students can reply or comment on the new state or log updated by other students.

Recommendation. The system can recommend the good learning resources which are downloaded many times by other users. It also can recommend friend, learning partner or pair-help partner to students.

Reminder. Pair-help partners can remind learning process and evaluate each other. The evaluation can be used to choose pair-help partner.

Sharing. Students can share their learning resources linked in their home page. Their friends can download the learning resources.

Learning space. The system provides a virtual study room as a virtual learning space. The learning group members can share resources and do cooperation homework in the learning space.

Intelligent Game. It provides amusement game to students during their learning process which can attract the student stay longer and study more in the learning system.

Awarding mcchanism. The integral mechanism promotes students to participate in activities. Students can obtain scores through the activities. Score activities: perfecting the personal data, writing learning diaries, upload resources, learning resources recommended and answer questions to others. The deducted activities: create group, download learning resources, advertising, etc.

Instant communication. The tool meets the requirement of real-time communication among users.

5 Conclusion

SNS is an emerging Internet social communication model which has good interactive features and can support cooperative learning in network learning. This paper presents a SNS communication model design in network education system including the roles and activities design. It can solve the problems of lack communication and learning lonely in the remote learning students. This design gives a valuable reference to apply SNS communication into network education system. The further research will focus on cooperative learning based on SNS in network education.

Acknowledgement. School of Network Education, Beijing University of Posts and Telecommunications (BUPT) support this work in funding (No.2009RC0902) and research environment.

References

1. Ping, W., Lixinp, L.: Research on Interactivity in Modern Distant Education System. Journal of Henan Normal University(Edition of Philosophy and Social Science) 30(6), 75–91 (2003)
2. Yao, L.: Rise and Prospect of SNS Web Site in China. Network Culture (2009)
3. Yongguang, C.: Analysis of Cooperative Learning Environment in Network Based on New SNS Web Site. Journal of Outside-school Education of China 12(3), 12–16 (2008)
4. Lijun, W., Fengjuan, L.: Preliminary Exploration in Exploitation and Application of Fictitious Classroom. Journal of Chinese Modern Educational Equipment 20(2), 25–28 (2007)
5. Honggang, Y., Yuwen, N., Donghuai, G., Xiajuan, S.: Research on Constructing Learning Community in Network Based on SNS. Journal of Modern Educational Technology 8(4), 40–45 (2010)
6. Tai, Y.: Searching on Strategies of Interactivity between Teachers and Students in Modern Distant Education. Journal of Distant Education 40(1), 30–36 (2009)

A New Energy Efficient Routing Scheme for Wireless Sensor Network

Linhai Cui

Software School, Harbin University of Science and Technology,
Harbin 150040, Heilongjiang, China
cuilinhai@hrbust.edu.cn

Abstract. Energy efficiency has always been a key factor in wireless sensor networks. So far, many energy-efficient routing protocols have been proposed and much attention has been attracted to cluster-based routing protocols. However, some of these protocols need location information of the sensor nodes in the network. A new energy efficient, self-organized and hierarchical routing approach by combining cluster-based routing protocol with Directed Diffusion protocols is proposed in this paper. By this scheme, sensor nodes are self-organized into clusters, and directed diffusion is used in Inter-cluster. This approach makes the energy consumption and the lifetime of the whole network more balanced and longer. The result of simulation shows that this approach is energy efficient and makes the lifetime of the network longer.

Keywords: Wireless Sensor Network, Energy Efficient, Cluster-based Routing, Directed Diffusion.

1 Introduction

Wireless sensor network is a network consisting of thousands of sensors spanning over a large geographical area. The sensors are able to communicate with each other, exchange information and perform tasks collaboratively. Each node in a sensor network is typically equipped with a radio transceiver or other wireless communications device, a small microcontroller, and an energy source, usually a battery.

Routing in sensor networks is very challenging due to several characteristics that distinguish them from contemporary communication and wireless ad hoc networks. First, it is not possible to build a global addressing scheme for the deployment of sheer number of sensor nodes. Second, almost all applications of sensor networks require the flow of sensed data from multiple regions to a particular sink. Third, generated data traffic has significant redundancy in it, and such redundancy needs to be exploited by the routing protocols to improve energy and bandwidth utilization. Fourth, sensor nodes are tightly constrained in terms of transmission power, on-board energy, processing capacity and storage. Due to such differences, many new algorithms have been proposed for the problem of routing data in sensor networks where energy awareness is an essential consideration.

Y. Yu, Z. Yu, and J. Zhao (Eds.): CSEEE 2011, Part II, CCIS 159, pp. 201–206, 2011.
© Springer-Verlag Berlin Heidelberg 2011

2 Related Work

In this section we review specific prior studies that dealt with the two main routing protocols that have been used in wireless sensor networks, Clustering and Directed Diffusion. Clustering is based on the distance between nodes and the number in a cluster for wireless sensor networks, while Directed diffusion is a data-centric routing protocol based on purely local interactions between individual network nodes.

LEACH (Low-Energy Adaptive Clustering Hierarchy) is a self-organizing, adaptive clustering protocol that uses randomization to distribute the energy load evenly among the sensors in the network. In LEACH, the nodes organize themselves into local clusters, with one node acting as the local base station or cluster-head.

LEACH includes randomized rotation of the high-energy cluster-head position such that it rotates among the various sensors in order to not drain the battery of a single sensor. In addition, LEACH performs local data fusion to "compress" the amount of data being sent from the clusters to the base station, further reducing energy dissipation and enhancing system lifetime.

LEACH leverages balancing the energy load among sensors using randomized rotation of cluster heads, but it does not guarantee good clustering head distribution. So, it is difficult for LEACH to make the whole network energy consumption more balanced.

Directed diffusion uses application-specific context for data aggregation and dissemination. Therefore, it can be completely matched to the application requirements in a large distributed sensor network. Many works have been recently done to improve the energy efficiency of this protocol. To perform a sensing task, a querying node creates an interest, which is named according to the attributes of the data or events to be sensed. When an interest is created, it is injected into the network by the sink node by broadcasting an interest message containing the interest type, duration, and an initial reporting rate to all neighbors. Nodes receiving the interest messages find the relevant interest entry in their caches and update the gradient field toward the node from which the message was received to the rate defined in the interest message. Each gradient also has expiration time information, which must be updated upon the reception of the interest messages. Interests are diffused throughout the network toward the sink node using one of a number of forwarding techniques.

Directed diffusion uses flooding to transfer interests. In the flooding scheme, sources flood all events to every node in the network. Flooding is a watermark for directed diffusion; if the latter does not perform better than flooding does, it cannot be considered viable for sensor networks. The drawbacks of directed diffusion are the invalid processing and the waste of energy.

To overcome the above drawback, a new energy efficient, self-organized and hierarchical routing approach by combining cluster-based routing protocol with Directed Diffusion protocols is proposed. A two level structure is used in this approach. The upper level is the inter-cluster which is used to transfer data to the sink via the rout between cluster-heads, and the bottom level is the intra-cluster which is used to collect required information. All the sensor nodes are divided into clusters, and the sensor node within a cluster can collect data and make data aggregation. In the inter-cluster level, directed diffusion is used to transfer data when sink node transfer interests or when the source node want to transfer data to sink node so as to reduce the number of flooding packet and the energy consumption.

3 Design and Implementation of the New Approach

This paper focuses on the clustering and directed diffusion routing algorithms. In the proposed scheme, all the sensor nodes are divided into clusters, and then in the directed diffusion the interest flooding scheme is improved so as to save lots of energy. So the main comparison parameters in the simulation program are the lifetime of the network and the average energy consumption.

3.1 Sensor Network Clustering

The basic idea of clustering is to estimate the energy level of a node according to the information of the cluster which the node within so as to confirm the threshold in the next rotation. Thus, the node with higher energy level may have more opportunity to be the cluster head and to make the network energy consumption more balanced. So, energy level of a node is the key to clustering algorithm.

3.2 Energy Level Estimation of the Node within a Cluster

Assume that there are N sensors nodes in an M x M (km) area, there are r clusters when in r rotation

Performing clustering on a sensor network deployment prior to localization has two advantages. First, it creates a regular pattern from which location information can be extracted. Second, it helps reduce the amount of communication overhead since only the cluster-heads need to be involved in the initial phase of the localization. Sensors elect themselves to be local cluster-heads at any given time with a certain probability. These cluster head nodes broadcast their status to the other sensors in the network. Each sensor node determines to which cluster it wants to belong by choosing the cluster-head that requires the minimum communication energy.

Once the interest reaches the desired region, sensor nodes within the region process the query and begin producing data at the specified rate (if more than one entry for the same interest type exist, data is produced at the maximum rate of these entries). After receiving low rate events from the source (recall that the initial reporting rate is set low), the data sink may reinforce higher quality paths, which might be chosen, for example, as those that experience low latency or those in which the confidence in the received data is deemed to be high by some application-specific measure. Reinforcement messages simply consist of the original interest messages set to higher reporting rates. These reinforced routes are established more conservatively than the original low rate interest messages so that only a single or few paths from the event to the sink are used.

Directed diffusion is a data-centric routing protocol based on purely local interactions between individual network nodes. Using this method, load-balancing is implemented to increase the life-time of the sensor nodes collaborating in the routing process. The proposed protocol, Multi-path directed diffusion (MDD), can produce more than one disjoint or braided paths and spread the data collected in the sources, properly between the paths. In this way, an efficient load balancing mechanism has been implemented. The simulation results show that through using MDD, the lifetime of the network connections between the sources and the sink will be increased and the interest flooding rate which is proved to be an expensive operation can be reduced.

Directed diffusion (DD) [4] is a data-centric routing protocol proposed for data gathering in wireless sensor networks. In DD, attribute-value pairs are used for describing the information. This algorithm in its basic form has two phases. In the first phase, the sink node floods a request packet called "interest" containing the desired attribute-value pairs. When this packet reaches a source node that has the requested information (second phase), the source node floods an "exploratory data".

Cluster-based approaches are suitable for habitat and environment monitoring, which requires a continuous stream of sensor data. Directed diffusion and its variations are used for event-based monitoring.

4   Analysis of the Simulation Results

To determine whether the energy consumption and network lifetime of the new routing approach is improved, a simulation was made by using the network simulation tool OPNET.

4.1   Simulation Environment

In order to evaluate the performance of the improved routing approach from the view point of average energy consumption and network lifetime, the network simulation tool OPNET is used to simulate the algorithm. 100 wireless sensor nodes and a fixed sink node are used in the simulation. Sink node is far from the sensing area and each wireless sensor node has the initial energy 2J. The size of the data packet is 500 byte and the size of the metadata is 25 byte. The energy consumption of a node is 0 W when it is idle and sleeps, send/receive circuit energy consumption is $50_{(E_{DA})}/nJ \cdot bit^{-1}$, free-space model magnification is $10 \zeta_{fs} / pJ \cdot (bit \cdot m^{-2})^{-1}$, Two-ray ground model magnification is $0.0013 \ \zeta_{fs} / pJ \cdot (bit \cdot m^{-2})^{-1}$, the energy consumption of data fusion is $5 (E_{DA}) / nJ \cdot bit^{-1}$.

4.2   Simulation Results

Figure 1 shows the curve of simulation for average energy rested (pre: LEACH; post: New approach). From figure 1 you can see the performance of the new approach is better than that of LEACH as the network area increased.

Figure 2 shows the network lifetime. The network nodes is distributed in the area of 10 km x10 km, the sink node is far away from sensing area, and the simulation environment is the same as above. From figure 2 you can see that the network lifetime of the new approach is longer than that of LEACH. At the time 300s, the number of left nodes are 78 and 82 for LEACH and the new approach respectively, that is to say, the new approach do increase the network lifetime.

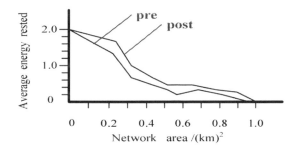

Fig. 1. Average energy rested

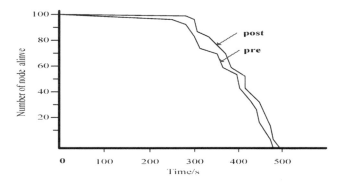

Fig. 2. Network lifetime

5   Conclusion

This paper proposed a new approach by combining cluster-based routing protocol with Directed Diffusion protocols. The cluster-based routing scheme uses the cluster head rotation, and the current energy is considered during the cluster head rotation. The new algorithm estimate the energy level of a node according to the local information of the cluster which the node within after the rotation, and so as to confirm the threshold in the next rotation. Thus, the node with higher energy level may have more opportunity to be the cluster head and to make the network energy consumption more balanced.

References

1. Heinzelman, W.R., Chandrakasan, A., Balakrishnan, H.: Energy-Efficient Communication Protocols for Wireless Microsensor Networks. In: Proceedings of the 33rd Annual Hawaii International Conference On Systems Sciences, Maui, pp. 3005–3014 (2000)
2. Younis, S.F.: A Hybrid, Energy-efficient, Distributed Clustering Approach for Ad-hoc Sensor Networks. IEEE Trans. On Mobile Computing. 3(4), 660–669 (2004)

3. Lindsey, S., Raghavendra, C.S.: PEGASIS: Power Efficient Gathering in Sensor Information Systems. In: Proc of IEEE Aerospace Conference, IEEE Aerospace and Electronic Systems Society, Montana, pp. 1125–1130 (2002)
4. Younis, O., Fahmy, S.: Distributed Clustering in Ad-hoc Sensor Networks: A Hybrid, Energy Efficient Approach. In: Proceedings of IEEE Infocom, pp. 629–640 (2004)
5. Al-Karaki, J.N., Kamal, A.E.: Routing Techniques in Wireless Sensor Networks: A Survey. IEEE Wireless Communications 6(11), 6–28 (2004)

Test and Feasibility Study of Heat Pump Instead of Oil Field Furnace in Daqing Oil Field

Yongying Jia[*], Lei Zhang, Xiaoyan Liu, and Zhiguo Wang

The College of Architecture and Civil Engineering, Northeast Petroleum University,
Hei Longjiang Daqing 163318, China
{jyy1219,fss_lbb}@163.com, liu-xydq@sohu.com,
wangzhiguo99@yahoo.com.cn

Abstract. According to the rich heat resources of sewage waste in Daqing Oil Field and the supply water temperature in the heater is over 70℃ in the building, proposed the idea of using heat pump that the high temperature sewage- source can be used in Daqing Oil Field, showed the technical feasibility of using the heat pump in stead of the oil field furnace, analyzed and evaluated of its energy usage of two schemes, The test and evaluation results indicate that heat pump instead of oil heater is feasible. The study provides a reliable scientific basis for heat pump in the oilfields.

Keywords: Sewage, Heat Pump, Oil Furnace, Energy Analysis, Test.

1 Introduction

Currently in the research of process heat pump, representative is parkas, Lawrence and so on [1-7], they study thermo economic of heat pump. The temperature of sewage discharge in Daqing Oil Field is between 40℃ and 50℃, its sewage discharge is about 6.01 million tons per hour, and this paper studied technical feasibility of using the heat pump in stead of the oil field furnace to conserve energy. The approach has more systematic and practical value.

2 Retrofit Scheme

The experimental site is chosen at the Combined-station, the basic situation of this station is the following:

The technical scheme of the heat pump instead of oil heater is shown in Figure1. The thermal source of the heat pump evaporator is the treated sewage, the temperature is about 43℃. In the condenser, because the oil-water mixture has not been treated, the condenser can easily lead to corrosion and scaling if it directly gets into the condenser, and influence the Heat transfer performance and lifetime of the consenter, so we set up intermediate heat exchanger which resistances.

---

[*] Corresponding author.

Y. Yu, Z. Yu, and J. Zhao (Eds.): CSEEE 2011, Part II, CCIS 159, pp. 207–211, 2011.
© Springer-Verlag Berlin Heidelberg 2011

Table 1. The basic value of the chosen station

Item	Area	The amount of Gathering and transfixing oil	Sewage treatment capacity	The temperature of treated sewage	Natural gas consumption
Unit	$10^4 m^2$	$10^4 t/a$	$10^4 t/a$	°C	$10^4 Nm^3/a$
Value	4.68	46.1	242.0	43	180.0

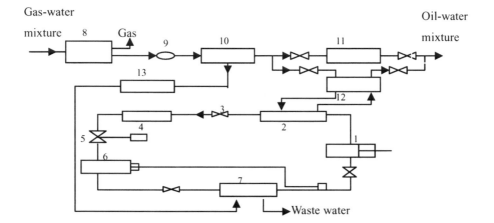

1-compressor   2-condenser   3-disconnecting valve   4-device for drying and filtering 5-magnet valve 6-expansion valve 7-evaporator   8-Electric heater 9-Pressure pump 10-edimentation tank 11-urnace section 12-intermediate heat exchanger 13- sewage treatment plant

Fig. 1. The heat pump transformation process diagram

To acid corrosion and is easy to clean. The recycled water in heat pump consenter is clean and exchange heat with the oil-water mixture. Before using the heat pump, the test is done to measure and calculate heating energy consumption, fluid flow and temperature, thermal losses and operating efficiency and so on. Then another test is done when heat pump instead of oil furnace and stable operation for some time.

3   Calculation Results

The data of the oil furnace operation is shown in Table 2; the data of heat pump operation is shown in Table 3.

According to the Table 2, we can get:

Operational efficiency:
$$(77.5\% + 74.2\%)/2 = 75.8\% \tag{1}$$

Gas consumption per hour:
$$(40.77 + 43.93)/2 = 42.25 Nm^3/h \tag{2}$$

Annual gas consumption:

$$42.25 \times 8000 = 33.88 \times 104 Nm3/a \text{ (Year terms by 8000 hours)} \qquad (3)$$

Industry Gas Prices:

$$1.25 yuan/Nm^3$$

Operation and management costs: Additional 10% of the operating management and maintenance costs. Annual operating costs:

$$(1+10\%) \times 1.25 \times 33.88 = 465800 yuan/a \qquad (4)$$

Table 2. The data of the oil field furnace operation

Item	Unit	Value The first time(summer)	The second time(winter)	Remark
Fluid drag-in flow	kg/h	45000.0	44800.0	
Fluid drag-in temperature	℃	45.0	45.0	
Fluid drag-out temperature	℃	56.5	57.0	
Gas consumption	Nm^3/h	40.77	43.93	
Calorific value of gas	kJ/Nm^3	41259.3	41259.3	
Mixture of heat absorption	MJ/h	1303.7	1344.9	
Heat supply	MJ/h	1682.2	1812.1	The average value of 3 times the value measured
Positive equilibrium Thermal efficiency	%	77.5	74.2	
Exhaust gas temperature	℃	216.0	178.0	
O2 content in flue gas (volume)	%	13.5	15.9	
N2 content in the flue gas (volume)	%	4.8	3.6	
CH4 content in the flue Gas (volume)	%	0.02	0.02	
Air factor	%	2.65	3.78	
Smoke losses	%	20.96	24.15	
Loss of incomplete combustion of gas	%	0.18	0.22	
Heat loss	%	3.3	3.0	
Anti-equilibrium thermal efficiency	%	76.6	72.6	

Table 3. The data of heat pump system operating

Item	Unit	Value The first time	The second time	Remark
Condenser Entering Water Temperature	℃	53.8	55.0	
Condenser Outlet Water Temperature	℃	58.3	60.0	
Circulating water flow	kg/h	72000.0	72000.0	The average value of 3 times the value measured
Evaporator inlet temperature	℃	43.2	43.2	
Evaporator water outlet temperature	℃	41.0	41.0	
Sewage flow	kg/h	123000.0	135000.0	
Inlet temperature of the mixture	℃	45.0	45.0	
Outlet temperature of the mixture	℃	56.5	57.0	
Compressor power consumption	kW	87.3	92.9	
Coefficient of performance	——	4.345	4.507	

According to the Table 3, we can get:
Coefficient of performance:

$$(4.345+4.507)/2=4.426 \tag{5}$$

Primary energy ratio:

$$4.426×0.3=1.328 \text{ (Power plant efficiency by 30\% of the total)} \tag{6}$$

Power consumption per hour:

$$(87.3+92.9)/2=90.1kWh/h \tag{7}$$

Annual power consumption:

$$90.1×8000=72.08×10^4kWh/a \tag{8}$$

Field of industrial electricity prices: 0.54 RMB/kWh
Annual operating costs:

$$(1+5\%)×0.54×72.08=431400Yuan/a \text{ (Additional 5\% of the management and maintenance costs)} \tag{9}$$

4 Conclusion

The key indicators of using the oil furnace system: operational efficiency: 75.8%, annual operating costs: 465800yuan/a. The key indicators of using the heat pump system: coefficient of performance: 4.426, primary energy ratio: 1.328, annual operating costs: 431400 yuan/a.

Coefficient of performance heat pump system operation --COP is 4.426, primary energy ratio—PER is 1.328.The thermal efficiency of the original furnace was 75.8%.Compared between the two results, the energy saving rate of the heat pump is 57.04%. But the economic point of view, the annual operating costs of the heat pump system is 431400 Yuan, and the oil furnace system is 465800 Yuan, the annual operating costs are almost the same. Engineering Test results show that the idea of using the heat pump instead of the oil furnace is feasible. There is no combustion process in heat pump system, so it does not produce any CO_2, SO_2 and other harmful gases, and no coal-fired dust. Therefore it cannot cause any pollution, and solve the traditional way of heating which causes the air pollution effectively. At the same time, reduce the circulating water pollution of the environment, environmental benefits are significant.

Acknowledgments

This project was funded partly by the National Natural Science Foundation (50776014), and Scientific and technological research project in Heilongjiang Province (GZ07A302).

References

1. Prakash, R.: Damshala: Thermoeconomic Analysis of OCHP System by Iterative Numerical Techniques. J. ASHRAE Transactions 106(1), 327–337 (2000)
2. Song, Z.-P.: Total Energy System Ana Ysis of Heating. J. Energy (25), 807–822 (2000)

3. Yantovsk, I.E.: Exergonomics in Education. J. Energy 25, 1021–1031 (2000)
4. Dentice, D., Accadia, M.: Optimal Operation of a Complex Thermal System: A Case Study. J. International Journal of Refrigeration 6(4), 290–301 (2001)
5. Lawrence, N., Kortekaas, H.Y.P.: DECSIM-A PC-based Diesel Engine Cycle and Cooling System Simulation Program. J. Mathematical and Computer Modelling 33(6), 565–575 (2001)
6. Smith, M.A., Few, P.C.: Second Law Analysis of an Experimental Domestic Scale Cogeneration Plant Incorporating a Heat Pump. J. APPlied Thermal Engineering 21, 93–110 (2001)
7. TaPio, P.: Industrial Ecology of the PaPer Industry. J. Water Science and Technology 40(11), 21–24 (1999)

Highway Road Accident Analysis Based on Clustering Ensemble[*]

Taoying Li, Yan Chen, Shengjun Qin, and Nan Li

Transportation Management College, Dalian Maritime University,
Dalian 116026, P.R.China
ytaoli@126.com, {chenyan_dlmu,qinsj1984}@163.com,
linan_dlmu@yahoo.cn

Abstract. We employ clustering ensemble to partition highway roads according to traffic accident information to avoid the occurrence of accidents in this paper. Above all, we use fuzzy k-means clustering to classify numerical data of accidents for producing numerical clustering membership, and produce categorical memberships using values of corresponding categorical attributes. Then we adopt clustering ensemble to merge all clustering memberships to solve the sole clustering. Finally, the clustering ensemble was used to group 16 highway roads and results show that it is effective and could be used to avoid occurrence of traffic accidents.

Keywords: Clustering ensemble; Fuzzy k-means; Accident analysis.

1 Introduction

The Global Status Report on Road Safety issued by World Health Organization showed that road traffic accident kills global 1.27 million people each year in 2009, and forecasted that the number of people killed by road traffic accident will reach to 2.4 million in 2030 if the road safety condition is left as there are not.[1]. Although every government has taken measures to protect drivers and passengers, which has gotten progress, however, number of people dying from traffic accidents increased because of the structural problems of Vehicular Traffic Laws.

Between years from 2000 to 2004, the loss caused by traffic accident in China was about 1 to 3 percent of gross domestic product, and amount of loss was $12.5 billion. During these five years, Chinese traffic accident killed more than 500 thousand people and caused 2.6 million people hurt, which means one person will be killed by traffic accident in every five minutes, and the mortality rate of China is the highest [2]. In 2008, the number of traffic accidents occurred in China was 265204, and 73484 people were dead and 204919 people were injured, and direct property damage was 1.01 billion [3]. In 2009, the number of traffic accidents occurred in China was

---

[*] This work was supported by the Doctoral Fund of Ministry of Education of China (Grant No. 200801510001), the Key Project of Chinese Ministry of Education (Grant No. 209030), and the National Science and Technology Supporting Plan of the eleventh five-year(2009BAG13A03).

Y. Yu, Z. Yu, and J. Zhao (Eds.): CSEEE 2011, Part II, CCIS 159, pp. 212–217, 2011.
© Springer-Verlag Berlin Heidelberg 2011

238351, and 67759 people were dead and 275125 people were injured, and direct property damage was 0.91 billion [4]. The occurrence rate and mortality rate of China are staying at the top, and how to change the traffic situation, prevent and decrease the occurrence of traffic accidents is an urgent problem in China.

We use clustering ensemble to analyze data of highway accidents, and partition highway into several different clusters or ranks, which will strengthen the public safety consciousness and assist managers of highway departments to make decision scientifically, and further avoid the occurrence of highway accidents effectively.

Existing algorithms for classifying categorical data or mixed numerical and categorical data have disadvantages of instability, stochastic and low precision. Besides, many algorithms transfer categorical data into numerical data or transfer numerical data into categorical data for clustering, or employ clustering ensemble to partition categorical data or mixed data [5]. Thus, we introduce clustering ensemble to analyze highway accidents.

Clustering ensemble is to integrate some results of some existing clustering partitions for higher quality and better robust, and Topchy proved that the ensemble method can be better than anyone clustering algorithm in [6-7]. We introduce clustering ensemble to analyze highway accident with mixed data, which will overcome the disadvantages of single algorithms and improve the effectiveness and precision of clustering [5]. Clustering ensemble was proposed by A. Strehl and J. Ghost in [8], and it was already studied by A. L. Fred in 2001 [9]. Recently, the study focused on two aspects, one is how to produce efficient clustering membership and another is how to design the mutual function for merging clustering membership [10].

2 Clustering Membership Production

It is difficult for us to enumerate all values of numerical data, and we use some algorithms to classify numerical data. We adopt fuzzy k-means algorithm to partition numerical data in this paper. Then we make use of values of different attributes to obtain their partitions for getting corresponding clustering memberships, and then use clustering ensemble to integrate all clustering memberships.

We eliminate the difference of units of numerical dimensions according to (1), which will make all data points be zero dimension.

$$x_{ji} = \left| \frac{x^{original}{}_{ji} - \min_t x_{ti}}{\max_t x_{ti} - \min_t x_{ti}} \right| \quad 1 \le i \le m \tag{1}$$

$$F(T,W,C) = [\sum_{j=1}^{n}\sum_{i=1}^{m}\tau_{lj}\omega_i(c_{li} - x_{ji})^2]/[\sum_{i=1}^{m}(c_{li} - \bar{x}_i)^2] \tag{2}$$

We can delete the units of attributes by (1), and at same time make the values of all attributes between 0 and 1. Then we use fuzzy k-means clustering to partition numerical data in this paper, the objective function of which was proposed by us in [11] and is shown as (2). Where, $\sum_{l=1}^{k}\tau_{lj} = 1, 1 \le j \le n, \tau_{lj} \in \{0,1\}$, $\sum_{i=1}^{m}\omega_i = 1, 0 \le \omega_i \le 1$, k, n and m are respectively the number of clusters, objects, and dimensions. x_{ji} is the value of the ith dimension of the jth object. $C=[c_{li}]$ is a k-by-m matrix, and c_{li} is the value of the ith dimension of the lth cluster center. $T=[\tau_{lj}]$ is a k-by-n matrix, and τ_{lj} is the degree of membership of the jth object belonging to the lth cluster. $W=[\omega_{li}]$ is a k-by-m matrix, and ω_{li} is the weight of the ith dimension in the lth cluster. $\overline{x}$ is the mean of all objects, $\overline{x}_i$ is the avarage of values of the ith dimension. If $n > 1$, the new algorithm is available, otherwise, the denominator is zero and the cost function $F(T,W,C)$ is not computable because it is a variable and is linear to the square sum of the distances from the mean of all objects to the means of all clusters.

Minimization of F in (2) with constraints forms a class of constrained nonlinear optimization problems whose solutions are unknown. Generally, the method for solving optimization is partial optimization for T, W and C. Thus we first fix C and get appropriate T, and then fix T to search appropriate C. We repeat the steps mentioned above until the value of objective function is least.

It is well known that the jth object will belong to the lth cluster if it is closest to the center of the lth cluster, which is same to (3).

$$\tau_{lj} = \begin{cases} 1, & if \sum_{i=1}^{m} \omega_{li}(c_{li} - x_{ji})^2 \leq \sum_{i=1}^{m} \omega_{zi}(c_{zi} - x_{ji})^2 \\ 0, & otherwise \end{cases} \tag{3}$$

$$c_{li} = \sum_{j=1}^{n} \tau_{lj} x_{ji} / \sum_{j=1}^{n} \tau_{lj} \tag{4}$$

Where $\tau_{lj} = 1$ denotes that the jth object belongs to the lth cluster absolutely, or it means it doesn't belong to the lth cluster. Fix T, we use method for obtaining average to get values of C. We can produce one clustering membership according to the method mentioned above, which means all numerical attributes would produce one clustering membership, and every categorical will produce one clustering memberships, and the number of total memberships is number of categorical attributes plus 1.

3   Clustering Ensemble

Now we suppose that there are m clustering memberships, $(m-1)$ of which are produced by categorical attributes and each has its own weight, one clustering membership was produced by numerical attribute and its weight is the sum of weights of all numerical attributes, $\sum_{i=1}^{m} \omega_i = 1$.

Then, we set the value of threshold θ ($0 < \theta \leq 1$), and search all possible groups, in each of which the sum of weights of all clustering memberships is large than θ. We set Π empty set, and suppose there are t clustering memberships in any group, and we fetch a cluster in each membership, and their conjunction is $\pi_i = \overbrace{C_{1 \cdot j_1} \cap C_{2 \cdot j_2} \cap \cdots \cap C_{t \cdot j_t}}^{t}$, if the number of objects in π_i is two or more, then put it in Π, or delete it. Then we search the next group until all groups have been traveled.

Weights of sets in Π equal to or are large than θ, at same time there may be conjunction among them. Let $\pi(i, j) = \pi_i \cap \pi_j \neq \Phi$, then get a triangle sparse matrix R with $s \times s$. For all those conjunction is nonempty set, arrange them in the sequence of value $|\pi(i, j)| / \max\{|\pi_i|, |\pi_j|\}$ from big-to-small. At this time, set the value of λ $(0 < \lambda \leq 1)$.

Step1. Let $\alpha = \max\left\{\dfrac{|\pi(i, j)|}{\max\{|\pi_i|, |\pi_j|\}}\right\}$ Step2. If $\alpha \geq \lambda$ then merge π_i and π_j to be π^*, and let $s = s-1$, then delete π_i and π_j from Π, and add π^* to Π, and at same time delete rows and columns of π_i and π_j from R, and add the row and column of π^*. Go to Step1. Else go to Step3. Step3. If $\exists \, \pi(i, j) \neq \Phi$ then If $|\pi_i| \triangleright |\pi_j|$ then $\pi_j = \pi_j - \pi(i, j)$ Else

$\pi_i = \pi_i - \pi(i, j)$, Else Go to Step4. Step4. If $\exists \, x_i \notin \bigcup\limits_{j=1}^{s} \pi_j$ then Add $\{x_i\}$ to Π, go to Step4. Else stop the computation.

Because Π contain all data points and conjunction of any two sets is empty, thus it is a partition of data set.

The time complexity of traditional clustering is O (mn^2), and its number is exponent to the change of number of data points, which means that the computation is very large and some time it is not obtainable for large data set. The time complexity of algorithm adopted in this paper is O(mnk) and similar to that of [11]. At the same time, we can use k-center to initial fuzzy k-means for decreasing the number of iteration.

4 Application of Highway Traffic Accidents

We analyze data of 16 highway roads accidents, and its attributes are number of overload vehicles, vehicle situation, number of vehicles carrying dangerous cargo, number of accidents, average number of vehicles, mortality rate, accident type, weather status, visibility, road surface condition and light condition, the former six attributes are numerical and the last five attributes are categorical. For numerical attributes, the value after eliminating unites of attributes according to (1) can be shown as Table 1.

Then we can obtain the minimum F according to (2) and its value is 12.38, and its partition for numerical data is $\{\{6\}, \{2,5,8,11,13\}, \{1,3,4,7,9,10,12,14,15,16\}\}$. We can obtain the clustering memberships according to categorical data in Table2 as follows:

For accident type, we obtain the clustering membership $\{\{2,6,11\}, \{1,3,15,16\}, \{4,5,7,8,9,10,12,13,14\}\}$. For weather, we obtain the clustering membership, $\{\{1,2,3,4,9\}, \{5,8,11,14,15,16\}, \{6,7,10,12,13\}\}$. For visibility, we obtain the clustering membership, $\{\{1,2,4,5,7,8,9,11\}, \{3,6,10,13,14\}, \{12,15,16\}\}$. For light condition, we obtain the clustering membership, $\{\{2,,3,4,6,7\}, \{1,5,8,9,10,12,13,14,16\}, \{11,15\}\}$. For road surface condition, we obtain the clustering membership, $\{\{1,2,5,6,7,11,14,15,16\}, \{3,8,9,10\}, \{4,12,13\}\}$.

Table 1. Numerical data

Road ID	Overload	Vehicle Situation	Number of Vehicles Carrying Dangerous Cargo	Number of Accidents	Average Number of Vehicles	Mortality Rate
1	0.101391	0.021983	0.001794527	8	278.625	0.125
2	0.101993	0.025791	0.002344666	3	284.3333	1.333333333
3	0.102343	0.019729	0.001849568	6	270.3333	0
4	0.101553	0.023297	0.001792115	6	279	0.333333333
5	0.101345	0.021973	0.001793722	8	278.75	0.875
6	0.101327	0.022187	0.001753583	47	279.0638	0.85106383
7	0.103107	0.021387	0.002701486	16	277.625	0.5
8	0.102698	0.024369	0.001740644	4	287.25	0.75
9	0.098233	0.022615	0.001413428	5	283	0.4
10	0.102094	0.022074	0.002110047	22	280.0455	0.5
11	0.108696	0.018116	0	1	276	1
12	0.094089	0.025332	0	3	276.3333	0.333333333
13	0.101958	0.022507	0.002025658	16	277.6875	1.125
14	0.097826	0.027174	0.001811594	2	276	0.5
15	0.102163	0.024639	0.001802885	6	277.3333	0
16	0.092527	0.017794	0	1	281	0

Table 2. Categorical data

Road ID	Accident Type	Weather	Visibility	Road Surface Condition	Light Condition
1	3/5,2/2,1/1	1/5,2/3	4	1	3/3,1/5
2	1/3,	1/2,2/1	4/2,3/1	1/2,2/1	3/2,1/1
3	2/6,	1/5,3/1	4/3,3/3	3/1,1/5	3/3,1/3
4	1/2,2/4	1/5,4/1	4/5,3/1	5/1,9/1,1/4	3/3,1/3
5	1/5,2/3	1	4	1	2/4,1/3,3/1
6	1/29,2/16	1/30,5/17	1/11,4/36	1/37,2/10	1/22,3/25
7	1/5,3/5,2/6	5/2,2/2,1/12	4	1	1/7,3/9
8	1/2,2/1,3/1	1	3/1,4/3	1/3,3/1	1/3,3/1
9	1/2,2/3	1/4,3/1	4	1/4,3/1	1/4,3/1
10	1/10,2/6,3/6	4/2,3/2,2/1,1/17	3/6,1/2,2/4,4/10	5/2,3/2,1.18	2/4,3/4,1/14
11	1	1	4	1	3
12	1/1,2/1,3/1	1/1,3/1,5/1	1/2,4/1	1/1,2/1,4/1	3/1,1/2
13	3/6,1/10	3/2,4/2,1/12	2/3,3/2,4/11	1/12,5/4	3/3,2/2,1/11
14	1/1,3/1,	1	3/1,4/1	1	1
15	3/5,2/1	1	1/4,4/2	1	3
16	2	1	1	1	1

Finally, we use clustering ensemble to merge all clustering memberships and then get the final partition {{1,3,4,7,9,10,12,14,15,16},{2,6,11},{5,8,13}}. {2, 6, 11} is the cluster that accidents occurring on these highway roads are most seriously. {5, 8, 13} is the clustering that accidents occurring on these highway roads are more seriously.{1, 3, 4, 7, 9, 10, 12, 14, 15, 16} is the clustering that accidents occurring on these highway roads are not that serious compared with other highway roads. Results of clustering are consisted with the facts.

5 Conclusions

More and more people were killed by traffic accidents in China and we lose the opportunity at the same time. In order to avoid the occurrence of traffic accidents, we use clustering ensemble to partition highway roads according to information of accidents on these roads, and results show that it is effective and can be used to assist people who work for highway road departments to make decision scientifically.

References

1. http://news.dayoo.com/world/57402/200906/30/57402_9657278.htm
2. http://news.163.com/05/1215/12/2510AHJ700011MS6.html
3. http://news.dayoo.com/china/57400/200901/04/57400_5065587.htm
4. http://www.167ok.com/shop/Normal/
   news_view.asp? CompanyMemberID =&ID =3653
5. Zhao, Y., Li, B., Li, X., Liu, W., Ren, S.: Cluster Ensemble Method for Databases with mixed Numeric and Categorical Values. J. Tsinghua Univ. (Sci & Tech) 46(10), 1673–1676 (2006)
6. Topchy, A., Jain, A.K., Punch, W.: A Mixture Model for Clustering Ensembles. In: The 4th SIAM International Conference on Data Mining, pp. 379–390. Society for Industrial and Applied Mathematics, Lake Buena Vista (2004)
7. Topchy, A., Jain, A.K., Jain, A.K., et al.: Adaptive Clustering Ensembles. In: The 17th International Conference on Pattern Recognition (ICPR 2004), vol. 1, pp. 272–275. IEEE Computer Society Press, Cambridge (2004)
8. Strehl, A., Ghosh, J.: Cluster Ensembles: A Knowledge Reuse Framework for Combining Multiple Partitions. Journal of Machine Learning Research 3(3), 583–617 (2003)
9. Fred, A.L.: Finding Consistent Clusters in Data Partitions. In: Kittler, J., Roli, F. (eds.) MCS 2001. LNCS, vol. 2096, pp. 309–318. Springer, Heidelberg (2001)
10. Li, T., Chen, Y.: Fuzzy Clustering Ensemble Algorithm for Partitioning Categorical Data. In: The 2009 International Conference on Business Intelligence and Financial Engineering, pp. 170–174. IEEE Press, Beijing (2009)
11. Li, T., Chen, Y.: An Improved k-means Algorithm for Clustering Using Entropy Weighting Measures. In: The 7th World Congress on Intelligent Control and Automation, Institute of Electrical and Electronics Engineers, pp. 149–153. Chongqing (2008)

Multi-Objective Evolutionary Algorithm Based on Improved Clonal Selection

Shaobo Li[1,2], Xin Ma[1], Qin Li[1], and Guanci Yang[2,*]

[1] Key Laboratory of Advanced Manufacturing Technology,
Guizhou Univ., 550003 Guiyang, China
[2] Chengdu Inst. of Computer Applications,
Chinese Academy of Sciences, 610041 Chengdu, China
lishaobo@gzu.edu.cn, maxincad@163.com,
helen850922@163.com, guanci_yang@163.com

Abstract. Evolutionary algorithm is widely used to search Pareto-optimal set. The paper proposes a kind of multi-objective evolutionary algorithm based on improved clonal selection (MOEAICS), which incorporates improved clonal selection algorithm (ICSA) into multi-objective evolutionary algorithms to replace genetic operators such as crossover and mutation in traditional evolutionary algorithms. ICSA characterized by 1) excavation of excellent gene fragments in antibody population and generating a memory antibody set, 2) packaging operation of excellent gene fragment, and 3) replacing low affinity antibody with high affinity antibody with probability from mutation antibody population during updating memory antibody population. The testing results show that MOEAICS is capable of maintaining population diversity, improving search efficiency, and accelerating convergence.

Keywords: Clonal Selection, Evolutionary Algorithm, Gene Excavation.

1 Introduction

It has become an attractive research to design better offspring-generated method to replace traditional operation, such as crossover and mutation [1]. The researcher introduced a constructive model into evolutionary algorithm [2] to extract excellent solutions' information, and then generate new individuals by probability distribution. In reference [3], the crossover and mutation operations were replaced by the Bayesian algorithm based on the theory of constructive probability model. The Bayesian network model [4] combined with SPEA2 and strength Pareto evolutionary algorithm based on decision map Bayesian network was put forward, which achieved good results. On the basis of clonal selection, this paper proposes a kind of multi-objective evolutionary algorithm based on clonal selection to replace crossover and mutation operations with antibody gene fragment extraction and antibody reuse mechanism.

---

* Corresponding author.

Y. Yu, Z. Yu, and J. Zhao (Eds.): CSEEE 2011, Part II, CCIS 159, pp. 218–223, 2011.

2 Principle and Algorithm of Clonal Selection

The basic idea of clonal selection theory is that clone is aimed at the cells which can recognize antigens, and these cells cloned are selected to survive and grow up, but these clonal cells which can't recognize antigens are replaced by other clonal cells. Denote Ab the initial antibody population and Ag the antigen set, and suppose $size$ is the size of Ab and $ilen$ is the length of each antibody, $i=0, 1, ..., size-1, j=0, 1, ..., ilen-1$. Let Fit be the antibody affinity, $Ab_i(j)$ be the antibody and $Ag_i(j)$ be antigen. Clonal selection algorithm (CSA) [5] can be detailed as follows.

Step 1 initialize antibody set. Antibody corresponds to candidate solutions
Step 2 calculate antibody affinity

In clonal selection algorithm, antibody's quality is embodied by affinity which is reflected by bond strength between antibody and antigen. The match of antibodies and antigens is decided by formula (1), and the antibody affinity Fit can be calculated by formula (2).

$$c_j = \begin{cases} 1, Ab_i(j) \neq Ag_i(j) \\ 0, Ab_i(j) = Ag_i(j) \end{cases} \tag{1}$$

$$Fit = \sum c_j + \sum 2^i \tag{2}$$

$$Ab_i(j) = \begin{cases} 1 - Ab_i(j), rand()\%1000 < p_m *1000 \\ Ab_i(j), rand()\%1000 \geq p_m *1000 \end{cases} \tag{3}$$

Step 3 clonal proliferations
Put antibodies into current excellent antibody population $CurrestAb$ after having finished sort descending according to antibody affinity, and then reproduce these antibodies in $CurrestAb$. The clonal scale of antibodies is proportional to the antibody affinity. Clonal rate, that is $Clonerate= \beta \times size/ (i+1)$, determines the number of antibody to clone, which is denoted by the clonal coefficient β and $size$. Place the clonal antibodies in clonal antibody population $CloneAb$.

Step 4 mutation
Mutate the clonal antibodies with the probability of P_m, and place these mutated antibodies in mutated antibody population $MutationAb$ according to formula (3).

Step 5 If satisfy termination conditions, then stop evolution, else update the antibody population and go to step 2.

3 Improved Clonal Selection Algorithm

Clonal selection algorithm achieves evolution through high-frequency variation operation. Although clonal selection algorithm tries to maintain the population diversity as far as possible during evolution process, the algorithm still have large risk to produce a large number of similar solutions, which will lead to premature convergence. Considering those risk, this paper has done some research to improve clonal selection algorithm, which includes the followings:

1) Excavation of excellent gene fragments in antibody set and generating a memory antibody set

After calculating affinity of antibody in step 2 of pre-section, discover the excellent gene fragments in antibody population based on the match of the antibody and antigen. It is obvious that excellent gene fragment plays a crucial role in evolution. In the process of excavation of excellent gene fragments of antibody population and generating a memory antibody, Firstly, match the antibody and antigen according to formula (1), and excellent gene fragment are stored as a temp memory antibody and some information is recorded as well. Secondly, 0\1 codes generated by a random generator are used to fill the temp memory antibody's vacancy except excellent gene fragment, and then we generate a complete memory antibody, and store it as a member of memory antibody set *MemoryAb*. Finally, refresh antibody affinity in memory set *MemoryAb*.

2) Packaging operation of excellent gene fragment

Package the discovered excellent gene into a fixed block with length of *n*. Then use 0/1 code to fill the fixed block to produce new antibody, and then store it as a member of antibody set *SubAb*.

3) Replacing low affinity antibody with high affinity antibody with probability from mutation antibody population during updating memory antibody population.

Based on those improvements above, the improved clonal selection algorithm (ICSA) can simply described as follows:

1) Generate an initial antibody set *Ab*.

2) Empty set of *MemoryAb、 MutationAb* and *SubAb*.

3) Calculate antibody affinity, and order it.

4) Minning the excellent gene fragments of antibody and then updates *MemoryAb* by new population composed of excellent gene fragment.

5) Calculate antibody affinity in *MemoryAb* and order it.

6) Clonal proliferate in antibody by the given strategy.

7) Apply crossover and mutation operators, and put the new group into *MutationAb*.

8) Replace high affinity antibody in *MutationAb* with the low affinity antibody in *MemoryAb* by given probability.

9) Assign *MemoryAb* to *Ab*, and then obtain a new population of antibody.

10) If terminal condition is reached, then stop, else go to 2).

4   MOEA Based on Improved Clonal Selection

Muti-objective evolutionary algorithm based on improved clonal selection (MOEAICS) adopts external and internal populations. Genetic copying is operated in internal population, and non-dominated solutions of each evolution are stored in external population, then we delete the dominant solutions and replace non-dominated solutions in external population, and decide the Pareto fitness of individuals according to dominance relationship of individuals in the two populations, and apply improved clonal selection algorithm to generate next generation. MOEAICS is described as follows:

1) Initialize evolutionary population Ab, and empty non-dominated external population P'.

2) Calculate individual fitness.

3) Determine non-poor individuals in Ab.

4) Apply similarity crowding algorithm [6] to update external population P'.

5) Apply ICSA to generate new population Ab.

6) If achieve maximal evolutionary generations, end the algorithm, else go to 2).

Replace the ICSA with CSA described in section 1 in step 5); it can obtain a kind of multi-objective evolutionary algorithm based on clonal selection (MOEACS). In the section of testing and analysis, we will compare MOEACS with MOEAICS to check their performance.

5   Testing and Analysis

We take multi-objective 0/1 knapsack problem as testing problem to check the performance of algorithms. Generally, a 0/1 knapsack problem consists of a set of items, weight and profit associated with each item, and an upper bound for the capacity of the knapsack. The task is to find a subset of items which maximizes the total of the profits in the subset, yet all selected items fit into the knapsack, i.e., the total weight does not exceed the given capacity. Formally, the multi-objective 0/1 knapsack problem considered here is defined in the following way: Given a set of n items and a set of m knapsacks, with p_{ij}(profit of item j according to knapsack i), w_{ij} (weight of item j according to knapsack i), c_i(capacity of knapsack i). Find a vector $x=(x_1,\ldots,x_n)$ $\in \{0,1\}^n$, such that $\forall i \in \{1,\ldots,m\}:\Sigma w_{ij}\cdot x_j \leq c_i, j\in \{1, ,\ldots,n\}$, and the vector $f=(f_1(x),\ldots,$ $f_m(x))$ is maximized, where $f_i(x)=\Sigma p_{ij}\cdot x_j$, and $x_j=1$ iff item j is selected. In order to obtain reliable and sound results, we use nine deferent test problems where both the number of knapsacks and the number of items are varied. Two, three, and four objectives are taken under consideration, in combination with 100, 150, and 200 items. Uncorrelated profits and weights are chosen, where p_{ij} and w_{ij} are random integers in the interval [10,100]. Table 1 shows the detailed information. The knapsack capacities c_i are set to half the total weight regarding the corresponding knapsack (see formula (4)).

$$c_i = \frac{1}{2}\times \sum_{j=1}^{m} w_{ij} \, (i = 1,2,\ldots n) \qquad (4)$$

A binary string s of n bits is used to encode the solution $x\in \{0,1\}^n$. Since many coding lead to infeasible solutions, a simple repair method is applied to the string s. The repair algorithm removes items from the solution coded by s step by step until all capacity constraints are fulfilled. The order in which the items are deleted is determined by the maximum profit/weight ratio per item. For item j the maximum profit/weight ratio q_j is given by $q_j=\max \{p_{ij}/w_{ij}\}$. The items are considered in increasing order of the q_j, i.e., those with the lowest ratio q_j are removed first.

Table 1. Parameters and Testing results

n	m	N′	N	I′	I	ΔI	n	m	N′	N	I′	I	ΔI	n	m	N′	N	I′	I	ΔI
	100	40	120	13.9	6.8	1.04		100	40	160	35	32.5	0.08		100	40	200	39.9	37.1	0.08
2	150	40	120	13.9	6.6	1.11	3	150	40	160	34.2	29.3	0.17	4	150	40	200	40	39.3	0.02
	200	40	120	19.9	6.6	2.02		200	40	160	35.2	20.7	0.70		200	40	200	40	39.4	0.02

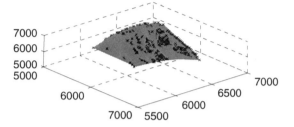

a) Pareto-optimal front distribution of 3 knapsacks using MOEAICS

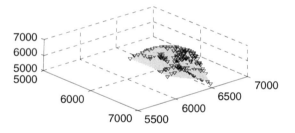

b) Pareto-optimal front distribution of 3 knapsacks using MOEACS

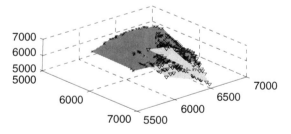

c) Dominant relationship between a) and b)

Fig. 1. Distributions of non-dominated solutions generated by MOEAICS and MOEACS for 0/1 knapsack problems with 3 knapsacks, the numbers beside the coordinate axes denotes the profit values of various knapsacks respectively.

In the testing, Single-point crossover and hyper-mutation is adopted. One-point crossover probability is set to 0.8 and mutation probability 0.01, and both are fixed. On all test multi-objective 0/1 knapsacks problems, 8000 generations were simulated per optimization run, and MOEAICS and MOEACS runs 20 times independently at the same initial population. We can measure the performance about algorithm relying on the increment of diversity of optimal solution and the number of optimal solution I' and I in external population by using MOEAICS and MOEACS, calculate the increment of diversity of optimal solution by $\Delta I = (I'-I)/I$. The size of population, number of bags and the number of goods refer to table 1(the size of population is N, the size of external population is N').

After 20 times running of each algorithm respectively, the statistics result about the increment of diversity of optimal solution refers to table 1, which strongly shows that MOEAICS can find out more optimal solutions compared to MOEACS. For the $n=3$, $m=200$, figure 1 is a curved surface which was composed of the final random sampling optimal solutions. Comparison between figure 1-a) and figure 1-b), MOEAICS' curved surface is wider than MOEACS', which shows that MOEAICS can search more and better distributed non-inferior solutions than MOEACS. From figure 1-c) we can see that the two curved surface is overlapping each other, from which can be known that the station of non-inferior solutions searched by MOEAICS is higher than MOEACS', and the Pareto front searched by MOEAICS be cover with MOEACS' mostly. According to figure 1, it shows that, compared to MOEACS, MOEAICS is more competent to get more evenly distributed Pareto optimal solutions with higher precision and convergence speed to Pareto front.

6 Conclusion

On the basis of traditional clonal selection algorithm, MOEAICS makes use of the excellent gene fragment of antibodies, and adds the operation of encapsulation of excellent gene fragment of antibodies, and replaces the low affinity antibody in memory antibody set with probability. By comparison of the performance of MOEAICS and MOEACS, the test shows that MOEAICS can not only avoid falling into local optimum, but also can obtain more accurate solutions.

Acknowledgments. This work is supported by Guizhou Provincial Natural Science Foundation of China (2010[2095]) and Program for New Century Excellent Talents in University of China (NCET09-0094).

References

1. Gong, M.J., Jiao, L.C., Yang, D.D., et al.: Research on Evolutionary Multi-Objective Optimization Algorithms. Journal of Software 20(2), 271–289 (2009)
2. Lozano, J.A., Zhang, Q.F., Larrinaga, P.: Special Issue on Evolutionary Algorithms Based on Probabilistic Models. IEEE Trans. Evolutionary Computation 13(6), 1197–1198 (2009)
3. Karshenas, H., Nikanjam, A.: Combinatorial Effects of Local Structures and Scoring Metrics in Bayesian Optimization Algorithm. In: 1st ACM/SIGEVO, New York, pp. 263–270 (2009)
4. Yao, J.T.: An Improving Strength Pareto Evolutionary Algorithm Based on Bayesian Network with Decision Graphs. Chinese J. of Computer 28(12), 1993–1999 (2005)
5. Mo, H.F.: The Principle and Application of Artificial of Artificial Immune System, Harbin (2004)
6. Li, S.B.: Genetic Programming and Creative Design of Mechatronic System, Beijing (2009)

Petroleum Contaminated Site Remedial Countermeasures Selection Using Fuzzy ANP Model

Wei Zhong Yang[1], Yu Hui Ge[1], Bin Xiong[2], and Xiang Bing Wang[1]

[1] School of Management, University of Shanghai for Science and Technology,
Shanghai, P.R. China, 200093
[2] College of Electric Engineering, Guangxi University,
Nanning, P.R. China, 530004
okyang95@gmail.com, viagrae@gmail.com, viagrae@sina.com,
viagre@163.com

Abstract. Advances in fuzzy analytic network process (ANP) are discussed to support decision making to derive the priorities of petroleum contaminated site remedial countermeasures (PCSRC). Conventional ANP approach seems to be incapable of capturing the vagueness and fuzziness during value judgment elicitation. The aim of this paper is to present an adaptive fuzzy ANP decision support method to make up for this shortcoming in the evaluation of PCSRC process. A numerical example is carried out to illustrate our approach.

Keywords: Site Remedial countermeasures, ANP, fuzzy number, prioritization.

1 Introduction

The selection of petroleum contaminated sites remedial countermeasures (PCSR) is inevitably a balancing act of many diverse factors such as social, economic, and technological issues. The decision making involving PCSR is a process characterized by complexity and uncertainty, as well as, multiple and conflicting criteria. The Analytic Network Process (ANP) have great potential for use in these practical evaluations its capability of modeling complex and the way in which comparisons are performed in complex situation [1]. Exact numerical values are replaced by fuzzy judgments for insufficiency and imprecision to incorporate the vagueness of human being [2], [3], and [4]. The fuzzy prioritization method purposed by Chang's extent analysis is ingrained into the ANP for the selection of PCSR [5], and [6].

2 Network Structure of PCSRC

A Chinese petroleum enterprise intended to select the most appropriate remedial alternatives for a contaminated site caused by a petroleum pipeline leak. Five remedial countermeasures are briefly described in Table 1. The evaluation criteria are defined as follows: c_1, the social acceptability of the countermeasures according to the perception of stakeholders; c_2, implementability in terms of administrative and

Y. Yu, Z. Yu, and J. Zhao (Eds.): CSEEE 2011, Part II, CCIS 159, pp. 224–229, 2011.

technological feasibility ; c_3, financial affordability with regards to the overall cost of the clean-up; c_4 , environmental effectiveness to protect public health and environment resources. Criteria are grounded into two clusters: external environment (including c_1 and c_4), and internal capabilities (including c_2 and c_3).

The network is formed as showed in Fig.1. All alternatives are influenced by above four defined criteria. The solid arrows represents the dependencies within the clusters resulted from criteria dependencies within the clusters, and the dotted arrows show connections between criteria within one cluster or two different clusters.

Table 1. The petroleum contaminated site remedial countermeasures

Remedial alternatives	Site remediation containing the waste layer	Prevention of contaminant spreading	Remediation of surrounding area
Alternative (A_1)	In situ disposal by incineration is a commercial possibility, if the volume of the petroleum-contaminated soil is large enough		In situ remediation (e.g. enhanced bioremediation, natural attenuation)
Alternative (A_2)	Complete removal of waste from the petroleum-contaminated soil from the site, and off-site treatment and disposal of the excavated waste		In situ remediation (e.g. enhanced bioremediation, natural attenuation)
Alternative (A_3)	In situ remediation (e.g., enhanced bioremediation, soil washing, etc.)	Capping and plume control (e.g., groundwater extraction)	In situ remediation (e.g. enhanced bioremediation, natural attenuation)
Alternative (A_4)	In situ remediation (e.g., enhanced bioremediation, soil washing, etc.)	Capping and vertical cut-off wall (e.g., sheet piling, chemical grout, etc.)	In situ remediation (e.g. enhanced bioremediation, natural attenuation)
Alternative (A_5)	In situ remediation (e.g. enhanced bioremediation, natural attenuation)		

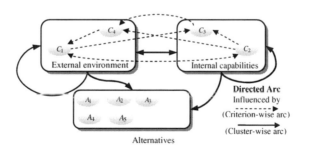

Fig. 1. Network of PCSRC selection

3   Fuzzy ANP Evaluation of PCSRC

3.1   Fuzzy Sets and Fuzzy Number

A triangular fuzzy number (TFN), $\tilde{M}$ is shown in Fig.2. A TFN is denoted simply as $(l.m.u)$. The parameters l , m and u , respectively, denote the smallest possible value, the most promising value, and the largest possible value that describe a fuzzy event. Each TFN has linear representation on its left and right side such that its membership function can be defined as the following:

$$\mu(x/\tilde{M}) = \{0, \ if \ x < l \ or \ x > u; \ (x-l)/(m-l), if \ l \le x \le m; \ (u-x)/(u-m), \ if \ m \le x \le u\}. \tag{1}$$

A fuzzy number can always be given by its corresponding left and right representation of each degree of membership (see Fig.2.):

$$\tilde{M} = (M^{l(\alpha)}, M^{u(\alpha)}) = (l + (m-l)\alpha, u + (m-u)\alpha) \ , \ \alpha \in [0,1] \ , \tag{2}$$

where $l(\alpha)$ and $u(\alpha)$ denote the left side representation and the right side representation of a fuzzy number, respectively.

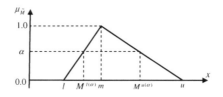

Fig. 2. A triangular fuzzy number $\tilde{M}$

3.2 Fuzzy Prioritization Method

Let $X = (x_1, x_2, \cdots, x_n)$ be an object set and $U = (u_1, u_2, \cdots, u_m)$ be a set of goal. According to the method of Chang's extent analysis, each object is taken and extent analysis for each goal is performed, respectively. Therefore, m extent analysis values for each object can be obtained with the following signs:

$$M_{gi}^1, M_{gi}^2, \cdots, M_{gi}^m \ , \ i = 1, 2, \cdots, n \ , \tag{3}$$

where all the M_{gi}^j ($j = 1, 2, \cdots, m$) are TFNs.

$$S_i = \sum_{j=1}^{m} M_{gi}^j \otimes \left[\sum_{i=1}^{n} \sum_{j=1}^{m} M_{gi}^j \right]^{-1}. \tag{4}$$

where $\left[\sum_{i=1}^{n} \sum_{j=1}^{m} M_{gi}^j \right]^{-1} = (\frac{1}{\sum_{i=1}^n u_i}, \frac{1}{\sum_{i=1}^n m_i}, \frac{1}{\sum_{i=1}^n l_i})$, as for fuzzy numbers.

Then, the degree of possibility for each convex fuzzy number s_i $i = 1, 2, \cdots, n$ to be greater than other convex fuzzy number s_j $j = 1, 2, \cdots, n$; is determined by next formula.

$$V(S_i \ge S_1, S_2, \cdots, S_n) = V[(S_i \ge S_1), and(S_i \ge S_2), and \cdots and(s_i \ge S_n)] = \min V(S_i \ge S_j) \ , \ i, j = 1, 2, \cdots, n \ ; i = j \ . \tag{5}$$

Assume that $s_1 = (l_1, m_1, u_1)$ and $s_2 = (l_2, m_2, u_2)$, then

$$V(S_2 \ge S_1) = \sup[\min(\mu_{s1}(x), \mu_{s1}(y))] = \left\{ 1, if \ m_2 \ge m_1; 0, if \ l_1 \ge u_1; \frac{l_1 - u_2}{(m_2 - u_2) - (m_1 - l_1)}, otherwise \right\}. \tag{6}$$

The non-fuzzy priority vector would be as the following:

$$\omega = (\min V(S_1 \geq S_j), \min V(S_2 \geq S_j), \cdots, \min V(S_n \geq S_j))^T .$$ (7)

The priority vector is normalized and used in the section 3.3.

3.3 Steps of the Proposed Fuzzy ANP for the Evaluation

The process of the fuzzy ANP for the evaluation comprises of the following steps:

Step 1: Construct pairwise matrices of the components with fuzzy judgments. The fuzzy scale regarding relative importance to measure the relative priorities is given in Fig. 3 and Table 2.

Step 2: Determine the local priorities from each matrix using the fuzzy prioritization method.

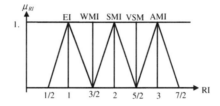

Fig. 3. Linguistic scale for relative importance

Table 2. Linguistic scale for relative importance

Linguistic scale for importance	Triangular fuzzy scale	Fuzzy reciprocal scale
Equally important	(1/2,1,3/2)	(2/3,1,2)
Weakly more important	(1,3/2,2)	(1/2,2/3,1)
Strongly more important	(3/2,2,5/2)	(2/5,2,2/3)
Very strongly more important	(2,5/2,3)	(1/3,2/5,1/2)
Absolutely more important	(5/2,3,7/2)	(2/7,1/3,2/5)

Step 3: Fill the super matrix with the elicited priorities to form limit weighted supermatrix in the SUPER DECISION SOFTWARE.

Step 4: Normalize the scores of alternatives from the limit weighted supermartrix into final priorities.

4 Calculation and Ranking

The comparison matrix of the alternatives with respect to social acceptability is demonstrated in Table 3, and the local priorities are calculated as follows:

$S_{A1} = (3.83.4.67,7.00) \otimes (1/18.50,1/25.50,1/36.00) = (0.106,0.183.0.378)$,

$S_{A2} = (4.00.6.00,7.00) \otimes (1/18.50,1/25.50,1/36.00) = (0.111,0.235,0.378)$

$S_{A3} = (3.17.5.00,7.50) \otimes (1/18.50,1/25.50,1/36.00) = (0.088,0.196,0.405)$

$S_{A4} = (3.83.5.50,8.50) \otimes (1/18.50,1/25.50,1/36.00) = (0.106,0.215,0.459)$,

$S_{A5} = (3.67.4.33,6.00) \otimes (1/18.50,1/25.50,1/36.00) = (0.101,0.170,0.324)$ are obtained.

Using this vector: $V(S_{A1} \geq S_{A2}) = 0.837$, $V(S_{A1} \geq S_{A3}) = 0.957$, $V(S_{A1} \geq S_{A4}) = 0.892$, $V(S_{A1} \geq S_{A5}) = 1.00$, $V(S_{A2} \geq S_{A1}) = 1.00$, $V(S_{A2} \geq S_{A3}) = 1.00$, $V(S_{A2} \geq S_{A4}) = 1.00$, $V(S_{A2} \geq S_{A5}) = 1.00$, $V(S_{A3} \geq S_{A1}) = 1.00$, $V(S_{A3} \geq S_{A2}) = 0.883$, $V(S_{A3} \geq S_{A4}) = 0.937$, $V(S_{A3} \geq S_{A5}) = 1.00$, $V(S_{A4} \geq S_{A1}) = 1.00$, $V(S_{A4} \geq S_{A2}) = 0.948$, $V(S_{A4} \geq S_{A3}) = 1.00$, $V(S_{A4} \geq S_{A5}) = 1.00$, $V(S_{A4} \geq S_{A1}) = 0.944$, $V(S_{A4} \geq S_{A2}) = 0.826$, $V(S_{A4} \geq S_{A3}) = 0.975$, $V(S_{A4} \geq S_{A5}) = 0.858$ are obtained.

Thus, the priority vector from Table 3 is calculated as $\omega = (0.41, 0.25, 0.15, 0.65, 0.56)^T$.

Table 3. Comparison matrix of alternatives with respect to social acceptability

Social acceptability	A_1	A_2	A_3	A_4	A_5	Local priorities
A_1	(1,1,1)	(1/2,2/3,1)	(2/3,1,2)	(2/3,1,2)	(1,1,1)	0.194
A_2	(1,3/2,2)	(1,1,1)	(1/2,1,3/2)	(1/2,1,3/2)	(1,3/2,2)	0.232
A_3	(1/2,1,3/2)	(2/3,1,2)	(1,1,1)	(1/2,1,3/2)	(1/2,1,3/2)	0.205
A_4	(1/2,1,3/2)	(2/3,1,2)	(2/3,1,2)	(1,1,1)	(1,3/2,2)	0.220
A_5	(1,1,1)	(1/2,2/3,1)	(2/3,1,2)	(1/2,2/3,1)	(1,1,1)	0.191

Table 4. Local priorities under five criteria for five alternatives

	A_1	A_2	A_3	A_4	A_5
Social acceptability	0.194	0.232	0.205	0.220	0.191
Implementability	0.299	0.042	0.046	0.362	0.251
Financial affordability	0.201	0.133	0.267	0.240	0.159
Environmental effectiveness	0.092	0.281	0.249	0.253	0.125

In Table 4, the local priorities of them are uniquely determined and there is no dominating alternative for all the criteria. Comparison matrices of the internal capability cluster under social acceptability, and the external environment cluster under implementability, are showed in Table 5-6. For cluster comparison, the resulted matrices together with their local priorities are shown in Table 7-8.

Table 5. Comparison matrix of the internal capabilities cluster under social acceptability

Social acceptability	IM	FA	local priorities
implementability	(1,1,1)	(2,5/2,3)	0.536
financial affordability	(1/3,2/5,1/2)	(1,1,1)	0.464

Table 6. Comparison matrix of the external environment cluster under implementability

Implementability	SA	EE	Local priorities
Social acceptability	(1,1,1)	(1/2,1,3/2)	0.500
Environmental effectiveness	(2/3,1,2)	(1,1,1)	0.500

Table 7. Comparison matrix of the clusters with respect to external environment

External environment	AL	IC	EE	Local priorities
Alternatives	(1,1,1)	(1/3,2/5,1/2)	(2/3,1,2)	0.059
Internal capabilities	(2,5/2,3)	(1,1,1)	(2/3,1,2)	0.588
External environment	(1/2,1,3/2)	(1/2,1,3/2)	(1,1,1)	0.353

Table 8. Comparison matrix of clusters with respect to internal capabilities

Internal capabilities	AL	IC	EE	local priorities
Alternatives	(1,1,1)	(1/2, 2/3,1)	(1/2,1, 3/2)	0.207
Internal capabilities	(1, 3/2,2)	(1,1,1)	(3/2,2, 5/2)	0.471
External environment	(2/3,1,2)	(2/5,1/2, 2/3)	(1,1,1)	0.322

The limit weighted supermatrix is formed by putting the local priorities in the SUPER DECISION SOFWARE. Its values in rows in front of the alternatives are final scores of the alternatives, and can be normalized into final priorities, $\omega_{final} = (0.198, 0.155, 0.213, 0.261, 0.174)$ of all alternatives, their ranking is followings: $A_4 \succ A_3 \succ A_1 \succ A_5 \succ A_1$. Based on this analysis, A_4 is identified as the most preferred alternative.

5 Conclusions

Petroleum contaminated land management is an important issue throughout China. The need for developing techniques and approaches to improve the decision making process for remediation of petroleum contaminated land is widely recognized. In this paper, a fuzzy ANP decision model has been systematically developed and applied to the evaluation of PCSRC in China. Fuzzy ANP could adequately handle the judgments derived from perception-based information which are intrinsically imprecise, reflecting the bounded ability of human mind to resolve detail and store information.

Acknowledgement

This research is funded by Research and Innovation Key Project (10ZS96) of Shanghai Education Commission, and Shanghai Key Disciplines (Phase III) (No.S30504).

References

1. Promentilla, M.A.B., Furuichi, T., Ishii, K., Tanikawa, N.: Evaluation of Remedial Countermeasures Using the Analytic Network Process. Waste Manage 26(21), 1410–1421 (2006)
2. Razmi, J., Sangari, M.S., Ghodsi, R.: Developing a Practical Framework for ERP Readiness Assessment Using Fuzzy Analytic Network Process. Adv. Eng. Software 40(1), 1168–1178 (2009)
3. Tuzkaya, U.R., Andönüt, S.: A Fuzzy Analytic Network Process Based Approach to Transportation-mode Selection between Turkey and Germany: a Case Study. Info. Sci. 178(14), 3133–3146 (2008)
4. Tang, Y.C.: An Approach to Budget Allocation for an Aerospace Company-Fuzzy Analytic Hierarchy Process and Artificial Neural Network. Neurocomputing 72(20), 3477–3489 (2009)
5. Chang, D.Y.: In Extent Analysis and Synthetic Decision. Optimization Techniques and Applications 1, 352–390 (1992)
6. Chang, D.Y.: Applications of the Extent Analysis Method on Fuzzy AHP. Eur. J. Oper. Res. 95(12), 649–655 (1996)

Palmprint Recognition Using Gabor Feature-Based Bidirectional 2DLDA

Feng Du[1,*], Pengfei Yu[1], Hongsong Li[1], and Liqing Zhu[2]

[1] School of Information Science, Yunnan University, Kunming, 650091, China
[2] Department of Electrical Engineering, Yunnan Vocational college of Mechanical and Electrical Technology, Kunming, 650203, China

Abstract. A procedure for palmprint recognition and a novel fusion strategy are proposed in this paper. The procedure is based on a combination of Gabor filters and bidirectional two dimensional linear discriminant analysis' (2DLDA) fusion. We apply horizontal 2DLDA (H2DLDA or 2DLDA) and vertical 2DLDA (V2DLDA) to extract two kinds of features of Gabor-based images (Gaborpalms): one kind is composed of horizontal discriminant features mainly extracted by 2DLDA, and the other is composed of the vertical features extracted by V2DLDA. To fuse these two kinds of features together, a distance-based adaptive strategy is designed. Finally the nearest neighbor classifier is used for classification. Using the Gaborpalms of a little higher dimension which means low computational cost, experimental results on PolyU Palmprint Database demonstrate that the proposed procedure delivers several desired results.

Keywords: Palmprint Recognition, Gabor Filters, Bidirectional 2DLDA, Adaptive Fusion Strategy.

1 Introduction

Palmprint recognition has rapidly developed in recent years and is becoming an active research field of biometrics recognition technologies. The methods of palmprint recognition are mainly classified to line-based, subspace-based and statistical methods [1]. LDA [2] and 2DLDA are two important subspace-based methods. LDA is to seek the discriminant vectors of making the ratio of the between-class distance to the within-class distance the larger. However, LDA often surfers from the small sample size problem, which leads the within-class scatter matrix to be singular. Matrix-based methods have been well developed to deal with the problem. Li et al. [3] proposed 2DLDA which directly calculates the scatter matrices from matrices and extracts the features based on Fisher's criterion. Moreover, compared with LDA, 2DLDA (or V2DLDA) has a lower computational cost because the dimensions of its scatter matrices are lower than LDA's and can well preserve the spatial structure information of the matrices. But 2DLDA is sensitive to variations caused by illumination and rotation of original images. Using the Gabor features of original palmprint images instead of the palmprint images themselves can overcome the problem. So many literatures [4,5] have reported using Gabor

* This work is supported by Science Research Foundation of Yunnan Provincial Education Department (Grant No. 08Y0032) and training program of key teachers in Yunnan University.

Y. Yu, Z. Yu, and J. Zhao (Eds.): CSEEE 2011, Part II, CCIS 159, pp. 230–235, 2011.

filters to extract the texture features of palmprint images. 2DLDA and V2DLDA, which complements each other, can extract two kinds of features from Gaborpalms. To get a higher recognition rate, some researchers fused these two kinds of features together, for example, Du Haishun et al. [6] proposed using bidirectional 2DLDA's fusion directly for face recognition.

In order to make the fusion more effective, a new fusion strategy, which computes the coefficients based on bidirectional distances, is developed in this paper. First, Gabor feature matrices of different scales and orientations are extracted by the convolution of Gabor filters and a palmprint ROI (Region Of Interest, the central part of the palmprint image with a size of 100×100). Features extracted from them are reorganized to a new feature matrix, which is called Gaborpalm of the original image. Subsequently, 2DLDA and V2DLDA are used respectively to reduce the dimension of Gaborpalm and to obtain bidirectional discriminant features. Third, a special strategy is applied to conduct the feature fusion. Finally, the nearest neighbor classifier is used for classification. Our experimental results present its stronger adaptiveness. The entire recognition procedure is shown in Fig. 1.

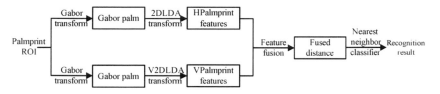

Fig. 1. Palmprint recognition procedure for the proposed method

2   Gabor Filters for Extracting Features

2.1   Gabor Filters

Choose the following Gabor form:

$$G_{f,\theta,\sigma_x,\sigma_y}(x,y) = \frac{1}{2\pi\sigma_x\sigma_y} \exp\left(\frac{x^2}{2\sigma_x^2} \quad \frac{y^2}{2\sigma_y^2} \right) \exp(2\pi f j(x\cos\theta + y\sin\theta)). \quad (1)$$

where f is the central frequency of each filter. θ controls their orientation. σ_x and σ_y are each filter's sharpness parallel to major axis and minor axis. j equals to $\sqrt{-1}$. To extract useful features of palmprint ROIs, a set of Gabor filters on various orientations and scales are required [7]:

$$\sigma_x = \sigma_y = 5.6179, \theta = \frac{u}{6}\pi, f = \frac{0.2592}{\sqrt{2^v}}. \quad (2)$$

where $u, v = 0, 1, 2, 3, 4, 5$. Thus we can get 36 filters which are denoted by $G_{u,v}$.

2.2   Gabor Representation of Palmprint ROI

Gabor features of a palmprint ROI I can be obtained by convolution of the image and $G_{u,v}$:

$$H_{u,v} = I * G_{u,v} . \tag{3}$$

Compute convolution magnitude $|H_{u,v}|$. So I can be represented by a set of Gabor coefficients. Normalize $|H_{u,v}|$ and then they are mapped to between $[0, 255]$ so that they can be stored and be shown as grayscale images which are denoted as $H'_{u,v}$. Concatenate 36 $H'_{u,v}$ by 6 results each row and each column to build a matrix. The dimension of this matrix (552×552) is quite high. To solve this problem, it is downsampled by one per four rows and four columns to get an 138×138 matrix denoted by Gaborpalm. Fig. 2 shows Gaborpalms of different samples of the same class.

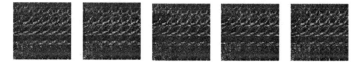

Fig. 2. Gaborpalms of different samples of the same class

3   Bidirectional 2DLDA for Dimensionality Reducing of Gaborpalms

Suppose there are n training Gaborpalms, denoted by $A_1, A_2, \cdots A_n$, who have uniform dimension $m \times n$. They can be classified to c classes. The jth class is denoted by w_j, which represents a set of samples of the jth class. A_k^j is the kth sample of w_j and n_j is the sample number owning to w_j.

Their between-class scatter matrix and within-class scatter are evaluated by

$$S_B = \frac{1}{n} \sum_{j=1}^{c} n_j (m_j - m)^T (m_j - m) . \tag{4}$$

$$S_W = \frac{1}{n} \sum_{j=1}^{c} \sum_{k=1}^{n_j} \left(A_k^j - m_j \right)^T \left(A_k^j - m_j \right) . \tag{5}$$

where

$$m_j = \frac{1}{n_j} \sum_{k=1}^{n_j} A_k^j, \; m = \frac{1}{n} \sum_{j=1}^{c} \sum_{k=1}^{n_j} A_k^j . \tag{6}$$

2DLDA aims to find some projection vectors to make the $J(x)$ the larger. These vectors organize a projection matrix which is optimal to separate subjects from different classes. $J(x)$ is defined as

$$J(X) = \frac{X^T S_B X}{X^T S_W X}. \tag{7}$$

which is named general fisher's criterion.

Based on the theory of matrices, the optimal projection vectors $XH_1, \cdots, XH_{d_1}$ are the normalized eigenvectors of $S_w^{-1} S_b$ corresponding to its d_1 larger eigenvalues. We select them to construct an optimal projection matrix $[XH_1, \cdots, XH_{d_1}]$ with a dimension of $n \times d_1$. Consequently, a discriminant matrix $YH_k^j(d_1)$ with lower dimension $m \times d_1$ is computed by

$$YH_k^j(d_1) = A_k^j[XH_1, XH_2, \cdots, XH_{d_1}]. \tag{8}$$

It is preserved as the horizontal discriminant characteristic of the kth palmprint image in the jth class.

Analogously, as for V2DLDA, the Gaborpalms' transposes $(A_1)^T, \cdots, (A_n)^T$ are obtained instead of $A_1, \cdots, A_n$, using the same algorithm:

$$YV_k^j(d_2) = (A_k^j)^T[XV_1, XV_2, \cdots, XV_{d_2}]. \tag{9}$$

with a size $n \times d_2$ can be extracted as the vertical discriminant characteristic.

4   The Design of Classifier

When having a palmprint image (ROI) I to be classified, its normalized Gaborpalm can be easily obtained, then $IH(d_1)$ and $IV(d_2)$, which can be got the same as getting $YH_k^j(d_1)$ and $YV_k^j(d_2)$. Using Frobenius norm, we have

$$DH_k^j = \| YH_k^j(d_1) - IH(d_1) \| . \tag{10}$$

$$DV_k^j = \| YV_k^j(d_2) - IV(d_2) \| . \tag{11}$$

Then

$$D_k^j = (min(DV_k^j)/min(DH_k^j))DH_k^j + DV_k^j. \tag{12}$$

is be computed as the discriminant distance, where $min(DV_k^j)$ and $min(DH_k^j)$ are the minimal values of DV_k^j and DH_k^j respectively, $min(DV_k^j)/min(DH_k^j)$ is the adaptive coefficient which can normalize DH_k^j to DV_k^j's level.

Finally, the nearest neighbor criterion is applied to decide which class it belongs to.

5   Experimental Results

The PolyU Palmprint Database contains 7752 grayscale images corresponding to 386 different palms in BMP image format. About twenty samples from each palm were collected in two sessions, where around 10 samples were captured in the first session and the second session respectively. The average interval between the first and the second collection was two months. 2000 images from 100 subjects are selected for our expriments. Each subject has 10 images in each session. Images from one session are selected randomly for training and the whole of the other session for testing. Taking into account the computational cost, we simply take $d_1 = d_2 = 2$ in the following experiments.

Table 1 presents comparison of recognition rates using 2DLDA, V2DLDA, FusionHV (Fusion of 2DLDA and V2DLDA) on original images and G2DLDA, GV2DLDA, GFusionHV (our method) on Gaborpalms. The comparison of Gabor-based methods is shown in Fig. 3.

Table 1. Comparison of recognition rates (%) of several methods

Method	Training sample number per class					
	2	3	4	5	6	7
2DLDA	39.48	41.05	42.17	43.19	45.05	44.87
V2DLDA	49.62	50.94	51.13	52.06	53.89	54.05
FusionHV	52.08	53.52	54.23	55.01	56.77	56.96
G2DLDA	93.58	95.14	95.32	96.51	96.76	96.92
GV2DLDA	88.44	91.29	91.77	92.73	92.88	93.29
GFusionHV	95.07	96.85	97.12	97.81	98.01	98.27

The experimental results show that the recognition rates of Gabor-based methods are much higher than those of direct recognition, which demonstrates the effectiveness of Gabor features' extracting. The proposed method GFusionHV obtains a series of highest recognition rates. You may find it is curious that the recognition rates of 2DLDA are lower than those of V2DLDA while those of G2DLDA are higher .In fact it is

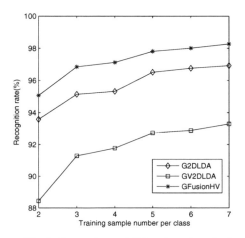

Fig. 3. Comparison of Gabor-based methods

reasonable because the recognition rates of G2DLDA and GV2DLDA don't have direct relationship with original palmprint images and they are only depened on the placing order of Gabor features.

6 Conclusion

This paper presents a novel fusion strategy and a better recognition procedure (than other listed methods). Because of the use of Gabor features which are robust to variations caused by illumination and rotation of original images and the use of the fusion strategy, the recognition rates are promoted greatly. Using a little higher dimension Gaborpalms and a very few projection vectors means not more computational cost is spent on. The better recognition performance shown by our experimental results testifies the effectiveness of the proposed method.

References

1. Kong, A., Zhang, D., Kamel, M.: A Survey of Palmprint Recognition. Pattern Recognition 42(7), 1408–1418 (2009)
2. Wu, X., Zhang, D., Wang, K.: Fisherpalms Based Palmprint Recognition. Pattern Recognition Letters 24(15), 2829–2838 (2003)
3. Li, M., Yuan, B.: 2D-LDA: A Novel Statistical Linear Discriminant Analysis for Image Matrix. Pattern Recognition Letters 26, 527–532 (2005)
4. Kong, W.K., Zhang, D., Li, W.: Palmprint Feature Extraction Using 2-D Gabor Filters. Pattern Recognition 36, 2339–2347 (2003)
5. Pan, X., Ruan, Q.-Q.: Palmprint Recognition Using Gabor-Based Local Invariant Features. Neurocomputing 72, 2040–2045 (2009)
6. Haishun, D., Xiuli, C., Fengquan, W., Fan, Z.: Face Recognition Using a Fusion Method Based on Bidirectional 2DLDA. Chinese Journal of Scientific Instrument 30(9), 1880–1885 (2009)
7. Pan, X., Ruan, Q.-Q.: Palmprint Recognition Using Gabor Feature-Based $(2D)^2$PCA. Neurocomputing 71, 3032–3036 (2008)

Combination of Key Information Extracting with Spoken Document Classification Based on Lattice

Lei Zhang, Zhuo Zhang, and Xue-zhi Xiang

Information and Communication Engineering College,
Harbin Engineering University, Harbin, China
{zhanglei,zhangzhuo,xiangxuezhi}@hrbeu.edu.cn

Abstract. Traditionally, the query words in spoken document classification are generated by manual. Here, based on CHI, TFIDF and maximum poster probability (MPP) features, key information extraction is combined with spoken document classification system, where different class has different topic. From the extraction, the weights of the same key word in each topic may be distinct. These weights which reveal the relationship between the word and topic can be taken part in spoken document classification system. Additionally, in the classification system, document length information is adopted when no query is found. The whole classification system is based on lattice, which has more information than 1-best result in speech recognition system. Among CHI, TFIDF and MPP, the system performance of MPP is a little worse than the others. CHI is a little better than TFIDF when the key words number is increasing. Experiments show that when the system is combined weight and document length information, the best performance can achieve 0.769 MAP.

Keywords: Spoken Document Classification, Key Information Extraction, Lattice.

1 Introduction

Text-based searching engine has been applied widely, and many technologies such as automatic summarization [1], semantic extraction [2] based on text documents are studied in detail. However, how to integrate these approaches into spoken document processing is still a challenge. Most studies of Spoken Document Retrieval (SDR) use the speech recognizer to generate approximate transcripts and just to apply the text-based information retrieval techniques directly [3]. But for broadcasts and conversation data, the low recognition rate can worsen the performance of classification system. Lattice can reduce the impact of the error rate to some extent by providing multiple hypothesis [4-5]. We have shown the improvement of lattice for spoken document classification system in [6]. Further, in that work, spoken document length is firstly applied in spoken document classification.

Another important problem in the spoken classification system is about the key information extraction, or definition of query words.

There are many approaches to extract the key information from text, such as document frequency (DF), the χ^2 statistics (CHI), term strength (TS), mutual

Y. Yu, Z. Yu, and J. Zhao (Eds.): CSEEE 2011, Part II, CCIS 159, pp. 236–241, 2011.
© Springer-Verlag Berlin Heidelberg 2011

information (MI), and information gain(IG). In [7], the performance of these features is compared in text categorization. [8] gives the comparison among IG, CHI and the maximum posterior probability(MPP) measure. Analysis in [7,8] reveals that DF, IG and CHI scores of a term are strongly correlated, and MPP can get better performance in topic identification. So here, CHI and MPP are selected as the baseline approaches to extract key information. Additionally, *TFIDF* as the most common feature in text-based retrieval is also adopted here. In fact, different feature can reflect different aspect of the key information.

However, in most classification systems, the queries of each topic are assigned by manual, and these queries play the same roles during classification. Different from the queries obtained by manual, the queries extracted automatically have distinct weights. This kind of weight can reveal the relationship between query and topic. In fact, for each topic, some queries may be more important than others. That means even for the same topic, different query word can play different role in the classification. The main contribution of this paper is to combine this kind of information into the classification system. The basic system is introduced in section 2, and some key information extraction approaches are described in section 3. In section 3, the weight which can reflect the different influence is combined in classification system. At last, the experiment and results are listed in section 4.

2 System Framework

The whole system has three parts. The first one named off-line part converts the speech signal into lattice. In the first part, HMM model is built by syllable. Since there exist more than 80,000 commonly used words and more than 10,000 commonly used characters in Mandarin Chinese, it is hard to construct the recognition model based on words or characters. Furthermore, all characters are monosyllabic, and for Chinese, there are many homophones, then the total number of phonologically allowed syllables with tone is only 1345 [9]. So here, combined with language model smoothed by modified Katz, the system output the syllable lattice.

The second part is the classification system based on lattice. According to different queries, the relevance between lattice and queries are computed. Combined with document length, the system performance can be further enhanced.

Given a query $\mathbf{q}^t = (q^t_1 q^t_2 ... q^t_m)$ belonging to a topic t, the relevance of document D to $\mathbf{q}^t$ can be defined by the probability $P(D|\mathbf{q}^t)$ in (1).

$$P(D|\mathbf{q}^t) = P(\mathbf{q}^t|D)P(D)/P(\mathbf{q}^t)$$

$$= \frac{P(q_1 q_2 ... q_m|D)P(D)}{P(\mathbf{q})} \approx \prod_{w^t} (\frac{P(w^t|D)}{P(w^t)})^{c(w^t,\mathbf{q}^t)} P(D) \qquad (1)$$

Where $c(w^t,\mathbf{q}^t)$ is the word count of w in $\mathbf{q}$ belonging to topic t, $P(w^t)$ is always considered as uniform, and can be dropped for document ranking purpose. The probability of word w occurring in document D is $P(w^t|D)$, which can be estimated by maximum likelihood algorithm. As for the priors probability of documents, $P(D)$

is equal for every document. In (1), it can be drawn that whether the query word occurring in document or not, $P(D)$ has no effect for classification. In text retrieval method, there is a view that document prior probabilities depend on document length. Many researches tried to establish a connection between the likelihood of relevance and document length. The results in [10] confirm that the prior probability is proportional to document length. That means, the longer documents span more topics and are more likely to be relevant, if no query has been seen. So in our classification system, the document length information is considered. Considering the size and computing speed of lattice, each speech document is divided into M segments, then (1) is turned into (2).

$$P(D|q') \approx \begin{cases} \prod_w \sum_{k=1}^{M} P(w' | D_k)^{c(w',q')} & \text{any query} \\[2em] \prod_w \sum_{k=1}^{M} P(w' | D_k)^{c(w',q')} \times \dfrac{\sum_{k=1}^{M} E[| D_k |]}{L} & \text{no query} \end{cases} \qquad (2)$$

Where D_k is represented as lattice, it is the k-th segment in spoken document D. L is the whole length of all documents. Since there are many candidates in lattice, the expectation must be considered here. In (2), the document length is combined in classification. When there is no query occurring in lattice, the longer document is more likely to relate to the topic. But if there are some queries happening in document, the effect of document length can be ignored. The expected document length $E[| D_k |]$ is computed as (3).

$$E[| D_k |] = \sum_p | p | P(p | O_k) \qquad (3)$$

Where $| p |$ denotes the number of words in path p. And O_k is the corresponding utterance.

In order to handle the zero probability, $P(w' | D_k)$ in (2) is computed by Jelinek-Mercer (JM) method [11], in which an interpretation of the maximum likelihood model with the priori probability of word w is adopted.

The last part in the whole system is for key information extraction. Furthermore, this kind of information extracted is combined into the spoken document classification system.

3   Key Information Extraction

Here, three features as CHI-square, MPP and *TFIDF* are selected to extract the key information related to topic.

Supposing that the relationship between term m and topic t is independent, CHI-square test can be acted as (4). Let $e_t = 0$ denote the document is not in topic, and $e_t = 1$ is the document in topic. Similarly, $e_m = 0$ means the document does not contain term m. Then according to the value of e_t and e_m, $k_{e_t e_m}$ denotes the number of document under all conditions. n is the sum of k_{11}, k_{10}, k_{01} ,and k_{00} .

$$weight_{CHI} = \frac{n(k_{11}k_{00} - k_{10}k_{01})^2}{(k_{11} + k_{10})(k_{01} + k_{00})(k_{11} + k_{01})(k_{10} + k_{00})} \tag{4}$$

For MPP, the larger $P(t \mid m)$ is, the more important the term m is. But if m does not occur in topic t, the probability will be zero. So here, $P_{map}(t \mid m)$ is adopted as follows, which can handle the zero probability.

$$weight_{mpp} = P_{map}(t \mid m) = \frac{N_{mlt} + \alpha_1 N_t P_{map}(t)}{N_m + \alpha_1 N_t} \tag{5}$$

Where N_{mlt} is the occurring number of term m in topic t, N_t is the number of distinct topics. N_m is the number of m in training corpus. Further, $P_{map}(t)$ is as

$$P_{map}(t) = \frac{N_{dlt} + \alpha_2}{N_d + \alpha_2 N_t} \tag{6}$$

Similarly, the N_{dlt} is the number of document in topic t, and N_d is the whole number of document.

TFIDF is a kind of widely used weight in text-document retrieval. The value is calculated as

$$weight_{tfidf} = N_{mlt} \times \log(N / W(m)) \tag{7}$$

Where N is the total number of documents, $W(m)$ is the whole number of document containing term m. (7) is a little different with traditional *TFIDF*. Here, $W(m)$ can also present the character of inverse information. If this value is large, that means this term also happens not only in topic t, but other classes. Thus, the effect of this term should be weakened

Further, the weight is combined into classification system, and (2) is changed as

$$P(D \mid q') \approx \begin{cases} \sum_w weight_w \times \log(\bar{P}) & \text{any query} \\ \sum_w weight_w \times \log(\bar{P}) \times \dfrac{\sum_{k=1}^{M} E[\mid D_k \mid]}{L} & \text{no query} \end{cases} \tag{8}$$

Where $\sum_{k=1}^{M} P(w' \mid D_k)^{c(w',q')}$ is represented as $\bar{P}$. And $weight_w$ means any weight as above of query word w.

4 Experiments and Conclusion

HTK is employed to train acoustic and language models based on "863" corpus and part of broadcast corpus as adaptation data by maximum likelihood linear regression and maximum a posterior approach. The training set consists of 90821 utterances from 156 speakers, and the testing set and adapt database include 5924 utterances from broadcast programs, which are classified into six topics, such as national defense, sport, countryside, law, economy and politics. The broadcast corpus for retrieval in our experiments is from the programs of CCTV, which include the conversation programs, news and so on. Here, the evaluation is depended on mean average precision (MAP).

Table 1. MAP of different features without document length and weight

	CHI	TFIDF	MPP
1 key word	0.1982	0.2240	0.2290
5 key words	0.4551	0.5125	0.4966
10 key words	0.6365	0.6358	0.6280
20 key words	0.7317	0.7314	0.6928
30 key words	0.7338	0.7544	0.7126
50 key words	0.7535	0.7546	0.7166

Table 1 is based on the classification system without the document length and weight. From Table I, it can be seen that no matter what kind of feature is adopted, with the increasing of the key words number, MAP is increasing too. Among CHI, *TFIDF* and MPP, the system performance of MPP is a little worse than those of the others. And CHI is a little better than *TFIDF* when the key words number is increasing. But when the query words are enough, *TFIDF* can get the best performance. So in next experiment, *TFIDF* is used to test the performance of weight and document length effect.

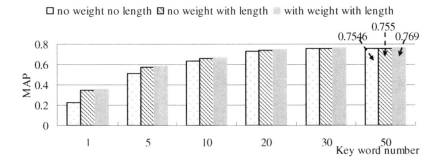

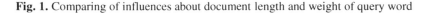

Fig. 1. Comparing of influences about document length and weight of query word

From this figure, it can be seen that considering the document length will obviously improve the performance of system, especially when the key word number is small. The reason is that only when the key word did not happen in the document, the document length can act. With the increasing of key word number, this kind of case seldom occurs. So the influence of document length weakens with the increasing of key word number.

As for the effect of weight, it can be seen in fig.1, the performance can be further enhanced when weight is considered. For 50 key words and combined with the weight and document length, the best MAP of 0.769 can be achieve, which has some improvement of only document length acting.

All in all, combining the key information extraction into classification can consider the different contribution of different key word, which can enhance the system performance further.

Acknowledgments. It is a project supported by National Natural Science Foundation of China #60702053, also supported by young teacher supporting plan by Heilongjiang province in China #1155G17 and young teacher supporting plan by Harbin Engineering University.

References

1. Chang, Y.-L., Chen, J.-T.: Latent Dirichlet Learning for Document Summarization. In: ICASSP 2009, pp. 1689–1692. IEEE Press, TaiWan (2009)
2. Malian, J.T., Throop, D.R.: Basic Concepts and Distinctions for an Aerospace Ontology of Functions, Entities and Problems. In: Aerospace Conf., pp. 1–18. IEEE Press, Big Sky (2007)
3. Chen, B., Wang, H.-M., Lee, L.-S.: Retrieval of Broadcast News Speech in Mandarin Chinese Collected in Taiwan Using Syllable-level Statistical Characteristics. In: ICASSP 2000, pp. 1771–1774. IEEE Press, Istanbul (2000)
4. Meng, C.-H., Lee, H.-Y., Lee, L.-S.: Improved Lattice-based Spoken Document Retrieval by Directly Learning from the Evaluation Measures. In: ICASSP 2009, pp. 4893–4896. IEEE Press, Taiwan (2009)
5. Mertens, T., Schneider, D.: Efficient Sub-word Lattice Retrieval for German Spoken Term Detection. In: ICASSP2009, pp. 4885–4888. IEEE Press, Taipei (2009)
6. Zhang, L., Gao, Y., Xang, X., Lu, D.: A New Syllable-lattice Based Approach for Mandarin Spoken Document Retrieval. In: Wireless Communications & Signal Processing, pp. 1–4. IEEE Press, ShangHai (2009)
7. Yang, Y.-M., Pedersen, J.O.: A Comparative Study on Feature Selection in Text Categorization. In: Proc. ICML-14, pp. 12–420 (1997)
8. Hazen, T.J., Richardson, F., Margolis, A.: Anna Margolis.: Topic Identification from Audio Recordings using Word and Phone Recognition Lattices. In: ASRU 2007, pp.659–664 (2007)
9. Chen, B., Wang, H.M., Lee, L.S.: A Discriminative HMM/N-gram-based Retrieval Approach for Mandarin Spoken Documents. ACM Trans. Asian Lang. Inform. Process. 3(2), 128–145 (2004)
10. Blanco, R., Barreiro, Á.: Probabilistic Document Length Priors for Language Models. In: Macdonald, C., Ounis, I., Plachouras, V., Ruthven, I., White, R.W. (eds.) ECIR 2008. LNCS, vol. 4956, pp. 394–405. Springer, Heidelberg (2008)
11. Zhai, C.-X., Lafferty, J.: A Study of Smoothing Methods for Language Models Applied to Information Retrieval. ACM Trans. Information Systems 22(2), 179–214 (2004)

Quickly Determining the Method of Chaotic Synchronization System and Its Application in Spread Spectrum Communication System

Jing Pang[1], Qun Ding[2], ShuangChen Su[1], and Guoyong Zhang[1]

[1] Hebei University of Technology, Langfang, 065000, China
[2] Heilongjiang University, Harbin, 150080, China
pangjing2002@sohu.com, ding-qun@263.net,
sushuangchen@sina.com, zgy660419@163.com

Abstract. According to the two situation of the same-structure chaos system, the paper demonstrates the method of confirming response equation of the system only by the way of observing rather than deducing. Then verifies the correction of the method through the simulation to lerenz and chen system. At last, the determinate synchronization system is applied to the spread spectrum communication: through the compression and coding, the solution to the drive and response equation is transferred into PN code of the spread spectrum communication. The code meet the requirements of spread spectrum communication through verify the feature of the code, and this can ensure the synchronism of the transmitting bit.

Keywords: Chaos, Synchronization, Linear Decomposition, Spread Spectrum Communication.

1 Introduction

The concept of chaos synchronization [1] is proposed in 1990, synchronization of two systems driven by uniform signal with circuit was achieved in the laboratory. Nowadays people have done already many in-depth studies in chaos synchronization, there are studies in the same-structure chaos system and there are also studies in different-structure chaotic system. And the chaotic synchronization of different types is achieved in the different chaotic systems. Such as, complete synchronization, generalized synchronization, phase synchronization, delay synchronization, projective synchronization [2].At present, some scholars have proposed the anti-synchronization of chaotic system for different-structures [3] [4], and this has been applied to the communication field [2].

This article describes a method to quickly determine the synchronization system response equation by observing. This method is that decomposing the chaotic system into the linear part and nonlinear part, and then overlaid. Moreover turning the vector of the linear part into two type :one is that the main diagonal elements which are all negative, and the other is not all negative, then define and illustrate respectively and prove its validity. Through two simultaneous system, lorenz and chen. At last, through

Y. Yu, Z. Yu, and J. Zhao (Eds.): CSEEE 2011, Part II, CCIS 159, pp. 242–246, 2011.
© Springer-Verlag Berlin Heidelberg 2011

compressing and coding the solution of the drive and response equations and transfer it into spreading PN code and dispreading PN code. Then it is proved that the bit error ratio, meet the requirements of spread spectrum communication.

Chaos system is widely used in research of secure communication and spread spectrum communication system, if the systems achieve synchronism by implementing in hardware, then the mathematical models of the two systems are closer, the circuitdesign is simpler. So studying the system has certain practical and theoretical signifi-cance.

2   Determination of Response System Equation and Examples

Autonomous system can be described:

$$\dot{x}_i = f(x_i) \tag{1}$$

x_i is the n-dimension state vector of the system, f is vector space domain[3,4] of n-dimension space. This system can be decomposed into[2,3]

$$\dot{x}_i = A_1 x_i + f_1(x_i) \tag{2}$$

Suppose that the driving system is decomposed into $\dot{x}_i = A_1 x_i + f_1(x_i)$, which matrix A_1 is the coefficient of linear term of chaos model, $f_1(x_i)$ is the nonlinear term.

2.1   Determination of System Response Equation Whose Main Diagonal Elements Are Not All Negative

Definition 1. If elements of main diagonal line on matrix A1 are not all negative, then we can continue to divide A1 into $A_{11}+A_{12}+A_{13}$.The driving system is:

$$\dot{x}_i = A_{11} x_i + A_{12} x_i + A_{13} x_i + f_1(x_i) \tag{3}$$

A_{11} is the term that only includes the main diagonal elements and the elements are all negative, A_{13} is the term whose values are twice for corresponding ones of positive diagonal elements for A_{11},and $A_{12}=A_1-A_{11}-A_{13}$, thus response system is

$$y_i = (A_{11} - A_{13}) y_i + (A_{12} + A_{13}) x_i + f_1(x_i) \tag{4}$$

Here, as long as the selection of suitable synchronous controller u_i would make state of (3)and (4) are under any initial state, when$(x(0),y(0))$, $\displaystyle \lim_{t \to \infty} \|y(t) - x(t)\| = 0$ is

established then, chaos systems (3) and (4) can achieve synchronization.

Prove 1. Driving system   is supposedas : $\dot{x}_i = A_{11} x_i + A_{12} x_i + A_{13} x_i + f_1(x_i)$.

whileresponse   system   is $\dot{y}_i = A_{11} y_i + A_{12} y_i + A_{13} y_i + f_2(y_i) + u_i$,   and $A_1 = A_{11} + A_{12} + A_{13}$, order system synchronization error of (3) and (4) is: $\dot{e}_i = \dot{y}_i - \dot{x}_i$ The dynamic system error is as below:

Order:
$$\dot{e}_i = \dot{y}_i - \dot{x}_i = A_{11}e + A_{12}(y_i - x_i) + A_{13}(y_i - x_i) + f_2(y_i) - f_1(x_i) + u_i$$
$$u_i = -A_{12}(y_i - x_i) - A_{13}(y_i - x_i) - f_2(y_i) + f_1(x_i)$$

$$\dot{e}_i = A_{11}e_i \tag{5}$$

To select function Lyapunov $V = 1/2(e_1^2 + e_2^2 + e_3^2)$, according to Eq.5, we can get $\dot{V} = e\dot{e} = A_{11}e^2$ .Moreover A_{11} only has the main diagonal elements and these elements are all negative, namely, $\lambda_1, \lambda_2, \cdots \lambda_i) < 0$ $\dot{V} \le 0$, namely when $e_i = 0$, $V = 0$ .So the drive system and response systems are synchronal. Response system equation is Eq.4

2.2   Determ Ination of Response System Equation Whose Main Diagonal Elements Are All Negative

Definition 2: If the main diagonal elements of A_1 are negative, you can continue decomposing matrix A_1 into $A_{11} + A_{12}$, and then the drive system is

$$\dot{x}_i = A_{11}x_i + A_{12}x_i + f_1(x_i) \tag{6}$$

A_{11} is the term whose main diagonal elements are all negative and only contains the main diagonal elements, $A_{12}=A_1-A_{11}$,Then response system is

$$\dot{y}_i = A_{11}y_i + A_{12}x_i + f_1(x_i) \tag{7}$$

Here, as long as the selection of suitable synchronous controller u_i would make state of Eq.6 and Eq.7 are under any initial state, when $(x(0),y(0))$,the $\lim_{t \to \infty} \|y(t) - x(t)\| = 0$ is established then, chaos systems Eq.6 and Eq.7 achieve synchronization.(prove omit).

2.3   Brief Summarization

Observing Method: determine the response equation of this system as Eq.3 by the method: using the variable xi instead of yi in the left of formula (chen) ,then changing the elements in the main diagonal of the coefficient matrix in linear items in the right one to a negative value (if the original one is negative value then it is unchanged; if the original is positive, then the size is unchanged and the signal is changed to a minus sign), and then change the variable of this linear array variable from xi to yi, while the others items in the right one are unchanged , but it is necessary to expand the positive value with the positive coefficient matrix 2 times in the linear item ($2 \times 28 x_2$), and then add this item to the system equation, namely, $\dot{y}_2 = (-7x_1 - 28y_2 - x_1x_3) + 56x_2$. This method has universal application.

Qualitative Analysis for Synchronized Kernel: observing the two systems error signals, for coefficient array of linear terms, whether the main diagonal elements are positive or negative, there are errors for the main diagonal elements of the linear terms by subtracting of two systems, the other terms are offset each other, the non-offsetting two signals are generated by the same system, so both of them are the synchronous system of zero error.

2.4 Examples

The system equations of Lorenz and chen are given according to literature[1], their response equations are respectively writed under by the method of observation refer to formula (4) and (7)

$$\begin{cases} \dot{y}_1 = 35(x_2 - y_1) \\ \dot{y}_2 = -7x_1 - 28\,y_2 - x_1 x_3 + 56\,y_2 \\ \dot{y}_3 = x_1 x_2 - 3\,y_3 \end{cases} \tag{8}$$

$$\begin{cases} \dot{y}_1 = 10(x_2 - y_1) \\ \dot{y}_2 = 28\,x_1 - y_2 - x_1 x_3 \\ \dot{y}_3 = x_1 x_2 - 8/3\,y_3 \end{cases} \tag{9}$$

3 Application to Spread Spectrum Communication

Design of Spread Spectrum Communication System[5],[6]. It has been studied that the traditional synchronization method of spread spectrum utilizes the periodicity of spread spectrum code in order to reach send-receive synchronization at both ends however chaos spread spectrum code is non-periodic. So it means that a code falls out step will result in failure to the whole system in chaos spread spectrum communication system. Therefore, this article presents a spread spectrum method based on sequence as spread spectrum code generated by chaos synchronous system. Its basic idea is: constructing two synchronization chaos systems, which assure synchronization of spreading code and dispreading code generated by receiver and sender and code in conveying as well.

Generate PN Code. Select one of the three variables of lorenz system, such as $x_1(t)$, and compress its number field from [-15 15] into [-1, 1]. Then transform $x_1(t)$ into PN code which include elements of $\{-1, 1\}$

Bit Error Ratio. (signal-to-noise ratio influence on bit error ratio) When signal-to-noise ranges from 1to 10 dB, as is known in Fig.2, the longer intercept the sequence, the greater the influence on bit error ratio of system, conversely, the shorter, the smaller. Comparing the bit error ratio of PN code between the chaos systems and the chaos synchronous system, find their bit error ratio is in accordance.

  Design notes: the compress ratio and coding of the driving and response system must be identical.

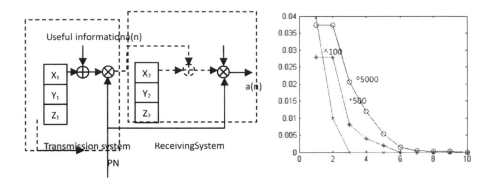

Fig. 1. Spread Spectrum Communication system functionalblock diagram

Fig. 2. The influence of sequence length on bit error ratio

4 Summary

We can find reasons for the synchronization of chaotic system through the analysis of the synchronized kernel of unified chaotic system: the non-linear elements and other elements on the non-diagonal line of two synchronized systems can be offset by each other in a certain way, while the diagonal elements can not be offset. They are generated by driving and response system, the two systems are generalized homology (Lorenz, Chen, Lu ect.)[1]. So the synchronous error is destined to be zero, namely system must be synchronized.

The application of lorenz system in the SSC (spread spectrum communication)field can enhance the synchronization and anti-noise ability of the system. The lorenz system can also be applied in multi-user spread spectrum system.

References

1. Yu, N., Ding, Q., Chen, H.: Different Constructure System Chaos Synchronization and Application in Secure Communication. Journal on Communications 28(10), 73–78 (2007)
2. Wang, X., Wang, Y.: Autonomous Chaos System's Projection Synchronization Based on Linear Separation. Physical Journal 56(8), 2498–2503 (2007)
3. Cai, N., Jing, Y.: Different Structure Chaos System's Auto-Adapted Synchronization and Counter-Synchronization. Physical Journal. 58, 802–813 (2009)
4. Xu, J., Zhang, X., Cai, G.: A New Chaos System's Different Structure and Counter-Synchronization. Jiang Su Scientific and Technical University Journal 23, 82–85 (2009)
5. Li, C., Zhou, J., Sun, K., Sheng, L.: Research and Simulation of Chaotic Sequence Spread-Spectrum Communication System Based on M Sequence Synchronization Method. Journal of System Simulation 21(4), 1198–1201 (2009)
6. Sun, K.-h., Zhou, J.-l., Mou, J.: Design and Performance Analysis of Multi-user Chaotic Sequence Spreac-Spectrum Communication System. Journal of Electronics & Information Technology 29(10), 2437–2440 (2007)

Mining the Astronauts' Health Condition Data: Challenges, Techniques and a Platform

Yi Han[1], Yan Zhao[2], Yan Jia[1], and Dongxu Li[2]

[1] College of Computer, National University of Defense Technology, China
[2] College of Aerospace and Materials Engineering,
National University of Defense Technology, China
yihan@nudt.edu.cn, regayaya@yahoo.com.cn,
jiayan@nudt.edu.cn, dongxuli@263.net

Abstract. In aerospace, stress and excessive workload could lead to astronauts in the physical, mental or psychological tension or disorder. In order to keep the safety of the astronauts, sensor network has been deployed in space craft for monitoring the physiological data. Such data can also be used for scientific research. In this paper, we discuss the background and challenges of mining the health condition data of astronauts, and propose a data processing platform to monitor the changing of astronauts' health conditions. We also briefly introduce the data processing methods we designed. In this paper, all the data processing units are implemented as portable services, and an interaction communication bus is introduced, which can support heterogeneous platforms simultaneously. We conduct the experiment on gathered data to verify the correctness of our processing solutions. The experimental results on real dataset indicated that our solutions are meaningful, and our methods are efficient in practice.

Keywords: Data Mining, Aerospace, Health Conditions, Framework.

1 Introduction

In aerospace, stress and excessive workload could lead to astronauts in the physical, mental or psychological tension or disorder, which may induce the disease. In order to keep the safety of astronauts and space crafts, the manned space flight control system monitors the health parameters of astronauts in real time, and continuously/ intermittently transfers the data to ground commanders. Sensor network, as an important system for gathering information, has been deployed on the body of astronauts for continuously monitoring the changing of the health conditions.

Body Sensor Network [1], BSN for short, is a wireless sensor platform for pervasive healthcare monitoring. BSN is widely used in early detection of disease treatment, health monitoring, telemedicine and other directions. In our experiments, we deployed BSN for keeping monitoring the astronauts' movements and vital signs, and the collected data has been transferred to ground computer systems by space communication channels. Such collected data can be used for diagnosis of astraunauts' health conditions, and also can be used for scientific research.

Y. Yu, Z. Yu, and J. Zhao (Eds.): CSEEE 2011, Part II, CCIS 159, pp. 247–252, 2011.

Our goal is to transform the gathered data into the form of human understandable knowledge, or to discover the interesting and unknown laws. In this paper, we introduce our work on analyzing and mining the gathered data by BSN, and illustrate some techiques applied on our experimental systems.

Preprocessing data is a necessary prerequisite for data mining. Many methods have been developed to handle explicitly missing values or noisy data sets [2-4]. However, in aerospace, data gathering cannot always be controlled precisely, resulting in out-of-range values, background noise, missing values, etc. Therefore, the data preprocessing settings are much more complicated than in traditional healthcare monitoring. We mainly describe two types of data uncertainty brought by space activities.

The contextual noise, as a kind of neglected data uncertainty, causes by the shakiness of spacecraft. Because of the shakiness of spacecraft on launching, returning, docking, or undocking, the sensor may not always fit the body tightly, and such shakiness brings a strong background noise on data; we call the uncertainty brought by environmental aspects as contextual uncertainty. Fortunately, since the space activities conducted by astronauts are controlled by ground commanders, and most of routines are predictable, contextual uncertainty is also predictable, that is, the strength and size the time windows of contextual uncertainty can be estimated before it really happens.

The sensor network has been deployed in the space suit. In aerospace, with the movement of the astronauts, the signal gathered by sensor could be unstable, or inconsistent with other sensors. For example, we deployed sensors on joint parts of astronaut space suit, and muscle's mechenic specificity is expected. However, the activity of arms may bring signal jitter, that is, defect sampling causes data points move irregularly in the sensible range. Such data uncertainty brought by the movement of body parts is unpredictabe, and we call it movement uncertainty.

Besides above 2 types of uncertainty, there still exists some uncertainty in our gathered data caused by the ability of the equipment itself or equipment malfunction. Since space experiments are not redoable, dirty data is not avoidable either. Therefore, our priority is to clean the dirty data as much as possible, or to develop the robust mining algorithms on data with errors.

The rest of the paper is organized as follows. We discuss the related work in Section 2, and introduce our platform structure in Section 3. We discuss the preprocessing techniques we conducted in Section 4, and introduce our data mining methods in Section 5. Section 6 concludes the paper.

2   Related Work

Our work is highly related to the previous studies on middleware techniques, data preprocessing and regression, and health-informatics. In this section, we review some representative work briefly.

A middleware platform consists of a set of services that allows multiple processes running on a distributed system to interact. Recently, some studies on distributed middleware platform focus on adding new features, like reflective action [5], or meeting some requirement, like real time reaction [6,7]. To the best of our knowledge, there is no previous research putting middleware architecture into space communication systems for meeting the requirement of resource management and continuous coming data mining. In this paper, all the data sources, diagnosis expert system interfaces, and data cleaners have been packed as the form of services. CORBA (Common Object Request Broker Architecture [8]) has been extended by adding reflective plugins, and an information and configuration repository has been added to coordinate the service interaction.

Data processing techniques, when applied prior to mining, can substantially improve the overall quality of the results and the running time. Data preprocessing techniques has been studied extensively in recent years. There are some existing work applied data cleaning to remove noise and correct inconsistencies [4]. Data integration [9] merges data from multiple sources into a coherent database. Since the astronauts' health condition diagnosis system mainly gathers the data from the sensors, and some redundant sensors will be deployed for backup and cross validation, the data cleaning and integration should be applied on multiple data streams. Moreover, data normalization [10, 11] has been applied for improving the accuracy and efficiency of mining algorithms involving distance measurements. Data reduction [12] can reduce the data size by aggregating, eliminating redundant features. In remote diagnosis system, the sample rate of some sensors is up to 512kbps, and some physical condition of astronauts might be highly related to the movement of the planet, the external factor should be considered for data mining. In order to reduce the data scale to fit the limited data channel, the task-related data features should be filtered in advance.

In recent years, mining clinical data and early diagnosis received a lot of attention. A retrospective study of the clinical data of infants admitted to a neonatal intensive care unit [13] found that the infants, who were diagnosed with sepsis disease, had abnormal heart beating time series patterns several hours preceding the diagnosis. Monitoring the heart beating time series data and classifying the time series data as early as possible may lead to earlier diagnosis and effective therapy.

3   Data Mining Platform

In this section, we study the architecture of DMAF. DMAF is based on the basic CORBA structure, and all the functions are packed as services. In DMAF, the control unit coordinates the data flow. If system received the commands from the control unit, the corresponding service will be activated by calling its interface.

We use STARBUS [14, 15] as the basic platform, which is a CORBA toolkit and running platform developed by our group, and all the services can be deployed on heterogeneous systems. Fig1 shows the services we deployed and interaction relationships between them.

Data processing part is implemented as several portable interceptors, which are packed as a single service with data sources. Since portable interceptors can be called transparently, the other data consumers, like data mining services, cannot feel the existence of them. Portable interceptors "hook" the dirty data, and smooth the data flow by adjusting the data values. Some parameters are used for control the smoothness and information loss. We will introduce these parameters later.

Feature selection services choose the data sources for different mining tasks. Since the communication channel is limited, only highly related data flows will be transferred to data mining service. Reflection and anomaly detection service monitors the changing of the data flows. Each data mining functions was implemented as a portable service. They won't be called until changing was detected by reflection services or a control signal was sent by control service.

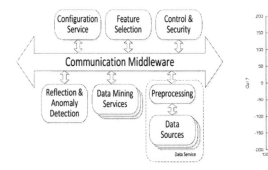

Fig. 1. The architecture of DMAF   Fig. 2. Movement trace recorded by a sensor

4   Data Preprocessing

Data preprocessing is performed on raw data flows for further processing procedures. Our goal is to use it to transform the data into a format that will be more easily and effectively processed. Defect sampling causes data points move irregularly. Fig2 shows the data points of the movement recorded by an arm muscle sensor. The movement is supposed to be a form the curves; however, because of the inevitable shakiness and precision of the sensor, the traces of the movement are blurred very much.

Local regression is adopted here for smooth the movement traces. In order to provide better perception of the pattern of dependence, a smooth curve was added to the scatter plot. To fit a loess curve, values need to be set for 2 parameters, a smoothing parameter α and the degree of the polynomials λ that are fitted by the regression. We use ϖ to measure the information loss of the local regression.

Definition 1 [Information Loss]: For a set of points $S = \{p_1, p_2, ..., p_n\}$ and its loess curve $S' = \{p_1', p_2', ..., p_n'\}$, the information loss of S', denoted by $loss(S', S)$, is the mathematical expectation of $|p_i - p_i'| (1 \le i \le n)$.

Since the movement trace after local regression has a diviation from the original curve, we use the information loss measures the discrimination between them. In information analysis, the clearer curve brings better presentation but larger information loss. Therefore, the configuration units are designed for controling the parameters.

We deployed hundreds of sensors monitoring the physical conditions of astronauts. The sensors continuously gather the samples with rate of 512kbps, and the data gathered by most sensors is just for cross validation. Since the channel band width is strictly limited, feature selection has been used in our system. That is, we consider using representative characteristics to replace the original data, to achieve the goal of reducing data size. In addition, by reducing $o(n)$ data size, data mining algorithms can save $o(n^k)$ running time (k is determined by the algorithm complexity); therefore, our goal is, for each potential mining target, to extract a subset of characteristic data from multiple threads of data. The feature selection units are designed for such purposes.

5 Data Mining Services

To determine the health condition of the astronauts, we adopt machine learning algorithms for generating the diagnosis system. In computer theory, classification is a supervised machine learning procedure, in which individual items are placed into groups based on statistical information on one or more characteristics inherent in the items and based on a training set of previously labeled items. We extended the idea of statistical classification to time series. A time series $TS = \{v_1, v_2, ..., v_n\}$ is a sequence of tuples with timestamps. The data values are ordered in timestamp ascending order. We denote by v_i the value of time series TS at timestamp i. We consider an early classifier should be able to affirm the earliest time of reliable classification so that the early predictions can be used for further actions. The main idea early classification is as follows.

For timestamp i, the data items before time i can be regarded as the training set for generating the classifier and the data after i can be considered as the test data. Such procedure can be repeated by move the timestamp i forward, until the accuracy of the classifier meets the requirement of the system.

For detecting the changing the of the data flows, the anomaly detection services are also deployed. Time series pattern matching is the key part of anomaly detection. An information repository has been used for storing the normal and abnormal patterns, and dynamic time wrapping based algorithms has been adopted here. For the limit of the space and some confidential reasons, we don't put the details here.

6 Conclusion

We conducted the experiments in the space simulator, which is provided by China's Aerospace Medical Research Laboratory. Limited by confidential agreement, we omit some details here.

In this paper, we discussed the background and challenges of mining the health condition data of astronauts, and introduced the method we are using currently. We proposed novel and interesting platform architecture to verify the correctness of our processing solutions. The experimental results conducted on real dataset gathered from volunteers indicated that our solutions are meaningful, and our methods are efficient in practice.

References

1. Lo, B.P.L., Thiemjarus, S., King, R., Yang, G.-z.: Body Sensor Network - A Wireless Sensor Platform for Pervasive Healthcare Monitoring. In: 3rd International conference on Pervasive Computing, pp. 77–80 (2005)
2. Fayyad, U., Piatetsky-Shapiro, G., Smyth, P., et al.: Knowledge Discovery and Data Mining: Towards a Unifying Framework. In: Proc. 2nd Int. Conf. on Knowledge Discovery and Data Mining, Portland, OR, pp. 82–88 (1996)
3. Fayyad, U., Piatetsky-Shapiro, G., Smyth, P.: The KDD Process for Extracting Useful Knowledge from Volumes of Data Communications of the ACM. Magazine Communications of the ACM 39(11), 27–34 (1996)
4. Famili, A., Shen, W.M., Weber, R., Simoudis, E.: Data Pre-processing and Intelligent Data Analysis. International Journal on Intelligent Data Analysis 1(1), 499–502 (2010)
5. Kon, F., Costa, F., Blair, G., Campbell, R.H.: The Case for Reflective Middleware. Communications of the ACM 45(6), 33–38 (2002)
6. Schantz, R.E., Loyall, J.P., Rodrigues, C., Schmidt, D.C., Krishnamurthy, Y., Pyarali, I.: Flexible and Adaptive Qos Control for Distributed Real-Time and Embedded Middleware. In: ACM/IFIP/USENIX 2003 International Conference on Middleware, pp. 374–393 (2003)
7. Schmidt, D.C.: Middleware for Real-Time and Embedded Systems. Communications of the ACM 45(6), 43–48 (2002)
8. Vinoski, S.: CORBA: Integrating Diverse Applications Within Distributed Heterogeneous. IEEE Communications Magazine 35(2), 46–55 (1997)
9. Lenzerini, M.: Data Integration: A Theoretical Perspective. In: Proceedings of The Twenty-First ACM SIGMOD-SIGACT-SIGART Symposium On Principles Of Database Systems, pp. 233–246 (2002)
10. Bolstad, B.M., Irizarry, R.A., Astrand, M., Speed, T.P.: A Comparison of Normalization Methods for High Density Oligonucleotide Array Data Based on Variance and Bias Bioinformatics, vol. 19(2), pp. 185–193. Oxford University Press, Oxford (2003)
11. Chang, J., Remmen, H.V., Ward, W.F., Regnier, F.E., Richardson, A., Cornell, J.: Processing Of Data Generated By 2-Dimensional Gel Electrophoresis For Statistical Analysis: Missing Data, Normalization, And Statistics. J. Proteome Res. 3(6), 1210–1218 (2004)
12. Labrie, K., Allen, C., Hirst, P., Holt, J., Allen, R., Dement, K.: The Gemini Recipe System: A Dynamic Workflow for Automated Data Reduction. In: Proceedings of SPIE - The International Society for Optical Engineering, vol. 7737, pp. 7737–7738 (2010)
13. Griffin, M.P., Moorman, J.R.: Toward The Early Diagnosis Of Neonatal Sepsis and Sepsis-Like Inllness Using Noval Heart Rate Analysis. Pediatrics 107(1), 97–104 (2001)
14. Yan, S.: Application Research of Data Mining Technology to Teaching Evaluation of Higher Education. Journal of Guizhou University of Technology 10(5), 164–166 (2008)
15. Ouyang, J.Q., Ding, B., Wang, H.M., Shi, D.X.: Component Based Context Model. In: The Ninth International Conference on Web-Age Information Management, pp. 569–574 (2008)

Research and Development of Granularity Clustering

Hong Zhu[1,2], Shifei Ding[1], Li Xu[1], and Liwen Zhang[1]

[1] School of Computer Science and Technology,
China University of Mining and Technology, Xuzhou, China
[2] Xuzhou Medical College, Xuzhou, China
zhuhongwin@sina.com, dingsf@cumt.edu.cn, xl412@126.com,
ji.wenji@163.com

Abstract. The rapid development of the Internet and information systems result in massive, high-dimensional, distributed, dynamic and complex data which often have characteristics of incompleteness, unreliability, inaccuracy, inconsistency and so on. Traditional cluster methods are very difficult to satisfy these data cluster demand. The granular computing has developed the cluster analysis research area, enhanced the use value further, and made theoretical significance closer to reality. The cluster may be carried on at different levels and from different angles through the granularity transformation. This will be beneficial to the solution of the problem. This article gives an overview of the essential relationship between granular computing and clustering, the advantages and scope of granule clustering, the research results about granule clustering based on rough sets, fuzzy sets and quotient space. Then the article discusses the necessity that three granular computing models integrate into each other to be used in granule clustering through comparative analysis, and make the analysis and the forecast to establish a unified granule clustering model at last.

Keywords: Clustering Algorithm, Granular Computing, Fuzzy Set, Rough Set, Quotient Space.

1 Introduction

Data clustering is to group a set of data (without a predefined class attribute), based on the conceptual clustering principle: maximizing the intraclass similarity and minimizing the interclass similarity [1]. Clustering is one of the important research contents in the field of pattern recognition, image processing data mining, machine learning and so on. It plays a vital role in aspect of identifying data's intrinsic structure. The study of cluster analysis is always the hot focus because of its importance, multiple application fields and (the application multi-domains as well as) cross-cutting features with other research direction. The conventional (hard) clustering methods restrict that each point of the data set belongs to exactly one cluster. But the rapid development of the Internet and information systems results in massive, high-dimensional, distributed, dynamic and complex data. These data often have characteristics of incompleteness, unreliability, inaccuracy, inconsistency and so on. Traditional cluster methods are very difficult to satisfy these demands.

Y. Yu, Z. Yu, and J. Zhao (Eds.): CSEEE 2011, Part II, CCIS 159, pp. 253–258, 2011.

Granularity is a powerful tool of describing uncertain objects. Granular computing (GrC) is a new research field in artificial intelligence, and it covers all research relate to granular, including the theory, method, technique and skill, majors in intelligent processing of uncertain, incomplete, fuzzy mass of information [2]. Granular computing extends clustering to the field of soft computing enhances the practical value of it. Clustering can be carried out at different levels and from different angles through transforming of granularity. It's beneficial to solve problem by using an optimal granularity to cluster.

2   Principle of Granularity in Clustering

Any subset, class, object, or cluster of a universe is called a granule [2]. Granules are composed of finer granules that are drawn together by distinguishability, similarity, and functionality. Of all human activities, granularity is omnipresent [2, 3]. Granules present in a particular level. They are subject to the level of study. Granular Computing is multi-level and multi-view structured thinking which inspired by the law of human cognition, the process of human problem solving. It designs algorithm to solve problems using granularity conception.

Clustering operation is an equivalence relation essentially which defined among the sample points. Many scholars have done research on the essential relation between granularity and clustering [4, 5]. No matter which clustering methods, there are two factors determine clustering results: similarity function and similarity threshold. The size and number of clustering results are determined by the threshold values. Different granules could be seen as different clustering which has different threshold. All clustering algorithms are uniformed by granular thought.

The size of clustering may be described by granularity. According to a series of threshold values, clustering makes a series of similarity relations from coarse to fine. The difference among sample objects will be discovered if the smaller threshold value is selected. On the other hand, the difference will be neglected. In the clustering process, the size of the granularity is always changing. The goal of cluster analysis is to find an "optimal" granularity among all the possible size based on the similarity function.

The research on the combination of clustering and granularity theory attracted widespread attention at home and aboard recent years because of the granularity thought hidden in clustering analysis. A. Bargiela, W. Pedrycz do lots of systematic research on granularity computing method, the description of granularity world under the sense of clustering [6-8]. Y. Xie [9] proposes fuzzy clustering algorithm based on the granularity analysis theory. Pu Dongbo [5] discusses the principle of granularity in clustering and proposes a new classification algorithm based on the theory of information granularity.

Granularity clustering is that with the help of granularity computing methods, such as fuzzy set theory, coarse set theory and quotient space theory, people can expand the scope of clustering analysis study in order to find an "optimal" granularity, get the best result of clustering and solve the problem better.

3 Advantages of Granularity Clustering

There are many advantages of granule clustering:

1) Approximate solution could be obtained by using coarse granularity of clustering when practical problems include incomplete, uncertain, imprecise or vague information
2) People can preprocess the data using coarse granularity of clustering when facing high-dimensional mass data.
3) When the exact solution is obtained from a high price to pay, or it's unnecessary to work out the exact solution, we can obtain approximate solution from a low price by adopting coarser granulation clustering.
4) To better understand these topics instead of submerging in the unnecessary details of the problem, we can abstract and simplify the problem using granulation clustering when the problem is too elaborate.
5) It's beneficial to the problem to discover or hide details of problems by changing the size of information granularity. In addition, the conversion from the details of space to the granularity of space can help us change the NP problem to the problem can be solved in polynomial time.
6) Clustering methods based on granularity are easy to be integrated, and easy to combine with soft algorithm such as neural network and evolution computation to improve performance of the algorithm.
7) Select the appropriate initial cluster size can reducing complexity and improve the correctness of clustering.
8) Eliminating incompatibility between clustering results and prior knowledge.

4 The Basic Methods of Granule Clustering

4.1 Fuzzy Clustering Analysis

Fuzzy set theory proposed by Zadeh [10] in 1965 gave an idea of uncertainty of belonging which was described by a membership function. The use of fuzzy sets provides imprecise class membership information. In order to make clustering analysis, Ruspinid [11] introduced conception of fuzzy partition. Because the fuzzy clustering expresses the fuzziness which class the sample belongs to, and reflect the real world objectively, it becomes the main trend of the research on clustering analysis.

Fuzzy clustering based on the objective functions is the most widely used method, in the literature of fuzzy clustering, the fuzzy c-means (FCM) clustering algorithms defined by Dunn and generated by Bezdek are the well-known and powerful methods in cluster analysis. There are two successful ideas to implement this algorithm:

(1) Introduce weight index of membership function

$$\left(U^*, P^*\right)\left\{J_m \overset{\min}{=} \sum_{i=1}^{c}\sum_{k=1}^{n}(\mu_{ik})^m \cdot D(x_k, p_i) + \zeta, s.t. f(\mu_{ik}) \in C\right\} \tag{1}$$

(2) Introduce entropy of information

$$J = \sum_{j=1}^{c}\sum_{i=1}^{n} u_{ij}d_{ij} + \lambda^{-1}\sum_{j=1}^{c}\sum_{i=1}^{n} u_{ij}\log u_{ij} \qquad (2)$$

4.2 Rough Clustering

Rough set was originally developed by Z.Pawlak [12] as a theory for data analysis and classification. It's an effective soft computing tool for dealing with data which have characteristics of incompleteness, unreliability and inaccuracy. Rough clustering is a simple extension of the notion of rough sets, involving two additional requirements – an ordered value set of attributes and a distance measure. The value set is ordered to allow a meaningful distance measure, and clusters of objects are formed on the basis of their distance from each other, in a similar manner to standard clustering techniques. In addition, Clusters are formed in a similar manner to agglomerative hierarchical clustering. However, an object can belong to more than one cluster. Clusters can then be defined by a lower approximation (objects exclusive to that cluster) and an upper approximation (all objects in the cluster which are also members of other clusters), in a similar manner to rough sets.

There are three ways to implement clustering algorithm based on rough set: (1) For data preprocessing; (2) Clustering with the concepts of rough set; (3) Clustering with the properties of rough set.

4.3 Clustering Analysis Based on Quotient Space

Quotient space theory [13] is proposed by china academy of sciences professor Zhang Ling and Zhang Bo. Suppose that the triple (X, F, T) describes a problem space, where X denotes the universe, T is the topology structure of X, and F indicates the attributes (or features) of universe. The coarser universe [X] can be defined by an equivalence relation R on X.

Unlike rough set, quotient space introduces granularity on topology. Quotient space theory can obtain the comprehension of the problem from different levels and different views through different quotient spaces.

Clustering analysis based on quotient space describes clustering structure using quotient space structure. Different quotient space can produce different clustering result. It can solve clustering problem using topology, fuzzy set, rough set and other mature theories because of the characteristic of quotient space.

Covering is the extension of partition from the view that equivalence relation is equated with partition. Covering clustering algorithm based on quotient space granularity[14] determines the number of clusters depending on the nature of the sample properties and actual needs in accordance with that the distance or similarity value is less than the threshold size to merge, and didn't know the number of clusters beforehand. So, it can avoid the effect of parameter settings on cluster results, and is easier to get the global optimal solution.

5   Clustering-Combination Methods Based on Granularity

Clustering-combination methods based on granularity have received considerable attentions in recent years, and many ensemble-based clustering methods have been introduced. There are three main models of granular computing, fuzzy set, rough set and quotient space. Analyses and compares among them are benefit for us to understand the relations and differences among them, and benefit for us to find out how to integrate them and how to construct a unified granularity clustering model [15].

5.1   Fuzzy Rough Set Clustering

The membership degree of objects in fuzzy set theory is given directly by the experts. It does not depend on other objects. So it's subjective and lack of precision. The membership degree of objects in rough set theory depend on Knowledge Base, the threshold attained from the data which are to be processed is objective. On the other hand, any subset of approximation space corresponds to a fuzzy set, the lower approximation and the upper approximation equal core and closure of fuzzy set. Fuzzy set and rough set are shown to complement each other.

5.2   Fuzzy Quotient Space Clustering

Fuzzy quotient space theory extends quotient space to the fuzzy granular world. There are three ways to realize it: introduce fuzzy set into universe; introduce fuzzy topology into structure; introduce fuzzy equivalence relation. The following four statements are equivalent: (1) a fuzzy equivalence relation given in universe X; (2) a normalized isosceles distance d given in quotient space [X]; (3) a hierarchical structure given in X. Based on these equivalence, fuzzy granular computing can be changed into computing which structure is ([X], d), so we can research the problem on quotient space [16].

5.3   Rough Quotient Space Clustering

Quotient space uses equivalence classes to describe granularity, rough set does it either. But quotient space which is the theory describing spatial relationship focus on researching inter-conversion and inter-dependence relations among different granularity world. Rough set focus on granular presentation and the inter-dependence between granularity and conception. So from this point of view, rough set is a special case of quotient space. Because rough set is microcosmic granularity computing while quotient is macroscopic, we often combine them together to cluster.

6   Prospect

Fuzzy set, rough set and quotient space are the main model of granularity computing. They all have merits and drawbacks. The trend of research is how to integrate them and how to construct unified granular clustering model based on the interinfiltration, integration, complement of them.

References

1. Han, J., Kamber, M.: Data Mining: Concepts and Techniques, 2nd edn. Morgan Kaufmann Publishers, Massachusetts (2006)
2. Yao, Y.Y.: Granular Computing: Basic Issues and Possible Solutions. In: Proc. of the 5th Joint Conf. on Information Sciences, vol. 5(1), pp. 186–189 (2000)
3. Zadeh, L.A.: Fuzzy logic=Computing with Words. IEEE Trans. on Fuzzy Systems 4(2), 103–111 (1996)
4. Yanping, Z., Bin, L., Yao, Y.Y.: Quotient Space and Granular Computing. Science Press, China (2010)
5. Dongbo, P., Shuo, B., Guojie, L.: Principle of Granularity in Clustering and Classification. Chinese Journal of Computers 25(8), 810–815 (2002)
6. Bargiela, A., Pedrycz, W.: Granular Computing: An Introduce. Kluwer Academic Publishers, Boston (2003)
7. Bargiela, A., Pedrycz, W.: Recursive Information Granulation: Aggregation and Interpation Issues. IEEE Transactions on Systems, Man and Cybernetics, Part B: Cybernetics 33(1), 96–112 (2003)
8. Pedrycz, W., Keun, K.C.: Boosting of Granular Models. Fuzzy Sets and Systems 157(22), 2934–2953 (2006)
9. Xie, Y., Raghavan, V.V., Dhatric, P., Zhao, X.Q.: A New Fuzzy Clustering Algorithm for Optimally Finding Granular Prototypes. International Journal of Approximate Reasoning 40(1-2), 109–124 (2005)
10. Zadeh, L.A.: Fuzzy Sets. Inform and Control 8(3), 338–353 (1965)
11. Ruspini, E.H.: A New Approach to Clustering. Inform and Control 15(1), 22–32 (1969)
12. Pawlak, Z.: Rough Sets. International Journal of Information and Computer Science 11(5), 341–356 (1982)
13. Bo, Z., Ling, Z.: The Solving Theory of Problem and Applications. Tsinghua University Press, China (2007)
14. Li-Li, Y., Yan-ping, Z., Bi-yun, H.: Covering Clustering Algorithm Based on Quotient Space Granularity. Application Research of Computers 25(1), 47–49 (2008)
15. Daoguo, L., Duoqian, M., Hongyun, Z.: The Theory, Model and Method of Granular Computing. Journal of Fudan (Natural Science) 43(5), 837–841 (2004)
16. Lin, Z., Bo, Z.: Theory of Fuzzy Quotient Space. Journal of Software 14(4), 770–776 (2003)

Graph-Based Semi-supervised Feature Selection with Application to Automatic Spam Image Identification

Hongrong Cheng, Wei Deng, Chong Fu, Yong Wang, and Zhiguang Qin

School of Computer Science and Engineering, University of Electronic Science and
Technology of China, Chengdu 611731, China
hongrong.c@sohu.com, {dengwei,fuc,cla,qinzg}@uestc.edu.cn

Abstract. In this paper, we propose a new spectral semi-supervised feature
selection criterion called s-Laplacian score. It identifies discriminate features
by measuring their capability of preserving both local and global geometrical
structure. To address the limitation for spectral feature selection which cannot
handle redundant features, we define Classification Information Gain degree
(CIG) to measure redundant features. Based on s-Laplacian and CIG, we
propose a graph-based semi-supervised feature selection algorithm (GSFS).
The experimental results on real-world image dataset for automatic spam
image identification problem show that GSFS can do well in utilizing small
labeled samples and a large amount unlabeled data to select discriminate
features.

Keywords: Semi-supervised Feature Selection, Laplacian Score, Conditional
Mutual Information, Spam Image Identification.

1 Introduction

The high dimensionality of data poses a challenge to real-world classification tasks.
Since many irrelevant and redundant features are introduced to the tasks, feature
selection is an important data processing step to improve a classifier's performance.

It is common in many real applications that there are a large number of samples but
only a small portion of them are labeled. Semi-supervised feature selection methods
[1] can incorporate information from labeled and unlabeled samples to select relevant
features. Recently, semi-supervised feature selection has received an increased
interest in the pattern recognition and several algorithms have been developed, such
as clustering-based methods [2] and algorithms using spectral graph theory [3, 4].

In clustering-based methods, each original feature is firstly grouped as a cluster.
The pair of clusters that minimize the increment intra-cluster variance are merged at
each step until the number of clusters reaches the pre-defined value. Every feature
representative for a cluster is picked out to form the selected feature subset. Since
irrelevant features can also be grouped as clusters, the selected feature subset may
include irrelevant features. Besides, redundant features may be selected while

Y. Yu, Z. Yu, and J. Zhao (Eds.): CSEEE 2011, Part II, CCIS 159, pp. 259–264, 2011.

important features may be ignored if the pre-defined number of clusters is set inappropriately.

The algorithms based on spectral graph theory demonstrate relatively reliable and promising performance in semi-supervised feature selection. They use the supervised information in constructing neighborhood relationship matrix of graph and identify relevant features through measuring their capability of preserving geometrical structure of graph. However, since the algorithms measure features individually, they cannot handle redundant features. In addition, the hybrid form in [3] may have the difficulty of choosing an optimal weight parameter λ.

2 S-Laplacian Score

The Laplacian Score (LS) [5] algorithm is a promising unsupervised feature selection criterion which measures features through their local preserving capability. We extend LS to semi-supervise feature selection by adding the consideration of features' capability of preserving global geometrical structure.

Let f_{ri} denote the i-th sample of the r-th feature, $i = 1, \ldots, m$ and $f_r = [f_{r1}, f_{r2}, \ldots, f_{rm}]^T$. We construct a with-class graph G^w and a between-class graph G^b. The matrix S^w evaluates the similarity between data points in G^w. Similarly to LS, a "good" feature is to minimize $\Sigma_{i,j}(f_{ri}-f_{rj})^2 S^w_{ij}$. The matrix S^b evaluates the dissimilarity between data points in G^b. We expect data points with different labels are as far apart as possible. Thus, the following objective function on G^b should be maximized:

$$\max_{f_r} \Sigma_i (f_{ri}-u_r)^2 D^b_{ii} \tag{1}$$

where u_r is the mean value of f_r. Since $D^b_{ii} = \Sigma_j S^b_{ij}$, we have

$$\Sigma_i (f_{ri}-u_r)^2 D^b_{ii} = \Sigma_i (f_{ri}-u_r)^2 \Sigma_j S^b_{ij} = \Sigma_{i,j} (f_{ri}-u_r)^2 S^b_{ij} \tag{2}$$

We know from (2) that by maximizing (1), the smaller S^b_{ij} gets, the bigger $|f_{ri}\text{-}u_r|$ becomes. So the graph structure G^b can be well preserved. Thus, a reasonable criterion for choosing a good feature is to minimize the following s-Laplacian object function:

$$Score(f_r) = \frac{\Sigma_{i,j}(f_{ri}-f_{rj})^2 S^w_{ij}}{\Sigma_i (f_{ri}-u_r)^2 D^b_{ii}} \tag{3}$$

We can remove the mean u_r in (3) by using a similar way to [5] and get

$$Score(f_r) = \frac{\tilde{f}_r^T L^w \tilde{f}_r}{\tilde{f}_r^T D^b \tilde{f}_r} \qquad (4)$$

where $\tilde{f}_r = f_r - \frac{f_r^T D1}{1^T D1}1$, $1 = [1, ..., 1]^T$.

In s-Laplacian Score, we define S^w and S^b as follows:

$$S_{ij}^w = \begin{cases} 1 & \text{if } x_i \text{ and } x_j \text{ share the same label or} \\ & \text{if } x_i \text{ or } x_j \text{ is unlabeled,} \\ & x_i \in KNN(x_j) \text{ or } x_j \in KNN(x_i) \\ 0 & \text{otherwise.} \end{cases}$$

$$S_{ij}^b = \begin{cases} 1 & \text{if } x_i \text{ and } x_j \text{ have different labels} \\ 0 & \text{otherwise.} \end{cases}$$

3 GSFS Algorithm

Redundant features increase dimensionality unnecessarily and make learning performance worse. It is shown empirically that removing redundant features can result in significant performance improvement [6].

Let Y denotes the class set. In [7], we give the definition of classification redundant feature based on conditional mutual information as follows:

Definition 1 (*Classification Redundant Feature*). A feature fi is a classification redundant feature of the selected feature subset S if $I(Y; fi) > 0$ and $I(Y; fi \mid S) = 0$.

Since approximate evaluation of classification redundant feature is acceptable for general applications, we define Classification Information Gain degree (CIG) to evaluate approximately the classification redundant features of a selected feature f_s.

Definition 2 (*Classification Information Gain degree*). A feature fi's classification information gain degree for fs is defined as:

$$CIG(f_i, f_s) = \frac{I(Y; f_i \mid f_s)}{I(Y; f_i)} \qquad (5)$$

We set a threshold δ to qualify the likeness of f_i being a classification redundancy feature of f_s. While $CIG(f_i, f_s) < \delta$, f_i can be treated as the classification redundant feature of f_s and removed from the original feature set.

Let X_L and Y_L denote the set of the labeled training samples and their corresponding labels, X_U denotes the set of unlabeled training samples. Based on s-Laplacian and CIG, our proposed GSFS algorithm is as follows:

Algorithm 1. GSFS

Input: X_L, X_U, Y_L, δ.
Output: S_{semi}, the features ranked in the selection order.
   1: $S_{semi} = []$.
   2: Construct graphs G^w and G^b from X_L and X_U.
   3: Caculate L^w, D^b from G^w and G^b.
   4: Compute each feature's score by using Eq. (4).
   5: S_{semi} = the set of features ranked according to their s-Laplacian Score in ascending order.
   6: Investigate S_{semi} for removing redundant feature f_i from S_{semi} while f_i, $f_s \in S_{semi}$, Score(f_i) > Score(f_s), and CIG(f_i, f_s) < δ.

---

4 Experiments

To evaluate the effectiveness of GSFS, we apply it to the automatic spam image identification problem and compare it with other two representative graph-based feature selection algorithms: LS and Fisher Score (Fisher) [8]. The experiments are conducted on personal spam/ham image datasets [9] which concludes 3298 spam images and 2021 ham images. We use the higher-order local autocorrelation approach, which we present in [10] to obtain the features of images. To verify the feature selection quality of GSFS in the datasets with redundant features and irrelevant features, we add 8 classification redundant features and 2 irrelevant features.

In the experiments, the δ value is empirically set to 0.01 and the neighborhood size k is set to 5. After feature selection, C4.5 classifier is employed for classification. To simulate the small labeled sample context, we set l, the number of labeled data, to be 50 and 100, respectively. The labeled data is randomly selected from the datasets. The process of randomly selecting labeled samples for each l is performed 5 times. All the classification experiments are conducted in the WEKA [11]. The obtained classification error rate is averaged and used for evaluating the quality of the feature subsets.

Fig. 1 shows the average classification error rate vs. different numbers of selected features on personal spam/ham datasets with one standard deviation error bars. As shown in the figures, as expected, the feature subset chosen by Fisher generally has lower classification accuracy than that obtained by LS, when there are only 50 or 100 labeled samples available for Fisher, but the whole dataset for LS. GSFS performs better than the other two algorithms when certain number of features is selected. The better performance of GSFS indicates that integration of supervised and unsupervised information can improve the selected features' classification accuracy even when the number of labeled samples is small.

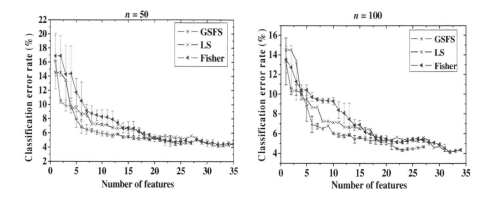

Fig. 1. Average Classification error rate on Personal Datasets

5 Conclusion

This paper presents a new semi-supervised feature selection criterion(s-Laplacian), based on which we propose a feature selection algorithm called GSFS. GSFS exploits spectral graph theory and conditional mutual information to select relevant features and remove redundant features. We compare GSFS with the well-known Fisher and LS algorithms which are graph-based supervised and unsupervised algorithms respectively. The experimental results show that GSFS achieves higher performance than Fisher on the small labeled samples, and outperforms LS on the whole sample data with labels removed.

Acknowledgments. This work is supported by National High Technology Research and Development Program of China (863 Program) under Grant 2009AA01Z422, National Science Foundation of China under Grant 60903157 and Sichuan Scientific Planning Project under Grant 2008GZ0120.

References

1. Zhu, X.: Semi-supervised Learning Literature Survey. Technical Report 1530, Computer Sciences, University of Wisconsin, Madison (2005)
2. Quinzán, I., Sotoca, J.M., Pla, F.: Clustering-based Feature Selection in Semi-supervised Problems. In: 9th international conference on Intelligent Systems Design and Applications, pp. 535–540. IEEE Press, Pisa (2009)
3. Zhao, Z., Liu, H.: Semi-supervised Feature Selection via Sepectral Analysis. In: 7th SIAM International Conference on Data Ming, pp. 641–646. SIAM, Minnesota (2007)
4. Yeung, D., Wang, J., Ng, W.: IPIC Separability Ratio for Semi-Supervised Feature Selection. In: 8th International Conf on Machine Learning and Cybernetics, pp. 399–403. IEEE Press, Baoding (2009)
5. He, X., Cai, D., Niyogi, P.: Laplacian Score for Feature Selection. In: Advances in Neural Information Processing System, pp. 507–514. MIT Press, Vancouverm (2005)

6. Duangsoithong, R.: Relevant and Redundant Feature Analysis with Ensemble Classification. In: 7th International Conference on Advances in Pattern Recognition, pp. 247–250. IEEE Press, Kolkata (2009)
7. Cheng, H., Qin, Z., Wang, Y., Li, F.: Conditional Mutual Information Based Feature Selection Analyzing for Synergy and Redundancy. ETRI Journal 33(2), 210–218 (2011)
8. Duda, R.O., Hart, P.E., Stork, D.G.: Pattern Classification, 2nd edn. John Wiley & Sons, New York (2001)
9. Dredze, M., Gevaryahu, R., Bachrach, A.E.: Learning Fast Classifiers for Image Spam. In: 4th Conference on Email and Anti-Spam, California (2007)
10. Cheng, H., Qin, Z., Liu, Q., Wan, M.: Spam Image Discrimination using Support Vector Machine Based on Higher-Order Local Autocorrelation Feature Extraction. In: IEEE International Conferences on CIS and RAM, pp. 1017–1021. IEEE Press, Chengdu (2008)
11. Witten, I.H., Frank, E.: Data Mining - Practical Machine Learning Tools and Techniques with JAVA Implementations, 2nd edn. Morgan Kaufmann, San Francisco (2005)

A Provenance-Aware Data Quality Assessment System

Hua Zheng[1,*], Kewen Wu[2], and Fei Meng[3]

[1] School of Management and Engineering, Nanjing University
[2] Department of Computer and Information Management,
GuangXi University of Finance and Economics,
Ming Xiu west road 100#, Nanning, Guang Xi, China, 530003
[3] Department of Information Management, Nanjing University,
Han Kou road 22#, Nanjing, Jiang Su, China, 210093
gxhuazheng@yahoo.com.cn, kewen-wu@163.com, fei.meng@foxmail.com

Abstract. The data quality assessment (DQA) process has the lack of sufficient attention on enterprise infomationization, and existing technologies and methods have their limitations. In order to solve data quality(DQ) problems from the source and realize the traceability of data, after research on data provenance technology and determining the idea of achieving the way data can be traced, the framework of data quality assessment based on data provenance and SOA is presented. Then the logical architecture is described, simultaneously core technology are focus to analyze. Finally, specific application is discussed and the direction of further work is given.

Keywords: Data quality management, Data quality assessment, Provenance, SOA.

1 Introduction

Data is the enterprise critical strategic resource, and reasonably, effectively using the correct data can guide business leaders make the right decisions to enhance the competitiveness of enterprises. Unreasonably using incorrect data (ie, poor DQ) can lead to the failure of decision-making. A survey of the total data quality management (TDQM) [1] project from Massachusetts Institute of Technology (MIT) is shown: only 35% of the companies trust the own data, only 15% of the companies trust the partner data. In the United States there is cost about 600 billion U.S. dollars annually to ensure DQ, or to compensate for DQ problems caused economic losses. It can be seen that the issues of DQ have been begun to attach importance by the enterprises.

Currently, most enterprise information technology projects are not starting from scratch, and they need to use the data that already resides in the enterprise and has the poor quality. In particular, because the current scope of data has been expanding and more widely shared and increasingly diverse forms of data have been, a large, complex, heterogeneous data environment has been formed. Therefore, the analysis of data

---

* Corresponding author.

Y. Yu, Z. Yu, and J. Zhao (Eds.): CSEEE 2011, Part II, CCIS 159, pp. 265–269, 2011.

generation and evolution process, then evaluate the quality and accuracy of data, as well as revise data result that is very important.

In this paper, firstly the background of DQA and basic ideas are introduced; In section 2, the related research works are given, which includes DQA and provenance; The framework of DQA based on provenance and SOA is designed in section 3; Provenance model is given in section 4; A concrete example is implemented in section 5; Finally a brief summary is given.

2   Related Works

Provenance is a current research focus, and how to use provenance technology to achieve DQA is a worthy research direction.

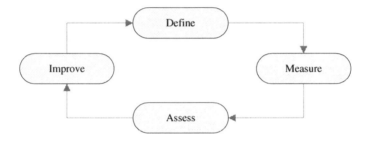

Fig. 1. Data Quality Management Process

DQ[2] is defined as "suitability for use", and this definition is now widely accepted. Much research in data quality management (DQM)[3] focuses on methodologies, tools and techniques for improving quality, and DQA [4] is an important part of DQM(shown in Fig.1). The current mainstream methods of DQA are summarized in literature[5], including: TDQM, DWQ, TIQM, AIMQ and so on, a total of 13 ways; Yang et al[6] designed a six-dimensional DQA model, and the quality situation of the data system can be assessed by the application of quantitative indicators.

Provenance[7] refers to the data are generated, and the information of the whole process of evolution over time. Data provenance contains static source data information and dynamic data evolution process, which characteristic is focused on to describe the variety application data sources and evolutionary process information, including richer metadata. After Simmhan et al[8] analyze and compare the exiting research achievements of provenance, it is given that data provenance is defined as derivative process information came from source data to data product.

3   Architecture of Provenance-Aware DQA

Realistic DQA must be carry for a specific environment and user, and there is no uniform standard. So in this paper, a framework of DQA is presented that enables the various functions of assessment to dynamically be added and in which the evolution of the data by data provenance can be technically analyzed, and the appropriate system functionality of assessment can be selected for DQA.

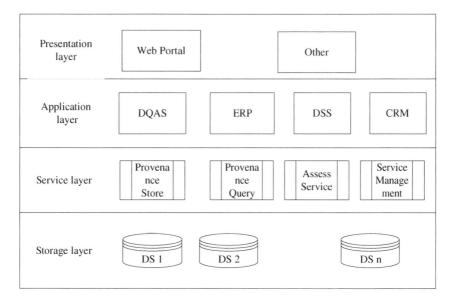

Fig. 2. Logic Architecture

SOA(Service-Oriented Architecture) is an abstract model, and represents a specific implementation which does not involve the software infrastructure and has direct service-oriented. Enterprise business logic can be achieved at lower cost for rapid reconstruction. Therefore, the assessment functions based on SOA will be abstracted in the form of service, and then the framework of DQA that can adapt to all kinds of requirement for assessment in a different environment is created. The logical architecture of the framework shown in Fig.2 is divided into four layers (storage layer, service layer, application layer, presentation layer). Its core is the service layer, in which all the functional components are packaged into the form of web service, here including the assessment of service components, provenance storage components, provenance query components, service management components. All of the services will be integrated by the ESB (Enterprise Service Bus). The entire solution is based on a typical SOA framework.

The specific process of DQA can be divided into three parts: (1)source system analysis. Data quality metrics are applied to determine data quality levels of source system; (2)target system analysis. The differences between the target system and the source system are analyzed, and then the recommendations for the elimination of differences are created; (3)alignment and harmonization requirements for each relevant data elements are assessed. The results of DQA indicators are interpreted, and translated into business terms. Detailed reports, charts and summary are created to describe DQ levels and provide recommendations.

4 Semantic Model of Provenance

In order to collect and query the provenance data in provenance management module, firstly the semantic model which can describe provenance is given. Detailed exposition is given below.

A provenance entity can be viewed as a static entity record or a activity record, which can be defined as(type, entity), type $\subseteq$ {S,A}. Suppose type=S, then the provenance entity is a static entity record; suppose type=A, then the provenance entity is a activity record. a provenance relationship is a relationship between the two provenance entities, which can be defined as(causal-entity, consequential-entity, role, annotation, relationship-id), causal-entity and consequential-entity both are provenance entity, role $\subseteq$ DICT, relationship-id $\subseteq$ NS.

There are two types of provenance entities: static entity records and activity records. A provenance entity can be accessed, so we can define the storage of the provenance entity as a physical object that can be accessed (called PES). Provenance relationship is actually the relationship between the two provenance entities, and we define a provenance relationship store here (called PRS) as a storage object of provenance relationship. Therefore, the provenance storage is actually to achieve storage of provenance entities and their relations.

An important operation on the provenance is the query, so we have developed a query model by which provenance entities and relationships can be operated. The query model is constituted by PES, PRS, and various queries operators. With comparing the general data query, the difference is that the results of a provenance query include not only the structure of complex content, but also includes the structure of complex relationship. The query of provenance information is mainly completed by defining operators of these two types of objects. TO query PES (type as a query parameter) as an example for a description: Suppose S is a object of PES, type=t, use s and t as input, the operator can be defined as:

$$\sigma_{type}(S,t) = \sigma(S, \Theta type = t) \tag{1}$$

You can use the operator to retrieve the records of all activities of PES. Because space is limited, on the other types of operator is not defined within the details.

It is known that the semantic model is constituted by provenance entities and their relationships from the above definitions. By defining the semantic model, the provenance information of all the static and dynamic elements in application system can be described by a flexible way.

5 System Implementation

Here, the DQ problems of the sugar factory are selected to study. In the infomationization process of sugar factory, a distributed control system (DCS), a equipment management system (EMS), a condition monitoring system (CMS), a enterprise resource planning (ERP) system and so on are respectively constructed. For the different business lines, channels or products categorization, the data is often stored and processed in different ways by using different technologies. Core enterprise information is distributed in multiple vertical systems with multiple copies. The maintenance information of each system is according to their own context, regardless of the context of the whole enterprise, which further exacerbates the inconsistencies in the process of business.

Based on the solution that was discussed in the previous sections, we developed a manufacturing-enterprise-oriented DQA system for the sugar factory, in which the data of management and control will be monitored by provenance management, and the problems of DQ caused by misuse and false behavior will be found. The dimension of evaluation can be customized by user. The major algorithm of evaluation is using simple ratio, maximum/minimum operation, weighted average, etc. This system provides strong support to enterprise DQM and contributes heavily to control/management integration.

6 Conclusions and Future Work

Because it is not feasible for universal dimensions and methods of DQA, so the dimensions should be analyzed for different application environments and individual, then the appropriate assessment methods are selected. In this paper, the proposed DQA framework based on SOA and provenance is a feasible solution, in which assessment function is dynamically adjusted by the form of web service.

The future works will be focused on further optimizing the details of this solution, including structural provenance model, automatically collecting and storing provenance data, and how to secure the provenance data, etc.

Acknowledgment

This work was supported in part by a grant from the Research Foundation of Philosophy and Social Science of GuangXi Province, China (No.08FTQ001).

References

1. Wang, R.Y.: A Product Perspective on Total Data Quality Management. Communications of the ACM 41(2), 58–63 (1998)
2. Wang, R.Y., Strong, D.M.: Beyond Accuracy: What Data Quality Means to Data Consumers. Journal of Management Information Systems 12(4), 33–50 (1996)
3. Evena, A., Shankaranarayananb, G., Bergerc, P.D.: Evaluating a Model for Cost-Effective Data Quality Management in a Real-World CRM Setting. Decision Support Systems 50(1), 152–163 (2010)
4. Pipino, L., Lee, Y., Wang, R.Y.: Data Quality Assessment. Communications of the ACM 45(5), 211–218 (2002)
5. Batini, C., Capplcllo, C., Francalancl, C.: Methodologies for Data Quality Assessment and Improvement. ACM Computing Surveys 41(3), 40–52 (2009)
6. Yang, Q.Y., Zhao, P.Y., Yang, D.Q.: Research on Data Quality Assessment Methodology. Computer Engineering and Applications 40(9), 3–4 (2004) (in Chinese)
7. Buneman, P., Khanna, S., Tan, W.C.: Why and Where: a Characterization of Data Provenance. In: 17th International Conference on Data Engineering, pp. 316–330. ACM Press, London (2001)
8. Simmhan, Y., Plale, B., Gannon, D.: A Survey of Data Provenance in E-Science. ACM SIGMOD Record 34(3), 31–36 (2005)

Prediction of Bacterial Toxins by Relevance Vector Machine

Chaohong Song

College of Science, Huazhong Agricultural University,
Wuhan 430070, China
chh_song@mail.hzau.edu.cn

Abstract. Using Relevance Vector Machine, and with an improved feature extraction method, a novel method was proposed here to predict bacterial toxins, the jackknife cross-validation was applied to test the predictive capability of the proposed method. Our method achieved a total accuracy of 95.95% for bacterial toxin and non-toxin, and a total accuracy of 97.33% for discriminating endotoxins and exotoxins, the satisfactory results showed that our method is effective and could play a complementary role to bacterial toxins prediction.

Keywords: Prediction; Bacterial toxins; Approximate entropy; Relevance vector machine.

1 Introduction

The bacterial toxins are a major cause of diseases because they are responsible for the majority of symptoms and lesions during infection [1]. But some of these powerful disease-causing toxins have been exploited to further basic knowledge of cell biology or for medical purposes. For example, diphtheria toxin has been used to treat many disorders including cervical and laryngeal dystonia, writer's cramp, hemi facial spasm, tremors, and tics [2], and also used cosmetically to reduce deep wrinkles caused by the contraction of facial muscles [3]. These showed that bacterial toxins could play the very vital role as long as we could identify them and make use of them, but with protein database growing, experimental method to identify bacterial toxins alone is more time-consuming and costly, so it is necessary to propose a precise and rapid theoretical method to identify bacterial toxins.

In fact, Saha and Raghava [4] currently had used support vector machines (SVM) based on amino acids and dipeptide composition to predict whether a protein sequence is a bacterial toxin or a non-toxin, and achieved an accuracy of 96.07% and 92.50%, respectively. For endotoxins and exotoxins, two classes of bacterial toxins, they discriminated them using the same method and achieved an accuracy of 95.71% and 92.86%, respectively. Using ID-SVM algorithm on the same dataset, Yang and Li [5] also predicted bacterial toxins and achieved higher MCC for entotoxins and exotoxins than that of [4]. Encouraged by their research, in this paper, we attempted to develop a new method to predict bacterial toxins and their class.

Y. Yu, Z. Yu, and J. Zhao (Eds.): CSEEE 2011, Part II, CCIS 159, pp. 270–274, 2011.
© Springer-Verlag Berlin Heidelberg 2011

2 Methods and Material

2.1 Dataset

From http://www.imtech .res.in/raghava/btxpred/supplementary.html, we could download all datasets we used in this paper, which were collected form Swiss-Prot database [6] and the dataset used by [4].

The first dataset is the classification of bacterial toxins and non-toxin. We used the cd-hit soft [7] to remove sequences that have more than 90% sequence identity and deleted the sequences which the length is less than or equal to 100, than we got the 141 bacterial toxins and 303 non-toxins.

The second dataset is the classification of exotoxins and endotoxins, we also used the cd-hit soft [7] to remove sequences that have more than 90% sequence identity, and finally the dataset contains 73 exotoxins and 77 endotoxins.

2.2 Approximate Entropy

The approximate entropy is a measure of system complexity [8] and has been widely used to deal with physiological signal [9]. In this paper, we used the approximate entropy to describe the complexity of protein sequences.

Suppose X is a protein sequence, we first translate the amino acid sequence of a protein into the corresponding original hydrophobicity value sequence, then the digitized expression is formulated by $X = (u(1), u(2) \cdots u(N))$, let $x(i) = (u(i), u(i+1) \cdots u(i+m-1))$, $i = 1, 2 \cdots N - m + 1$ are m-D (dimensional) vector which is composed by sequence $u(i)$ according to its order, then the approximate entropy of the protein sequences can be expressed as the following equation:

$$ApEn(m, r) = \varphi^m(r) - \varphi^{m+1}(r) \tag{1}$$

$$\varphi^m(r) = \frac{1}{N-m+1} \sum_{i=1}^{N-m+1} \ln C_i^m(r) \tag{2}$$

$$C_i^m(r) = \sum_j \text{sgn}(r - d(x(i), x(j))), 1 \le i \le N - m + 1 \tag{3}$$

$$\text{And, } d(x(i), x(j)) = \max_{k=1,2\cdots m} |u(i+k-1) - u(j+k-1)| \tag{4}$$

Where N is given data of samples, r is filter parameter, which describe the similarity of the protein sequences and m is mode dimension, different m expresses different length of amino acids pairs. Here we select $m = 2, 3, 4$ and $r = 0.1, 0.15, 0.2, 0.25$, then we can get 12 approximate entropies.

2.3 Feature Vector

In this research, every protein is represented as a point or a vector in a 32-D space, the first 20 components of this vector were supposed to the occurrence frequencies of the 20 amino acids in the protein sequence, and the last 12 components of this vector were the a times approximate entropy of the protein sequence. The best value of a is obtained by numerical experiments, Here we choose $a = 0.022$ for the prediction of bacterial toxins and non toxins, and $a = 0.3$ for the prediction of exotoxins and endotoxins.

2.4 Relevance Vector Machine

Relevance vector machines (RVM) is proposed by Tipping [10] on the Bayesian framework, with the same function form as support vector machine (SVM). Based on kernel function, Relevance vector machine maps non-linear problems of low-dimensional space into linear problems of high-dimensional space; it solves the weight of associated vector by maximizing a posteriori probability.

Given sample training set $\{ t_n, x_n \}$, $x_i \in R^d$, the output of RVM model is

$$y(x,w) = \sum_{i=1}^{n} w_i \phi_i(x_i) = \phi(x_i)w \tag{5}$$

Where $\phi(x_i) = (1, K(x_i, x_1), K(x_i, x_2) \cdots K(x_i, x_n))$, $K(x, x_i) = \exp(-\dfrac{\| x - x_i \|^2}{2\sigma^2})$ is

kernel function, n is the number of samples, $w = (w_1, w_2 \cdots w_n)^T$, w_i is the weight of the model.

Let $t = (t_1, t_2 \cdots t_n)$, where $t_n = y(x_n, w) + \varepsilon_n$, the noise ε_n subject to Gaussian distribution with mean 0 and variance σ^2, t_i independent and identically distributed, so the training sample likelihood function should be

$$p(t \mid w, \sigma^2) = (2\pi\sigma^2)^{-\frac{1}{2}} \exp(-\frac{\| t - \phi w \|^2}{2\sigma^2}) \tag{6}$$

In order to improve the model's generalization ability and avoid too much relevant vectors in models, which could cause over-fitting problems, RVM algorithm define a priori sub-Gaussian distribution for the weight w :

$$p(w_i \mid a_i) = \sqrt{\frac{a_i}{2\pi}} \exp(-\frac{a_i w_i^2}{2}) \tag{7}$$

Where, a_i is hyper parameter of a priori distribution, By Bayesian inference, we can obtain the posterior distribution of w and t :

$$p(w \mid t, a, \sigma^2) = N(\mu, \textstyle\sum) , \; p(t \mid a, \sigma^2) = N(0, C) \tag{8}$$

Where $\mu = \sigma^{-2} \sum \phi^T t$, $\sum = (\sigma^{-2} \phi^T \phi + A)^{-1}$ and

$A = diag(a_0, a_1 \cdots a_n)$, $C = \sigma^2 I + \phi A^{-1} \phi^T$

By maximizing the likelihood distribution of super parameters, we can obtain the optimal parameter values a_{MP} and σ_{MP} , then we can make forecast for the new observational data x_* and calculate its predictive distribution.

$$p(t_* \mid t, a_{MP}, \sigma_{MP}^2) = N(\mu_*, \sigma_*^2) \tag{9}$$

Where, μ_*, σ_*^2 is the mean and variance of the prediction output.

$$\mu_* = \mu^T \phi(x_*) , \quad \sigma_*^2 = \sigma_{MP}^2 + \phi(x_*)^T \sum \phi(x_*) \tag{10}$$

3   Results and Discussion

Here we also use the sensitivity (Sn), specificity (Sp), Matthew's correlation coefficient (MCC) and the overall prediction accuracy (Ac) to evaluate the correct rate and reliability, and use Jackknife test on the dataset [5].

The performances of various methods developed for discriminating the bacterial toxins from non-toxins and exotoxins from endotoxins had been shown in Table 1and Table 2 respectively. (in two tables, a comes from [4], b coms from [5])

Table 1 and table 2 showed that the performance of our method was satisfactory, the total accuracy for predicting toxins was 95.95%, which were close to the best prediction of the previous results, and the total accuracy and MCC of our method for predicting exotoxin and endotoxin were 97.33% and 0.9466 respectively, which were higher than that any other existence results. Moreover RVM need shorter calculation time than that of SVM, this because RVM is more sparse than SVM, although the vector number of RVM increases with the number of training samples, but the relative growth rate is slower than that of SVM, RVM is more suitable for on-line detection. Our method could play a complementary role to bacterial toxins prediction.

Table 1. The performances of various methods in predicting bacterial toxins

	Sn	Sp	MCC	Ac
Our method	93.62%	93.62%	0.9065	95.95%
Amino Acids[a]	92.14%	100%	0.9293	96.07%
Dipeptides[a]	86.43%	98.57%	0.8612	92.50%

Table 2. The various performances in discrimination of exotoxins and endotoxins

	Sn	Sp	MCC	Ac
Our method	97.26%	97.26%	0.9466	97.33%
ID[b]	92.91%	99.24%	0.9428	
Amino Acids[a]	100%	91.43%	0.9293	95.71%
Dipeptides[a]	94.29%	91.43%	0.8612	92.86%

References

1. Böhnel, H., Gessler, F.: Botulinum Toxins – Cause Of Botulism and Systemic Diseases? Vet. Res. Commun. 29, 313–345 (2005)
2. Kessler, K.R., Benecke, R.: Botulinum Toxin: From Poison to Remedy. Neurotoxicology 18, 761–770 (1997)
3. Carter, S.R., Seiff, S.R.: Cosmetic Botulinum Toxin Injections. Int. Ophthalmol. Clin. 37, 69–79 (1997)
4. Saha, S., Raghava, G.P.: BTXpred: Prediction of Bacterial Toxins. In Silico Biol. 7, 405–412 (2007)
5. Yang, L., Li, Q.Z., Zuo, Y.C., Li, T.: Prediction of Animal Toxins Using Amino Acid Composition and Support Vector Machine. Inner Mongolia University 40, 443–448 (2009)
6. Boeckmann, B., Bairoch, A., Apweiler, R., Blatter, M.C., Estreicher, A., Gasteiger, E., Martin, M.J., Michoud, K., O'Donovan, C., Phan, I., Pilbout, S., Schneider, M.: The Swiss-Prot Protein Knowledgebase and Its Supplement TrEMBL in 2003. Nucleic Acids Res. 31, 365–370 (2003)
7. Wei, Z.L., Godzik, A.: Cd-Hit: A Fast Program for Clustering and Comparing Large Sets Of Protein or Nucleotide Sequences. Bioinformatics 22, 1658–1659 (2006)
8. Sci., USA, vol. 88, pp. 2297–2301 (1991)
9. Richman, J.S., Moorman, J.R.: Physiological Time-Series Analysis Using Approximate Entropy and Sample Entropy. Am. J. Physiol. Heart Circ. Physiol. 278, 2039–2049 (2001)
10. Tipping, M.E.: Sparse Bayesian Learning and the Relevance Vector Machine. JMLR 1, 211–244 (2001)

Simulation of Spatial Distribution Evolution of Plants in Forest

Liang Shan, Gang Yang*, and Xinyuan Huang

School of Information Science and Technology,
Beijing Forestry University, Beijing, China
Stephanie581@126.com, yanggang@bjfu.edu.cn,
hxy@bjfu.edu.cn

Abstract. As the impact of natural environment, resource competition, propagation of trees and other factors, the spatial distribution of trees in forest is changing continuously. In this paper, we put forward the method to simulate this dynamic evolution of forest based on the theory of forest ecology. In the method, the spatial distribution information of trees is recorded and visualized in a form of two-dimensional distribution diagram, and three important evolution rules of forest: self-thinning, species succession and clustering growth, are simulated with iterated deduction manner. The method can be used to provide the dynamic distribution data of forest for further realizing the realistic simulation of dynamic forest scenes. Furthermore, it also has potential value in forestry research and applications.

Keywords: Dynamic Evolution of Forest, Visual Simulation, Spatial Distribution of Trees.

1 Introduction

Forest is one of the most important biological communities on the Earth, it covers a large area of land and contains rich biological resources and has markedness influence on environment and climate. Because of these reasons, forest and its related issues have always been important research topics for reseachers. Forest is an ecological system with dynamical evolution. In a forest, thousands of plants propagate, grow and die constantly. An individual plant may compete for resources with the other plant that is adjacent to it, thus will have a significant impact on their growth. One kind of plant community may die out gradually and be substituted by another kind of plant community since the competition between two plant communities or others environmental factors. For the above natural phenomena, the distribution of plant communities in the forest will change continuously. The research and simulation of these natural phenomena that take place in the evolution process of forest may reveal the pattern, the speed and the degree of stability of the dynamic evolution in forest, thus can simulate and predict the past and the future of the forest. If we could use computer

---

* Corresponding author.

Y. Yu, Z. Yu, and J. Zhao (Eds.): CSEEE 2011, Part II, CCIS 159, pp. 275–280, 2011.
© Springer-Verlag Berlin Heidelberg 2011

technology to simulate and visualize the forest evolution based on the correlative theory of forest ecology, it would be very significant and valuable for forestry research and forest management. In addition, for the research of computer graphics and virtual reality, the simulation of forest evolution process is also the key issue for representing the time-varying landscape of forest realistically.

There are many factors affecting the forest evolution, and the theory of it is also complex. In forest ecology, researchers have observed and researched on the dynamic changes of forest for a long time. They found several natural phenomena, which can affect the change of forest plants distribution remarkably. One is Self-thinning phenomenon, the other one is species succession, the third one is clustering growth. Many plants adopt the method of propagating, which sow their seeds around themselves. In this paper, we based on the theoretical research on forest ecology to analyze and model these three natural phenomena. We use iterative calculation method to deduce and calculate these three phenomena, and adopt two-dimensional distribution graph to express simulation results.

In this paper, section1 gives previous work. In section 2.1, we introduce two-dimensional graph and the basic idea of iterative methods. Then we introduce how to simulate the natural phenomena of self-thinning, species succession and clustering growth by the iterative methods in section 2.2 - 2.4. Conclusions are presented in section 3.

2 Previous Work

Many researchers of forest ecology have been concerned about the pattern of plant growth and plant distribution in forest. Yoda [1] proposed the -2/3 self-thinning rule at 1963, which applied the quantitative method to describe the self-thinning phenomenon. [2] and [3] applied the FON model to simulate the distribution change of plants, which due to the resources competition.

In computer graphics, one of the earliest researches on forest evolution simulation is the work of Deussen in [4]. In this paper, Deussen represented individual plant as a circle with certain growth properties. He iterative calculated these circular areas based on the -2/3 self-thinning rule, and hence deduced the plants distribution in different evolution stages. Lane [5] improved Deussen's method and use Multiset L-system to simulated the self-thinning phenomenon.

In this paper, we also adopt the circular area to represent individual plant, but put forward the iterative method and the accelerating algorithm to simulate these three phenomena.

3 Simulation of Forest Evolution Phenomena

3.1 Iteration Method

In this paper, We use a kind of two-dimensional distribution graph to represent the plants in forest, each plant can be represented as an circular area, the expansion of circular area represent the growth of plant, the intersection of two circular areas means the competition for resource between the two plants. Based on the two-dimensional distribution graph, we adopt the iterative method to simulate three kinds of natural phenomena in forest evolution. The work flow of the iterative method is clearly, and we can add and modify a variety of growth pattern in its work flow conveniently.

The growth of plant can be regarded as a process that growing by time unit. In growing period, the individual plant in the forest will go through one time unit after another unit. In each time unit, every plant has its own state, and the state then decide how the plant should grow in the next time unit. So the growing process of plants can be described as an iterative computation process by time unit.

For iterative method, each plant is expressed as a structure with multiple properties, these properties determine the state of plant in each time unit. In each round of iteration, all the individual plants in forest will be traversed, and for each individual plant, its property values in the next time unit will be determined by its current property and surrounding environment conditions according to certain growing rules. Thus the space distribution state of plants in the forest will be simulated by the time unit iteratively.

3.2 Self-thinning

In an ecosystem, as the natural resources, such as space, water, sunlight are always limited, adjacent individual plants will compete for the limited natural resources. Larger and stronger plants often hinder the development of surrounding smaller and weaker plants, leading the death of part of the weaker plants, resulting in the decline of plants density. This is called self-shinning phenomenon.

In simulation of self-thinning, we divide the state of plant into three categories: dominated state, ripe state and regular growth state. A plant that is dominated means that the plant is hindered and dominated by other larger, stronger plants. The dominated plants will die with a certain probability. Ripe state is the state that the size of a plant has reached the maximum growing limit. If a plant has reached ripe state, it will not grow any more but maintain the current status. If a plant is in the regular growth state, the plant will grow continuously with a certain speed.

The main properties of plant are listed as follows:

```
Tree:
{ centre_x      X coordinate of the location of plant
  centre_y      Y coordinate of the location of plant
  radi          radius of the plant
  dom           dominated identifier
  ripe          live identifier
  speci         species identifier
  speed         growth increment
}
```

In the two-dimensional distribution, each plant is described by a circle with the radius radi, the property radi represent the size of plant. The property dom represent whether the plant is dominated or not. When one circle is intersected with another, we set the smaller one is in dominated state. The property mature represent if the plant is ripe. If the radius of a plant has reached a preset value R, the plant is in the ripe state and it will stop to grow. The property live represent whether the plant is live or not. When a plant is in the dominated state in current time unit, this plant will die in the next time unit with a certain probability. The value of live is 0 means the plant is dead, and the dead plant will be removed from the distribution graph.

In the process of iteration, every plant will be judged and decided which state it will be in the next step according to its current properties and surrounding environment conditions. Thus realize the simulation of resources competition.

The essence of self-thinning can be captured using iteration method shown in Figure.1.

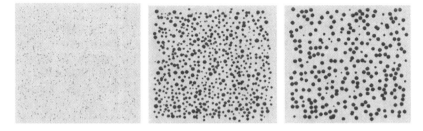

Fig. 1. Three stages of simulation of the self-thinning process

3.3 Species Succession

In the growth of forests, as the different species of plant have different growing abilities, one kind of plant community may be replaced by another kind gradually, and the latter will become the dominant species of the forest. We call this evolution phenomenon as species succession. The species succession can be simulated based on the basis of self-thinning, but compared with the phenomenon of self-thinning, the focus of species succession is on the different characteristics among different species. During the plants development, different species will perform different growing processes. In this paper, we use iteration method to simulate the succession between two species successfully.

In the simulation of species succession, we also divide the state of plant into three categories: dominated state, ripe state and the regular growth state. The meaning of each state is the same as the meaning in self-thinning.

For each species, we define several species properties as listed below.

```
Species:
{ sp      plant species
  grow    the speed of growth
  shaded  the probability of survival when plant at the state
          of dominated
  oldage  the probability of survival when plant at the state
          of ripe
}
```

During the simulation, the plant model should be added with the properties of species. The value grows represent of the speed of plant growth. The value shaded, called the shade tolerance of the plant, is a measure of how likely it is to survive in the shadow of resources. If a plant is under dominated state, it will die with the probability of shaded. If the plant is under ripe state, it will survive with the probability of oldage.

A plant that does not survive will be removed from the community. Because of the difference of the value of shaded, old age and grow of different kinds of species, different species of plants will show different variation over time.

Assuming that species 1 has a higher growth rate but lower shade tolerance and old-age survivorship than species 2 (grow (1) > grow (2), shaded [1] < shaded [2], oldage [1] < oldage [2]). At first, the field will be dominated by species 1. With the increase of plants density, largest members of species 1 will die, while the smaller members of species 2 will survive due to their greater shade tolerance. Spcecies 2 will have a size advantage over young seedlings of species 1 gradually. Finally species 1 will be replaced by species 2, and the latter become the dominant species of the whole forest. This process is illustrated in Figure 2.

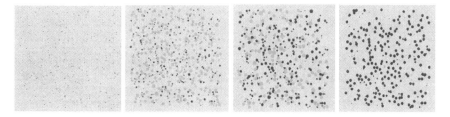

Fig. 2. Four stages of simulation of species succession. Using green circles to indicate the early-succession plants (species 1), and blue circles to indicate the late-succession plants (species 2).

3.4 Clustering Growth

In nature, many kinds of plants sow their seeds in surrounding earth for propagation.

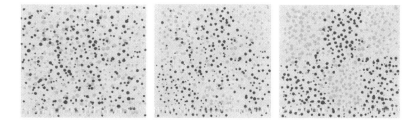

Fig. 3. Three stages in the simulation of plant are clustering. At the beginning, plants distribute by the form of dispersion, and with the growing of plants, it will appear the phenomenon of clustering.

By this method, with the propagation of plants, the clustering of same species plants will be formed gradually. The clustering phenomenon has remarkable influence on the distribution of plants in forest.

In order to simulate the clustering phenomenon, we set that if one plant is at the state of growing, it will not only grow, but also will sow their seed around itself at the same time. Figure.3 show the simulation of plant clustering. We can see that with the growing and propagating of plants, the clustering phenomenon will appear.

4 Conclusions

In this paper, we put forward the method to simulate the plants distribution change in forest evolution. We adopt iterative method, and proposed corresponding evolution rules to simulate the self-thinning, species succession and clustering growth phenomena successfully. Our method can not only be used in the forestry research and forest management, but also have significance and application values in the realistic simulation of time-vary forest scenes in computer graphics and virtual reality.

Acknowledgment. This work was supported in part by a grant from Central University special fund basic scientific research and operating expenses (Project Number: YX2010-24) and National 948 Program (Project Number: 2009-4-41).

References

1. Yoda, K., Kira, T., Ogawa, H.: Self Thinning in Overcrowded Pure Stands Under Cultivated and Natural Conditions. J. Biol. Osaka City Univ. 14, 107–129 (1963)
2. Berger, U., Hildenbrandt, H.: A New Approach to Spatially Explicit Modeling of Forest Dynamics: Spacing, Ageing and Neighbourhood Competition of Mangrove Trees. Ecological Modeling 132(3), 287–302 (2000)
3. Berger, U., Hildenbrand, H.T., Grimm, V.: Towards a Standard for the Individual Based Modeling of Plant Populations. Self-Thinning and the Field of Neighborhood Approach. Net Resource Model 15(1), 39–54 (2002)
4. Deussen, O., Hanrahan, P., Lintermann, B., Mech, R., Pharr, M., Prusinkiewicz, P.: Realistic Modeling and Rendering of Plant Ecosystems. In: Proceeding of SIGGRAPH. ACM SIGGRAPH, pp. 275–286 (1998)
5. Lane, B., Prusinkiewicz, P.: Generating Spatial Distributions for Multilevel Models of Plant Communities. In: Proceeding of Graphics Interface, Calagary University, pp. 69–80 (2002)

Trends and Development on Green Computing

Jiabin Deng[1,*], Juanli Hu[1], Wanli Li[1], and Juebo Wu[2]

[1] Zhongshan Ppolytechnic, Computer Engineering, Boaiqi Road.
25, 528404 Zhongshan, China
[2] Shenzhen Angelshine Co., LTD, South area.
25, 518057 Shenzhen, China
hugodunne@yahoo.com.cn, hjlfoxes@163.com,
{hugodunne24,wujuebo}@gmail.com

Abstract. Mainly from the angle of computer system, this paper analyzes the negative influence of natural environment caused by the computer system. It presents a novel concept for green computing with generalized aspect and expands the traditional tasks and relationships of green computing research. Firstly, the deep discussion of generalized green computing using hardware and software technology and method is carried out, and the indicative meaning behind it is analyzed, including all kinds of the scheme for research institutions and enterprises. Then, some basic ideas of green computing and generalized methods are put forward in order to establish the necessary foundation for the methods and tools of green computing.

Keywords: Green Computing, Power Saving, Portable Computing, Sustainable Computing.

1 Introduction

In recent years, energy conservation, emission reduction and scientific development are not only the key task for energy-intensive heavy industry such as steel, chemical engineering and manufacturing, but also the urgent problem which IT has to confront with. Thus, green computing can play an important role in dealing with such problems in terms of coordinating the relationship between IT technology and environment. Currently, the main emphasis of green computing is comprehensive utilization of advanced technology as various soft/hardware, which can lower the workload of computer system and improve the operation efficiency in order to reduce energy consumption and decrease the number of computer systems. Meanwhile, it could make a better design for computer system, and it can also improve resource utilization, recovery and recycling utilization. By using the new energy and reducing carbon dioxide emissions or greenhouse gases, energy-saving, environmental protection and saving can be realized.

---

* Corresponding author.

Y. Yu, Z. Yu, and J. Zhao (Eds.): CSEEE 2011, Part II, CCIS 159, pp. 281–286, 2011.
© Springer-Verlag Berlin Heidelberg 2011

The paper will focus on the following aspects: portable computing [1], electronic waste [2], hardware power savings modes [3], new energy application [4], etc. It will also briefly cover the meaning and positive effects in green computing of what technologies are available in the race to meet green computing requirements.

2 The Environmental Impact by Computer System

The computer system can be divided into the humanistic environment and the natural environment. As the computer system widely used in daily life, it leads to some negative impact to environment. The computer system can be divided into the humanistic environment and the natural environment. As the computer system widely used in daily life, it leads to some negative impact to environment. The famous company Smart [5] cooperated with the Climate Group and the Global e-Sustainability Initiative (GESI) issued SMART 2020 in 2008: enabling the low carbon economy in the information age. The study highlighted the significant and rapidly-growing footprint of the ICT (Information and communications technology) industry and predicted that because of the rapid economic expansion in places like India and China, among other causes, demand for ICT services will quadruple by 2020, as shown in Table 1.

Table 1. The emission of carbon dioxide of the ICT

	Emissions 2007 (MtCO2e)	Percentage 2007	Emissions 2020 (MtCO2e)	Percentage 2020
World	830	100%	1430	100%
Server farms/ Data Centres	116	14%	257	18%
Telecoms Infrastructure and devices	307	37%	358	25%
PCs and peripherals	407	49%	815	57%

3 Green Computing Analysis

Monitor is the big energy consumption in computer system. Take early flat display (CRT Monitor) for example, it is a kind of display device using cathode ray tube which mainly has five parts: electron gun, defection coils, shadow mask, phosphor and glass shell [6]. The core component of CRT is the tube which works the same as the principle of television. The consumption of such monitor using in the past can reach 80w-100w on average differing from manufacturers and brands.

Computer Graphics is one of the key parts in computer system. The traditional graphics are generally independent components, and multi-core, the high frequency and the high volume display emerge in endlessly along with the needs of the users. Though they have different functions, the power and heat are ascending accompanied by such functions. Because it is a small proportion in the people who occupy the professional image work, scientific research and pursue the high display effect, the majority are still in a common state. So then it appears the motherboard integrates the graphics, even local video memory. The latest integrated graphics techniques have been integrated into the CPU, which can greatly reduce power consumption, such as Intel® Atom™ Processor N410, integrated Intel® multimedia accelerator with integrated storage 3150 controller. TDP is only 5.5 W and it also reduces calculating stress of mainboard chipset. Although it exits some differences in performance compared with independent video, it still caters to the users' demands.

VISION is a novel concept presented by AMD in 2009, aiming to help consumers simplify the buyer experiences. The earliest version is based on AMD laptop microprocessor with ATI chipset collocation, but now it has transited to desktop. It is no mention of technical parameters, but it is composed of processor, graphics, chipset and software etc. It is divided into basic version, premium version and black version for meeting the needs of the users', and the user just needs to choose corresponding grade products according to their application. In addition, AMD emphasizes the new energy technology into the platform specially. It also includes multi-core power management, the optimal PowerNow 2.0, CoolCore, and graphics PowerXpress support chipsets. The memory supports the low voltage DDR3L 1.35 V, Vari-Bright and software support of power management. Among them, PowerXpress 2.5 can realize integrated graphics card, independent of the dynamic switch, convenient to satisfy different needs, and extend battery application of time. (detail in 4.6.1) Using the appropriate integration, it can be not only more reasonable collocation of hardware and stable performance, but also save cost, energy use and more reasonable, as shown in Fig.1.

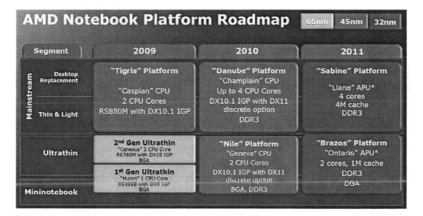

Fig. 1. The Roadmap of AMD Notebook Platform

4   Portable Computing

The rapid development of Internet technology is another reason that the portable computing is widely applied in daily life. Currently the popular mobile Network is mainly 3G Wireless and Wi-fi (Wireless Local Area Network). The acceptance of 3G standard by ITU mainly contains WCDMA, CDMA2000 and TD-SCDMA. The speed can reach 3.1 Mbit/s downloading and 1.8 Mbit/s uploading close to the cable broadband speed. Some companies launch Worldwide Interoperability for Microwave Access (WiMAX) [7] with the covering area of 1km2. Its safety and stability get a considerable progress than 3G and wireless technology with Wi-fi. With the time passing, with the further decline of wireless internet access prices, it can provide favorable conditions for mobile computing.

As the above content, it can be clearly seen that the portable computing is of great advantage to the average daily data processing. It can also break away from the earth to natural environment and limit the low consumption and low emission, bringing a vast space to the development of green computing.

5   Electronic Waste

E-Waste (Electronic waste) may be defined as all secondary computers, entertainment device electronics, mobile phones, and other items such as television sets and refrigerators, whether sold, donated, or discarded by their original owners.

There are two ways to solve E-waste problems as following:

(1). The hardware equipments of computer system can be made by using nontoxic and harmless, convenient and recycling materials, and establish guidelines to treatment and recycling. Currently, some companies like Apple have already used a virulent, environmental degradation and recycle materials to manufacture related hardware. Through the way such legislation, it can force enterprise and individual shall to take reasonable treatment and recycling strategies, in order to reduce the discharge of wastes.

(2). To increase the utilization ratio of the computer system hardware. According to the research, most of the discarded hardware is not because of their physical damage, such as the durable consumer goods of CPU, Mainboard, Memory, but due to the changing needs. Take Hard Disk for example, due to the improvement of speed and storage, now all the hard disks are using the lower materials of energy and system complexity, namely SATA (Serial Advanced Technology Attachment). In order to save cost and performance, large enterprises and individuals replaced hard disks and auxiliary equipments. Though in the reading speed and the energy consumption, the traditional ATA technology can not better than SATA technology. If the rest of the Disk can be linked through RAID (Redundant Array of Independent Disk), it can reduce operating costs for the enterprise and reduce the discard of E-waste by using Backup Dataset and Data Warehouse.

6   Hardware Power Savings Modes

Hardware Power Savings is the most commonly modes for reduce energy consumption, and its methods and strategies can involve all the Hardware in the computer system. Take CPU for example, Intel released core i CPU in 2010 which has advanced micro structure and intelligent technology, including farsighted frequency speed technology, hyper-threading technology and smart caching techniques, etc. These technologies are called intelligence CPU, so the users need not to interfere. In series I5 and I7, Turbo Boost is used into them based on the power management technology with Nehalem framework. Through the analysis of the current situation of the CPU load, it can close some useless function intelligently, and let energy apply into the running core in order to carry out higher frequencies and further improve performance. Conversely, when the program requires multiple cores, it will open the corresponding core to readjust frequency. So, it can put the core work to higher frequency without influencing the CPU' TDP.

The proper use of Graphics is a very positive significance to energy saving and emission reduction. The current popular Graphics have the ability of adjusting performance and power consumption automatically. As AMD Puma platform, the lack of the traditional graphics processing capacity can be handled by joint graph processing components to the mobile platform. Puma platform introduced several new technologies in graphics processing technologies, such as: Hybrid CrossFireX and Power Xpress. Hybrid CrossFireX means that two graphics can work together. In the condition of cooperating work, the display performance can dramatically increase by 70%. According to the testing data, it shows that after opening the Hybrid CrossFireX in the two mainstream products 3450 and 3470 of HD300 series, the performance can reach to high-end independent graphics level which can reduce the investment for high-tech laptop.

Hard Disk is the necessary software system in computer system, the technology has converted from SCSI (Small Computer System Interface) to ATA and SATA which can reduce the power consumption while improve the performance. Along with the development of technology and demand, Hard Disk has been gained a substantial improvement in recent years. Solid State Disk was introduced and it is thorough separation from the restrictions of magnetic read, motor and disc by FLASH memory chip and DRAM storage technology. It not only can access speed increased by about 40 percent, but also has zero noise, low energy consumption advantages.

The energy-saving models of hardware are varied, including hardware shielding modes, etc. For example, the hardware shielding by Founder Company can reduce power consumption when a user is not running some hardware equipments, such as CD-ROM wireless network and bluetooth devices, etc. Like hard disk, many hardware can start the modes of sleep modes and Hibernate, which can reduce power consumption, large energy saving and emission reduction.

7 Solar Power Equipment

The solar power equipment has the characteristic of inexpensive and easy-to-use, so it is widely applied and spread. But it mostly relies on environment and instability which brings great limitations to the solar power equipment. Currently, the solar power equipments do not always refer to the solar battery, but supply battery directly from solar power.

The power supply device of computer system using solar energy is mainly portable devices in domestic, including laptop and smart phone. At present, many manufacturers have been introduced various solar power equipments, such as solar charger and solar notebook computer (pack), which aims to the devices with less power and the computer system working outdoors. The power supply volume is restricted between 100mAh and ten 1,000 mAh while voltage from 3.0V to 19V. And it can meet the demands of all the popular portable devices recently. The specific technologies includes silicon solar battery, multiple compound film solar cells, polymer layers modified electrodes type solar battery, nanocrystalline solar battery, organic solar battery, where silicon solar battery is the most mature with leading position in the application.

8   Conclusions

Green computing as a new calculation model and technology, has turned the design strategy from the technology and the user as the center of IT products and competition to environmental for the center of IT products and competition strategy, and it is also the important way of sustainable development in the design of a computer system.

Although people have realized that the environment for human development plays an important role and more and more people have actively involved in the movement of green computing, the green computing is still in the initial stage when many methods, technologies or standards are not mature. And the computer system products of green computing are still immature with high price, and all these have become resistance of green computing.

References

1. Wernsing, J.R., Stitt, G.: Elastic Computing: A Framework for Transparent, Portable, and Adaptive Multi-core Heterogeneous Computing. J. ACM SIGPLAN Notices 45, 115–124 (2010)
2. Widmera, Heidi, O.K., Deepali, S.K., Max, S., Heinz, B.: Global Perspectives on E-waste Rolf. J. Environmental Impact Assessment Review 25, 436–458 (2005)
3. Schmidt, K., Härder, T., Klein, J., Reithermann, S.: Green computing - A Case for Data Caching and Flash Disks. In: 10th International Conference on Enterprise Information Systems, pp. 535–540. IEEE Press, Portugal (2008)
4. Dieter, H.: The New Energy Paradigm. Oxford University Press, United Kingdom (2007)
5. Enabling the Low Carbon Economy in The Information Age,
   http://www.smart2020.org/_assets/files/
   03_Smart2020Report_lo_res.pdf
6. Roy, S.B., Ricardo, J.M., Mark, E.G.: Theory and Practice. Color Research & Application. J. CRT Colorimetry 15, 299–314 (2007)
7. Jeffrey, G., Andrews, Arunabha, G., Rias, M.: Fundamentals of Wimax, Understanding Broadband Wireless Networking. Prentice Hall PTR, USA (2007)

Behavioral Consistency Analysis of the UML Parallel Structures

Huobin Tan[1,2], Shuzhen Yao[1], and Jiajun Xu[2]

[1] Computer School, BeiHang University, China
[2] College of Software, BeiHang University, China
{thbin,szyao}@buaa.edu.cn, jiajunm.xu@gmail.com

Abstract. In this paper, we analyze the parallel structural features of UML state machine diagrams and sequence diagrams, and put forward the technique used to describe parallel regions in a state machine diagram and parallel fragments in a sequence diagram through the parallel structure units of Petri nets. Then, we validate the common behavior consistency of state machines and sequence diagrams by introducing the Petri net language and equivalence theory, and present UML model's consistency analysis technique based on Petri net processes and occurrence nets. The consistency analysis technique is also extended to consistent validation of other UML diagrams.

Keywords: UML, Parallelism, Consistency, Petri net.

1 Introduction

When modeling a system with UML, it is necessary to build multi-perspective models. These models split the whole system up into the different and interrelated views. From the viewpoints of system behaviors, the state machine and interaction model respectively describe the difference behavior facets, and describe completely the system's behaviors only when they can keep pace with each other.

The model consistency is a key issue for the system model and implementation. Because of parallel mechanism involved, the model consistency validation becomes more difficult. The paper will analyze the parallel structural features in the state machine diagram and sequence diagram (referred to UML dynamic models), and put forward a consistency analysis method based on Petri nets.

2 Parallel Structural Features of UML Dynamic Models

The state machine diagram provides three kinds of transitions associated with parallel activities, namely *fork*, *join* and *syn*. A *fork* divides an object's control into several parallel parts, and each part corresponds to one region. A *join* merges object's parallel controls coming from parallel sub-state machines. A *syn* transition is used to specify parallel processes' synchronized point. In Fig 1, the transitions T_1 and T_2 respectively play roles of a *fork* and a *join* control. The composite state A_2 contains two parallel

Y. Yu, Z. Yu, and J. Zhao (Eds.): CSEEE 2011, Part II, CCIS 159, pp. 287–292, 2011.
© Springer-Verlag Berlin Heidelberg 2011

regions separated by a dashed line. The synchronized point "*Syn*" (a circle with "***" mark in the centre) expresses a constraint relation between the parallel regions. It is used to synchronize the transition t_1, t_2 and t_3. Similar to a synchronized clock, it assures that only a transition is executed at a clock cycle.

The sequence diagram describes the interactions among the objects, and provides the interaction fragment for a reusable structure unit. A parallel fragment expresses that the interaction sequences inside two or more parallel regions execute in parallel. In Fig 2, the sequence diagram depicts the interaction relations between the object A and B, and the object A sends sequentially the message a and b to B, meanwhile A sends the message c to B in parallel.

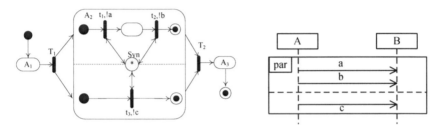

Fig. 1. An example of parallel behaviors   **Fig. 2.** An example of the parallel fragment

3   Petri Net Language and Equivalence

It is well known that a set of strings meeting some specific conditions in an alphabet is called a language of the event alphabet. When given a mapping from the set T of transitions in a Petri net to an alphabet Γ, all transition firing sequences of the Petri net may be mapped to the event sequence set of Γ^* and regarded as a language over T, where T^* and Γ^* are respectively any transition combination and event combination. It is given four different types and twelve classes Petri net languages in [1]. We use L-type with empty-label, language L^λ to express the event sequence set corresponding to state transitions in a Petri net.

Let $\Sigma = (P, T; F, M_0)$ be a Petri net, Γ be a labeled symbol set, and $\varphi: T \to \Gamma$ be a labeled function, then (Σ, φ) is a labeled Petri net. If Q_t is a specific subset of $R(M_0)$, namely $Q_t \subseteq R(M_0)$, $L = \{\varphi(\sigma) \in \Gamma^* | \sigma \in T^*: M_0[\sigma>M, M \in Q_t]$ is the L-type language produced by Σ through φ and expressed as $L(\Sigma, \varphi)$. If $\exists t \in T$, $\varphi(t) = \lambda$ (λ means an empty string), the L is the L^λ-type language and expressed as $L^\lambda(\Sigma, \varphi)$.

For the convenience of use, the equivalence based on the L^λ-type Petri net language is divided into the complete equivalence and partial equivalence.

Let both $\Sigma_1 = (P_1, T_1; F_1, M_{01})$ and $\Sigma_2 = (P_2, T_2; F_2, M_{02})$ are labeled Petri nets. There exist two mappings whose labeled functions are $\varphi_1: T_1 \to \Gamma$ and $\varphi_2: T_2 \to \Gamma$ to the same labeled symbols and $L_1(\Sigma_1, \varphi_1) \subseteq L_2(\Sigma_2, \varphi_2)$, then it is called that Σ_1 is an implementation of Σ_2 or Σ_1 includes Σ_2. If they implement each other, Σ_1 and Σ_2 are completely language equivalent.

Let both $\Sigma_1 = (P_1, T_1; F_1, M_{01})$ and $\Sigma_2 = (P_2, T_2; F_2, M_{02})$ be labeled Petri nets. There exist two mappings whose labeled functions are $\varphi_1: T_1 \rightarrow \Gamma$ and $\varphi_2: T_2 \rightarrow \Gamma$ to the same labeled symbols and for $\sigma_1: \varphi_1(\sigma_1) \in \Gamma^* |\sigma_1 \in T_1{}^*$ there must exist $\sigma_2: \varphi_2(\sigma_2) \in \Gamma^* |\sigma_2 \in T_2{}^*$ and $\varphi_1(\sigma_1) = \varphi_2(\sigma_2)$, then it is called that Σ_1 is an branch implementation of Σ_2 or Σ_1 branch includes Σ_2. If they are branch implementation each other, Σ_1 and Σ_2 are partially equivalent.

As can be seen above, although both the language equivalence and partial equivalence are used to analyze the transition sequences, their semantics are slightly different. The former emphasizes the whole transition sequences from the initial state to the final state, and the latter is used to check any potential sequences between any two states. The language equivalence may detect the event's relationships of sequential occurrences, whereas the partial equivalence may detect the labeled event's relationships in the different branches.

4 Mapping Dynamic Models to Labeled Petri Nets

As relatively independent reusable units, parallel regions and parallel fragments may be regarded as abstract transitions. Such abstract transitions can be analyzed when being mapped to parallel structures units in Petri nets.

4.1 Mapping a State Machine Diagram to a Labeled Petri Net

A state machine diagram is generally expressed as a 8-tuple $SM=(S, Init, SH, Labs, \delta, EE, CE, AE)$, where S, $Init$, $SH: S \rightarrow 2^s$; $Labs$, $\delta \subseteq 2^s \times Labs \times 2^s$; EE, CE and AE respectively represent the nonempty finite set of states, initial state, hierarchy function, labeling set, transition functions, event set, condition set and action set. If flatting of SM is adopted (see [3]) and implementation features like events, conditions and actions, are not considered, SM may be simplified as a 4-tuple $SM=(S, Init, Labs, \delta)$. The mapping rules from a simplified SM to a labeled Petri net $\Sigma=(P, T, F, M_0, \Gamma, \varphi)$ are as follows:

a. Mapping S to P.
b. Defining M_0 according to $Init$.
c. Mapping $Labs$ to Γ. In order to distinguish the difference, sending a message adds a prefix "!" whereas receiving a message adds a prefix "?". For example, "!a" d. expresses sending a message a and "?a" expresses receiving a message a.
Constructing T, F and φ according to δ.

Besides, the following rules are also necessary:

a. Mapping a composite state to a parallel unit.
b. Mapping a synchronized point in a parallel region to a synchronized place.

Fig 3 is the labeled Petri net mapped from Fig 1, where S_6 is a synchronized place and used to synchronize t_1, t_2, t_3.

4.2   Mapping a Sequence Diagram to a Labeled Petri Net

A sequence diagram is defined as a 4-tuple $SD=(O, I, Mess, \lambda)$, where O is a set of object identities, I is a set of all location sequence numbers (are natural numbers) and $max(I)$ represents the maximal number, $Mess$ is a set of message's identities, and λ: $I \rightarrow O \times O \times \{!,?\} \times Mess$ is a set of location-message mapping functions, explains the corresponding relationships between messages and locations and symbols "!" and "?" respectively represent sending and receiving the message.

The mapping relationships from a SD to a labeled Petri net $\Sigma=(P, T, F, M_0, \Gamma, \varphi)$ are as follows (see [5]):

Setting initial place p_0 to express the interaction beginning.

Mapping $Mess$ to Γ.

Constructing P, T, F according to λ. If λ is a single-valued mapping, each i is corresponding to one p_i, t_i, and arc $f_{i-1}=<p_{i-1}, t_i>$ and $f_i=<t_i, p_i>$; If λ is a multi-valued mapping, it is necessary to add a composite transition t_i corresponding to a parallel unit.

Fig 4 is the labeled Petri net mapped from Fig 2.

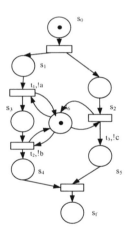

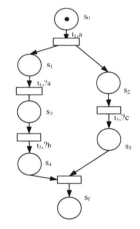

Fig. 3. The labeled Petri net of Fig 1   **Fig. 4.** The labeled Petri net of Fig 2

5   The Consistency Analysis Based on Object Communications

From the viewpoint of consistency, the responsive order of message requests in the state machine diagram should be consistent with the message sequence in the interaction diagram. For example, if the object A sends messages to B by "a, b" and the responsive order is "b, a", their communication behaviors are inconsistent. The consistence discussed above can be validated by means of Petri net's language equivalence according to two theorems as follows (see [6]):

They are behavior consistent if their Petri net languages are equal or language equivalent for the state machine and sequence diagram.

They are branch consistent if their Petri net languages which are produced respectively by the corresponding initial identities that are equal or partial equivalent for the state machine and sequence diagram.

Considered the language equivalence of Petri nets in Fig 3 and Fig 4, their sets of language are equals, namely $\{a, c, ab, ca, abc, acb, cab\}$, then their state machine and sequence diagram are behavior consistent. Because their prefix expressions are same, e.g. $a{\to}b$, $b{\to}c$, $c{\to}b$, $a{\to}c{\to}b$, $a{\to}b{\to}c$ and $c{\to}a{\to}b$, they are also branch consistent.

It is necessary to discuss the relations between events as smaller execution units in order to analyze the parallel structures. It is a kind of previous-next relationship in a sequence structure between two events, whereas there might be no any constraint relation between them in a parallel structure. This relationship needs to be expressed by the partial order based on processes.

The net theory containing parallel relations doesn't describe event sequences with *totally ordered* time axes, and holds that each resource's running track in the system forms a line and is a total order. Their running tracks for two resources refer two different time systems. Only when they intersect each other and participate together in the same variation, the two time systems are "simultaneous" (see [7]). This idea in the net theory can be expressed by the partial order relation. The partial order relation is built on occurrence nets and Petri net processes. The occurrence net and process are defined as follows (see [7]).

Let $N = (P, T; F)$ be a Petri net, if
$$\forall p \in P: |{}^\bullet p {\le} 1| \wedge |p^\bullet {\le} 1|, \forall x, y \in P \cup T: (x,y) \in F^+ \Rightarrow (y,x) \notin F^+$$

N is an occurrence net where F^+ is the transitive closure of the arc F.

Let $\Sigma = (P, T; F, M_0)$ be a Petri net, Let $N=(B, E; G)$ be an occurrence net, $\rho: N{\to}\Sigma$ satisfies the conditions as follows:

$$\rho(B) \subseteq P \wedge \rho(E) \subseteq T \wedge \forall (x,y) \in G: \rho(x,y)=(\rho(x), \rho(y)) \in F$$
$$\forall e \in E: \rho({}^\bullet e)= {}^\bullet \rho(e) \wedge \rho(e^\bullet)=\rho(e)^\bullet$$
$$\forall b_1, b_2 \in B: \rho(b_1)=\rho(b_2) \Rightarrow {}^\bullet b_1 {\neq} b_2{}^\bullet \wedge b_1{}^\bullet {\neq} {}^\bullet b_2$$
$$\forall p \in P: |\{b| {}^\bullet b = \varnothing \wedge \rho(b)=p\}| {\le} M_0(p)$$

Then (N, ρ) is called a *process*.

A Petri occurrence net is an observable record or actual execution process, whereas a process reflects an event net with labeled events. The partial order relations based on processes are a set of event's ordered pairs.

Let $\Sigma_1 = (P_1, T_1; F_1, M_{01})$, $\Sigma_2 = (P_2, T_2; F_2, M_{02})$ be labeled Petri nets. If there exist the partial orders: $(\Sigma_1, <)$, $(\Sigma_2, <)$ and labeled functions: $\varphi_1: T_1{\to}\Gamma$, $\varphi_2: T_2{\to}\Gamma$, there must exist t_2, $t_2' \in T_2$: $\varphi_2(t_2)<\varphi_2(t_2')$ and $\varphi_1(t_1)=\varphi_2(t_2)$, $\varphi_1(t_1')=\varphi_2(t_2')$ for all t_1, $t_1' \in T_1$: $\varphi_1(t_1)<\varphi_1(t_1')$. Then Σ_1 is a partial order implementation of Σ_2, or Σ_1 is partial order included by Σ_2. If both are partially ordered implementation each other, Σ_1 and Σ_2 are the partially ordered equivalence.

Though Fig 3 and Fig 4 are language equivalent, their event relations are essentially different because of the differences on structures. The a and c may in parallel execute in Fig 4, whereas the a and c can't in parallel execute because of the conflict in the P_2 in Fig 3. The partially ordered set based on processes is

$PO_1= \{a{\to}b,\ c\}$ in Fig 4, whereas $PO_2= \{a{\to}b{\to}c,\ b{\to}c{\to}a,\ c{\to}a{\to}b\}$ in Fig 3 and $PO_1{\neq}PO_2$, so their events' relations in the sequence and state machine diagram are inconsistent. This kind of consistency checking problem is ascribed to the Petri net's partially ordered equivalence. Namely, the sequence and state machine diagram are consistent if there exits partially ordered equivalence between their labeled Petri net.

The parallel structure is a kind of basic structure in UML dynamic behavior models, and the partially ordered equivalence theory provides a good foundation for the consistency analysis between models containing parallel structures.

6   Conclusion

This paper starts with parallel regions in UML state machine diagrams and parallel fragment in UML sequence diagrams, analyzes the UML parallel mechanisms and constructs Petri net's parallel structure units. Through mapping UML dynamic models to labeled Petri nets, the paper puts forward a kind of method of analyzing UML model consistency by Petri net language. For the sake of clarity, the paper analyzes communication interaction between two objects. Likewise, there exist two kinds of methods used to validate the model consistency among more objects. The first method is building labeled Petri nets for each object in the sequence diagram and checking the partially ordered equivalence with the labeled Petri nets of its state machine diagram. Another one is building labeled Petri net expressing the whole system interaction, extracting the partial orders for each object, comparing with the partial orders of its state machine diagram, and checking the dynamic model's consistency.

Acknowledgement. The research is supported by Science and Technology on Avionics System Integration Laboratory under grant No. 20085551042, and Commission of Science Technology and Industry of China.

References

1. Peterson, J.L.: Petri Net Theory, the Modeling of Systems. Prentice-Hall, Englewood Cliffs (1981)
2. Störrle, H.: Analysis and Design with UML and Petri-nets. Ludwig Maximilians University, Munich (2000)
3. Yao, S.Z., Shatz, S.M.: Consistency Checking of UML Dynamic Models Based on Petri Net Techniques. In: Proceedings of the 15th International Conference on Computing, CiC, pp. 289–297 (2006)
4. Hu, Z., Shatz, S.M.: Explicit Modeling of Semantics Associated with Composite States in UML Statecharts. Journal of Automated Software Engineering 13(4), 423–467 (2006)
5. Li, G.Y., Yao, S.Z.: Research on Mapping Algorithm of UML Sequence Diagram to Object Petri Nets. In: 2009 WRI Global Congress on Intelligent Systems, GCIS, vol. (4) (2009)
6. Yao, S.Z., Tan, H.B., Wang, C.M.: Consistency Analysis Among Models Based on Equivalence Theory of Petri Net. The Proceeding of the 12th Petri Net Theory and Application 38(1-2), 195 (2009)
7. Yuan, C.Y.: Petri Net Principle and Application. Publishing House of Electronics Industry (2005)
8. Zeng, Q.T., Wu, Z.F.: Process Net System of Petri Net. Chinese Journal of Computers 25(12) (2002)

Utilizing Morlet Spectral Decomposition to Detect Oil and Gas

Huixing Zhang[1,2], Bingshou He[1], and Xiaodian Jiang[1]

[1] Key Lab of Submarine Geosciences and Prospecting Techniques,
Ministry of Education, Ocean University of China, Qingdao, China
[2] State Key Laboratory of Coal Resources and Safe Mining,
China University of Mining and Technology, Beijing, China
{zhxing,hebinshou,xdjiang}@ouc.edu.cn

Abstract. Underground media consists of solid and liquid. Dual-phase medium theory can describe underground media more accurately than single-phase medium theory. Therefore, oil and gas detection method based on dual-phase medium theory will have higher accuracy. Through solving the seismic equations in dual-phase media, we get the results that seismic energy will decrease in dual-phase media and the attenuation coefficient is proportional to frequency. This means that lower frequency energy will be relatively enhanced and higher frequency energy will be relatively weakened when seismic propagating in hydrocarbon region. To extract this characteristic accurately can detect oil and gas. By using Morlet wavelet's advantage of showing signal's local properties both in temporal and spectral domain, we decomposed the seismic signals and extracted the characteristic of dual-phase media to detect oil and gas. The results of filed data show that Morlet wavelet spectral decomposition can be used for detecting oil and gas effectively.

Keywords: Morlet Wavelet, Oil And Gas Detection, Seismic Exploration, Dual-Phase Medium, Spectral Decomposition.

1 Introduction

Single-phase medium assumes that underground medium consists of pure solid. The traditional seismic exploration theory is based on single-phase medium theory, which plays an important role in detecting structural reservoirs. Currently, it is hard for single-phase medium theory to adapt to the requirements of seismic exploration because of without considering the influence of fluid. We must establish a new theory in order to improve the accuracy and reliability of exploration. Dual-phase medium assumes that underground medium consists of solid and fluid, where solid refers to the framework of the rock and fluid refers to liquid or gas filling in the pores. There is a relative movement between solid and fluid. Obviously, when underground medium contains oil and gas, it shows the characteristic of dual-phase medium. The petrophysical hypothesis of seismic wave propagation based on dual-phase medium theory is more reasonable, and it can describe seismic wave propagation mechanism in fluid-saturate porous solid more accurately. Therefore, theoretically, it can solve

Y. Yu, Z. Yu, and J. Zhao (Eds.): CSEEE 2011, Part II, CCIS 159, pp. 293–298, 2011.

the problem of oil and gas exploration better. To develop the method and technology based on dual-phase medium theory can improve the accuracy and reliability of oil and gas detection.

Dual-phase medium theory was established by Biot in his two papers [1, 2]. Later, many scholars studied on dual-phase medium theory from different aspects [3-6]. In order to describe the relative motion between solid and fluid delicately, Mavko and Nur proposed "squirt flow" theory in 1979 [7], and Dvorkin and Nur proposed Biot-Squirt (BISQ) model in 1993[8]. In the aspect of describing wave's attenuation, the squirt flow theory has the same conclusion with Biot's theory at low frequencies. However, it is difficult to put squirt flow theory into practice because it needs to know the information of rock structures, while these details are not easy to be obtained. BISQ model is the most close model to the real nature, but it is so complex that it hardly to be used. Therefore, Biot's theory is the relatively appropriate dual-phase theory that can be applied to practice currently.

In this paper, through solving Biot's equations, we get the analytic solutions of the equations and study the dynamic properties of seismic waves caused by the relative motion between solid and fluid.Using Morlet's advantage of showing signal's local characteristic both in temporal and spectral domain, we intend to apply Morlet spectral decomposition to oil and gas detection.

2   Dynamic Properties of Seismic Waves in Dual-Phase Medium

Biot's dilatational wave equations [1] are:

$$
\begin{cases}
\nabla^2[(A+2N)\theta + Q\varepsilon] = \dfrac{\partial^2}{\partial t^2}(\rho_{11}\theta + \rho_{12}\varepsilon) + b\dfrac{\partial}{\partial t}(\theta - \varepsilon) \\
\nabla^2(Q\theta + R\varepsilon) = \dfrac{\partial^2}{\partial t^2}(\rho_{12}\theta + \rho_{22}\varepsilon) - b\dfrac{\partial}{\partial t}(\theta - \varepsilon)
\end{cases}
\tag{1}
$$

where A and N correspond to the familiar Lame coefficients in the theory of elasticity. The coefficient R is a parameter related to fluid. The coefficient Q is of the nature of a coupling between the volume change of the solid and that of the fluid. The parameters ρ_{11}, ρ_{22}, ρ_{12} are mass coefficients. θ and ε are the strain in the solid and that in the fluid, respectively. The parameter b is dissipation coefficient which is related to Darcy's coefficient of permeability and the fluid viscosity and the porosity of the material. The parameter t represents time.

Consider a plane wave propagating in the x-direction:

$$
\begin{cases}
u = u_0 e^{i(k'x - \omega t)} = u_0 \cdot e^{-\alpha x} \cdot e^{ikx - i\omega t} \\
U = U_0 e^{i(k'x - \omega t)} = U_0 \cdot e^{-\alpha x} \cdot e^{ikx - i\omega t}
\end{cases}
\tag{2}
$$

where u is the displacement of solid, U is the displacement of fluid, and u_0 and U_0 is the initial value of the solid and that of fluid respectively. k' represents complex wave number, and $k' = k + i\alpha$, where k is the wave number, and α represents the attenuation coefficient. The parameter ω represents angular frequency and i is imaginary unit. The parameter x represents the distance of wave propagating.

After a series of derivation, we obtain:

$$\begin{cases} k = \sqrt{\dfrac{\sqrt{(\rho_{11}\rho_{22} - \rho_{12}^2)^2 H^2 \omega^4 + \rho^2 b^2 H^2 \omega^2} + (\rho_{11}\rho_{22} - \rho_{12}^2)H\omega^2}{2\rho(PR - Q^2)}} \\[3ex] \alpha = \sqrt{\dfrac{\sqrt{(\rho_{11}\rho_{22} - \rho_{12}^2)^2 H^2 \omega^4 + \rho^2 b^2 H^2 \omega^2} - (\rho_{11}\rho_{22} - \rho_{12}^2)H\omega^2}{2\rho(PR - Q^2)}} \end{cases}, \tag{3}$$

where $H = P + R + 2Q$, $\rho = \rho_{11} + \rho_{22} + 2\rho_{12}$.

Since attenuation coefficient α is not equal to zero, attenuation of seismic waves appears when they propagate in dual-phase media, and the attenuation coefficient is approximately proportional to angular frequency. Eq.(1) is not the viscoelastic, so the attenuation is not caused by the viscosity of the rock but caused by the existence of fluid, for, as we know, plane seismic wave energy will not attenuate propagating in single-phase perfect elastic medium. Therefore, if we neglect the viscosity of the rock and take it as perfect elastic medium, there will be seismic wave attenuation in dual-phase medium, that is, the lower frequency components of the energy decrease less while the higher frequency components of that decrease more, which leads to the lower frequency energy is relatively enhanced and the higher frequency energy is relatively weakened comparing with the single-phase medium. This conclusion can be used to detect whether underground media is dual-phase or filled with oil and gas by using seismic data.

3 Oil and Gas Detection Method and Principle by Using Morlet Wavelet

The attenuation coefficient is approximately proportional to frequency in dual-phase medium, so the energy distribution of seismic wave will change after propagating in oil and gas region. As to a certain target layer, if we can extract the lower and higher frequency energy of seismic wave accurately, we will be able to judge whether underground media contain oil or not by the characteristic of lower frequency energy enhanced and higher frequency energy weakened.

Using spectral decomposition technology can extract a certain frequency range of seismic information. Morlet is a kind of wavelet that has pretty good local properties both in temporal and spectral domain. It is likely to extract the low and high frequency energy of seismic waves accurately so as to realize oil and gas detection.

Propose $x(t)$ is square integratiable function, $\varphi(t)$ is basic wavelet or mother wavelet function, then

$$WT_x(a, \tau) = \frac{1}{\sqrt{a}} \int_{-\infty}^{\infty} x(t)\varphi^*(\frac{t - \tau}{a})dt \tag{4}$$

is called wavelet transform [9], where $a > 0$ refers to scale factor, τ shows the displacement, superscript* represents conjugate. If we select morlet $\varphi(t) = e^{-\frac{t^2}{T}} e^{j\omega_0 t}$ as basic wavelet, then we call it morlet wavelet transform.

The procedures utilizing morlet wavelet to decompose seismic signals are as follows:

(1) Select at least one seismic record respectively from oil-bearing wells and dry wells. For the target layer, do spectrum analysis and calculate the optimal sensitive frequencies fl1 - fl2 and fh1- fh2 of hydrocarbon, where fl1 and fl2 represents for the begining and the ending of the lower sensitive frequencies respectively, while fh1 and fh2 respresents for that of higher frequencies respectively;

(2) Apply morlet wavelet transform to seismic data according to fl1, fl2, fh1 and fh2. For any seismic trace, adopting a certain frequency step length df, calculate sensitive frequency parameters: low frequency $f_i = FL1 + i*df$, high frequency $f_i = FH1 + i*df$, where f_i represents the ith frequency; according to $a_i = \dfrac{2}{dt \cdot f_i}$, calculate relevant scales a_i , where dt means sample interval; according to a_i , apply morlet wavelet transform to the seismic data of target layer;

(3) If the number of frequency in lower or higher frequency range is n , then n results will be acquired through wavelet transform. Find out the maximum amplitude from n filtering results and take it as lower frequency energy or higher frequency energy;

(4) Repeat procedure (2) and (3) until all seismic data are calculated;

(5) For the calculated seismic wave energy in lower and higher frequencies, according to whether the characteristic that the lower frequency energy becomes relatively stronger and the higher frequency energy becomes relatively weaker appears or not to map oil and gas regions.

4   Applying Example

Selecting a 3D seismic data of G work area, the target layer is Shasanzhong. Shasanzhong is a set of turbidite sandstone reservoir consists in a large set of mudstone in Shasan. The main type of reservoir in Shasanzhong is lithologic reservoir controlled by sandstone updip pinch out trap.

There are five wells in this work area. From the oil testing data, we have known that three of them (well1, well2, well3) are oil wells and the other two (well4 and well5) do not contain any oil. We use Morlet wavelet transform to extract lower and higher frequency energy oil and gas sensitive to, as shown in Fig. 1. Time window length we used is 50ms. The lower and higher frequency parameters used are: 1-10Hz and 35-45 Hz. In Fig.1, lower and higher frequency seismic wave energy values are normalized, and locations of the wells are marked: oil wells marked in red, dry wells marked in white.

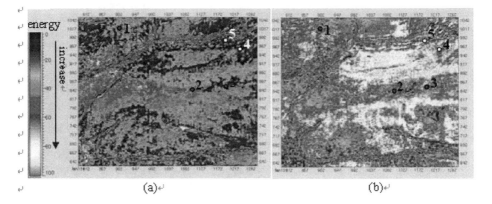

(a)   (b)

Fig. 1. Oil and gas detection results in G work area: (a) lower frequency energy(1-10Hz), (b) higher frequency energy(35-45Hz)

From Fig.1 (a) and Fig.1 (b), we can see that we ll 1, well 2, and well 3 are mainly located in areas where lower frequency energy values are larger while higher frequency energy values are smaller, which illustrated that the three wells have a large possibility of containing oil. In fact, they are oil wells as we have already known. Well 4 and well 5 are located in areas that low-frequency energy is smaller, high-frequency energy is larger, which shows a less possibility of containing oil. The two wells are dry wells actually. Therefore, the oil and gas detection results of the 5 wells by Morlet wavelet transform are in accordance with the actual facts. For the convenience of explanation, Fig.1 (a) and Fig.1 (b) are combined to Fig.2 through algorithm. Fig.2 is the final result of oil and gas detection of this work area. In Fig.2, the greater the energy value, the larger the possibility of containing oil and gas. It can be more clearly seen that oil wells are located in the larger energy area and dry wells are located in the smaller energy area, which shows the accuracy and effectiveness of this method.

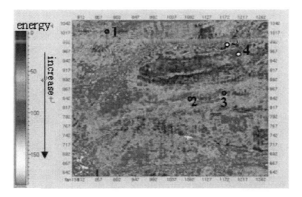

Fig. 2. Final result of oil and gas detection in G work area

5 Conclusions

In dual-phase medium, there exists seismic wave attenuation because of the influence of the fluid. The attenuation coefficient is approximately propotional to frequency, which leads to lower frequency energy is relatively enhanced and higher frequency energy is relatively weakened compared with the same layer of surrounded rocks after seismic wave propagating in oil and gas reservoirs. To extract this characteristic of dual-phase medium accurately and effectively can detect the existence of fluid or hydrocarbon.

Using the advantages of Morlet wavelet describing local properties of signals both in temporal and spectral domain, we can achieve the purpose of extraction of the required information. The results of Morlet transform to field data and hydrocarbon detection show that Morlet wavelet can realize the accurate extraction of different frequency energy of seismic waves and can be used for oil and gas detection.

Acknowledgement. We thank our colleagues at Shengli Oil Company (SLO) for providing the seismic data for our scientific research project within the framework of a scientific cooperation agreement between Ocean University of China (OUC) and SLO. This research is co-supported by National High Technology Research and Development Program 863 - 2008AA09Z302, P. R. China and State Key Laboratory of Coal Resources and Safe Mining (2007-03).

References

1. Biot, M.A.: Theory of Propagation of Elastic Waves in A Fluid-Saturated Porous Solid. Low-Frequency Range. Journal of the Acoustical Society of America 28(2), 168–178 (1956)
2. Biot, M.A.: Theory of Propagation of Elastic Waves in a Fluid-Saturated Porous Solid. II. High-Frequency Range. Journal of the Acoustical Society of America 28(2), 179–191 (1956)
3. Plona, T.J.: Observation of a Second Bulk Compressional Wave in a Porous Medium at Ultrasonic Frequencies. Appl. Phys. Lett. 36, 259–261 (1980)
4. Nagy, P.B., Adler, L., Bonner, B.P.: Slow Wave Propagation in Air-Filled Porous Materials and Natural Rocks. Appl. Phys. Lett. 56, 2504–2506 (1990)
5. Boyle, F.A., Chotiros, H.P.: Experimental Detection of a Sow Acoustic Wave in Sediment at Shallow Grazing Angles. J. Acoust. Soc. Am. 91, 2615–2619 (1992)
6. Zhang, H.X., He, B.S., Ning, S.N.: High-Order Finite Difference Solution of Dilatational Wave Equations in Two-Phase Media. Geophysical and Geochemical Exploration 28(4), 307–309 (2004)
7. Mavko, G., Nur, A.: Wave Attenuation in Partially Saturated Rocks. Geophysics 44, 161–178 (1979)
8. Dvorkin, J., Nur, A.: Dynamic Poroelasticity: A Unified Model with the Squirt and the Biot Mechanisms. Geophysics 58(4), 524–533 (1993)
9. Yang, F.S.: Engeering Analysis and Application of Wavelet Transform. Science Press, Beijing (2001)

Improving SVM via Local Geometric Structure for High Dimensional Data Classification

Yanmin Niu[1] and Xuchu Wang[2]

[1] College of Computer and Information Science, Chongqing Normal University,
Chongqing 400050, P.R. China
[2] Key Laboratory of Optoelectronic Technology and Systems of Ministry of
Education, Chongqing University, Chongqing 400044, P.R. China
niuym@cqnu.edu.cn, seadrift.wang@gmail.com

Abstract. The support vectors to maximize between-class margins in traditional Support Vector Machines (SVMs) are usually exclusive from with-class partners and hardly reflect the class distribution information, which easily makes the separating hyperplane less consistent to the intrinsic manifold structure in high dimensional feature space. In this paper a modified class of SVMs with local geometric structure preserving is proposed by introducing the local manifold structure based distance metric. Specifically, a graph matrix is firstly defined to represent the local geometric structure and then introduced to reformulate the objective of distance metric with small sample size consideration for high dimensional data, finally the proposed objective is deduced for efficient computation. Experimental results on known data sets validate the effectiveness and superiority of the proposed method.

Keywords: supervised machine learning, support vector machine, manifold learning, local geometric structure preserving.

1 Introduction

Support vector machine (SVM) is a powerful machine learning method arisen from Vapnik's statistical learning theory. Different from traditional empirical risk minimization approaches that usually attempt to minimize the misclassification errors on the training set, SVM essentially finds a linear or kernel-based nonlinear separating hyperplane which achieves the maximal margin among two different classes of data [1]. This structural risk minimization strategy makes it obtain good generalization and computationally efficient convex optimization.

Although the support vectors to maximize between-class margins can determine the separating hyperplane in SVM, they are exclusive from with-class partners and hardly reflect the distributions of class. To overcome this drawback, the minimum class variance SVM [2] and its variants [3,4] were introduced to modify the traditional SVMs and the improvements have been validated by various experiments, however, these methods consider the with-class samples equivalently due to the scatter matrix can just recover globally linear structure

Y. Yu, Z. Yu, and J. Zhao (Eds.): CSEEE 2011, Part II, CCIS 159, pp. 299–304, 2011.

of data. To incorporate with the local nonlinear structure, a Laplacian SVM (LSVM) method by combining SVM and LPP is proposed [5], but just aims at the semi-supervised learning tasks. Furthermore, another modified SVM by combining SVM and LPP is recently presented [6], in which the locality preserving with-class scatter matrix is introduced to explicitly consider the data manifold structure and experimental results indicates its improvements on traditional SVM. However, this method ignores the constraints of the summarized entries of LPP, in some sense, the lack of normalization of the entries makes it lose discriminative such as clustering property. It is necessary to directly consider both the data manifold structure and the traditional SVM from other way.

In this paper we propose a modified SVM method called LGS-SVM which adds constrain of neighbors to preserve with-class manifold structure on marginal distance. We firstly define a graph matrix to represent local geometric structure, then reformulate distance metric by introducing this matrix with SSS consideration for high dimensional data, and finally deduce an efficient computation. We also validate it by experiments on UCI and PIE data sets.

2   SVM with Local Geometric Structure Preserving

2.1   Building Within-Class Local Geometric Structure Matrix

The local geometric structures of samples play key role in nonlinear feature extraction and classification tasks. For example, recent investigations have shown that the underlying assumptions of global un-linearity but local linearity provide much help for discovering the nonlinear structure of the manifold and extracting nonlinear features [7]. The process of unfolding nonlinear structures of manifold are usually by means of mapping nearby points in the high-dimensional space to nearby points in a low-dimensional feature space, so we propose to extend the within-class scatter matrix $\mathbf{S}_w$ as the following form [8]:

$$\mathbf{S}_{nw} = \sum_{i=1}^{N}(\mathbf{x}_i - \sum_{\Phi(\mathbf{x}_i,\mathbf{x}_j)\neq\Gamma} A_{ij}\mathbf{x}_j)(\mathbf{x}_i - \sum_{\Phi(\mathbf{x}_i,\mathbf{x}_j)\neq\Gamma} A_{ij}\mathbf{x}_j)^{T}, \qquad (1)$$

where Γ denotes an empty set and $\Phi(\mathbf{x}_i, \mathbf{x}_j)$ a set by two samples $\mathbf{x}_i$ and $\mathbf{x}_j$, A_{ij} the normalized weight to measure the difference between $\mathbf{x}_i$ and $\mathbf{x}_j$. There are two ways to build $\Phi(\mathbf{x}_i, \mathbf{x}_j)$: (1)$\Phi(\mathbf{x}_i, \mathbf{x}_j) = \{\mathbf{x}_j | \mathbf{x}_j \in \Omega(\mathbf{x}_i) \text{ or } \mathbf{x}_i \in \Omega(\mathbf{x}_j)\}$; (2) $\Phi(\mathbf{x}_i, \mathbf{x}_j) = \{\mathbf{x}_j | \mathbf{x}_j \in \Omega(\mathbf{x}_i) \text{ and } \mathbf{x}_i \in \Omega(\mathbf{x}_j)\}$, where $\Omega(\mathbf{x}_i)$ denotes the set of k nearest (or $\varepsilon-$) within-class neighbors of $\mathbf{x}_i$. They both shrink the within-class relationship of $\mathbf{x}_i$, $\mathbf{x}_j$ from globality to locality. The stricter case (2) is taken. By putting Φ on A_{ij} to preserve the local structures, $\mathbf{S}_{nw}$ can be reformulated as

$$\mathbf{S}_{nw} = \sum_{i=1}^{N}(\mathbf{x}_i - \sum_{j=1}^{N} A_{ij}\mathbf{x}_j)(\mathbf{x}_i - \sum_{j=1}^{N} A_{ij}\mathbf{x}_j)^{T} = \mathbf{X}\mathbf{M}\mathbf{X}^{T} \qquad (2)$$

where $\mathbf{M} = (\mathbf{I}\text{-}\mathbf{A})^{T}(\mathbf{I}\text{-}\mathbf{A})$ and $\mathbf{A}$ is a matrix with entries A_{ij}. So, $\mathbf{S}_{nw}$ is positive and semi-positive definite. The resulting form in Eq.(2) is equivalent to the reconstruction error in Locally Linear Embedding (LLE) and in its linear approximation Neighborhood Preserving Embedding (NPE) [7]. This helps to investigate $\mathbf{S}_{nw}$ more flexibly by means of the data manifold. From the viewpoint of graph, $\mathbf{A}$ constructs a weighted k- nearest or $\varepsilon-$ neighbor graph which models explicitly the within-class data topology.

A_{ij} should be potentially determined by imposing the constraint that each sample in the reduced space is reconstructed from its neighbors by the same weights used in the input space after linear projection. In our case, A_{ij} under this reconstruction sense can be chosen as follows. For each point $\mathbf{x_i}$ find k nearest neighbors to build the neighborhood matrix $\mathbf{X}_i$, where each column is a vector from the set of k nearest neighbors of $\mathbf{x_i}$, then use singular value decomposition (SVD) to write $\mathbf{X}_i = \mathbf{ULV}'$, build $\mathbf{U}_2 = \{\mathbf{u_{d+1}}, \cdots, \mathbf{u_k}\}$ as the basis of null space from $\mathbf{U}$ corresponding to the zero eigenvalues in $\mathbf{L}$ and compute $\mathbf{A}_{i:} = \mathbf{U}_2\mathbf{U}_2'\mathbf{1}/\mathbf{1}'\mathbf{U}_2\mathbf{U}_2'\mathbf{1}$, where $\mathbf{1}$ denotes a vector with all entries of one. A_{ij} is estimated to satisfy certain optimality properties, i.e., invariance to rotations, scalings, and translations to make the affinity graph preserves intrinsic geometric characteristics of each neighborhood.

2.2   Local Geometric Structure Constrains for SVM

SVM differs from other hyperplane-based classifiers by virtue of how the hyperplane is selected. By defining the distance from the separating hyperplane to the nearest expression vector as the margin of the hyperplane, the SVM selects the maximum-margin separating hyperplane in an Euclidean distance sense. We shall improve the margin as Mahalanobis distance metric by incorporating $\mathbf{S}_{nw}$, considering the "soft margin" directly and the objective and constraint functions becomes

$$\arg\min_{\mathbf{w},b} \left(\frac{1}{2}\mathbf{w}^T\mathbf{S}_{nw}\mathbf{w} + C\sum_{i=1}^{N}\xi_i \right),$$ (3)

$$\text{s.t.} \quad y_i(\mathbf{w}^T\mathbf{x}_i + b) \geqslant 1 - \xi_i; \xi_i \geqslant 0; i = 1, \cdots, N,$$

where $\boldsymbol{\xi}$ denotes the non-negative slack variables for sample $\mathbf{x}_i$, C is a given regularization constant denoting the cost of classification errors. Larger C corresponds to higher penalty assigned to errors. The linearly separable case, i.e., "hard margin", can be achieved when choosing $C = +\infty$. As in SVM and MCVSVM, this problem can be transformed into its corresponding dual problem by introducing the following primal Lagrangian

$$\mathcal{L}(\mathbf{x}, b, \boldsymbol{\alpha}, \boldsymbol{\beta}, \boldsymbol{\xi}) = \frac{1}{2}\mathbf{w}^T\mathbf{S}_{nw}\mathbf{w} + C\sum_{i=1}^{N}\xi_i - \sum_{i=1}^{N}\alpha_i[y_i(\mathbf{w}^T\mathbf{x}_i + b) - 1 + \xi_i] - \sum_{i=1}^{N}\beta_i\xi_i,$$ (4)

where $\boldsymbol{\alpha}, \boldsymbol{\beta}$ are Lagrangian multipliers for the constraints. Suppose the matrix $\mathbf{S}_{nw}$ is nonsingular (the singular problem of $\mathbf{S}_{nw}$ will be discussed in next subsection), by differentiating with respect to $\mathbf{w}$, b and $\boldsymbol{\xi}$ and using the KKT conditions, we have $\mathbf{w} = \mathbf{S}_{nw}^{-1}\sum_{i=1}^{N}\alpha_i y_i \mathbf{x}_i$. Putting it into Eq.(4) and using the KKT conditions, Eq.(3) is reformulated to the Wolfe dual functional

$$\arg\min_{\boldsymbol{\alpha}} \left(f(\boldsymbol{\alpha}) = \mathbf{1}_N^T \boldsymbol{\alpha} - \frac{1}{2} \boldsymbol{\alpha}^T \mathbf{Q} \boldsymbol{\alpha} \right),$$

$$\text{s.t.} \quad 0 \leqslant \alpha_i \leqslant C; i = 1, \cdots, N; \boldsymbol{\alpha}^T \mathbf{y} = 0; \tag{5}$$

where $\mathbf{1}_N$ is a N-dimensional vector of ones and $Q_{ij} = y_i y_j \mathbf{x}_i^T \mathbf{S}_{nw}^{-1} \mathbf{x}_j$. Eq.(5) is a typical convex quadratic programming problem easy to be solved. Suppose $\boldsymbol{\alpha}^*$ is a solution vector, then the optimal weight vector is $\mathbf{w}^* = \mathbf{S}_{nw}^{-1} \sum_{i=1}^{N} \alpha_i^* y_i \mathbf{x}_i$. If $\alpha_i^* \in (0, C)$, the corresponding sample $\mathbf{x}_i$ is called a support vector (SV). Although threshold b^* can be determined by an arbitrary SV, its optimization is better found by exploiting the fact that according to the KKT condition, for all SVs their corresponding slack variables are zero. By averaging over all support vectors yields usually a numerically stable solution. So we can calculate b^* as

$$b^* = \frac{1}{N_{sv}} \sum_{i=1}^{N_{sv}} y_i (1 - \sum_{j=1}^{N} \alpha_j^* Q_{ij}); \mathbf{x}_i \in \Psi_{sv}, \tag{6}$$

where Ψ_{sv} is the set of SVs and N_{sv} the number of SVs. As a result, the corresponding decision surface of LGS-SVM is:

$$g(\mathbf{x}) = \text{sgn}(\mathbf{w}^T \mathbf{x} + b) = \text{sgn} \left(\sum_{i=1}^{N} \alpha_i^* y_i (\mathbf{x}_i^T \mathbf{S}_{nw}^{-1} \mathbf{x}) + b^* \right). \tag{7}$$

2.3   LGS-SVM for Small Sample Size Problem

The small sample size problem occurs whenever the number of samples (N) is smaller than the dimensionality of the samples (D) [9]. In some high-dimensional sample cases such as image appearance-based face recognition and gene pattern analysis, $D \gg N$. Considering the rank of $D \times D$ matrix $\mathbf{S}_{nw}$ not larger than $N - C$, $\mathbf{S}_{nw}$ usually encounters the heavy singularity. To address this, the optimization problem of LGS-SVM is reformulated into an equivalent one in a lower dimensional space through dimensionality reduction using Principle Component Analysis (PCA), where the within-class scatter matrix $\mathbf{S}_w$ is nonsingular to make an efficient solution, i.e. all the samples in the original space is first transformed to a low-dimensional space $\mathbf{X}_{pca} = \mathbf{W}_{pca}^T \mathbf{X}$, where $\mathbf{W}_{pca} \in \mathbb{R}^{D \times B}$ is the transform that projecting $\mathbf{X}$ to its PCA range subspace, due to the rank of $\mathbf{S}_t$ is not larger than $N - 1$, $B \leq N - 1$ (the equality holds in case that the training samples are linearly independent). Then $\mathbf{S}_{nw}$ should be modified as

$$\mathbf{S}_{nw,pca} = \mathbf{W}_{pca}^T \mathbf{S}_{nw} \mathbf{W}_{pca}. \tag{8}$$

Intrinsically, This PCA preprocessing itself does not lose any optimal discriminatory information for solving the problem in Eq.(3). This claim is achieved because during PCA, let $\mathbf{W}_{pca}$, $\mathbf{W}_{pca}^{\perp}$ be the range space and null spaces spanned by the eigenvectors of $\mathbf{S}_t$ that correspond to nonzero eigenvalues and to zero eigenvalues, the following theorem holds:

Theorem I. Let $\mathbf{w} = \mathbf{v} + \mathbf{v}^\perp$; $\mathbf{w} \in \mathbf{R}^D$, $\mathbf{v} \in \mathbf{W}_{pca}$, $\mathbf{v}^\perp \in \mathbf{W}_{pca}^\perp$, the optimization problem Eq.(3) is equivalent to $\underset{\mathbf{v},b}{\arg\min} \left(\frac{1}{2}\mathbf{v}^T \mathbf{S}_{nw,pca}\mathbf{v} + C \sum_{i=1}^{N} \xi_i \right)$, s.t. $y_i(\mathbf{v}^T \mathbf{W}_{pca}^T \mathbf{x}_i + b) \geqslant 1 - \xi_i; \xi_i \geqslant 0; i = 1, \cdots, N$.

Its proof is omitted due to space limit. Discarding the vector corresponding to the smallest nonzero eigenvalue in all vector basis in PCA, now the optimal weight vector is $\mathbf{w}_{pca}^* = \mathbf{S}_{nw,pca}^{-1} \sum_{i=1}^{N} \alpha_i^* y_i \mathbf{W}_{pca}^T \mathbf{x}_i$, and Eq.(5) has $Q_{ij,pca} = y_i y_j \mathbf{x}_i^T \mathbf{W}_{pca} \mathbf{S}_{nw,pca}^{-1} \mathbf{W}_{pca}^T \mathbf{x}_j$, thus,

$$b_{pca}^* = \frac{1}{N_{sv}} \sum_{i=1}^{N_{sv}} y_i(1 - \sum_{j=1}^{N} \alpha_j^* Q_{ij,pca}); \mathbf{x}_i \in \Psi_{sv}, \tag{9}$$

finally the decision hyperplane becomes

$$g(\mathbf{x}) = \mathrm{sgn}\left(\sum_{i=1}^{N} \alpha_i^* y_i (\mathbf{x}_i^T \mathbf{W}_{pca} \mathbf{S}_{nw,pca}^{-1} \mathbf{W}_{pca}^T \mathbf{x}) + b_{pca}^* \right). \tag{10}$$

3   Experimental Results and Conclusions

We evaluated the proposed LGS-SVM method and the related SVM, MCVSVM, MLPCVSVM as well as LDA on the following representative data sets: (1)Four subsets on UCI; (2)CMU PIE. More details about them refer to websites and [8]. For each classifier, the recognition rate was estimated with 5-fold cross validation to ensure good statistical behavior. This procedure is repeated five times and then the average recognition rate and the standard deviation across all trials is computed. Considering parameters determination for SVM-based methods is still an open problem, we use grid searching strategy to settle them. The common slack variable C is from $\{0.001, 0.01, 0.1, 1, 10, 100\}$; k for MCLPVSVM is the class number and t is from $\{0.5, 1, 1.5, 2, 2.5\}$; and the only neighbor number for LGS-SVM is from $\{3, 6, 9, 12, 15, 18, 24\}$. The evaluation are performed on UCI without SSS consideration as well as on PIE and FERET with SSS consideration. Their experimental results are reported in Table 1. From where we can start with two general observations: First, the LGS-SVM classifier performs, in general, better for most data sets. The one exceptions are Breast data sets, where

Table 1. Average recognition rate (%) and standard dev. of the cross validation on selected UCI PIE, and FERET data sets in non-SSS and SSS cases

Data set	LDA+NN	SVM	MCVSVM	MCLPVSVM	LGS-SVM
Heart	84.07±.048	84.44±.045	85.19±.051	85.58±.048	**85.92±.043**
Breast	96.57±.022	**96.86±.021**	**96.86±.021**	**96.86±.021**	**96.86±.021**
Wdbc	95.61±.011	97.19±.007	96.82±.008	96.82±.008	**97.21±.010**
Ionosphere	86.00±.034	87.71±.030	84.86±.037	89.12±.054	**89.68±.046**
PIE	84.56±.049	84.85±.051	85.45±.056	86.19±.052	**86.97±.058**

the SVM-based methods obtain very similar recognition rates. This may indicate that in most data sets, there is indeed separation information present in manifold structure constraints of the class distributions. Second, we see that SVMs with class distribution consideration indeed can improve the accuracy of the classifier in most cases. Note, although, that this is not always the case (for example, SVM performs also best on Wdbc) and wether it does hold the improvements are sometimes not very convincing. However, the error rate does not drop considerably and we can strength the claim by investigating the higher recognition rates in the SSS case, where LGS-SVM method remarkably outperforms other methods, e.g. on the high dimensional PIE data sets.

This paper presents a new modified SVM for high-dimension data classification. This method introduces neighborhood-based manifold structure to adjust the locality property of support vectors by regulating the objective to enforce locality on the separating hyperplane. In SSS case, PCA is introduced to reduce the original dimensionality without loss of information. As a result, the proposed method is more robust and homogeneous for the bias and SSS problem. The experimental results clearly show its advantage over SVMs, especially when input samples are heavy bias and with high dimensionality. Of course, the proposed method can be extended by kernel trick, feature pre-selection and etc. They are our next main research interests.

Acknowledgments. This work was partially supported by the Natural Science Foundation of Chonqing Science & Technology Committee (CSTC 2009BB3192).

References

1. Vapnik, V. (ed.): The Nature of Statistical Learning Theory. Springer, Heidelberg (1995)
2. Tefas, A., Kotropoulos, C., Pitas, I.: Using support vector machines to enhance the performance of elastic graph matching for frontal face authentication. IEEE Trans. Pattern Anal. Mach. Intell. 23, 735–746 (2001)
3. Xiong, T., Cherkassky, V.: A combined SVM and LDA approach for classification. In: Proc., IEEE Int'l Joint Conf. Neural Networks, pp. 1455–1459 (2005)
4. Zafeiriou, S., Tefas, A., Pitas, I.: Minimum class variance support vector machines. IEEE Trans. Image Process. 16, 2551–2564 (2007)
5. Belkin, M., Niyogi, P., Sindhwani, V.: Manifold regularization: A geometric framework for learning from labeled and unlabeled examples. Journal of Machine Learning Research 7, 2399–2434 (2006)
6. Wang, X., lai Chung, F., Wang, S.: On minimum class locality preserving variance support vector machine. Pattern Recognition 43, 2753–2762 (2010)
7. Kokiopoulou, E., Saad, Y.: Orthogonal neighborhood preserving projections: A projection-based dimensionality reduction technique. IEEE Trans. Pattern Anal. Mach. Intell. 29, 2143–2156 (2007)
8. Wang, X., Niu, Y.: Locality projection discriminant analysis with an application to face recognition. Optical Engineering 49, 077201 (2010)
9. Niu, Y., Wang, X.: Extraction of discriminative manifold for face recognition. In: King, I., Wang, J., Chan, L.-W., Wang, D. (eds.) ICONIP 2006. LNCS, vol. 4233, pp. 197–206. Springer, Heidelberg (2006)

Analysis and Modeling on the GHG Emissions in Dyeing and Finishing Processes

Yingxiang Fan[1], Ming Du[2], and Hui Song[2]

[1] College of Computer Science and Technology Donghua University
Shanghai, China
[2] College of Computer Science and Technology
Donghua University,Shanghai, China
fanyx8201@hotmail.com,
{duming,songhui}@dhu.edu.cn

Abstract. This paper analyzes the GHG emissions from each process in dyeing and finishing and finds the key points. The finding shows that the energy using and industrial activities contribute the main GHG emissions. And a model is created to calculate the GHG emissions in dyeing and finishing industry. With the given emission factor, an application was done for a real dyeing and finishing factory.

Keywords: GHG Emissions, Dyeing And Finishing, Carbon Emissions, Software Model, Consumption.

1 Introduction

In recent years, the world increasingly concerns about global warming issue, especially in the greenhouse gas (GHG) emissions. If nothing to do, greenhouse effect will threat the human's life.

The Chinese government, in 2009 United Nations Climate Change Conference in Copenhagen, announced that China will reduce the intensity of carbon dioxide emissions per unit of GDP in 2020 by 40 to 45 percent compared with the level of 2005 [1]. The GHG emissions from the dyeing and finishing industry can not be discounted [2]. Up to now, many enterprises still lag behind. They are high energy and high water consumption.

Carbon footprint' is a term used to describe the amount of GHG emissions caused by a particular activity or entity, and thus a way for organizations and individuals to assess their contribution to climate change [3]. "Kyoto Protocol" Annex-A highlight the six kinds of gas, such as carbon dioxide (CO_2), methane (CH_4), nitrous oxide (N_2O) and so on. Further, the proportion of carbon dioxide is more than 60%. Usually through the use of Global warming potential (GWP), the greenhouse gases are converted into carbon dioxide equivalent emissions.

This paper analyzes the GHG emissions from every process in dyeing and finishing and finds that the energy using and industrial activities contribute the main GHG emissions. And a model is created to calculate the GHG emissions in dyeing and finishing industry.

Y. Yu, Z. Yu, and J. Zhao (Eds.): CSEEE 2011, Part II, CCIS 159, pp. 305–309, 2011.
© Springer-Verlag Berlin Heidelberg 2011

2 Analysis on the GHG Emissions

Dyeing and finishing processes, according to the different kinds of fabrics, are very different, especially pretreatment. For example, pretreatment of cotton fabric includes original cloth preparation, singeing, desizing, scouring, bleaching, open rate, rolling water, drying and mercerization, and so on; pretreatment of wool includes select wool, scouring and carbonization process; pretreatment of silk fabric, mainly to degumming, remove most of silk sericin, pigments and other impurities. Most of these processes consume gas, natural gas to contribute the 'direct carbon emissions'; meanwhile they consume electricity and water to contribute the 'indirect carbon emissions' [5]. The whole process is divided into three processes: transportation, production and environment control. Then the GHG emissions of the three processes could be calculated through the energy, water consumption. The sum of three results is the total GHG emissions of the whole process.

The transportation process includes the supply and distribution logistics in the transport and production logistics in the pipeline, conveyor belts and other forms of transport. It is divided into two parts: transportation in the factory and out of the factory. The former includes the materials transportation from a workshop to another. The latter includes raw materials in, the finished product and the waste shipped out. The factories mainly use gasoline trucks or kerosene trucks, which gain momentum by burning the fuel and discharge greenhouse gases into the air. Then the GHG emissions should be calculated by the model 2 in the next section.

The production process mainly consumes electricity, gas, water and so on. And the electricity and water contribute the 'indirect carbon emissions', while the gas or other fuels contribute the 'direct carbon emissions'. The model also can calculate these GHG emissions together. The electrical energy needs to be specially treated, because the emissions from electrical energy are different according to different regions.

The environment control process is very similar with the production process. The GHG emissions from above three processes will be put together to make the whole result out.

3 The Models for GHG Emissions Calculation

Greenhouse gases consist of many kinds of gases. However, the GWP can be used to help us to treat with the complex greenhouse gases simply. The equivalent carbon dioxide emission is obtained by multiplying the emission of a GHG by its Global Warming Potential (GWP) for the given time horizon [4]. For a mix of GHGs it is obtained by summing the equivalent CO_2 emissions of each gas [4].

A basic model is created to calculate the GHG emissions with GWP. The model 1 is as follows:

$$M = \sum_{i}^{n} (V_i \times G_i) \tag{1}$$

Where M: the whole GHG emissions in the whole production process, Kg, V_i: the amount of the i type of GHGs, Kg; G_i: the GWP of the i type of GHGs.

However, it's hard to get the amount of each greenhouse gases. So the basic model is not suitable for practical application. Fortunately, emission factor can help to calculate the emissions. Emission factors of various raw materials unify the greenhouse gas emissions into carbon dioxide emissions. The other model 2 is as follows:

$$M = \sum_{i=1}^{n}(M_i \times EF_i)$$ (2)

Where M: the whole GHG emissions in the whole production process, Kg; M_i: the amount of the i type of raw materials consumed, Kg; EF_i: the emission factor of the i type of raw materials, $KgCO_2e/Kg$.

4 Software Model and Calculation by Computer

The software model is developed to calculate the GHG emissions. As the Fig.1 shows, the software structure includes three parts: user interface layer, calculation model and database layer.

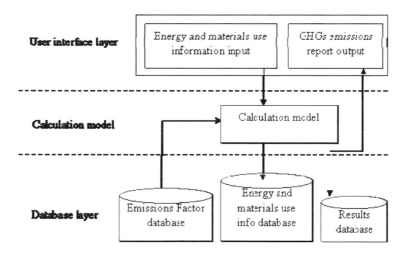

Fig. 1. Software structure

User interface layer is provided to the users to input the energy and materials use information, transmit the data to the calculation model and show the results in report. The calculation model applies the model A and model B in Section II to calculate the GHG emissions according to the data from user interface and deliver the data and result to the database layer. The database layer is responsible for storing the data, including the energy and materials use information and the GHG emissions result from the calculation model, and providing the emission factor to the calculation model. The graphical user interface (GUI) is showed by Fig.2.

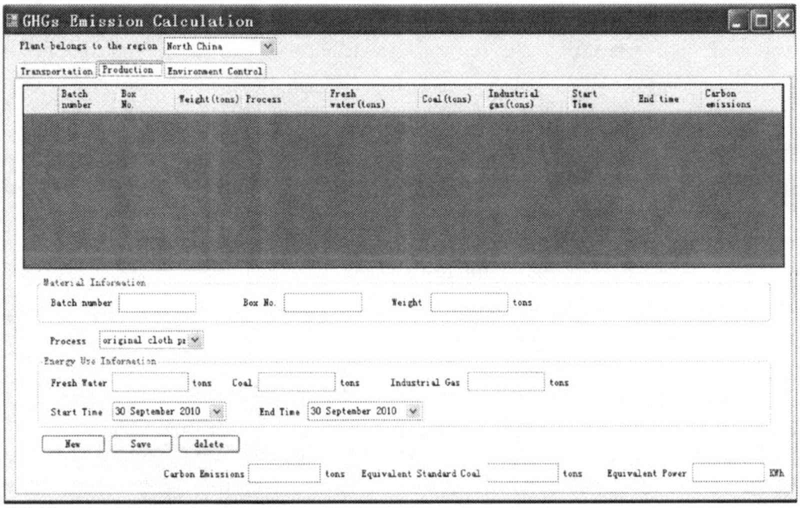

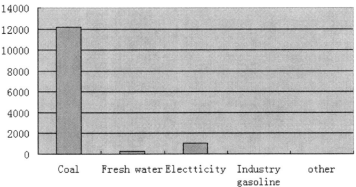

Fig. 2. The graphical user interface

An application has been done for a dyeing and finishing factory in Jiangsu province. The main production is Advanced Yarn-dyed Fabric. The factory produced 11,820,000 meters of the Advanced Yarn-dyed Fabric. This factory consumed fresh water 285279 t a year, coal 5500t a year, electricity 1492400 KWh a year, industrial gasoline 9.5 t a year and so on.

The emission factors of these activity data are shown in table 1, which is obtained from secondary data sources [6].

Table 1. Emissions Factor

name	value
Fresh Water	0.91 Kg CO_2/t
Coal	96 Kg CO_2/Gj
Electricity	0.6826 Kg CO_2/KWh
Industry gasoline	69.2 Kg CO_2/Gj

Fig. 3. The GHG emission from the factory

Fig.3 shows the GHG emissions from the factory. The GHG emission caused by the use of coal is the main point. 6.503 t greenhouse gases have produced when 1 t Yarn-dyed Fabric done. 0.1138 t greenhouse gases have produced when one hundred meters Yarn-dyed Fabric done. That is about equivalent to 42.15 Standard coals.

5 Conclusions

In this paper, we analyzed the processes in the dyeing and finishing industry and proposed a calculation method of the GHG emissions. According to the analysis, we developed a software model, which is used to an application for a dyeing and finishing factory in Jiangsu province. However, the model is not suitable for all kinds of dyeing and finishing processes. How to expand application scope of the model will be an important work in the future.

References

1. Li, H.Z.: Ambitious Emissions Cut Target Poses Challenges to China,
   http://news.xinhuanet.com/english/2009-11/27/
   content_12546756.htm
2. Daneshvar, N., Oladegaragoze, A., Djafarzadeh, N.: Decolorization of Basic Dye Solutions by Electrocoagulation: An Investigation of the Effect of Operational Parameters. J. Hazard. Mater 129, 116–122 (2006)
3. BSI: Guide to PAS 2050: How to Assess the Carbon Footprint of Goods and Services. British Standards Institute, London (2008)
4. Pachauri, R.K., Reisinger, A.(Core Writing Team): Climate Change 2007: Synthesis Report; Reisinger, A.: IPCC, Geneva, Switzerland (2008)
5. Schils, R.L.M., Verhagen, A., Aarts, H.F.M., Šebek, L.B.J.: A Farm Level Approach to Define Successful Mitigation Strategies for GHG Emissions from Ruminant Livestock Systems. Nutr. Cycl. Agroecosyst 71, 163–175 (2005)
6. BSI: PAS 2050:2008: Specification for the Assessment of the Life Cycle Greenhouse Gas Emissions of Goods and Services. Technical report, British Standards Institute, London (October 2008)

GPR Simulation for the Fire Detection in Ground Coal Mine Using FDTD Method

Yanming Wang, Deming Wang, Gouqing Shi, and Xiaoxing Zhong

School of Safety Engineering, China University of Mining and Technology,
Xuzhou 221116, China
{cumtwangym,wdmcumt,shiguoqing2001,zhongxx_2003}@163.com

Abstract. As the radar wave is a kind of electro-magnetic wave and the spread of the wave accords with Maxwell equation, the wave field simulation method has been used in forward simulation of ground-penetrating radar (GPR) profile. The finite-difference time-domain (FDTD) scheme for simulation is derived from the Maxwell's curl equations. The method is applied to simulate the (GPR) profile of fire detection in ground coal mine. The numerical results show that this method is correct and feasible.

Keywords: Ground-penetrating Radar, Finite-difference Time-domain Method, Numerical Simulation, Ground Coal Mine.

1 Introduction

Over the past decade, borehole ground-penetrating radar (GPR) has become an increasingly popular tool for non-invasive, highresolution imaging of the shallow subsurface. Applications of this technique include delineation of ore bodies [1], location of underground tunnels and voids [2], mapping fractures in bedrock [3], and estimation of subsurface lithology and hydrogeological properties [4]. Due to the strong correlation of electromagnetic (EM) wave velocity with water content in the subsurface, the technique is commonly used to detect differences in porosity in the saturated zone, as well as difference in soil water retention and thus grain size in the vadose zone [5].

Near-surface, environmental investigations often require monitoring of the spatial distribution of water content. Water content estimates are needed to model and predict pollutant transport through the vadose zone, and to bsequently design an efficient and reliable remediation plan. Time-lapse GPR has been used successfully to image mass transport such as vegetable oil emulation [6] and saline water [7]. The timelapse imaging at field sites can be divided into three main modes of operation: surface-based or single-borehole reflection surveying [8], surface-to-borehole surveying [9], and cross-borehole surveying [10]. Among these modes, time-lapse crossborehole GPR has gained popularity for monitoring water content changes, thanks to its high-resolution capabilities [11].

Y. Yu, Z. Yu, and J. Zhao (Eds.): CSEEE 2011, Part II, CCIS 159, pp. 310–314, 2011.

Although the ZOP mode is rapid in data acquisition and straightforward in the processing and interpretation, multiple travel paths through the subsurface can give rise to measurement errors. Specifically, in layered media with sharp changes in water content with depth, critically refracted waves may arrive before direct waves at some depths. In this paper, the wave field simulation method has been used in forward simulation of ground-penetrating radar (GPR) profile. The finite-difference time-domain (FDTD) scheme for simulation is derived from the Maxwell's curl equations.

2 Methods

For the forward modeling of EM waves propagating in heterogeneous media, we employ a FDTD solution of Maxwell's equations. The modeling scheme is used to simulate EM fields radiated from a transmitter antenna located in a source borehole.

In this formulation, rotational symmetry about the vertical z-axis is assumed so that Maxwell's equations can be separated into transverse electric (TE) and transversemagnetic (TM) modes, which are two sets of coupled partial-differential equations involving (EΦ, Hρ, Hz) and (Eρ, Ez, HΦ) components, respectively. For crosshole GPR modeling, where antennas are oriented parallel to the z-axis, only the TM-mode equations are required. Maxwell's equations are reduced in TM-mode to

$$\nabla \times H = \frac{\partial D}{\partial t} + J$$
$$\nabla \times E = -\frac{\partial B}{\partial t} - J_m \tag{1}$$

where D is the electric flux density, E is the electric field, H is the magnetic field. In the cartesian coordinate system, Eq.1 can be rewrited as

$$\begin{cases} \dfrac{\partial E_y}{\partial z} - \dfrac{\partial E_z}{\partial y} = \mu \dfrac{\partial H_x}{\partial t} + \sigma_m H_x \\[2mm] \dfrac{\partial E_z}{\partial x} - \dfrac{\partial E_x}{\partial z} = \mu \dfrac{\partial H_y}{\partial t} + \sigma_m H_y \\[2mm] \dfrac{\partial E_y}{\partial y} - \dfrac{\partial E_y}{\partial x} = \mu \dfrac{\partial H_z}{\partial t} + \sigma_m H_z \end{cases} \tag{2}$$

and

$$\begin{cases} \dfrac{\partial H_y}{\partial z} - \dfrac{\partial H_z}{\partial y} = -\varepsilon \dfrac{\partial E_x}{\partial t} - \sigma E_x \\[2mm] \dfrac{\partial H_z}{\partial x} - \dfrac{\partial H_x}{\partial z} = -\varepsilon \dfrac{\partial E_y}{\partial t} - \sigma E_y \\[2mm] \dfrac{\partial H_x}{\partial y} - \dfrac{\partial H_y}{\partial x} = -\varepsilon \dfrac{\partial E_z}{\partial t} - \sigma E_z \end{cases} \tag{3}$$

For the 2D simulation, the Maxwell's equation can be reduced as

$$\begin{cases} \dfrac{\partial H_x}{\partial t} = \dfrac{1}{\mu}\dfrac{\partial E_y}{\partial z} \\[2mm] \dfrac{\partial H_z}{\partial t} = -\dfrac{1}{\mu}\dfrac{\partial E_y}{\partial x} \\[2mm] \dfrac{\partial E_y}{\partial t} = -\sigma\dfrac{1}{\varepsilon}E_y + \dfrac{1}{\varepsilon}\left(\dfrac{\partial H_x}{\partial z} - \dfrac{\partial H_z}{\partial x}\right) + I_y \end{cases} \tag{4}$$

Eq.4 can be solved numerically in time-domain using a leapfrog staggered-grid approach, which offsets the electric- and magnetic-field components so that the FD approximations of all partial derivatives are centered in both space and time. And the magnetic-field components can be given by

$$\begin{aligned} \dfrac{\partial H_x^{n+1/2}(i,k)}{\partial z} &= \dfrac{1}{\Delta z}\left[H_x^{n+1/2}\left(i,k+\dfrac{1}{2}\right) - H_x^{n+1/2}\left(i,k-\dfrac{1}{2}\right)\right] \\[2mm] \dfrac{\partial H_z^{n+1/2}(i,k)}{\partial x} &= \dfrac{1}{\Delta x}\left[H_z^{n+1/2}\left(i+\dfrac{1}{2},k\right) - H_z^{n+1/2}\left(i-\dfrac{1}{2},k\right)\right] \end{aligned} \tag{5}$$

As shown in Fig.2, stepping forward in time is accomplished by alternately updating the electric and magnetic fields using explicit updates. Field components are calculated in space identically to avoid singularity problems.

The electric- and magnetic-field components are located at the edges and the center of a cell, respectively. Because the electric field is set to a cell edge, an averaged complex permittivity at the edge is evaluated as a weighted sum of the permittivities of the two adjoin cells, where the weighting is based on the area of each cell, so the field components can be solved as follows:

$$\begin{aligned} E_y^{n+1/2}(i,k) &= A_1 E_y^n(i,k) + A_2\left[H_x^{n+1/2}\left(i,k+\dfrac{1}{2}\right) - H_x^{n+1/2}\left(i,k-\dfrac{1}{2}\right)\right] \\[2mm] &- A_3\left[H_z^{n+1/2}\left(i+\dfrac{1}{2},k\right) - H_z^{n+1/2}\left(i-\dfrac{1}{2},k\right)\right] + A_4 I_y^{n+1}(i,k) \end{aligned} \tag{6}$$

$$\begin{aligned} H_x^{n+1/2}\left(i,k+\dfrac{1}{2}\right) &= H_x^{n-1/2}\left(i,k+\dfrac{1}{2}\right) + A_5\left[E_y^n(i,k+1) - E_y^n(i,k)\right] \\[2mm] H_z^{n+1/2}\left(i+\dfrac{1}{2},k\right) &= H_z^{n-1/2}\left(i+\dfrac{1}{2},k\right) - A_6\left[E_y^n(i+1,k) - E_y^n(i,k)\right] \end{aligned} \tag{7}$$

3   Results

Using the FDTD scheme as discussed in this paper, we get the GPR simulation profile for the mine fire area. As shown in Fig.2, the circle area is the underground hole caused by mine fire.

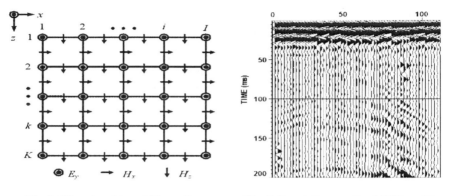

Fig. 1. Discrete grids and Yee cell   **Fig. 2.** Simulation results of GPR profile

4   Conclusion

As the radar wave is a kind of electro-magnetic wave and the spread of the wave accords with Maxwell equation, the wave field simulation method has been used in forward simulation of ground-penetrating radar (GPR) profile. The finite-difference time-domain (FDTD) scheme for simulation is derived from the Maxwell's curl equations. The method is applied to simulate the (GPR) profile of fire detection in ground coal mine. The numerical results show that this method is correct and feasible.

Acknowledgement. This work was supported by the State Key Laboratory Research Foundation of Coal Resources and Mine Safety (No. SKLCRSM08x06), the Fundamental Research Funds for the Central Universities (No. 2010QNA01), and the Scientific Research Foundation for the Introduction of Talent of China University of Mining & Technology.

References

1. Alumbaugh, D., Chang, P.Y., Paprocki, L., Brainard, J.R., Glass, R.J., Rautman, C.A.: Estimating Moisture Contents in the Vadose Zone Using Cross-borehole Ground Penetrating Radar: A Study of Accuracy and Repeatability. Water Resources Research 38(12), 1309–1320 (2002)
2. Bano, M.: Modeling of GPR Waves for Lossy Media Obeying a Complex Power Law of Frequency for Dielectric Permittivity. Geophysical Prospecting 52(1), 11–26 (2004)
3. Binley, A., Winship, P., Middleton, R., Pokar, M., West, J.: High-resolution Characterization of Vadose Zone Dynamics Using Cross-borehole Radar. Water Resources Research 37(11), 2639–2652 (2001)
4. Binley, A., Cassiani, G., Middleton, R., Winship, P.: Vadose Zone Flow Model Parameterization Using Cross-borehole Radar and Resistivity Imaging. Journal of Hydrology 267(7), 147–159 (2002)
5. Cassiani, G., Strobbia, C., Gallotti, L.: Vertical Radar Profiles of Deep Vadose Zone. Vadose Zone Journal 3(1), 1093–1105 (2004)

6. Chang, P.Y., Alumbaugh, D., Brainard, J., Hall, L.: Cross-borehole Ground Penetrating Radar for Monitoring and Imaging Solute Transport within the Vadose Zone. Water Resources Research 42(10), 10413–10425 (2006)
7. Davis, J.L., Annan, A.P.: Ground-penetrating Radar for High-resolution Mapping of Soil and Rock Stratigraphy. Geophysical Prospecting 37(5), 531–551 (1989)
8. Day-Lewis, F.D., Lane, J.W., Harris, J.M., Gorelick, S.M.: Time-lapse Imaging of Saline Tracer Transport in Fractured Rock Using Difference Radar Attenuation Tomography. Water Resources Research 39(10), 1290–1302 (2003)
9. Ellefsen, K.J., Wright, D.L.: Radiation Pattern of a Borehole Radar Antenna. Geophysics 70(6), K1–K11 (2005)
10. Fullagar, P.K., Livelybrooks, D.W., Zhang, P., Calvert, A.J.: Radio Tomography and Borehole Radar Delineation of the McConnell Nickel Sulfide Deposit, Sudbury, Ontario, Canada. Geophysics 65(6), 1920–1930 (2000)
11. Hammon III, W.S., Zeng, X., Corbeanu, R.M., McMechan, G.A.: Estimation of the Spatial Distribution of Fluid Permeability from Surface and Tomographic GPR Data and Core, with a 2-D Example from the Ferron Sandstone, Utah. Geophysics 67(5), 1505–1515 (2002)

Extraction of Chinese-English Phrase Translation Pairs

Chun-Xiang Zhang[1], Ming-Yuan Ren[1], Zhi-Mao Lu[2], Ying-Hong Liang[3],
Da-Song Sun[4], and Yong Liu[5]

[1] School of Software, Harbin University of Science and Technology,
Harbin, China
[2] College of Information and Communication Engineering,
Harbin Engineering University, Harbin, China
[3] School of Computer Engineering, Vocational University of Suzhou City,
Suzhou, China
[4] Computer Center, Harbin University of Science and Technology,
Harbin, China
[5] School of Computer Science and Technology, Heilongjiang University,
Harbin, China
zcxbysj2006@yahoo.com.cn, 15846527327@126.com,
lzm@hrbeu.edu.cn, liangyh7036@126.com,
z6c6x6@yahoo.com.cn, acliuyong@sina.com

Abstract. Phrase translation pairs are very useful for bilingual lexicography, machine translation system, cross-lingual information retrieval and many applications in natural language processing. In this paper, we align Chinese-English bilingual sentence pairs based on parsing tree of Chinese sentence and word alignment results, which is called as the tree-string alignment. From tree-string alignment, phrase translation pairs are extracted.

Keywords: Phrase Translation Pairs, Natural Language Processing, Parsing Tree, Word Alignment.

1 Introduction

Acquisition of phrase translation pairs is a task where phrases in source language and phrases in target language, which can be translated from and to each other, are extracted from bilingual sentence pairs.

Phrase translation pairs are very important translation knowledge in natural language processing, which can be used in a variety of applications such as bilingual lexicography [1], machine translation system [2] and cross-lingual information retrieval. Many methods have been proposed for acquisition of phrase translation pairs. John proves that finding optimal phrase alignment is NP-hard, and the problem of finding an optimal alignment can be cast as an integer linear program [3]. Parse-parse-match method is adopted firstly to extract phrase translation pairs [2]. Its main idea is that each language of bilingual corpus is parsed independently by a monolingual grammar, and then corresponding constituents are matched based on word alignment results. Melamed has proposed a fast and greedy algorithm called

Y. Yu, Z. Yu, and J. Zhao (Eds.): CSEEE 2011, Part II, CCIS 159, pp. 315–318, 2011.

competitive linking in order to find word-to-word equivalences [4], which provides aligning anchors for extracting phrase translation pairs. Zhang builds a two-dimensional matrix to represent a bilingual sentence pair where the value of each cell corresponds to the point-wise mutual information between source word and target one. Box-shaped region whose mutual information values are similar with each others is looked upon as a phrase translation pair [5]. Zhang uses individually the monolingual language model to identify phrases in Chinese corpus and phrases in English corpus. Alignments are built on Chinese phrases and English phrases in order to extract phrase translation pairs which are applied to an example-based machine translation system [6]. Venugopal utilizes an improved IBM model to create knowledge sources in phrase level that effectively represent local phrasal context and global phrasal context, which can be applied to the process of phrase alignment [7]. Philip uses a widely practised approach to get word alignments from two directions including source to target and target to source. With this refined word alignment, target candidate phrases will be extracted for a given source phrase in the target sentence by searching the left and right projected boundaries [8]. Kenji uses translation literality to evaluate literality of bilingual sentence pairs and cleans the corpus in order to improve the quality of phrase translation pairs [9]. Zhao proposes an algorithm for extracting phrase translation pairs, which do not need explicit word alignment results. For each phrase translation pair, a bilingual lexicon-based evaluation score is computed to estimate the translation quality between source phrase and target phrase. A fertility score is computed to estimate how good the lengths are matched between source phrase and target phrase. A center distortion score is computed to estimate the relative position divergence between source phrase and target phrase [10].

Wong proposes the annotation schema of translation corresponding tree (TCT) on bilingual sentence pairs, from which phrase translation pairs are extracted for constructing the example base [11]. Each TCT represents syntactic structure of source language sentence, and specifies the correspondence between source parsing tree and target string. In order to get TCT, source parsing tree and target sentence are aligned based on word alignment results. TCT can be viewed as the tree-string alignment of a bilingual sentence pair. The method can decrease the impact of the grammar disagreement between source language and target language. Phrase translation pairs can be acquired from TCT. This partly solves the problem that the alignment process is restricted by grammar incompatibility between source language and target language. From tree-string alignment, we can extract phrase translation pairs. There are only parsing information in source parts of phrase translation pairs, and target parts do not include any parsing knowledge.

In this paper, we extract phrase translation pairs based on word alignment results of bilingual sentence pairs and parsing trees of source language sentences.

2   Chinese-English Phrase Alignment

A Chinese parser tool and a Chinese–English word alignment tool are only used here. Firstly, Chinese sentence is analyzed by Chinese parser. Secondly, we use word alignment tool to align the bilingual sentence pair. At last, the tree-string alignment between Chinese and English is built according to word alignment results, from which phrase translation pairs can be extracted. The process of extracting Chinese-English phrase translation pairs from tree-string alignment is shown in Figure 1.

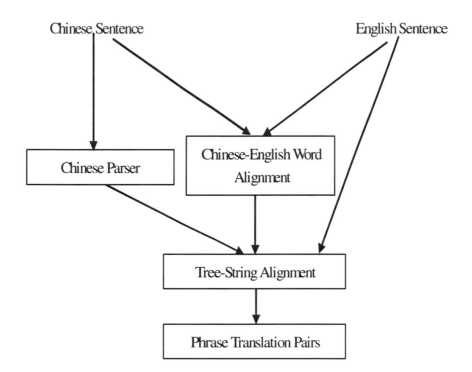

Fig. 1. Extracting Chinese-English phrase translation pairs from tree-string alignment

For a bilingual sentence pair (C, E), the algorithm of tree-string alignment is shown as follows:

1. The words in C and E are assigned with their positions respectively.
2. Parse Chinese sentence C and T is the parsing tree of C.
3. Align words between C and E by word alignment tool. Extract word links between C and E from word alignment results.

When the algorithm is applied to Chinese-English bilingual sentence pairs, the tree-string alignments will be gotten. From the tree-string alignments, we can extract phrase translation pairs when the TCT is post-traveled.

3 Conclusions

In this paper, we align Chinese-English bilingual sentence pairs based on parsing tree of Chinese sentence and word alignment results. From tree-string alignment, Chinese-English phrase translation pairs are extracted.

Acknowledgments

This work is supported by National Natural Science Foundation of China under Grant Nos. 60903082, 60975042, Science and Technology Research Funds of Education Department in Heilongjiang Province under Grant Nos. 11541045, Chun-Hui Cooperated Project of the Ministry of Education of China under Grant Nos. S2009-1-15002, Top-Notch Talent Funds of Harbin University of Science and Technology, and Jiangsu Province Support Software Engineering R&D Center for Modern Information Technology Application in Enterprise under Grant Nos. SX200907.

References

1. Gale, W.A., Church, K.W.: Identifying Word Correspondences in Parallel Texts. In: Proceedings of the 4th DARPA Workshop on Speech and Natural Language, California, pp. 152–157 (1991)
2. Imamura, K.: Application of Translation Knowledge Acquired by Hierarchical Phrase Alignment for Pattern-Based MT. In: Proceedings of the 9th Conference on Theoretical and Methodological Issues in Machine Translation, Japan, pp. 74–84 (2002)
3. John, D.N., Dan, K.: The Complexity of Phrase Alignment Problems. In: Proceedings of the 46th Annual Meeting of the Association for Computational Linguistics, Columbus, pp. 25–28 (2008)
4. Melamed, I.D.: A Word-to-Word Model of Translational Equivalence. In: Proceedings of the Eighth Conference on European Chapter of the Association for Computational Linguistics, USA, pp. 490–497 (1997)
5. Zhang, Y., Vogel, S., Waibel, A.: Integrated Phrase Segmentation and Alignment Model for Statistical Machine Translation. In: Proceedings of International Conference on Natural Language Processing and Knowledge Engineering, Beijing, pp. 567–573 (2003)
6. Zhang, Y., Brown, R.D., Frederking, R.E.: Adapting An Example-Based Translation System to Chinese. In: Proceedings of the First International Conference on Human Language Technology Research, Californian, pp. 1–4 (2001)
7. Venugopal, A., Vogel, S., Waibel, A.: Effective Phrase Translation Extraction from Alignment Models. In: Proceedings of the 41st Annual Meeting of the Association for Computational Linguistics, Japan, pp. 319–326 (2003)
8. Philip, K., Kevin, K.: Feature-Rich Statistical Translation of Noun Phrases. In: Proceedings of the 41st Annual Meeting of the Association for Computational Linguistics, Japan, pp. 311–318 (2003)
9. Imamura, K., Sumita, E.: Bilingual Corpus Cleaning Focusing on Translation Literality. In: Proceedings of the 7th International Conference on Spoken Language Processing, USA, pp. 1713–1716 (2002)
10. Zhao, B., Vogel, S.: A Generalized Alignment-Free Phrase Extraction. In: Proceedings of the ACL Workshop on Building and Using Parallel Texts, Ann Arbor, pp. 141–144 (2005)
11. Wong, F., Hu, D.C., Mao, Y.H., Dong, M.C.: A Flexible Example Annotation Schema: Translation Corresponding Tree Representation. In: Proceedings of the 20th International Conference on Computational Linguistics, Switzerland, pp. 1079–1085 (2004)

The Study of One Case Leading to the Whole Course in Teaching of Operating System

Guoxia Zou and Jianqing Tang

GuiLin College of Aerospace Technology Guilin, China, 541004
zouguoxia@163.com, tjq2007_cn@sina.com

Abstract. The operating system is a very important professional foundation course of computer, but it is very difficult to study, because its knowledge is much and abstract. To enable students to understand the role of the operating system and master its principle, to make good bedding for the subsequent course, to train students' abstract thinking ability and problem-solving skill, in the teaching process of operating system, a teaching method of one case leading to the whole course was developed. The teaching practice shows that the method has some inspiration, coherence. It makes students easily understand the operating principle and think the role of operating system with a certain high degree.

Keywords: Operating System; Teaching Methods; One Case Leading to the Whole Course.

1 Introduction

Operating system is the main course of computer, students must master its basic principle and understand the internal structure, then can design, develop and maintain complex communications software. Operating system is a profound masterpiece, its content is rich and deep thought [1-4], it always is difficult for computer students to learn [5-6]. With the science progresses, operating system has rapidly changed in the concept, technology and the architecture. In order to make students happily learn and master the course, prepare for the subsequent course, with the personally teaching process in recent years, the author developed a teaching method of operating system that is one case leading to the whole course.

2 The Introduction of One Case Leading to the Whole Course Preparation

One case leading to the whole course is that, using a simple source code known by everyone to lead to a series of related knowledge points of operating system during the process of the source code writing to running. In short word, it is using a simple example to lead to the full knowledge points of the text. Because the operating system is a very strong practical significance theory course, using a program that all

Y. Yu, Z. Yu, and J. Zhao (Eds.): CSEEE 2011, Part II, CCIS 159, pp. 319–324, 2011.

know to lead to the principle, it is so accessible, and also exercise the capability of thinking and exporting problems.

2.1 The Example of One Case Leading to the Whole Course

1) Writing a Source Program Hello. C, the Code Is as Follows.

```
#include <stdio. h>
  int main(int argc, char avgv[])
{
  char ch;
  printf("please put in a char:\n");
  scanf("%c", &ch);
  printf("the char is: %c", ch);
  return 0;
}
```

2) Saving the Source Program

To save the file, you need to know the location to store the file and the filename. By the filename, we can lead to file system management section about the rule of filename and the meaning of common extension name; by the document saving location can lead to the logical location and path, document storage physical structure; then in turn lead to file system management section about the logical structure and physical structure. The order of the structure shown in Fig. 1.

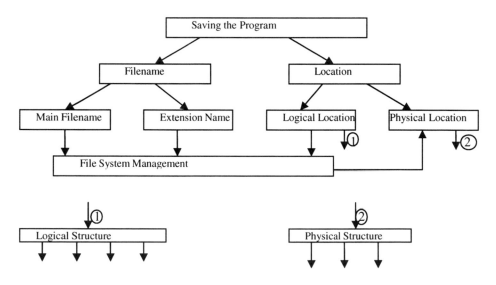

Fig. 1. Knowledge Frame Of Saving The File

3) Compiling and Linking the Source File

After the file is saved well, you can compile and link. During the compiling, the compiler is needed, and the compiler needs source files location, we can leads to the file system how to search the physical location of the file, thus the catalog tables, index node and FCB (File Control BLOCK) can be leaded. Source file compiled under compiler into object files, but we also use the printf, scanf and other library functions, for which, we need to link the object files and library files to generate an executable binary file. The process of leading to knowledge shown in Fig. 2.

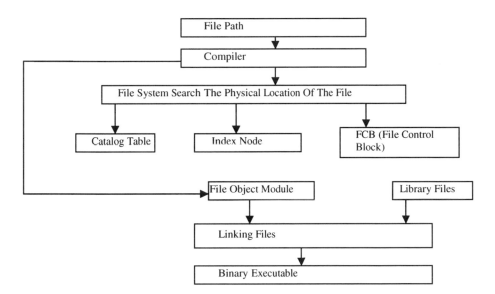

Fig. 2. The Process Of Compiling And Linking The Source File

4) Executing the File

Executing file need to use the file system management functions to search the file physical location, and then use the loading program to load the file into memory. There are two methods to load the file; they are dynamic loading and static loading.

During the loading, we can lead to the device management section. To control the device, we require device drivers, and device controller; to convey signal and data between device and CPU, or between device and memory, we need I/O control. To locate the data, we need algorithms, such as FCFS, SSTF, SCAN, CSCAN etc.

When the file is loading into memory, the memory distribution need to be considered, at this time, we can lead to the memory management. Memory management is mainly to allocate and recover the user available memory. There are three kinds of memory allocation: continuous distribution, discrete distribution, virtual storage. Every kind of memory allocation has its algorithms. The process of leading to knowledge shown in Fig.3.

5) Process Scheduling

After the executable file into memory, the process is created. Then the process is in the ready state. Why we need the process? What's the difference between process and program? To answer those question, we can lead the development of operating system, the characteristics of the process, to solve the question that when is the process executed, we can lead to the process state. Because there are many processes in the memory, they maybe some relations, and they are competing the CUP, so we need to understand the process scheduling, process communication, and scheduling deadlock. The process of leading to knowledge shown in Fig 4.

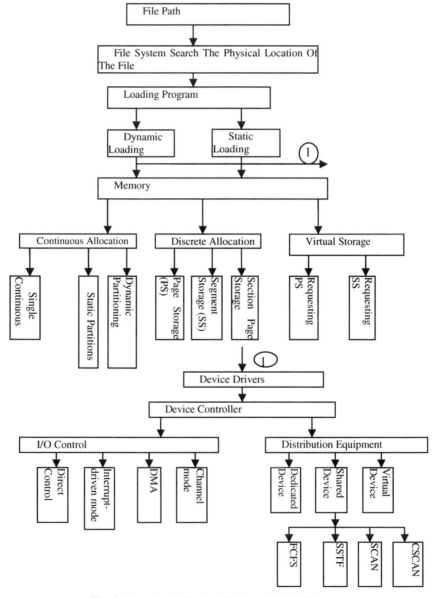

Fig. 3. Knowledge Frame Of Executing The File

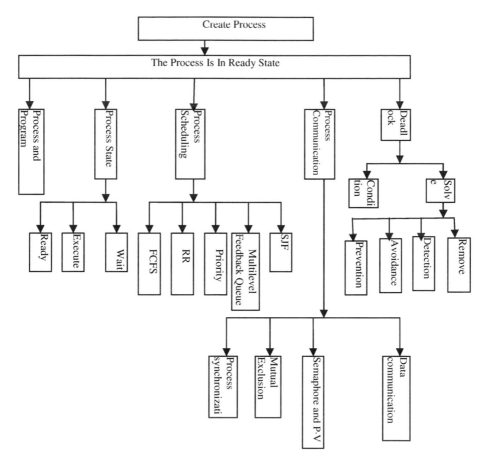

Fig. 4. Knowledge Frame Of Process Scheduling

3   The Contrast of Teaching

Using nearly years teaching effectiveness to analyze the teaching method—one case leading to the whole course. In test paper, the paper of 2010 includes all content that appeared in 2009 paper. What's more, 2010 paper is more content and difficulty than 2009 paper. In foundation of students, for the expansion of enrollment, entry requirements relative decline, 2010 students' foundation is relatively worse than 2009 students'. The same teacher used the same textbook and the same teaching time, the final examination results are distributed as table 1.

It is can be seen from the table that 2010 teaching effectiveness is better than 2009.

4   Summary and Outlook

Through the above example, the whole textbook knowledge can be leaded only by four knowledge frames. It is not a list of knowledge; the design is according to the way of human thinking-knowledge and solving-problem. The teaching method —one case leading to the whole course , can make students deeply understand the importance of the operating system, at the same time train students solving-problem. The method can make students see the operating system as a whole, students can remember the front chapter when they learn the behind knowledge.

Table 1. The final examination results contrast

Year	Highest	>=85	>=75	>=60	< 60	Lowest
2009	88	20	40	30	10	45
2010	97	27	50	17	6	46

References

1. Chung-nin, L.: Operating System Tutorial. Beijing, Tsinghua University Press (October 2000)
2. Yao-Xue, Z., Mei, S.: Computer Operating System Tutorial (2nd Edition). Beijing, Tsinghua University Press (2002)
3. Tsao, S.-L.: A Practical Implementation Course of Operating Systems. Curriculum Design and Teaching Experiences. In: Parallel and Distributed Systems ICPADS 2008, pp. 768–772 (2008)
4. Ge, J., Ye, B., Fei, X., et al.: A Novel Practical Framework for Operating Systems Teaching. In: Eighth International Conference on Embedded Computing, pp. 596–601 (2009)
5. Dou, J., Cao, J., Jiang, Y., et al.: Research on the Teaching Reform of Operating System Courses. In: The 9th International Conference for Young Computer Scientists, pp. 2402–2406 (2008)
6. Qingqiang, W., Langcai, C.: Teaching Mode of Operating System Course for Undergraduates Majoring in Computer Sciences. In: 4th International Conference on Computer Science & Education, pp. 1412–1415 (2009)

A Combining Short-Term Load Forecasting System for Abnormal Days

Ming Li and Junli Gao

College of Automation, Guangdong University of Technology
510006 Guangzhou, Guangdong, P.R. China
mingli4@mail.ustc.edu.cn, jomnygao@163.com

Abstract. A data mining and ENN combining short-term load forecasting system is proposed to deal with the weather-sensitive factors' influence on the power load in abnormal days. The statistic analysis showed that the accuracy of the short time load forecasting in abnormal days has increased greatly while the actual forecasting results of AnHui Province's total electric power load and the comparative analysis have validated the effectiveness and the superiority of the strategy.

Keywords: short-term load forecasting, combining forecasting, data mining, ENN, abnormal days.

1   Introduction

In the past few decades, a variety of power load forecasting algorithms have been proposed such as neural networks[1], expert systems[2], fuzzy systems approach[3], SVM[4], data mining[5], etc. However, due to the unusual weather conditions and holiday, here we call "Abnormal Days"; it is hard to model the relationships between the power loads and the variables that influence the power loads. To handle the major factors which make the modeling complicated, a combining forecasting system is proposed whose working flow is shown in Fig. 1.

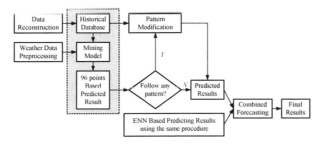

Fig. 1. The working flow of the combining forecasting system

The proposed system mainly consists of two modules. The Data Processing Module is responsible for converting the load data and meteorological data into the

Y. Yu, Z. Yu, and J. Zhao (Eds.): CSEEE 2011, Part II, CCIS 159, pp. 325–330, 2011.
© Springer-Verlag Berlin Heidelberg 2011

specific form of training data[6]. The Load Forecasting Module's task is to call the combining algorithm and gives the loads prediction[7]: first, an improved C4.5[8] algorithm and an ENN are used respectively to predict the power load. Second, for the exceptional changes in the abnormal days, special patterns are recognized with highly similar features; so a second procedure is applied to the data in these patterns to find the hidden law which determines the power load, corresponding modification is then applied to the initial results respectively. The last step is to combine the above two results to get a better prediction.

2 Algorithm

Suppose that there are m kinds of forecasting methods for the event F, if we can express the ith mapping in the form of $x \in X \subset R \xrightarrow{\varphi_i} y \in Y \subset R$, the nonlinear combination can be described in the form of $y = \Phi(x) = \Phi(\varphi_1, \varphi_2, \ldots \varphi_m)$. A three-layer BP neural network is chosen to determine the nonlinear mapping. In the offline training phase, the input is the individual predicting value $f_1(t)$, $f_2(t)$ and the output is the actual power load value recorded in the historical database where $f_1(t)$ is the data mining result while $f_2(t)$ is the ENN result. In the forecasting phase: the input of the BP network is the predicting value of the power load based on the weather forecasting data, and the output is the final desired predicting value.

2.1 Data Mining Prediction

The improved C4.5 algorithm builds decision trees from a set of training data using the concept of information entropy[6]. The training data is a set $S = s_1, s_2, \cdots$ of already classified samples. Each sample $s_i = x_1, x_2, \cdots$ is a vector where $x_1, x_2, \cdots$ represent attributes of the sample. The training data is augmented with a vector $C = c_1, c_2, \cdots$ where $c_1, c_2, \cdots$ represent the class to which each sample belongs. At each node of the tree, the algorithm chooses one attribute of the data that most effectively splits its set of samples into subsets enriched in one class or the other. Its criterion is the normalized information gain (difference in entropy) that results from choosing an attribute for splitting the data.

Suppose a training set S, we can calculate the the information entropy $I(S)$. After S has been partitioned in accordance with the n outcome of a test X. The expected information requirement $I_X(s)$ can be found as the weighted sum over the subsets. So $G(S) = I(S) - I_X(S)$ measures the information that is gained by partitioning S in accordance with the test X. The gain criterion selects a test to maximize this information gain. But this gain criterion has strong bias in favor of tests with many outcomes. It can be rectified by the potential information generated by dividing S into n subsets as $Split\_I(X)$. So a new gain criterion expressing the proportion of information generated by the split that appears helpful for classification is $Gain\_R(S) = G(S) / Split\_I(X)$. And then, The decision tree should be pruned through discarding one or more sub-trees and replacing them with a leave node.

2.2 ENN Prediction

The ENN prediction is part of the combining forecasting. In our system an ensemble learning enhancing ANN-based forecasting model has been applied to make the prediction, which is validated in many works[9] as shown in Fig. 2.

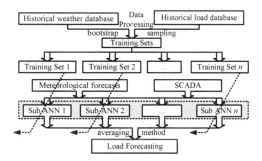

Fig. 2. The ENN prediction

The ENN is composed of multiple NNs, whose outputs are combined to be the ENN's final outputs. Here, a three-layer feedforward network has been used. The hyperbolic tangent function $f(x)=\tanh(1.5h)=(e^{1.5h}-e^{-1.5h})/(e^{1.5h}+e^{-1.5h})$ is used for hidden neurons and output neurons.

2.3 The Modification

Several kinds of the mutation patterns are summarized here for modification. The historical data following these patterns is submitted for the secondary learning to obtain the hidden correction rules of the highly similar features. Therefore, when the forecasting date is consistent with one of these patterns, the corresponding rule will be incorporated to the predicting process for better accuracy.

Temperature reconstruction within a day. It can be found that the sensitivity of predictive value varies greatly due to the different times in a day. Meanwhile, the temperature parameters' impact on the load forecasting under different conditions also changes a lot. So the weighted daily maximum temperature is used to reconstruct historical data, which is treated as part of the input as $T_w(t) = T(t)\times(1-w)+T_{max}\times w$ where $T_w(t)$ is the weighted temperature, $T(t)$ is the dry bulb temperature at time t, T_{max} is the highest daily temperature, w is the weighting coefficient. During the system design, only the historical data from June to September is reconstructed using the rule.

The temperature mutation. The patterns are summarized as follows: (1)Mutation point. A load-temperature mutation point always exists to cause the enormous changes on both sides of the point. According to the behavior and the magnitude when the actual temperature goes through the point, the corresponding condition can be classified into four cases: Minor warming through mutation point, rapid warming through mutation point, slight cooling through mutation point and rapid cooling through mutations. (2) High temperature and relative low temperature. When summer temperatures rise to a certain degree, even minor temperature changes will result in a large load change, At the same time there is also a load-temperature change saturation point, above this temperature, the ordinary power consumption will be on full load. To deal with the condition when the temperature of the base date and the forecasting temperature both are above the mutation point, five kinds of situations are considered: sustained high temperature, minor heating under the high temperature, rapid heating under high temperature, minor cooling under the high temperature and rapid cooling under high temperature. But when the temperature of the base date and the day to be forecasted are both below the mutation point, the modification is not significant. So the necessary corrections are also much smaller compared to the previous two patterns. (3) Continuous cooling. Five cooling patterns can be attained in accordance with the following factors: cooling rate in the daytime, cooling rate in the night time and average daily temperature change.

Towards the above patterns, modification of the initial prediction is as follows:

$$L_{predict}(i) = \Phi\left(L_{base}(i) \times \left(1+f_k\left(\Delta p_i, \Delta p_i'\right)\right)\right) \quad 1 \le i \le 96 \tag{1}$$

$$f_k\left(\Delta p_i, \Delta p_i'\right) = \beta_k \times \Delta p_i + \left(1-\beta_k\right) \times \Delta p_i' \quad 1 \le k \le 15 \tag{2}$$

where i refers to the 96 sampling time sequence, k is the representative of the 4 groups of 15 kinds of mutations, while $L_{base}(i)$ is the base load and $L_{predict}(i)$ is the forecasting load.

3 Application and Results

The accuracy formula used to evaluate performance of the forecasting is defined as follows:

$$R_j = \left[1 - \sqrt{\sum_{i=1}^{n} E_i^2 / n}\right] \times 100\% \tag{3}$$

where E_i^2 is the relative error of the forecasting points given in [10]. As the 96 points methods is adopted to get the predicting curve, so n equals to 96. In the modeling phase, the historical power and meteorological data from May 2005 to May 2008 is used as the training data. In the predicting phase, the obtained model is used to predict the power load from the June, 2008 to September, 2008.

To verify the performance of the proposed method, two comparisons is carried out, the first is the comparison between the forecasting power load and real-load of Anhui power load network as shown in Fig. 3; the second is the comparison of the performance between the improved system and the currently using ELPSDM system in Anhui Power Dispatching and Communication Center as shown in Fig 4.

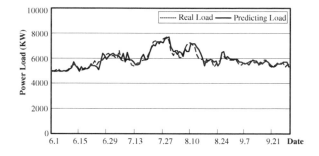

Fig. 3. Comparison between forecasting and real-load

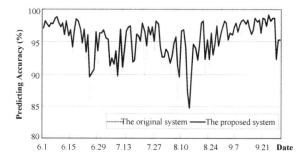

Fig. 4. Comparison between the performances of improved system and the current using one

It can be seen from Fig. 3 and Fig. 4 that the improved system will not only be able to maintain high accuracy of the load prediction throughout the summer, but also greatly improved the accuracy of the prediction when there exits frequently unusual weather conditions or rapid climate change. It is worth mentioning that because of the algorithm's dependence on the weather forecast to some extent, the forecasting error will have a considerable bad influence on the predicting results. So in the Fig. 4, the serious error forecasting of the cold spell in 13 August cause the prediction accuracy down to be slightly lower than 85%. However, the statistics of the forecasting accuracy over the entire summer shows that the improved system can keep highly accurate prediction to achieve an average prediction accuracy of 96.6% even when there are many anomalies in the weather conditions.

Analyzing the comparison between the currently using system and the proposed system, in the abnormal days when the currently using system is difficult to achieve accurate predicting, the average prediction accuracy has been improved by 1.3% while the monthly average accuracy throughout the year of the proposed system has reached 97.9%.

4   Conclusions

In this paper, a data mining and ENN based combining forecasting system has been proposed to achieve high predicting accuracy especially in abnormal days. A variety of abnormal patterns have been recognized and corresponding modification is given to improve the predicting accuracy. The actual prediction results have proved that the strategy has greatly improved the prediction accuracy in abnormal days while ensuring the overall prediction accuracy and enhanced the system's ability to adapt to the abnormal conditions. Future work will be focused on the following aspects, the first is how to make the system adaptive to other common abnormal situations such as political events, contingencies, and holiday etc. The second is how to redesign the system to improve the feedback performance of the system, and how to make the system robust to the weather forecasting.

Acknowledgements. This work is supported by the 211 Foundation, Guangdong Development and Reform Commission [431] and the Dr. Start Fund in Guangdong University of Technology (No. 405105028).

References

1. Osman, Z.H., Awad, M.L., Mahmoud, T.K.: Neural Network Based Approach for Short-Term Load Forecasting. In: Power Systems Conference and Exposition, pp. 1–8. IEEE Press, New York (2009)
2. Rahman, S., Bhatnagar, R.: An Expert System Based Algorithm for Short Term Load forecast. IEEE Trans. on Power Systems 3(2), 392–399 (1988)
3. Sachdeva, S., Verma, C.M.: Load Forecasting Using Fuzzy Methods. In: Joint International Conference on Power System Technology, pp. 1–4. POWERCON (2008)
4. Escobar, A.M., Perez, L.P.: Application of Support Vector Machines and ANFIS to the Short-Term Load Forecasting. In: Transmission and Distribution Conference and Exposition: Latin America, pp. 1–5. IEEE Press, New York (2008)
5. Lu, Y., Huang, Y.N.: Research on Analytical Methods of Electric Load Based on Data Mining. Intelligent Computation Technology and Automation 2, 1085–1088 (2010)
6. Hong, L., Zhang, H.Q., et al.: An Electric Load Prediction System Based on Data Mining. Mini-Micro Systems 125(3), 434–437 (2004)
7. Yang, Y.: Combining Forecasting Procedures: Some Theoretical Results. Econometric Theory 20(1), 176–222 (2004)
8. Information on, http://en.wikipedia.org/wiki/c4.5_algorithm
9. Fan, S., Chen, L.N., Lee, W.J.: Short-Term Load Forecasting Using Comprehensive Combination Based on Multimeteorological Information. IEEE Transactions on Industry Applications 45(4), 1460–1466 (2009)
10. Niu, D.X., Cao, S.H., et al.: The Methods and Application of Power System Load Forecasting. China Electric Power Press, Beijing (1998)

Numerical Simulation on Gas-Liquid Flow Field in Tee Tube Separator

Hanbing Qi, Jiadong Sun, Guozhong Wu, and Dong Li

The College of Architecture and Civil Engineering, Northeast Petroleum University,
Hei Longjiang Daqing 163318, China
{qihanbing,sunjiadong}@sina.com.cn,
{wgzdq,lidonglvyan}@126.com

Abstract. The water content measuring of the crude oil with gas influencing is serious. The key of popularizing the gas-liquid separator is improving the gas-liquid separating efficiency and decreasing the energy consumption. The phase distribution and the pressure characteristic of the two-phase flow in tee tube were analyzed in this paper, and the phase separation principle of tee tube was given, then the physical and mathematical models of the tee tube separator were established. The phase and pressure distribution of the two-phase flow in tee tube were simulated using FLUENT 6.3, and analyzed the influence of the number of the branch pipe of the same side in gas-liquid separation. The result shows that the influence of the number of the branch pipe of the same side was greatly urgent. And with 3 branch pipes, the gas content at the throttle orifice was from 0.2 of the entrance to 9.18×10^{-5}. The conclusions have provided some basis for the further application of the tee tube gas-liquid separator.

Keywords: Tee Tube, Gas-Liquid Separation, Branch Pipe, Numerical Simulation.

1 Introduction

Gas-liquid flow is a common flow pattern in the petroleum and the chemical industrial field. In petroleum chemical industry, the water content of the crude oil is the one of the important measuring parameters in the production and machining process of crude oil. Water content measuring of the crude oil is greatly influenced by the gas in the crude oil.

The technologies of gas-liquid separation mainly focus on gravitational separation and centrifugal separation. In 1970s and 1980s, the main separation method was gravitational separation, centrifugal separation developed quickly in 1990s. With the reason of its skilled technology, gravitational separation is still used in practical production. Gravitational separation is basic on the density difference and the incompatibility between gas and liquid. Considering the practical aspect, the process of gravitational separation not need to have additional power and reagent, there is no secondary pollution in this process and the maintenance cost is little [1]. It often needs to optimize the form of the component inside to improve the flow characteristic of the

Y. Yu, Z. Yu, and J. Zhao (Eds.): CSEEE 2011, Part II, CCIS 159, pp. 331–335, 2011.
© Springer-Verlag Berlin Heidelberg 2011

separator for obtaining a better separation effect [2]. To strengthen the separation process and improve the technical economy characteristic, scholars have studied the separation characteristic of gas-liquid separation [3, 5].

Tee tube is the simplest multi-channel element and used in industry and agriculture pipe system. It can uniformly distribute the liquid from the main pipe into the two branch pipe and also can the liquid flow into the main pipe from the branch pipes. For single-phase flow, the distribution and the pressure characteristic of the tee tube are simple, and flux of the every tube is assigned completely depending on the resistance characteristic of the branch pipe. When the two-phase flow flows through the tee tube, dryness appears significant difference at the exit of the main branch pipe and the other, and this phenomenon called the phase separation of the tee tube. By this phenomenon, part of the single gas or liquid and all of the single gas or liquid can be separated from the gas-liquid flow [6, 7]. For two-phase flow, the distribution characteristic of tee tube is complicated. The physical model and mathematical model of the separator on the basis of the means of gravitational separation and the distribution characteristic of two-phase flow in tee tube were established this paper, analyzed the separation models with different forms and simulated the flow using FLUENT 6.3.

2 Model Establish

In order to analyze the effect of gas-liquid separation with the different branching pipe number of the tee tube, three physical models were established, whose number of branching pipe is respectively one, two and three. The three branch pipe of gas-liquid separator as an example is as shown in Fig. 1. The interior diameter of branching pipe is the same with the main pipe. The branching pipe is arranged by the vertical-type and has sufficient height. Branching pipe connect with gas gathering tube by a pin hole, the diameter of pin hole is less than one-fifth of the interior diameter of branching pipe. The function of pin hole is to control the flow by increasing the resistance, and also prevent some drops of the larger size to rush into gas gathering tube. Orifice plate's role is to adjust the loop resistance and to improve the characteristics of shunt coefficient.

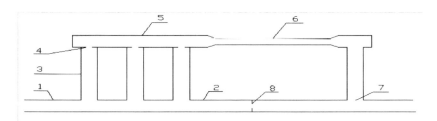

1-head; 2-through branch; 3- branch pipe; 4-holes; 5-Collecting pipe; 6-gas pipeline ;7-confluence of three links ;8-Orifice plates

Fig. 1. Schematic diagram of tee tube separator

Given the flow in tube is a process of steady state, and the fluid density keep constants. The energy conservation equation and momentum conservation equations are satisfied in the selected region. And also assuming flow in the pipe is smooth, without considering the friction heat of liquid viscous in the pipe.

Continuity equation

$$\frac{\partial u}{\partial x} + \frac{\partial u}{\partial y} = 0 \tag{1}$$

Nervier-Stocks equation

$$\cdot \ \rho \frac{dv}{dt} = \rho F - \mathrm{grad}p + \mu \nabla^2 v \tag{2}$$

Energy conservation equation

$$\frac{\partial(\rho T)}{\partial t} + \frac{\partial(\rho u T)}{\partial x} + \frac{\partial(\rho V T)}{\partial y} = \frac{\partial}{\partial x}\left[\frac{k}{c_p}\frac{\partial T}{\partial x}\right] - \frac{\partial}{\partial y}\left[\frac{k}{c_p}\frac{\partial T}{\partial y}\right] + \frac{\partial}{\partial z} + S_T \tag{3}$$

Where, C_P-specific heat; T-temperature; K-heat transfer coefficient of fluid; S_T-viscous dissipation.

The two-equation model is k-ε viscosity of turbulent flow.

$$\frac{\partial}{\partial t}(\rho k) + \frac{\partial}{\partial x_i}(\rho k u_i) = \frac{\partial}{\partial x_j}\left[\left[\mu + \frac{\mu}{\sigma_k}\right]\frac{\partial k}{\partial x_j}\right] + C_k + C_B - \rho \varepsilon - Y_M + S_k \tag{4}$$

$$\frac{\partial}{\partial t}(\rho \varepsilon) + \frac{\partial}{\partial x_i}(\rho \varepsilon u_i) = \frac{\partial}{\partial x_j}\left[\left[\mu + \frac{\mu}{\sigma_\varepsilon}\right]\frac{\partial \varepsilon}{\partial x_j}\right] + C_{1\varepsilon} \times \frac{\varepsilon}{k}(G_k + C_{3\varepsilon}C_B) - c_{2\varepsilon}\frac{\varepsilon^2}{k} + S \tag{5}$$

Where $C_{1\varepsilon} = 1.44$, $C_{2\varepsilon} = 1.92$, $C_{3\varepsilon} = 0.09$, $\sigma_\varepsilon = 1.3$, $\sigma_k = 1.0$.

3 Results and Discussion

Liquid is gas-liquid mixture, containing large amount of liquid, gas is air, with air content 20%. The velocity of entrance is 3m/s. Density of the air is 1.247 Kg/m^3, viscosity is 1.7894e^{-5} kg/(m•s), density of crude oil is 898 Kg/m^3. Volume fraction contours of gas and liquid of the three tube-types are obtained by numerical simulation, as shown in Fig. 1, 3.

It can be seen from Fig. 2, with the presence of branch pipe, part of the gas flows into the branching pipe, but large number gas accumulation in the orifice plates leads to higher gas rate in rear throttle plate because of throttle effect. The throttle effect results in the gas accumulation in the rear throttle plate, at last leads to higher contain rates of gas.

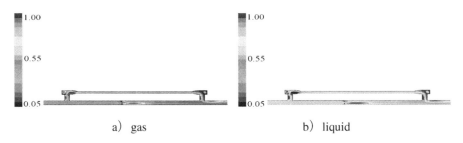

a) gas   b) liquid

Fig. 2. Gas and the liquid contours with one branch pipe

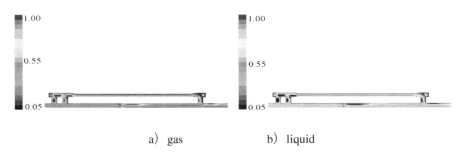

a) gas   b) liquid

Fig. 3. Gas and the liquid contours with two branch pipe

As shown in Fig. 3, two branching pipe contain gas rate significantly less than one, which indicates that the main channel containing gas rate begins to decrease. And because presence of two branching pipe, most of the gas flow into the branching pipe, while at the throttle plate, the impact of the front branching pipe that leads to gas accumulation is slow down at the rear throttle plates.

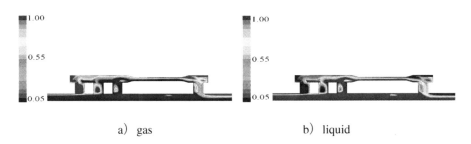

a) gas   b) liquid

Fig. 4. Gas and the liquid contours with three branch pipe

As shown in Fig. 4, the gas rate of the three branching pipe is significantly less than the others, which indicates that the gas rate of main channel is significantly reduced, at throttle orifice of the three branching pipe, the gas content reduced from 0.2 to 9.18×10^{-5}. It is shown that the existence of three branching pipe, most of the gas flows into the branching pipe, and at the throttle plate, the impact of the front branching pipe reduce at the rear throttle plate, however the reducing amount is significantly weaker.

4 Conclusion

The model of the tee tube separator was established in this paper, and simulated the flow in the tee tube with FLUENT, obtained the phase distribution of the gas-liquid flow. As the presence of branch pipe, part of the gas flow into the branch pipe. And at the throttle orifice, because of its throttling, the number of gas gathers at the rear throttle plate, so the gas content here increases. The separation effect of separator with three branching pipe is better than the other two.

References

1. Wan, C., Huang, F., Li, L.: Research Development in the Gravity Oil Water Separation Technology. Industrial Water Treatment 28(7), 13–16 (2008) (in Chinese)
2. Yuan, G., Zhang, L.: Study on Flow Field in Gravitational Oil-Water Separators with Regulating Components. Petro-Chemical Equipment 37(1), 13–15 (2008) (in Chinese)
3. Emelb, Bowman, R.W.: Residence Time Distribution in Gravity Oil-Water Separators. Journal of Petroleum Technology 9(18), 275–282 (1978)
4. Zhang, B., Liu, D., Wang, D.: Equal Quality Distribution of Gas-Liquid Two-Phase Flow Using Phase Separation Method. Hsi-An Chiao Tung Ta Journal of Xi'an Jiao Tong University 7(44), 106–110 (2010) (in Chinese)
5. Lu, Y., Xue, D.: Investigation on the Flow Dynamics of Gravity Separator Used For Oil-Water Emulsion (In Chinese). Acta Petro Lei Sinica 16(6), 23–29 (2000) (in Chinese)
6. Wang, D., Zhang, X.: The Split of Two-Phase Flow at Horizontal T-Junctions of Unequal Diameters. Journal of Engineering Thermophysics 16(6), 23–29 (2003) (in Chinese)
7. Willems, G.P., Kroes, J.P.: Performance of a Novel Rotating Gas-Liquid Separator Journal of Fluids Engineering. Transactions of the ASME 13(132), 313011–3130111 (2010)

Information Fusion Implementation Using Fuzzy Inference System, Term Weight and Information Theory in a Multimode Authentication System

Jackson Phiri[1], Tie Jun Zhao[1], and Jameson Mbale[2]

[1] Machine Intelligence and Natural Language Processing Group, School of Computer Science, Harbin Institute of Technology, Harbin, 150001, China
[2] Department of Computer Science, University of Namibia, Windhoek, Namibia
jackson.phiri@gmail.com, btjzhao@mtlab.hit.edu.cn,
cjmbale@unam.na

Abstract. Fuzzy inference systems have been used in a number of systems to introduce intelligence behaviour. In this paper we attempt to address an area of security challenges in identity management. The Sugeno-Style fuzzy inference is envisaged in the implementation of information fusion in a multimode authentication system in an effort to provide a solution to identity theft and fraud. Triangular and Sigmoidally shaped membership functions are used in the fuzzification of the three inputs categories namely biometrics, pseudo metrics and device based credentials. Term weight from text mining and entropy from information theory are used to compose the identity attributes metrics. Three corpora are used to mine the identity attributes and generate the statistics required to develop the metrics values from the application forms and questionnaires.

Keywords: identity attributes metrics, information fusion, fuzzy logic, authentication, term weight, entropy.

1 Introduction

In this information era, information technology is right at the core of our activities and almost every form of business. However, because of the security challenges, most business and government organizations are now grappling with identity theft, identity fraud, virus attack, espionage and many other similar challenges [1]. In this paper we attempt to address an area of security challenge in identity management called authentication through multimode authentication. We propose that a user is required to submit a biometric, pseudo metric and use a device based credential such as a smart card to be effectively authenticated [2]. Each of these identity attributes is assigned a weight and the user needs to meet a specified threshold to be authenticated in a multimode authentication system. The weights are combined or fused together in a technique of information fusion using Sugeno-Style fuzzy inference system and are composed by using the text mining technique called term weight and Shannon's information theory called entropy. Application forms for the various services offered

Y. Yu, Z. Yu, and J. Zhao (Eds.): CSEEE 2011, Part II, CCIS 159, pp. 336–341, 2011.

both in the real and cyberspace are used as the source of the identity attributes. In addition we roll out a questionnaire to obtain the identity attributes based on the user's opinion. Three corpora namely the AntConc, ConcApp and the TextSTAT are then used to mine the identity attributes from these two sources and generate the statistical information required to compute the term weight and entropy as the metrics values of these identity attributes [3][4][5].

2 Related Works

Information theory introduced by Claude Shannon in 1948 is used to quantify information and is based on probability theory and statistics [6]. The most important quantities of information theory are entropy and mutual information. In this paper, we use the statistical information generated by the three corpora to compute the empirical probabilities of the identity attributes and then (1) to compute their entropy using [6];

$$H(p) = \sum_{i=1}^{n} p_i \log_2 (1/p_i)$$

(1)

Where p_i is the probability of the identity attribute in the ith sample space.

In text mining, term weight is composed of term frequency and document frequency [7]. The term frequency $tf_{t,d}$ of the term t and document d is defined as the number of times that t occurs in d. Document Frequency (df) is the measure of the informativeness of the term t. The most useful component of df is the *Inverse Document Frequency* (idf). *Collection Frequency* (cf) of the term t is the number of occurrences of t in the collection, counting multiple occurrences [7]. In this paper we use the term weight to compose the identity attributes metrics. It is given by the following equation where N is number of documents in the corpus [7];

$$w_{t,d} = (1 + \log tf_{t,d}) \times \log_{10} \frac{N}{df_t}$$

(2)

Two commonly used fuzzy inferences are Mamdani and Sugeno-Style inference system. In FIS inference process, there are four major steps, which include fuzzification of the input variables, rule evaluation, aggregation of the rule outputs, and finally defuzzification [8]. Sugeno-Style fuzzy inference is used in this paper because it is computationally effective and works well with optimisation and adaptive techniques. *Triangular membership function* (trimf) and the *Sigmoidally shaped membership function* (sigmf) are used as fuzzification functions. Sigmoidally function depends on two variables a and c and is used in fuzzification of biometrics inputs. It is represented by the following function [8];

$$f(x,a,c) = \frac{1}{1 + e^{-a(x-c)}}$$

(3)

Triangular membership function (trimf) is a function of a vector x and depends on three scalar parameters a, b and c. The parameters a and c locate the feet of the triangle while the parameter b locate the peak as follows [8];

$$f(x:a,b,c) = \max\left(\min\left(\frac{x-a}{b-a}, \frac{c-x}{c-b} \right), 0 \right) \tag{4}$$

Information fusion has seen a lot of applications in the areas of robotics, geographical information systems and data mining technologies. For example, [9] uses a combination of artificial neural networks, Dempster-Shafer evidence theory-based information fusion and Shannon entropy to form a weighted and selective information fusion technique to reduce the impact uncertainties on structure damage identification. In [10] the quantitative metrics are proposed to objectively evaluate the quality of fused imagery. Information fusion in biometric verification for three biometrics at the matching score level is provided in [11] while [12] provides a two-level approach to multimodal biometric verification. In this paper, we introduce the pseudo metrics and device based metrics in addition to the biometrics and use Sugeno-Style fuzzy inference system to implement information fusion during multimode authentication.

3   Methodology

In this paper the services offered by both the public and private sectors which include the services such as the passport, birth certificate, VISA or permit issued to foreign nationals, the national identity card, citizen certificate, voter's card, insurance services, financial services, health care services and education services are used as the source of the identity attributes. Application forms in PDF format from the service providers' websites are downloaded and then converted into the text file for analysis using three corpora [3][4][5]. The online application forms are captured using the *Webspider* integrated into the TextSTAT corpus. Secondly, we roll out questionnaires targeting international respondents for the opinion based responses on the identity attributes deemed as important and can uniquely identify the user. Social networks and email addresses are used as the vehicle for obtaining responses from the electronic copy of the questionnaire. A total of *200 application* forms and *100 questionnaires* are used in this paper and each of the three corpora is then used to generate the *word frequency*, *document frequency* and *collection frequency* from the two sources hence creating *six sample spaces* of the identity attributes. Fig. 1 shows a total of 200 documents (application forms) in the corpus with 156 hit (collection frequency) for the *family name* using AntConc corpus. Also shown is the number of hits (term frequency) for each of the 156 documents. For example the file named *47ch.txt* has a term frequency of *3*. These generated statistical results are then used to compute the entropy yield and term weights of the identity attributes using (1) and (2). Fig. 2 shows the term weight results of 10 selected identity attributes while Table 1 shows both the computed term weight and entropy yield. Using Sugeno-Style FIS, an information fusion technique is then designed and implemented with three inputs, five rules and one output.

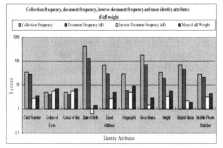

Fig. 1. AntConc corpus search results for family name showing term and document frequencies

Fig. 2. Collection frequency, document frequency, inverse document frequency, and the mean computed tf-idf weight

We use the three categories that constitute multimode authentication to design and implement an information fusion technique. These are pseudo metrics such as password or personal identification number (PIN), biometrics such as fingerprint or iris scan and device metrics such as international mobile equipment identifier [2].

Table 1. Computed entropy and term weight metrics values

#	Identity Attribute	Mean Identity Attributes Entropy	Mean Identity Attributes Term Weight
1	Card ID	0.048644	3.50358
2	Date of Birth	0.263487	1.44031
3	Email	0.085400	5.06057
4	Fingerprint	0.042741	8.97436
5	Given Name	0.217056	2.94966
6	Height	0.056102	5.57110
7	Mobile Phone ID	0.047691	2.76955
8	Passport Number	0.074050	2.08616
9	Sex	0.215595	1.82216
10	Surname	0.379362	1.34112

Three examples used in this paper are the *fingerprint* for the biometrics, *PIN* for the pseudo metrics and the *Card-ID* for the device metrics. Equation 2 is used as the membership function for the fingerprint biometrics and the parameters a and b are assigned 0.5 and 5.0 respectively. Equation 3 is used as the membership function for both pseudo metrics and device metrics. For the pseudo metrics the values of a, b and c are 0.5, 2.5 and 4.5 respectively while that for device metric the respective values of a, b and c are 2.5, 3.5 and 4.5. The computed values of the term weights of the identity attributes as shown in the last column of Table 1 help in coming up with the above values. Since we are using the Segeno-Style, we assign the output linguistic variables the following values where entropy values help to give the indication of the required range for each identity attribute. These are *all* with a score of 0.9, *VeryHigh* with a score of 0.5, *High* with a score of 0.4, *Medium* with a score of 0.3, *Low* with a score of 0.2 and *VeryLow* with a score of 0.1. We then formulate the five rules as shown in Fig. 3. The weighted average is used to come up with the defizzification crisp output value which is then used to implement multimode authentication.

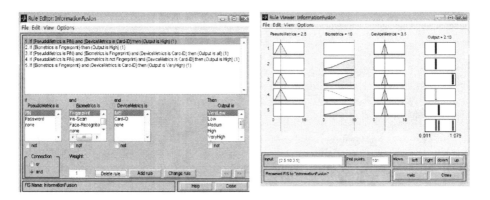

Fig. 3. Five fuzzy rules and linguistic variables   **Fig. 4.** Sugeno-Style FIS with 3 inputs and five rules

4   Results and Discussion

Table 2 shows a sample of eight selected input combination and the respective computed output of the FIS information fusion technique. When all the submitted copies matched the copies in the databases then the maximum output of the information fusion is *2.18* as shown also in Fig. 4. However, when none of the submitted copies matched those in the database then the system gives *0.03*. The rest of the combinations of the inputs lie in between this range as shown in Table 2. Fig. 5 shows the surface viewer for the biometrics, devices metrics and corresponding output. Using this range, it is possible to implement a multimode authentication where a user is required to submit a given range of identity attributes to enable the user meet the threshold value set in a given authentication system. A combination of biometrics, token based credentials and pseudo metrics will most likely form a very effective defence against imposters and will most likely help to reduce the cases of identity theft and fraud seen on most online services. A similar example using artificial neural networks with four input, four neurons in the hidden layer, one neuron in the output layer, sigmoid function as the transfer function and back propagation algorithm for learning, the network successfully yielded values between 0.131854 and 0.912294 to six decimal places. In the future, an additional fourth input for the name, date of birth, race and information fusion technique such as neural-fuzzy inference, Dempster-Shafer, Hidden Markov Model will be used and compare the results with our implementation.

Table 2. Inputs and corresponding outputs of a Sugeno-Style FIS information fusion technique

P	B	D	Y
0.0	0.0	0.0	0.03
0.0	10.0	0.0	0.37
2.5	0.0	0.0	0.03
0.0	0.0	3.5	0.07
2.5	0.0	3.5	0.91
2.5	10.0	0.0	0.37
0.0	10.0	3.5	0.83
2.5	10.0	3.5	2.18

P = Pseudo Metrics, B = Biometrics, D = Device Metrics and Y = Final Output.

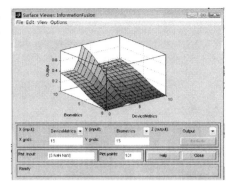

Fig. 5. Surface viewer of the biometrics, device metrics and the respective output of the FISReferences

References

1. Mike, S.: Unify and Simplify: Re-thinking Identity Management. Network Security 2006(7), 11–14 (2006)
2. Geoff, B.: The Use of Hardware Tokens for Identity Management. Information Security Technical Report (2004)
3. Anthony, L.: AntConc: Design and Development of a Freeware Corpus Analysis Toolkit for the Technical Writing Classroom. In: Professional Communication Conference Proceedings, pp. 729–737. IEEE Press, New York (2005)
4. ConApp Corpus, http://www.edict.com.hk/PUB/concapp/
5. TextSTAT Corpus, http://neon.niederlandistik.fu-berlin.de/en/textstat/
6. Togneri, R., DeSilva, S.J.C.: Fundamentals of Information Theory and Coding Design. Chapman & Hall Press, Florida (2002)
7. Manning, C.D., Raghavan, P., Schutze, H.: Introduction to Information Retrieval. Cambridge University Press, New York (2008)
8. Fuzzy Toolbox, http://www.mathworks.com/access/helpdesk/help/pdf_doc/fuzzy/rn.pdf
9. Hassan, R.M., Nath, B., Kirley, M.: A Fusion Model of HMM, ANN and GA for Stock Market Forecasting. J. Expert Systems with Applications 33(1), 171–180 (2007)
10. Zheng, Y., Essock, A.E., Hansen, C.B., Haun, M.A.: A New Metric Based on Extended Spatial Frequency and its Application to DWT Based Fusion Algorithms. J. Information Fusion 8(2), 177–192 (2007)
11. Cheung, M., Mak, M., Kungi, S.: A Two-Level Fusion Approach to Multimodal Biometric Verification. In: ICASSIP IEEE Conference, pp. 485–488. IEEE Press, New York (2005)
12. Ross, A., Jain, A.K., Qian, J.: Information Fusion in Biometrics. In: Bigun, J., Smeraldi, F. (eds.) AVBPA 2001. LNCS, vol. 2091, pp. 354–359. Springer, Heidelberg (2001)

Investment Efficiency Analysis of Chinese Industrial Wastewater Treatment Based on Expansion-Type DEA

Shuang Zhao[1,2], Hui-Min Wang[1,2], and Lei Qiu[1,2]

[1] State Key Laboratory of Hydrology Water Resource and Hydraulic Engineering of Hohai University, 210098 Nanjing, China
[2] Management Science Institute of Hohai University, 210098 Nanjing, China
a28129988@163.com, {hmwang,qiulei}@hhu.edu.cn

Abstract. Based on a shortage of an investment efficiency scientific analysis about Chinese industrial wastewater treatment, from the relationship between environment and economy, construct an assessment indicator system, and then targeted propose the DEA-based method with extended decision-making units, to solve the lack of the standard DEA method. The results showed that: At present, in each region the investment efficiency is low, especially northern and northwest regions are most serious. So increasing capital investment is no longer a good promoting role to improving the pollutant. Analyze the reasons, and then introduce the market rules, establish environmental supervision system with incentive compatibility constraint penalty, so that the problem of industrial wastewater treatment can be shoveled fundamentally.

Keywords: Industrial Water Pollution, DEA, Expansion-Type Decision-Making Units, Pollution Control Investment Efficiency.

1 Introduction

Water environment is seriously polluted as industrial wastewater discharged in China. Face this situation, government increases investment to treat it year by year, but if it is efficiency. Many experts from domestic and foreign have done the study. In domestic, Zhongkui Li(2003)[1], Kai Huang(2007)[2] research on environmental economic value. Hui Luo(2005)[3] study the environment treatment efficiency with property right theory. At abroad, Timo Kuosmanena (2007)[4], Hajkowicz Stefan (2008)[5] research on economics theory of cost-benefit to analyze the validity. By contrast and Summarizing what the domestic and foreign scholars have studied, DEA method used by Timo Kuosmanena is a easier method, which is most typical that whether put into effect due to produce results, and great significance of productive practice. However the traditional DEA method is limited to solving the effectiveness of investment. While the number of decision-making units and input and output indicators must meet Cooper condition [6]. For this reason, proposes the DEA-based method with extended decision-making units to research on the effectiveness of investment.

Y. Yu, Z. Yu, and J. Zhao (Eds.): CSEEE 2011, Part II, CCIS 159, pp. 342–347, 2011.
© Springer-Verlag Berlin Heidelberg 2011

2 Theory Methods and Model Construction

DEA (Date Envelopment Analysis) is a method that assess of relative efficiency on multi-input and multi-output decision-making units (DMUs) [7]. As it needn't assume the DMUs' production function, and avoid the complex and difficult task of estimating cost is often used to assess the effectiveness of many cases, technological progress, returns to scale and enterprise efficiency. The advantage of standard DEA model is determining flexibility the weight of indexes (v_i, u_r), but when DMUs number is limited relatively, if the number of input and output indicators is not met for cooper condition n≥max{m*s,3(m+s)}, then its identified capable will be low. This paper proposes an extended DMUs method to analysis investment efficiency.

There are n areas with m input and s output. $X_j = (x1_j, \ldots xs_j)$ is an input vector and $Y_j = (y1_j, \ldots ys_j)$ is an output vector in a set from (X_j, Y_j). The production possibility set of evaluating DMUs' size and technology is as follows.

$$T = \left\{ X, Y \mid \sum_{j=1}^{n} \lambda_j X_j \leq X, \sum_{j=1}^{n} \lambda_j Y_j \geq Y, \lambda_j \geq 0 \qquad j = 1, 2, \cdots n \right\} \tag{1}$$

In the condition that DMUs is limited, by the theory foundation of input and output analysis that general equilibrium theory of Walrasian that economic can automatically balanced development with supply and demand and the price fluctuation. Adjusting the input and output of DMUs the same direction, won't affect the reference value of assessment results of fictitious DMUs. The detailed method is as follows.

As shown in Table 1, expand each DMU. First, determine a adjustment value of each input or output. Fixed i and centered $DMU_{j(0)}$, adjust each date. The number of DMUs is determined by cooper condition. The adjustment value is as follows.

$$\frac{\max\{x_{j(0)}\} - \min\{x_{j(0)}\}}{(n-1) \times 2p} = \varepsilon \tag{2}$$

$$x_{j(q)} - x_{j(q-1)} = \varepsilon; \varepsilon = \{\varepsilon_i, i = 1, \cdots, m+s\}; q \in (-p+1, p); i = 1, \cdots, m+s; j = 1, \cdots, n$$

$DMU_{j(0)}$ is one of practical DMUs. $DMU_{j(k)}$ is one of fictitious DMUs, $k \in (-p,p)$ and $k \neq 0$. p is the number of expanding bilateral sides respectively of DMUs. $x_{ij(k)}$ is input value and $y_{ij(k)}$ is output value. ε is adjusting value of fictitious DMUs.

There are much similar points between C2R model and improved model as follows.

$$(P_0) \begin{cases} \min \theta \\ \sum_{j=1}^{n(2p+1)} \lambda_j X_j + s^- = \theta X_0 \\ \sum_{j=1}^{n(2p+1)} \lambda_j Y_j - s^+ = Y_0 \\ \lambda_j \geq 0, s^+ \geq 0, s^- \geq 0, j = 1, \cdots, n \end{cases} \tag{3}$$

For planning problems P_0, the necessary and sufficient condition of weakly DEA efficient of DMUs is $\theta=1$. The necessary and sufficient condition of DEA efficient of DMUs is $\theta=1$, and its optimal solutions $\lambda^0 = (\lambda_1^0 ;\cdots , \lambda_n^0)^T, s^{-0}, s^{+0}$ is met for $s^{-0} = s^{+0} = 0$.

Table 1. Expended decision unit

	DMU_j						
	DMU_{jr}	..	$DMU_{j(-1)}$	$DMU_{j(0)}$	$DMU_{j(1)}$	..	$DMU_{j(p)}$
v_1 $1\rightarrow$	$x_{1j(-p)}$	..	$x_{1j(-1)}$	$x_{1j(0)}$	$x_{1j(1)}$	..	$x_{1j(p)}$	
v_2 $2\rightarrow$	$x_{2j(-p)}$	..	$x_{2j(-1)}$	$x_{2j(0)}$	$x_{2j(1)}$	..	$x_{2j(p)}$	
$\vdots$ $\vdots$	$\vdots$		$\vdots$	$\vdots$	$\vdots$		$\vdots$	
v_m $m\rightarrow$	$x_{mj(-p)}$	..	$x_{mj(-1)}$	$x_{mj(0)}$	$x_{mj(1)}$	..	$x_{mj(p)}$	
	$y_{1j(-p)}$	..	$y_{1j(-1)}$	$y_{1j(0)}$	$y_{1j(1)}$	..	$y_{1j(p)}$	$1\ u_1 \rightarrow$
	$y_{2j(-p)}$	..	$y_{2j(-1)}$	$y_{2j(0)}$	$y_{2j(1)}$	..	$y_{2j(p)}$	$2\ u_2 \rightarrow$
	$\vdots$	$\vdots$	$\vdots$	$\vdots$	$\vdots$	$\vdots$	$\vdots$
	$y_{sj(-p)}$	..	$y_{sj(-1)}$	$y_{sj(0)}$	$y_{sj(1)}$	..	$y_{sj(p)}$	$s\ u_s \rightarrow$

Table 2. Indexes of industrial wastewater treatment efficiency evaluation

Target layer	Element layer	Index layer
Evaluation index system of investment in industrial wastewater treatment	Economical index	X1, investment amount of industrial wastewater treatment per unit [yuan/t]
	Industrial wastewater features index	Y1, COD setback ratio of industrial added value per unit; Y2, NH_3-N setback ratio of industrial added value per unit; Y3, petroleum contaminants setback ratio of industrial added value per unit; Y4, cyanide setback ratio of industrial added value per unit; Y5, Total mercury setback ratio of industrial added value per unit; Y6, Total cadmium setback ratio of industrial added value per unit; Y7, Cr^6+ setback ratio of industrial added value per unit; Y8, total lead setback ratio of industrial added value per unit;

3 Investment Efficiency Analysis of Chinese Industrial Wastewater Treatment

By considering the drainage distribution and the economic development of each province city comprehensively, divide our country into seven typical parts, which are north, northeast, east, central, south, southwest and northwest of China. Make them be DMUs to research their investment efficiency analysis.

3.1 Constructing Index System of Input and Output

Investment efficiency is a problem of relationship between economic development and environmental quality. It is explained by Environmental kuznets curve. When economic development in a lower level, environmental pollution will be worse with it, and vice versa. So, construct and evaluation index system of investment from the relationship between economic development and environmental quality. (Table 2)

3.2 Sampleing Data Choosing

With the data from China statistic year book, calculate investment amount per unit. The amount is on the basis of constant price for 2000. And contamination discharge setback ratio of industrial added value per unit from 2003 to 2007. The amount and setback ratio are used in DEA model analysis. (Table 3).

Table 3. Raw data of Chinese industrial wastewater treatment efficiency evaluation

Regions	Input indicators X_1	Output indicators							
		$Y_1(\%)$	$Y_2(\%)$	$Y_3(\%)$	$Y_4(\%)$	$Y_5(\%)$	$Y_6(\%)$	$Y_7(\%)$	$Y_8(\%)$
North	0.86	56.19	62.82	63.83	69.16	92.31	89.09	63.02	92.47
Northeast	0.71	37.92	40.12	23.46	59.97	99.93	99.67	72.98	90.64
East	0.63	51.18	58.49	69.44	74.85	89.03	85.41	62.31	80.38
Central	0.56	61.48	64.82	75.03	71.58	85.60	66.34	81.95	87.76
South	0.37	37.19	45.94	68.40	66.12	93.08	76.30	11.87	58.35
Sorthwest	0.52	67.03	66.61	78.65	91.17	94.88	79.62	75.05	62.69
Northwest	1.12	39.06	64.99	87.15	62.19	80.85	78.40	73.50	51.44

Note: data is according to China statistic year book 2004.

Table 4. DEA validity evaluation results of expended decision unit

Regions	DEA scale and technique validity (θ^*)	Technique validity (θ')	Returns to scale
North	inefficacy (0.641)	inefficacy (0.96)	degression
Northeast	inefficacy (0.835)	efficacy (1.00)	degression
East	inefficacy (0.815)	efficacy (1.00)	degression
Central	inefficacy (0.962)	efficacy (1.00)	degression
South	inefficacy (0.936)	efficacy (1.00)	degression
Southwest	inefficacy (0.959)	Efficacy(1.00)	degression
Northwest	inefficacy (0.475)	inefficacy (0.72)	degression

3.3 Calculation and Analysis of the Result

Expand the existing DMUs using the DEA model of expanded DMUs. Calculate the data in table 3, with DEAp2.1 Microsoft. The output result is in table 4.

Table 4 shows that from 2003 to 2007, none of seven regions meets DEA scale and technique available at the same time. And return to scale is all descending. So, it is useless for increasing contamination discharge setback ratio of industrial added value per unit, by investment more funds.

It is important to note that northern and northwest still can't satisfy DEA pure technology efficient and relative efficiency are lower than 0.7. It indicates that investment efficiency in north and northwest china are lowest, need to be improved urgently. These two regions are inputs redundancy and outputs scarcity of DEA pure technology efficient. Their target values are in table 5 as an improvement basis to treat industrial wastewater. From the output, the key point of industrial wastewater treatment is petrochemical enterprises and plating enterprise in north China. Investment efficiency in northwest is more serious than north china. Setback ratio of industrial added value per unit of COD, cyanide, Total mercury and total lead are all over 10%, so many industry should be improved expect petroleum contaminants.

Table 5. Technique validity improvement analysis of industrial wastewater treatment investment in North and Northwest of China

Regions	Input indicators X_1	Output indicators							
		Y_1 (%)	Y_2 (%)	Y_3 (%)	Y_4 (%)	Y_5 (%)	Y_6 (%)	Y_7 (%)	Y_8 (%)
North (θ=0.96)	0.86	56.19	62.82	63.83	69.16	92.31	89.09	63.02	92.47
s	0	0.70	0.00	5.30	3.60	5.30	0.00	8.30	0.00
North target value	0.83	56.89	62.82	69.13	72.76	97.61	89.09	71.32	92.47
Northwest (θ=0.72)	1.12	39.06	64.99	87.15	62.19	80.85	78.40	73.50	51.44
s	0	19.90	3.20	0.00	20.70	14.50	3.50	6.80	10.40
Northwest target value	0.81	58.96	68.19	87.15	82.89	95.35	81.90	80.30	61.84

There are some differences in technical and scale relative efficiency value in the five regions. Central, South and Northwest China are all over 90%, in a superior to Northeast China and east China. So divide the investment efficiency into three grades. High-grade is Central China, South China and Northwest China. Medium-grade is Northeast China and East China. And low-grade is North China and Northwest China.

In conclusion, the investment efficiency is low from 2003 to 2007 for various regions in China, and the scale revenues are all in the decline stage. The main reasons as follows. First, historical reasons are due to serious water environment damage. As backward country situation in early period, six industrial bases established without wastewater treatment facilities. Second, irrational industrial structure leads to structural pollution. The proportion of the secondary sector (industry) is too high. But the high energy consumption and high pollution industries such as power industries, steel industries, building materials industries and energy industries still control the economic lifelines of the country. Low efficiency of environmental protection facilities, ineffective of the environmental pollution control and the financial mismanagement are the third reasons. Fourth, there are multiple difficulties of the water environment quality supervision, with local governments underpowered supervision and enterprises underpowered law-abiding. Therefore, the key of the fundamental solution to the problem of industrial wastewater treatment is introduction the market rules under the guidance of Central Government, and establishing environmental supervision system with incentive compatibility constraint penalty.

4 Conclusion and Insufficiency

(1) This paper proposes the DEA-based method with extended decision-making units, to research that if the investment in Chinese industrial wastewater treatment is efficient. The DEA method with extended decision-making units can make up for the useless of standard DEA model.

(2) Construct and evaluation index system of investment in industrial wastewater treatment from the angle of the relationship between economic development and environmental quality. This index system can be applied to investment efficiency analysis of industrial wastewater treatment.

(3) The result of this paper shows that the lowest is North China and Northwest China of investment efficiency, where are the work in the future. Medium-grade is Northeast China and East China. The result is a scientific reference for enhancing industrial wastewater treatment efficiency in China.

There are some shortages in this paper need to be improved. The index system only contains environment and economics indicators. But determinants of investment efficiency also contain some policy and human factors. Adding policy and human factors to index system can make the evaluation more meaningful, and guides the nation and local government to make a significant environmental protection strategy.

Acknowledgments. We are grateful to national social science fund (08CJY022), national social science fund key (10AJY005), ministry of science and technology research key (108064), Jiangsu "333" project fund.

References

1. Li, Z., Song, R., Yang, M., et al.: Approach to the Environmental Economics Analysis of Benefit from Watershed Management. J. Sci. Soil Water Cons. 1(3), 56–62 (2003)
2. Huang, K., Guo, H., Yu, Y., et al.: Ecological Value Analysis of Watershed Water Pollution Control. J. China Pop. Resour. Environ. 17(20), 109–112 (2007)
3. Luo, H., Zhong, W., Liu, Y., et al.: Ecological Environment Treatment Efficiency in Plateau of North Shanxi Province: Property Damaged Theory in an Analytical View. J. China Pop. Res. Environ. 15(8), 50–54 (2005)
4. Timo, K., Mika, K.: Valuing Environmental Factors in Cost–benefit Analysis Using Data Envelopment Analysis. J. Ecol. Econ. 62(9), 56–65 (2007)
5. Stefan, H., Rachel, S., Andrew, H., et al.: Evaluating Water Quality Investments Using Cost Utility Analysis. J. Environ. Manage. 88(50), 1601–1610 (2008)
6. Cooper, W.W., Seiford, L.M., Tone, K.: Date Envelopment Analysis, A Comprehensive Text with Models, Applications, References and DEA-Solver Software. Kluwer Academic Publishers, Boston (2003)
7. Charnes, A., Cooper, W.W., Rhodes, E.: Measuring the Efficiency of Decision Making Units. Eur. J. O. R. 2(5), 429–444 (1978)

Electrospinning Polyurethane (PU) /Inorganic-Particles Nanofibers for Antibacterial Applications

Xiao-jian Han[1,*], Zheng-ming Huang[2], Chen Huang[1], and Chuang-long He[3]

[1] College of light-textile Engineering and Art, Anhui Agricultural University,
130 West Changjiang Rd., 230036 Hefei, China
[2] School of Aerospace Engineering & Applied Mechanics, Tongji University,
1239 Siping Road, 200092 Shanghai, China
[3] Institute of Biological Sciences and Biotechnology, Donghua University,
1882 West Yanan Rd, 201620 Shanghai, China
xiaojianhan@gmail.com, huangzm@tongji.edu.cn,
silkhch@163.com, hcl@dhu.edu.cn

Abstract. A blend-electrospinning technique was applied to transfer polyurethane (PU)/inorganic particles solutions into nanofibers, in which titanium dioxide (TiO2), cuprum (Cu) or/and silver (Ag) nanoparticles were packed. The resultant nanofibers were subsequently characterized by means of scanning electron microscope (SEM), fourier transform infrared spectroscopy (FTIR) and anti-bacterial test. Experimental results have shown that when the ratio of PU solution and titanium dioxide sol was in 3.5:1, 89.55% of the escherichia coli and 82.35% of the staphylococcus aureus bacteria were killed after a light activation treatment. Without any photocatalysis, however, an introduction of Ag or/and Cu nanoparticles into the PU/TiO2 (in the ratio of 3.5:1) nanofibers led to a significant improvement in their anti-bacterial ability without photocatalysis.

Keywords: Nanofibers, Electrospinning, Anti-Bacterial Performance.

1 Introduction

Recently, preparation of polymer nanofiber films containing inorganic nanoparticles has drawn great attention because the resultant nanocomposites can exhibit the outstanding characteristics of both inorganic nanoparticles (e.g., the quantum size effect and high ratio of surface atoms to innersphere atoms etc.) and polymer nanofibers (e.g., the high specific surface area and good interpenetrating capacity in other materials etc.) [1]. Titanium dioxide is an important material for a variety of applications such as catalytic devices [2], solar cells [3], and other optoelectronic devices [4].

Conventionally, it is extremely difficult to disperse TiO_2 powder homogeneously into the polymer solution due to the easy agglomeration of nanoparticles. There are a few investigations of electrospinning polymer/TiO_2 sol-gel for application in

---

* Corresponding author.

Y. Yu, Z. Yu, and J. Zhao (Eds.): CSEEE 2011, Part II, CCIS 159, pp. 348–353, 2011.
© Springer-Verlag Berlin Heidelberg 2011

photocatalysts [5], solar cells [6] and other optoelectronic devices[4] reported. He etc al. [5] have investigated the electrospinning parameters of electrospun polyvinyl alcohol (PVA)–Pt/TiO$_2$ composite nanosize fiber and the degradation rate of PVA by photocatalytic oxidation under UV irradiation. Viswanathamurthi etc al. [6] reported the electrospun poly (vinylacetate)/titanium dioxide hybrid fibers and obtained the pure ceramic metal oxide fibers through high temperature calcination of the composite.

In this work we fabricated PU/TiO$_2$ ultrafine fibrous membranes using a sol-gel electrospinning technique. In order to carry out their fine anti-bacterial ability without photocatalysis, a proper content of Ag or/and Cu was added into the PU/TiO$_2$ ultrafine fibers. The resulting composite ultrafine fibers are designated as PU (TiO$_2$, Cu), PU (TiO$_2$, Ag), or PU (TiO$_2$, Cu, Ag) respectively. The nanofibrous mats obtained by this technique can be used for filter media with high filtration efficiency (FE) and low air resistance, or bacterial protective clothing, depending on the PU polymer and TiO$_2$, which was also doped by silver or (and) copper, being used.

2 Experiments

2.1 Materials and Electrospinning

Polyurethane (PU, Elastollan 1185A) was purchased from Jinjiang plastic Ltd. (Shanghai), whereas Tetrahydrofuran (THF), dimethylformamide (DMF) and acetic acid were from Chemical Reagent Co. Ltd (Shanghai). Tetrabutyl titanate, was acquired from Runjie Chemical Reagent Co. Ltd (Shanghai). Copric nitrate (Cu(NO$_3$)$_2$) and Silver Nitrate (AgNO$_3$) were obtained from Zhenxin Reagent Co. Ltd (Shanghai) and NO. 1 Reagent Co. Ltd (Shanghai), respectively.

The devised experimental setup basically includes a spinneret apparatus with an inner steel capillary (inner diameter, 0.8 mm) coaxially located inside an outer plastic nozzle (inner diameter, 2.0 mm), which is fixed no a grounded hob, made a co-axial configuration at the tip. A syringe pump (model WZ-50C2, Zhejiang University Medical Instrument Co., Ltd) is for delivering solution through Teflon tube connecting the syringe with the inner capillaries (needles), a high voltage DC generator (Beijing Machinery & Electricity Institute) exerts a 20-30 KV voltage to spinning solutions through inner needle, and the negative terminal is attached to a aluminium foil covered collector (cathode) with ability to move to and fro in horizontal direction to collect the random fibers.

2.2 Morphology and Fourier Transformed Infrared Spectroscopy (FTIR)

The morphology of collected fibers was observed under a environmental scanning electron microscopy (ESEM, PHILIPS XL 30, Netherland) with a accelerating voltage of 20 kV.

Infrared spectroscopies of the nanofiber membranes, whose spectral range was from 4000 to 500 cm^{-1} with a resolution of 4 cm^{-1}, were obtained in terms of EQUINOX 55 (Bruker, Germany) under a 6.10 kHz scanner velocity.

2.3 Antibacterial Test

The in vitro antibacterial activities of PU fibers with nanoparticles were examined according to AATCC100 standard. The following microorganisms were used: Gram-positive taphylococcus aureus (S. aureus, ATCC 6538)) and Gram-negative Escherichia coli (E. coli, ATCC 2592). The counts were used to calculate the surviving number of bacteria by the GB/T 4789.2-2003 standard (Microbiological examination of food hygiene Detection of aerobic bacterial count). The anti-bacterial efficacy (ABE in %) of the specimen was calculated according to the following equation.

$$ABE(\%) = (A - B) / A \times 100 \tag{1}$$

where A (cfu/g)and B (cfu/g) stood for the average numbers of viable bacterial colonies of the blank control in the electrospun PU nanofibrous membrane and PU/inorganic specimen, respectively.

3   Results and Discussion

3.1   Fiber Morphology and FTIR Analysis

Fig.1 shows the SEM images of PU (TiO_2, Cu), PU(TiO_2, Ag) and PU (TiO_2, Cu, Ag) fibers, electrospun by the solution made from Cu (NO_3)$_2$ and (or) $AgNO_3$, PU/TiO_2 solution (weight ratio, 3.5:1) in the molar ratio (Cu/Ti, 2/100, Ag/Ti, 2/100), respectively. Even though the content of Cu and (or) Ag introduced into PU/TiO_2 solution is small, there is a strong interaction between $AgNO_3$ (or Cu(NO_3)$_2$) and PU molecular chains, which promoted phase separation of PU . This would change many parameters of mixture, such as solution concentration and surface tension. Therefore, the fine morphology of PU (TiO_2, metal particle) could be obtained in more harder electrospinning parameters (voltage, flow rate, the distance between the tip and the collector), than that of PU/TiO_2. For example, in the process of electrospinning PU (TiO_2, Ag), the environmental condition of the voltage (26-27kv), the distance (13-14cm) and the flow rate (1.6ml/h) could be distinguished from the respective parameters of PU/TiO_2. Even the same flow rate of PU (TiO_2, Cu) solution was used as that of PU/TiO_2 solution in electrospinning, the other parameters should be different (the voltage, 25kv, the distance, 13-14cm) for successfully fabricating PU (TiO_2, Cu) membrane without flaw on the resultant membrane. In the present study, the average diameter of the bead on the PU (TiO_2, metal particle) fibers corresponding to Cu, Ag, Cu and Ag, as shown in Fig.1 (b)- (d), were 320 nm, 461 nm, and 1350 nm, respectively. The average diameter of PU/TiO_2 (3.5:1) fibers was 862 nm (Fig.1 (a)). This is because the strong interaction between $AgNO_3$ (or Cu(NO_3)$_2$) and PU molecular chains slows the viscosity and the higher voltage applied in electrospinning, which increases the charge density of polymer jet and favors stretching or drawing of the jet, leads to in smaller diameter fibers. So Figs.1 (b) and (c) favours more thinner fibers without or less beads than Figs.1 (a). But too much metal particles in the PU (TiO_2, Cu, Ag) solution will weaken the electrospinning capability of mixture. So the average diameter of bead in the electrospun PU (TiO_2, Cu, Ag) fibers (Fig.1 (d)) is the most and dispersed less symmetrically in PU (TiO_2, metal particle) system.

Fig. 1. (a) SEM images of PU/TiO2 (3.5:1), (b) PU (TiO2, 3.5:1, Ag), (c) PU(TiO2, 3.5:1, Cu), (d) PU(TiO2, 3.5:1, Cu, Ag)

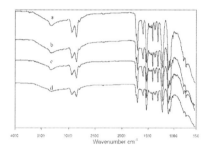

Fig. 2. FT-IR spectra of fibers: (a) PU/TiO2 (3.5:1), (b) PU (TiO2, 3.5:1, Ag), (c) PU (TiO2, 3.5:1, Cu), (d) PU (TiO2, 3.5:1, Cu, Ag)

From Fig.2, the FT-IR spectra of the PU polymer, we can see the following characteristics: a strong absorbing peak appeared in the 1700-1670 cm^{-1} region owing to the stretching vibration of C=O bond in carbamate (R-O-C (=O)-NHR). In general, the high frequency absorbing peaks corresponding to a mono-substituent and to a disubstituted compound may occur in the region of 1736-1700 cm^{-1}. A combination of the both gave rise to the peak of 1730 cm^{-1}. Attributed to C=O stretch of amido ester, there was also a 1701cm^{-1} region. The characteristic bands of alkyl aether produced the wide 1106 cm^{-1} (1150-1060 cm^{-1}) and 1079 cm^{-1} regions because of the C-O-C asymmetrical flexing vibration. Howerever, less difference from the FTIR of the PU (TiO$_2$, metal particle) (Fig.2 (b)-(d)) specimen and PU/TiO$_2$ (3.5:1) (Fig.2 (a)) could be found resulted from small amount of Ag and (or) Cu in PU (TiO$_2$, metal particle) system, and the characteristic absorption peaks might not be reflected in this point. Hence, we may conclude that the structure of the TiO$_2$, encapsulated by the PU polymer fibers, was formed in this regard.

3.2 Antibacterial Test

It is believed that the antimicrobial activity TiO_2 nanoparticles is dependent on their free radical of hydroxide radical, produced by the process of photooxidation after absorbing the light with the wavelength (below 385 nm), whose oxidizing potential energy could kill almost all the microorganism. The fluorescent light will activate the bonding electron on the conduction band. So the photoelectrons (e-) on the valence band and holes (h^+) on the conduction band was produced, and both of them could be separated and migrated to the surface of TiO_2 nanoparticles. Morever, the photoelectrons (e-) on the valence band, containing high reducibility with adsorbent oxygen molecule (O_2) on the surface to form superoxide anion ($\cdot O_2$-), can induce the further reaction with water (H_2O) to produce super-hydroxide radical ($\cdot OOH$) and oxyful (H_2O_2), which could be synthetized by the reciprocity of the active hydroxide radical ($\cdot OH$). The resultant O_2- , which contains the strong reducing property, can decompose the organotrophic bacteria and the materials that is necessary for them to survive. Morever, the $\cdot OH$, H_2O_2 and $\cdot OOH$ might produce many articulated oxidation reaction along with biomacromolecule, such as the esters, protein, enzyme and nucleic acid, and so on, to destroy the construction of biological cell. Table 1 shows the ABE of various electrospun samples for escherichia coli and staphylococcus aureus. It shows that the process of electrospinning PU/TiO_2 sol-gel produced the superfine fibers and intercept the agglomeration of TiO_2 nanoparticles, successfully. When these bacteria were incubated on the nanofibers, the ABE of PU/TiO_2 (3.5:1) specimen cultured in the fluorescent light, was 82.35% and 89.55% for Gram-positive S. aureus and Gram-negative E. coli, respectively, that is a good example to testify the function of TiO_2 particles.

Table 1. Antibacterial Efficiency of Various Electrospun Samples

Sample		Antibacterial Efficiency (%)	
		S. aureus	E. coli
PU/TiO_2 (3.5:1)	Incubating in sunlight for about 24h	82.35	89.55
PU (TiO_2, 3.5:1, Cu)	Incubating in camera obscura for about 24h	75.05	91.00
PU (TiO_2, 3.5:1, Ag)		92.37	93.20
PU (TiO_2, 3.5:1, Cu, Ag)		93.54	95.13

Besides, in this study, the antimicrobial activity of the PU/TiO_2 (3.5:1) nanofibers containing Cu and (or) Ag nanoparticles was tested against S. aureus and E. coli in the camera obscura. The free electron should be inclined to migrate to Ag^+ and (or) Cu^{2+} in the process of experiment, which would separate electrons (e-) and holes (h^+) that can be realized by the special environment of fluorescent light for single TiO_2 nanoparticles, and the necessary $\cdot OH$, H_2O_2 and $\cdot OOH$ for killing microbe can be produced. So, without the factor of illumination, all the PU(TiO_2, 3.5:1, metals) shows better function in the antimicrobial activity than PU/TiO_2 system does even in the fluorescent light, except the PU (TiO_2, 3.5:1, Cu, Ag) membrane against S. aureus (75.05%). The number of bacteria from the PU (TiO_2, 3.5:1, Ag) and PU (TiO_2, 3.5:1, Cu, Ag) membranes, was reduced more than 90% after 24 h incubation, indicating that the TiO_2 nanoparticles, modified by Ag^+ (Ag^+ and Cu^{2+}) ions, successfully inhibited the growth of these bacteria, that is also caused by good antimicrobial performance of Ag^+ in itself. Because of the poorer modifying performance from Cu^{2+} ions, the antimicrobial activity of PU (TiO_2, 3.5:1, Cu) nanofibers (75.05% (S. aureus), 91.00% (E. coli), respectively) was less than that of PU (TiO_2, 3.5:1, Ag) and PU (TiO_2, 3.5:1, Cu, Ag) one.

4 Conclusion

In this paper, we realized blend-electrospinning of PU/ inorganic particles system, in which TiO_2, Cu or (and) Ag nanoparticles were packed by sol-gel way. Experimental results have shown that when the ratio of PU solution (7wt%) and TiO_2 sol was 3.5:1, 89.55 percent of the escherichia coli bacteria and 82.35 percent of staphylococcus aureus were killed with light activation. Morever, the attachment of Ag or (and) Cu nanoparticles into PU/ TiO_2 (3.5:1) nanofibers leads to a significant antibacterial performance without photocatalysis.

Acknowledgements. The authors are grateful for the financial support of the natural science item of Anhui university and college research (Grant KJ2009A029Z), item to stable and subsidize talent's research (yj 2008-8), Anhui provincial young talents fund (2011SQRL048ZD) for their financial support.

References

1. Hai, Y.W., Yang, Y., Xiang, L., Li, L.J., Wang, C.: Preparation and Characterization of Porous TiO_2/ZnO Composite Nanofibers Via Electrospinning. Chinese Chemical Letters 21, 1119–1123 (2010)
2. Yu, Q.Z., Mang, W., Chen, H.Z.: Fabrication of Ordered Tio_2 Nanoribbon Arrays by Electrospinning. Materials Letters 64, 428–430 (2010)
3. Rowan, R., Tallon, T., Sheahan, A.M., Curran, R., McCann, M., Kavanagh, K., Devereux, M., McKee, V.: 'Silver Bullets' in Antimicrobial Chemotherapy: Synthesis, Characterisation and Biological Screening of Some New Ag (I)-Containing Imidazole Complexes. Polyhedron 25, 177–1778 (2006)
4. Madhugiri, S., Zhou, W.L., Ferraris, J.P., Kenneth, J., Balkus, J.: Electrospun Mesoporous Molecular Sieve Fibers. Microporous and Mesoporous Materials 63, 7–84 (2003)
5. He, C.H., Gong, J.: The Preparation of PVA– Pt/TiO_2 Composite Nanofiber Aggregate And the Photocatalytic Degradation of Solid-Phase Polyvinyl Alcohol. Polymer Degradation and Stability 81, 117–124 (2003)
6. Viswanathamurthi, P., Bhattarai, N., Kim, C.K., Kim, H.Y., Lee, D.R.: Ruthenium Doped Tio_2 Fibers by Electrospinning. Inorganic Chemistry Communications 7, 67–682 (2004)

Effects of Injection Parameters on Size Distribution of Particles Emitted from Diesel Engine

Haifeng Su[1,2], Xinli Bu[1], Ruihong Wu[1], and Xiaoliang Ding[2]

[1] Shijiazhuang Vocational Technology Institute, Hebei, 050081, China
[2] Beijing Institute of Technology, Beijing , 100081, China
suhf123@sina.com, {buxl,wurh}@sjzpt.edu.cn

Abstract. Number concentration and size distribution emitted from high pressure common rail diesel engine are investigated on the test rig, which are measured using electrical low pressure impactor (ELPI). Particle size distributions were measured at different injection pressures and different injection advanced angles. The results of the number concentration of particles distribution at A 25% engine condition are analyzed. The coarse mode particles varied little with injection pressure and injection advanced angle. To lower the accumulation mode particles emission, the rail pressure and the injection advanced angle increase were found to be an effective means. With the increase of the injection pressure, the dominant part of the total number emission shifted to nuclei mode particles. Both injection pressure and injection advanced angles increase, the nuclei mode particles emission would be increased obviously.

Keywords: particle emission, size distribution, injection pressure, injection advanced angle.

1 Introduction

The mass concentration of particles in the exhaust of diesel engines has reduced steadily over the past years due to the application of both engine design and after-treatment new techniques. However, some researchers have found that the mass reduction may have been accompanied by the increase in number of small particles emission, especially the number of ultrafine particles with diameter smaller than $0.1 \mu m$ and nanoparticles smaller than 50 nm. Studies have linked environmental exposure to ultrafine particles and nanoparticles with adverse health effects [1-6]. Now there is a growing concern over the impact on human health of ultrafine particles emitted from diesel vehicles.

In addition, some efforts have been devoted to the size distribution characterization of particle emitted from diesel engines. To realize particle distribution character and find the optimal injection parameters at engine operating conditions, an experimental investigation was performed on a small direct injection diesel engine equipped with a high pressure common-rail injection system. The particle size distributions were measured at different injection pressures and different injection angles with ELPI. The results of measurements conducted to determine the number and mass

Y. Yu, Z. Yu, and J. Zhao (Eds.): CSEEE 2011, Part II, CCIS 159, pp. 354–359, 2011.

concentration of particles emitted from the engine at A@25% condition were presented. Effects of injection parameters on particle distribution character were investigated.

2 Experimental Setup

The particulate emission measurement experiments were carried out on the engine dynamometer Test at the Lab of Low Emission Vehicle Research of Beijing Technology Institute. The exhaust was channeled to a dilution tunnel where it was turbulently mixed with filtered dilution air, and then was measured by the electrical low-pressure impactor (ELPI).

2.1 Engine and Working Condition

The four-stroke turbocharged diesel engine 4JB1 equipped with solenoid injectors high pressure common rail system, was used for the experiment without after-treatment system. The engine has four cylinders and a 2.8 L of displacement, with a peak power output of 80 kW at 3600 rpm. The output of the engine was coupled with a dynamometer for load control. In this paper, the engine was operated at 2130 rpm under 25 % load (A@25%). Injection pressure set 40MPa, 70 MPa and 100 MPa, injection advanced angle are 0°CA, 5°CA and 10°CA.

2.2 Dilution and Sampling System

The main factors considered in the design of the sampling port were exhaust temperature and concentration. The schematic diagram includes an engine bench, sampling system, a two-stage-air-ejector dilution tunnel, and a particle seizer. The diesel combustion exhaust was sampled at 25 cm downstream of the turbocharger. The engine exhaust was directed to the first diluter, then the second one. The diluters served two purposes. Firstly, the diluted exhaust was cooled to temperatures at which it could be sampled for further analysis, and secondly, particles concentrations were decreased to levels suitable for the particle sizing instrumentation. The exhaust gas should not be cooled down before it was diluted, as rapid cooling of a concentrated aerosol may result in significant changes of its physical-chemical characteristics. The dilution ratio was 64 in this experiment.

2.3 Particle Sizing Instrumentation

Particle size distributions were measured by the electrical low-pressure impactor, the product of DEKATI LTD. ELPI was equipped with oil-soaked sintered collectors, as shown in Fig.1 ELPI measures the aerodynamic-equivalent diameter of the soot particles. ELPI is capable of detecting the real time number size distribution with a size range of 7 nm-10μm and concentration measurement and enables the subsequent chemical analysis on the collected sample in 13 size fractions. ELPI software calculates the average number of charges on particles based on the aerodynamic particle diameter. So ELPI can also give some information about mass size distribution through computation which has a kind of value of analysis [7]. The formula is

$$M = \frac{\pi}{6} \rho D^3 n \tag{1}$$

Where M is the mass of particles; ρ is the density of particle; D is the diameter of particles; n is the number of particles.

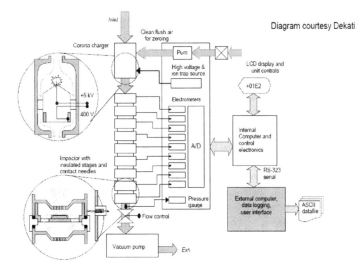

Fig. 1. Schematic of ELPI

3   Experimental Results and Discussion

According to particle formation mechanisms and particle size range, engine particles emission can be classified into three modes: nuclei mode; accumulation mode and coarse mode. The nuclei mode particles with geometric mean diameter were 5-50 nm, likely forming by condensation of carbon moieties. The nuclei mode particles may be solid or droplets. The accumulation mode with diameter between 100-300 nm and formed by coagulation of nuclei mode particles. The coarse mode particles with diameter greater than 1.0 μm often contained solid agglomerated material (elemental carbon, ash).

3.1   Effects of Injection Pressure on Particle Size Distribution

The particles with diameter greater than 0.5 μm often contained solid agglomerated material (elemental carbon, ash). The results of the whole experiment showed that the particles with diameter greater than 0.5μm were insensible to injection pressure. The coarse mode particles numerical concentrations didn't vary with injection pressure. Fig.2 shows the particle distributions varying with different injection pressures at injection advanced angle Ai=10°CA. At low injection pressure 40MPa, Size distributions of diesel particles have a single peak distribution character. The accumulation mode particles with diameter between 50 nm and 100 nm ('1' in Fig 2), were the dominant part of the total number emission which maximum number concentration was 4632650.944 cm-3. As injection pressure increased, the particles numerical concentrations of diameter between 50 nm and 150 nm decreased. The numerical concentrations of diameter 80nm decreased from 4632650 cm-3 to 621759 cm-3 as injection pressure increased from 40 MPa to 100 MPa. With the increase of the injection pressure, the well-established single peak character of Size distributions is disappeared. At high injection pressure 100 MPa, the dominant part of the total number emission was the nuclei mode particles, the particle diameter of maximum number concentration was 20 nm ('2' in Fig.2).

The test results were also obtained about the particles distributions varying with different injection pressures at injection advanced angle Ai=0°CA, 5°CA. The number emission showed the similar tendency as that under injection advanced angle Ai=10°CA condition.

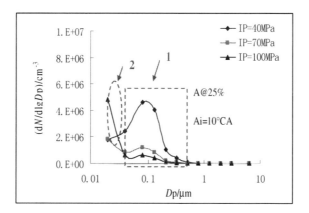

Fig. 2. particle size distribution at different injection pressures Ai=10°CA

It can be indicated that the rail pressure increase was an effective means to decrease the accumulation mode particles. With the increase of the injection pressure, the total particle number emission decreased significantly and the dominant part of the number emission shifted to nuclei mode particles.

3.2 Effects of Injection Advanced Angle on Particle Size Distribution

The results of the whole experiment showed that the particles with diameter greater than 0.5 μm were insensible to injection advanced angles too. The coarse mode particles numerical concentrations didn't vary with injection advanced angles. Fig.3 shows the particles distributions varying with different injection advanced angles at injection pressure IP=40 MPa. Size distributions of diesel particles have a single peak distribution character. At low pressure 40 MPa, the accumulation mode particles with diameter between 50 nm and 150 nm was the dominant part of the total number emission which maximum number concentration would be 8832036 cm^{-3}. As injection advanced angles increased, all particles numerical concentrations of diameter decreased, Particles between 50 nm and 150 nm decreased more seriously.

The test results were also obtained about the particles distributions varying with different injection advanced angle at injection pressures 70 MPa and100 MPa. With the increase of the injection pressure, the well-established single peak character of Size distributions is disappeared. As injection advanced angles increased, accumulation mode particles numerical concentrations decreased , but no more seriously than pressures 40 MPa. The dominant part of the total number emission shifted from accumulation mode to nuclei mode particles as injection advanced angle increase at injection pressures 70 MPa. Weight of diameter 20 nm Particles in total number increased with injection advanced angles at every pressure.

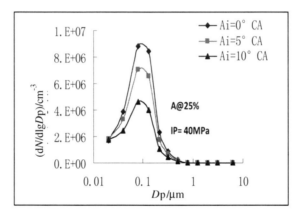

Fig. 3. particle size distribution at different injection angle IP=40MPa

4 Summary

This paper presents the results of the number concentration of particles emitted from high pressure common rail diesel engines at working condition A@25%, at three injection pressures and three injection advanced angles. The following conclusions can be achieved:

Size distributions of particles from the diesel engines, which have been designed for low PM mass emissions, are shifted towards the small accumulation and nuclei mode particles.

The coarse mode particles numerical concentrations didn't vary with injection pressure or injection advanced angle.

The accumulation mode particles number decreased with the increase of injection pressure or injection advanced angle. The affect of injection pressure in decrease particles number is more sensitively.

Increase injection pressure or injection advanced angle lead to nuclei mode particles number increase. The combined effects of High injection pressure and big injection advanced angles made nuclei mode particles to be the predominant part of total emission particle in number concentration.

Injection pressure and injection advanced angle are effective means to change particle distribution character and lower particles emissions.

References

1. Hongbin, M., Heejung, J., Kittelson: Investigation of Diesel Nanoparticle Nucleation Mechanisms. J. Aerosol Science and Technology 5, 335–342 (2008)
2. Villeneuve, P.J., Goldberg, M.S.: Fine Particulate Air Pollution and All-Cause Mortality within the Harvard Six-Cities Study. J. AEP 12(8), 568–576 (2002)
3. Theo, M.C.M., Hermen, A.L.D.: Toxicological Assessment of Ambient and Traffic-Related Particulate Matter: A Review of Recent Studies. J. Mutation Research 613, 103–122 (2006)
4. Steinvil, A., Levana, K.B.: Short-Term Exposure to Air Pollution and Inflammation-Sensitive Biomarkers. J. Environmental Research 106, 5–61 (2007)
5. Neubergera, M., Rabczenkob, D., Moshammer, H.: Extended Effects of Air Pollution on Cardiopulmonary Mortality in Vienna. J. Atmospheric Environments 41, 8549–8556 (2007)
6. Gao, J.D., Song, C.L., Zhang, T.C.: Size Distributions of Exhaust Particulates from a Passenger Car with Gasoline Engine. J. Journal of Combustion Science and Technology 13(3), 248–252 (2007)
7. Junfang, W., Yunshan, G., Chao, H.: Size Distribution of Particles Emitted from Liquefied Natural Gas Fueled Engine. J. Jounal of Beijing Institute of Technolgy (4), 410–414 (2008)
8. Catania, A.E., d'Ambrosio, S., Ferrari, A., Finesso, R.: Experimental Analysis of Combustion Processes and Emissions in a 2.0L Multi-Cylinder Diesel Engine Featuring a New Generation Piezo-Driven Injector. J. SAE Technical Paper. 2009-24-0040 (2009)

Effectiveness Evaluation about Self-Screening Jamming of Assaulting Aircraft in Complex EM Environment

Yanbin Shi[1], Yunpeng Li[2], and Junfeng Wang[3]

[1] Dept. Electronic Engineering, Aviation University of Air force,
Changchun, 130022, China
[2] Dept. Information Countermeasure, Aviation University of Air force,
Changchun, 130022, China
[3] The 62th unit, 94032 troop of PLA, Wuwei 733003, China
shiyanbin_80@163.com, yunpengli@126.com, wangjunfeng@sina.com

Abstract. The aviation electronic countermeasures effectively reduce the sensitivity of assault aircraft and enhance its fighting effectiveness and survivability in battlefield. The survivability of assaulting aircraft is researched when it breaks through the enemy side missile defense system under the SSJ supported, and the countermeasures effectiveness of airborne radar during assault is calculated out. Attention is concentrated on the detection probability of long-distance guard radar, missile direct hit damage probability, and penetration survival probability as well as resistance effectiveness of airborne radar to missile defense system. Studies have shown that the viability of aircraft obtains prominent enhancement under the radar electronic countermeasure support.

Keywords: Aerial Assault, SSJ, Electronic Countermeasures, Effectiveness Evaluation, Complex Electromagnetic Environment.

1 Introduction

The sensitivity is an important factor which affects the survivability of military aircraft [1, 2]. Along with lots of remote sensing technologies applied in aerial defense warning system and fire control system, it is more complex and urgent to reduce sensitivity of the assaulting aircraft. As the main tactics method, aerial assault plays the vital role on the decision of the advancement and result of war during modern local war. The goals of aerial assault are not only located on the depth of the strategy and tactics, but also safeguarded by the strict aerial defense systems [3~5]. Therefore breakthrough enemy's aerial defense systems and close the prearranged aerial assault targets are the decisive factors to complete the task. During this process, the fighter plane must been threatened by the search radar, the enemy intercepts aircraft group, the antiaircraft missile and the antiaircraft artillery, these weapon systems effectiveness depend on the electronic equipments. Each kind of electronic countermeasure means such as active jamming and passive jamming all can effectively reduce the sensitivity and improve the fighting effectiveness of the fighter

Y. Yu, Z. Yu, and J. Zhao (Eds.): CSEEE 2011, Part II, CCIS 159, pp. 360–365, 2011.
© Springer-Verlag Berlin Heidelberg 2011

plane. The work presented in this paper focuses on the evaluation about the disturbance effectiveness of airborne radar to ground-to-air missile system.

2 SSJ (Self-Screening Jamming)

Radar is important electronic equipment in the electronic warfare, and the radar countermeasure is the summarization of many tactical measures which weaken or destroy the using effectiveness of enemies' radar and make sure our radar effectiveness normally working. Radar countermeasure is the key content of electronic countermeasure. According to the difference of the origin of the disturbance energy, the radar jamming ways can be divided into the active jamming and the passive jamming. And according to the anthropogenic factor of the jamming, the radar jamming way also can be divided to intentional jamming and accidental jamming. According to the principles of the jamming signal working, it can be divided to enveloped jamming and fraudulence jamming. According to the interspaces dubiety of the radar and the goal, it can be divided to the SFJ (Stand Forward Jamming), ESJ (Escort Jamming), SSJ (Self Screening Jamming) and SOJ (Stand off Jamming), just like showed in Fig.1.

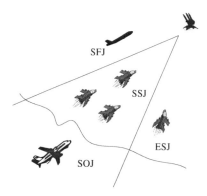

Fig. 1. Aviation EW battle disposition

Self-Screening jamming is used when the assaulting aircraft accomplish the missions such as break-through assaulting, bomb blockage, closing support and suppressing the enemy aerial defense weapon system and so on. Its goal is to avoid the assaulting aircraft be intercepted and captured by the defense system. The jamming must have the ability to deal with much menace when it counters the ground-to-air missile system in the broad electromagnetism frequency band simultaneous. That is impossible on a certain degree that the transmitters of aircraft has enough power to jam all the latent menace from radars and weapons because the load is limited by the aircraft, so it is generally using the fraudulence jamming to the enemy radars in order to reduce the demand of jamming power and make the jamming effectively focusing on the main lobe of the enemy radars [6, 7].

3   Effectiveness of Aerial Defense Weapon Systems

The effectiveness of weapon system refers to the degree to achieve the anticipated goal when the weapon system is used to carry out the task under the specific conditions, and the fighting effectiveness of the weapon system refers to the operational capacity when the weapon system is used at the combat [8] .The missions of the ground-to-air missile weapon include destroy the air target, district air defense or important position air defense. Its effectiveness is tactics index which closely-related not only by launch platform, guidance mode, radius of damage, damage mode, aerial defense district, but also by the interfere mode, the power of interfere signal, armada scale of the assaulting aircraft and so on. In order to simplify the research, on the assumption that the single assaulting aircraft carries out the scheduled assaulting task when it through interception area by the enemy side ground-to-air missile. At the same time the ground-to-air missile weapon system executes interception duty independently, which includes two processes include searching the defense aerial zone and vectoring radar navigates the missile to intercepts the assaulting aircraft when the assaulting aircraft enters the fire scope of the missile weapon system [9].

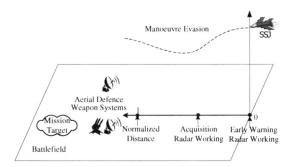

Fig. 2. the process of the assaulting aircraft through the enemy's missile defense system

In order to solve the contradiction between the ability of rapidly discover and accurately tracking the goal of anti-aircraft weapon system, the ground-to-air missile weapon system disposes two kinds of radar which have the different functions, respectively named the acquisition radar and the guidance radar. The primary mission of the acquisition radar is surveys many goals in the defense aerial zone, measures their parameters continuously, carries on the enemy identification and evaluates the threat degree, transmits the target information to the guidance radar and surveillance combat process.

The fighting effectiveness of the acquisition radar may express by the detection probability, and the detection probability is

$$P_d = P_{AEW}[P_{info} P_{P-S} + (1 - P_{info})P_{search}] + (1 - P_{AEW})P_{search} \tag{1}$$

Where $P_{AEW} = \exp\left(-\dfrac{4.75}{\sqrt{nS_n}}\right)$ is the detection probability of early warning

radar. $P_{search} = \exp\left(-\dfrac{9.5}{\sqrt{nS_n}}\right)$ is the detection probability of acquisition radar when it

is working as routine searching mode. P_{info} means the guided missile battlefield and warn radar whether to correspond expedite or not. If it is expedite, $P_{info} = 1$, otherwise it is 0. $P_{P-S} = 1 - (1 - P_{search})^{2\pi/\theta}$ is the detection probability of acquisition radar when it is working as emergency searching mode. S_n is the signal to noise ratio of single pulse. n is the accumulation number of radar pulse during a single scanning cycle. θ is the fan-sweeping angle of the acquisition radar when it is working in the emergency searching mode. The mode of the ground-to-air missile destroys the aerial assaulting aircraft can be divided into the direct hit destroy and the broken pieces destroy, the attention is concentrated on the destroy probability of direct hit in the paper. It is supposed that the random error of missile fire obeys the round normal distribution.

$$f(x,y) = \frac{1}{2\pi\sigma^2}\exp\left(-\frac{x^2 + y^2}{2\sigma^2}\right) \tag{2}$$

Missile undershoots error r obeys the Rayleigh distribute:

$$f(r) = \frac{1}{\sigma^2}\exp\left(-\frac{r^2}{2\sigma^2}\right) \tag{3}$$

σ^2 is mean-square error which the point of dispersal zone, which can be get by equation (4):

$$\sigma^2 = \frac{1}{2\ln 2}\left[\frac{(c_1 R_{ave}^2 + c_2)}{(S/J)} + c_3\right] \tag{4}$$

Where S/J is the jamming to signal ratio, if there is not have the disturbance, it can be replaced by the signal to noise ratio. $R_{ave} = 0.619 R_{missle}$ is the average firing range, where R_{missle} is fire field of missile; c_1, c_2 and c_3 are the fitting parameters. If assigned equivalent the aircraft is a circle, r_0 is radius, then the direct hit probability of single shot missile to aircraft is:

$$P_{1m} = 1 - \exp(-\frac{r_0^2}{2\sigma^2}) \tag{5}$$

The destruction probability of ground-to-air missile to assaulting aircraft.

The probability of the single shot missile wrecks the assault aero plane is $P_n = P_d \cdot P_{1m}$, supposing the number of times of the missile intercept the aircraft is l, may calculate the missile position to the single frame aircraft destruction probability is:

$$P_H = P_d \cdot \left[1 - \left(1 - P_{kk} P_{lm} \right)^{l \cdot k} \right] \tag{6}$$

Where P_{kk} is the reliability probability of missile launcher. k is the number of the missile position.

4   The Resistance Effectiveness of SSJ to Ground-to-Air Missile

The main purpose of this paper is to research disturbance effectiveness evaluation of the single aerial assaulting aircraft counterwork to the guided missile system when it break-through the enemy ground-to-air missile position support by the radar countermeasure. Through the above analysis, we can calculate the disturbance effectiveness using the ratio of penetration probability between the radar active jamming working and not working, and we can explains this process through the example in succession. $P_t = 1 - P_H$ is the penetration probability of aerial assaulting aircraft. The validity of radar countermeasure η_v is the ratio of penetration probability between the radar active jamming working and not working, we can look this ratio as the validity of radar countermeasure.

$$\eta_v = \frac{P_t^{'}}{P_t} \tag{7}$$

Where $P_t^{'}$ is the penetration probability of aerial assaulting aircraft when the radar active jamming is working. P_t is the penetration probability of aerial assaulting aircraft when the radar active jamming is not working.

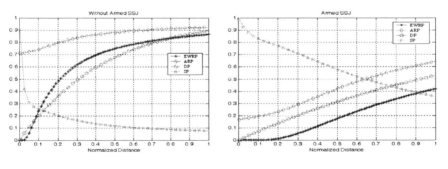

Fig. 3. The SSJ is not working

Fig. 4. The SSJ is working

It is not be considered the correspondence resistance in the equation (1), on the assumption that the missile position and the radar warning set correspondence is unimpeded, then $P_{info} = 1$, $P_{P\text{-}S} = 0.95$. According to reference [3], the signal to noise ratio in electromagnetism space is $S_n = (R_0/R)^4$ when the distance between the radar and the aerial assaulting aircraft is R, R_0 is the farthest coverage range of the warring radar, and $n = 30$, $R_{missile} = 1$, the fitting parameters $c_1 = 0.003$, $c_2 = 0$, $c_3 = 0$, $r_0 = 10m$, $P_{kk} = 0.95$, $l = 1.4$, $k = 20$ .After the correlative parameters have been ascertained, we can simulate the effectiveness evaluation using these parameters. It can be seen from the Fig.3 and Fig.4 (EWRP is the detection probability of early warring radar; ARP is the detection probability of acquisition radar; DP is destroy probability single missile) the penetration probability of aerial assaulting aircraft have been improved obviously.

5 Conclusion

This paper studies the survivability of the assaulting aircraft with self-screening jamming breaking through the enemy missile weapon system. And the survival probability and the detective probability are analyzed in the process of breaking through the enemy missile system. It can be seen from the modulation chart and the analytical results, with the SSJ, the survivability of the assaulting aircraft is enhanced obviously.

References

1. Yong-shun, Z., Ning-ning, T., Guo-qing, Z.: Theory of Radar Electronic Warfare. National Defense Industry Press, Beijing (2006)
2. Lee, D.S., Gonzalez, L.F., Srinivas, K., Auld, D.J., Wong, K.C.: Aerodynamic/RCS Shape Optimisation of Unmanned Aerial Vehicles using Hierarchical Asynchronous Parallel Evolutionary Algorithms. 8AUST, AIAA–3331 (2006)
3. Guo-yu, W., Lian-dong, W., Guo-liang, W., et al.: Radar Electronic Warfare System Mathematical Simulation and Evaluation. National Defense Industry Press, Beijing (2004)
4. Tao, Z., Ping-jun, W., Xin-min, Z.: Fighting Efficiency Evaluation Method of Surface-to-Surface Missile Armament, vol. 30(9). National Defense Industry Press, Beijing (2005)
5. Jiao, L., Zhong-liang, J., An, W., et al.: The Research of Airborne Radar's Basic Efficiency Model. J. Fire Control & Command Control 12(3), 12–14 (2003)
6. Na, W., An, Z.: Electronic Countermeasures of Airborne Weapon System. Electronics Optics & Control 10(2), 9–10 (2003)
7. Prevot, T., Lee, P., Smith, N., Palmer, E.: ATC Technologies for Controller-Managed and Autonomous Flight Operations. AIAA–6043 (2005)
8. Jian, Z.: Military Equipment System Effectiveness Analysis. Optimization and Simulation. National Defense Industry Press, Beijing (2000)
9. Xiao-hui, G., Bi-feng, S., Xu, W.: Evaluation of Survivability to an Aircraft with Electronic Countermeasures System. Systems Engineering Theory & Practice 25(6), 71–75 (2005)

Research on WEB Server Cluster System Based on Jini

Jun Xie, Minhua Wu, and Shudong Zhang

College of Information Engineering, Capital Normal University,
100048, Beijing, China
xiejunsoftware@tom.com, wuminhua@163.com,
sdz@mail.cnu.edu.cn

Abstract. WEB server cluster system is considered as an effective scheme to improve performance of network service. Now the research on this system is focused on flexible expansibility, high usability and reasonable load balance. Considering that Jini has good dynamic, scalability and self-healing ability, this paper put forward a WEB server cluster system based on Jini. The total architecture of this system was described in this paper, and at the same time the implementation of critical technology was analyzed.

Keywords: WEB Server Cluster, Jini, Lookup Service, Proxy.

1 Introduction

With the rapid development of Internet/Intranet, web services have become a major way to share information. As a mainstream solution [1] to improve the performance of network services, web server cluster system can dynamically request adaptation server to resolve the problem that network services are exposed to overload for a visit from the browser/client, and at the same time propose a series of challenges for construction of network services. First of all, a Web server cluster system must have high availability. Cluster consists of a number of web servers which interconnect Internet or intranet. Each server has the equivalent status and provides external services individually without assists of other servers. And if a server fails, the cluster system can mask out the failure server for the client, users will automatically visit the server which is running normally. Secondly, the cluster system must have good scalability. The performance of web server cluster system cannot be confined to a single server; new server must be able to join the cluster dynamically to enhance the performance of the cluster. Finally, the cluster system must be able to achieve load balancing. The browsing tasks received by cluster can be evenly distributed to each server in the cluster, making the load on each web server roughly the same.

In response to these challenges, we present a web server cluster system based on the Jini in this paper. Jini a new distributed computing system [1], launched by Sun Company, based on Java and RMI, can mask the details of communication of network

Y. Yu, Z. Yu, and J. Zhao (Eds.): CSEEE 2011, Part II, CCIS 159, pp. 366–370, 2011.

service resources object, provide a basic structure for transparent sharing and use of each other of network services, with a dynamic, scalable and self-repair capacity [2]. These features of Jini make it able to provide a good solution for constructions of web server cluster system.

2 Web Server Cluster System Based on the Jini

2.1 Jini

Jini system consists of infrastructure, programming model and services of three parts, and these three parts cooperate with each other to build a dynamic, distributed service network, that is Jini alliance. Among them, the infrastructure defines the minimal Jini technology core systems, including the discovery, join, and from the Jini service network agreement; programming model provides a set of programming interfaces for the Jini, including the lease interfaces, two-phase commit interface and event interface, infrastructure and services is to use this set of interfaces implemented; service is an entity of the Jini league, can provide services for the other participants in the alliance.

2.2 Architecture of Web Server Cluster System Based on the Jini

Figure 1 is the architecture model of the web server cluster system based on the Jini. Based on the traditional web server cluster infrastructure (usually consists of distributor and the network connection scheduling server pool), this model add a Jini lookup service and agent service platform.

The whole cluster system work flow is:

Start
a. Receive a request. Web server cluster system provides a single external IP address or domain name of the image, the request packets from the client / browser, will first be submitted to the cluster connection scheduling distributor;
b. Forwarding the request. After receives the request packet, connection scheduling distributor will traverse the registered agent information which is stored in the Jini lookup service, from which it obtain the latest load parameter information of each server in the servers pool. Then, based on a certain degree of load-balancing scheduling strategy, it chooses the lightest load server from the server pool, forwarding the request.
c. Processes the request and respond. The selected web server receives the request, processing, and respond directly back to the requesting client/browser.
End.

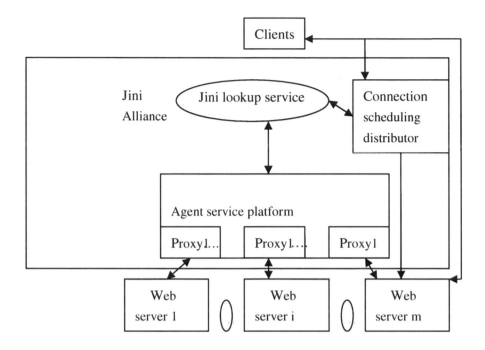

Fig. 1. The Jini-based Web Server Cluster System Architecture

2.3 Implementation of Agent Service Platform

As the web server may not be able to run a Java virtual machine, we spin off the web server from the Jini Union, while using a proxy service platform release agent on behalf of these web servers to participate in Jini alliance.

Proxy service platform specific work flow is:

Start

a. The initialization settings. Proxy service platform starts the socket server-side program to monitor and control connection requests from the Web server;

b. Requests. When a Web server want to join the cluster to provide services, it will apply for agency services to proxy service platform by the HTTP GET ;

c. Generate the proxy. When agent service platform receives a request, it will generate a new agent. The agency first found available lookup service in the Jini alliance, register (Registration information includes the name of the service agent, the information of web server the agent represented as well as web server load parameters, including static load parameters and dynamic load parameters), followed by response and send a executable socket client program to the web server who sends the request ;

d. Capture data. Web server executes of the socket client program to establish a connection with the proxy service platform, and non-stop obtains load parameter data from the server and transmits to the proxy service platform.

e. Update data. When the proxy of agent service platform obtain the latest load parameters message of web server it represents, it will promptly notify the Jini lookup service updates the relevant registration information.
End.

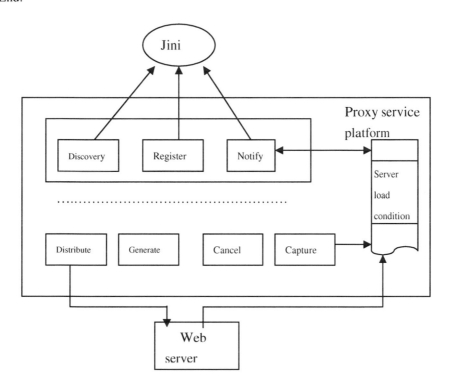

Fig. 2. Agent service platform implementation

2.4 Load Balancing Scheduling Algorithm

In the web server cluster system, load balancing scheduling strategy is directly dependent on two key factors: load balancing scheduling techniques and load balancing scheduling algorithm. Here, using the HTTP redirection approach as load-balancing scheduling technologies. The following shows focus on the design of load-balancing scheduling algorithm.

In order to compare the load condition of each web server valid comparison, we need to define a parameter which can intuitively reflect the load condition of web server. Referring to literature [5], use the formula (1) to calculate the load condition reflected value of each web server. Select any one server which works normally (assuming the No. M server) as the benchmark, then the load conditions reflected value of the other server (assuming the number i server) is shown as follows:

$$p_i = \alpha \times \frac{\lambda 1_i \times C_i}{\lambda 1_M \times C_M} + \beta \times \frac{\lambda 2_i \times R_i}{\lambda 2_M \times R_M} + \gamma \times \frac{\lambda 3_i \times D_i}{\lambda 3_M \times D_M} + \theta \times \frac{W_i}{W_M}$$

α : CPU comparison weights.
β : Memory comparison weights.
γ : Hard disk comparison weights.
θ : Network comparison weights.

The initial value of α, β, γ, θ is set to 1, its specific value can be increased or decreased to emphasize or reduce the load performance of a certain areas according to the actual situation of the cluster running.

3 Conclusions

As a result, the withdrawal or failure web server was shield, thus ensuring the high availability of web server cluster. (2) Scalable implementation. In the web server cluster this article refers to, the cluster consists of web server pools and Jini Union which consists of connection scheduling distributor , Jini lookup service and agent service platform. Jini providers a basic structure which makes the services in the networks can look for each other and take advantage of each other, so the scalability of various components of alliance can be guaranteed, and for the web server pool, using the HTTP GET Way release agent services to ensure the system scalability and resources to plug and play. (3) The load balancing implementation. In this article, we use the existing, more sophisticated load-balancing scheduling techniques and load-balancing scheduling algorithm to achieve load balancing.

References

1. Wang, J.: A Survey of Web Caching Schemes for the Internet. ACM Computer Communication Review 29(5), 25–27 (1999)
2. Jiang, J.-W., Xu-Fa, W., Guo-Rui, H., Quan, Q.: Multi-database System Based on Jini Construction. Computer Engineering 32(9), 54–56 (2006)
3. Jiang, Z., Heping, P., Tong, Z.: Jini Technology in Distributed Computing Applications. Computer Engineering 29(9), 24–26 (2003)
4. Chen, H.L.: Developing a Dynamic Distributed Intelligent Agent Framework Based on the Jini Architecture. University of Maryland Baltimore County (2000)
5. Ziang, H., Li, W.: Algorithms, Network Topology and the Relationship Between the Degree of Frequency and Dynamic Load Balancing. Computer Engineering and Science 1(22), 104–107 (2000)

The Study of the Knowledge Model Based on Event Ontology for the Occurrence of Paddy Stem Borer (Scirpophaga Incertulas)

Lin Peng[1,2] and Limin Zhang[2]

[1] College of Computer Engineering and Science, Shanghai University,
Shanghai, P.R. China
[2] College of Basic Science & Information Engineering, Yunnan Agricultural University,
Kunmin, P.R. China
penglin781026@shu.edu.cn, limin0789@126.com

Abstract. A paddy stem borer occurrence is a very complex natural phenomenon. The paddy stem borer occurrence is vividly, wholly, systematically described with digital ways, which is a basic to study the paddy stem borer occurrence with modern techniques. Firstly, based on the research on conventional methods of knowledge representation, this paper presents a new knowledge representation method of paddy stem borer occurrence by event ontology. Second, the new knowledge model of paddy stem borer occurrence is constructed with the method of knowledge representation. In comparison with the conventional method, event ontology represents knowledge with a higher granularity, meanwhile, it will more accurate and accord with reality, better extensibility and highly amalgamation. This paper provides new methodology and theory for pest disaster knowledge representation.

Keywords: Event ontology, Rice paddy stem borer (Scirpophaga incertulas), Knowledge representation, Knowledge model.

1 Introduction

According to the census of the FAO (Food and Agriculture Organization of the United Nation), food produce decreases 14% suffering from insects damages all over the world every year. And the total economic losses from the food produce decreasing are inestimable [1]. The average economic losses from pest insects in China outreach 10 billion every year. There are more than 300 pest insect species, including 30 key pest insect species. And rice paddy stem borer (Scirpophaga incertulas) broke frequently and severely in recent 20 years. Thus, rice paddy stem borer was becoming another crucial pest in China, the occurrence features of which include high population density, large damage extent, stricken damage degree, long occurrence time and enormous economic losses. The average occurrence area of paddy rice borer is about 15 million hectares and the average control cost is about RMB 4.57- 6 billion Yuan. The economic costs of the residual pest nearly reach RMB 6.5 billion Yuan[2,3].

Y. Yu, Z. Yu, and J. Zhao (Eds.): CSEEE 2011, Part II, CCIS 159, pp. 371–376, 2011.

In order to control the rice paddy stem borer efficiently, such as decreasing the damages on plant and cutting down the economic losses, there should be fundamental to build a knowledge model for studying the occurrence of paddy stem borer. And the new model should describe the occurrence of the pest sufficiently. But since that seldom researches were done on knowledge models in the knowledge management system for the insect, so the whole management system for the occurrence of paddy stem borer was incomplete.

This paper aimed at the knowledge model of the paddy stem borer occurrence, describing the insect occurrence by Event ontology, built a knowledge model based on event ontology for the occurrence of paddy stem borer.

2 Event Ontology

Ontology was originally a terminology in philosophy, which is an actual systematized interpretation in order to describe the innate characters of the objects. And it was introduced into computer science field about ten years ago and was applied into knowledge representation, knowledge acquisition, knowledge excavating, knowledge retrieval, knowledge sharing and reusing etc.

Various definitions for ontology were done by different researchers, but Studer et.al. (1998) defined the ontology as a specific, canonical, formalized explanation for sharing conceptual model, which was accepted by majorities. This definition emphasized that ontology is a semantic basis to describe the relationships between different conceptions. But with the development of the studying on ontology, there exist open defects in traditional definition which is focus on conceptions.

1. The world is motive, and this motive world is made of events, so realities in the world should be memorized and apprehend by "event", which is hardly described by static conception.
2. "Ontology" was an objects set with common attributes, and the objects were described by attributes in tradition. This definition is fit in static conception, but ignores the dynamic conditions of the objects, and results in neglecting other factors inevitably.
3. Lacking of advanced semantic preference over the conception hierarchy, traditional "Ontology" hardly express the complicated relationships in realities, and impossibly abstract and inference more advanced semantic mode.

Faced to above problems, LIU Zong-tian, HUANG Mei-li, ZHOU Wen et al.[5-7] proposed a new concept "Event ontology", which is a specific, canonical, formalized explanation for systematize model with sharing and objective events. Event ontology, as a knowledge representation method event oriented, is more coincident with existence regularity of the world and with cognition regularity of people on the present world. The realities in the world changed with the "event" changing. Today's world was formed with lots of "event" happening. Thus, describing the realities in the world by event units means studying the relations and regulations between the event units.

LIU Zong-tian, HUANG Mei-li, ZHOU Wen et al.[5-7] gave out the following definition:

Event: an event presents out some behavior characters with some participation in some particular temporal spatial circumstance. And denote "e" to express "Event", "e" includes six variables: A, O, T, V, P, L. e = (A, O, T, V, P, L). These six variables were denoted as six elements of each event.

A means action to describe the changing processes and features;

O means object to describe all studying subjects;

T means time to define the temporal intervals;

V means environment to describe the happening surroundings;

P means Peroration including precondition, middle peroration and postcondition;

L means language to describe the linguistic rules, which includes words set of kernel string, expression of kernel string and collocation of kernel string etc.

3 Knowledge Model Based on Event Ontology for the Occurrence of Paddy Stem Borer (*Scirpophaga Incertulas*)

3.1 Regulations between the Event Ontology for the Occurrence of Paddy Stem Borer

According to the relative research, the occurrence and happening time of paddy stem borer is depended on the climate, geography, water quality, control measurements and occurrence in the last generation, among these factors the climate information data is a special one. This paper built a set of event ontology corresponding to the above influent factors, and analyzed the relationships between the event ontology (fig.1).

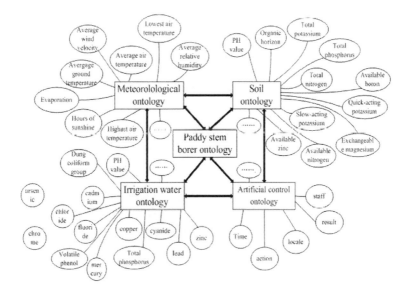

Fig. 1. Relations of the event ontology for paddy stem borer occurrence

3.2 Modeling

This paper built the Ontology model according to the modeling primitive for Paddy stem borer. There were a hexatomic tuple: E-Ontology: =(E-Action, E-Object, E-Time, E-Environment, E-Peroration, E-Language). E-Action means the state of the event; E—Object means the studying subjects; E—Time means the time intervals of the event; E-Environment means the surroundings of the event, focusing on the location information; E-Peroration means the peroration information; E-Language means the homonymy information which could be omitted if there were nonexistence.

An E-Ontology for paddy stem borer was shown below.

Ontology paddy stem borer {
E-Action:{ the first generation }
E-Object:{paddy stem borer}
E-Time:{2007-4-23}
E-Environment:{ 102°33′N and 24°12′E; XiZhuang Town of JianShui County }
E-Peroration:{58 }
E-Language:{null}
}

The meteorological information influencing the paddy stem borer occurrence includes 9 main factors, average ground temperature, average wind velocity, average air temperature, average relative humidity, hours of sunshine, evaporation, highest air temperature, lowest air temperature and precipitation. And An E-Ontology for meteorological information was shown below.

Ontology meteorology {
E-Action:{ null }
E-Object:{ average ground temperature }
E-Time:{2007-4-23}
E-Environment:{ 102°33′N and 24°12′E; XiZhuang Town of JianShui County }
E-Peroration:{ 19.6℃ }
E-Language:{null}
}

The soil information influencing the paddy stem borer occurrence includes 14 main factors, PH value, organic horizon (g/kg), total nitrogen (N) g/kg, available nitrogen (N) mg/kg, total phosphorus (P) g/kg, available phosphorus (P) mg/kg, total potassium (K) g/kg, quick-acting potassium (K) mg/kg, slow-acting potassium (K) mg/kg, exchangeable magnesium (Mg) g/kg, available molybdenum (Mo) mg/kg, available zinc (Zn) mg/kg, available manganese (Mn) mg/kg and available boron (B) mg/kg. And An E-Ontology for soil information was shown below.

Ontology soil {
E-Action:{ null }
E-Object:{ available nitrogen }
E-Time:{2007-4-23}
E-Environment:{ 102°33'N and 24°12'E; XiZhuang Town of JianShui County }
E-Peroration:{118.48 }
E-Language:{null}
}

The irrigation water information influencing the paddy stem borer occurrence includes 17 main factors, PH value, cadmium, lead, copper, zinc, mercury, arsenic, chrome, dung coliform group, fluoride, chloride, petroleum, COD, cyanide, total phosphorus, volatile phenol and total salt. And An E-Ontology for irrigation water information was shown below.

Ontology irrigation water {
E-Action:{ null }
E-Object:{ zinc }
E-Time:{2007-4-23}
E-Environment:{ 102°33'N and 24°12'E; XiZhuang Town of JianShui County }
E-Peroration:{0.016}
E-Language:{null}
}

The artificial control influencing the paddy stem borer occurrence includes cultural control, artificial control and chemicals control etc. And An E-Ontology for artificial control was shown blow.

Ontology artificial control {
E-Action:{ 150-200 gram 50%l60 emulsifiable concentrate per mu }
E-Object:{agricultural technical personnel; farmer}
E-Time:{2007 4 23}
E-Environment:{ 102°33'N and 24°12'E; XiZhuang Town of JianShui County }
E-Peroration:{ null}
E-Language:{artificial control}
}

The E-Ontology model for paddy stem borer occurrence built in this paper includes 42 classes, which contains 26 parent child relations and 18 non- parent child relations; 65 attributes; 782041 instances which contains 1015 event instances, 781026 other instances.

4   Conclusions

First, An E-Ontology model was studied with paddy stem borer occurrence knowledge in this paper. Second, the knowledge structure of paddy stem borer occurrence was analyzed and described carefully. Third, the knowledge model based on E-Ontology for the occurrence of paddy stem borer (Scirpophaga incertulas) was accomplished. Above all, the knowledge model based on E-Ontology integrated the knowledge structures of paddy stem borer and make the knowledge sharing come true, and lay a strong foundation for knowledge management system. The knowledge inference mechanisms in this type model are the research focus in the next step, and the knowledge sharing platform for paddy stem borer occurrence would be framed based on the E-Ontology model in the coming days.

Acknowledgments. The findings and the opinions are partially supported by projects of the Natural Science Foundation of Yunnan Province of China, NO. 2008ZC050M. This work is supported by Shanghai University, China and Yunnan Agricultural University, China.

References

1. Ma, F., Xu, X.-F., Zhai, B.-P., Cheng, X.-N.: The Framework of Complexity Theory in Insect Pest Disaster Research. Entomological Knowledge 40, 307–312 (2003)
2. Chen, H.-X., Chen, X.-b., Gu, G.-h.: Studies on the Injury and the Economic Threshold of (Walker). Entomological Knowledge 36, 322–325 (1999)
3. Sheng, C.-f., Wang, H.-t., Sheng, S.-y., Gao, L.-d.: Pest Status and Loss Assessment of Crop Damage Caused by the Rice Borers, Chilo Suppressalis and Tryporyza Incertulas in China. Entomological Knowledge 40, 289–294 (2004)
4. Li, H.: Research on Ontology-based Emergency Domain Knowledge Expression and Reuse. TianJin University (2008)
5. Liu, Z.-t., Huang, M.-l., Zhou, W., Zhong, Z.-m., Fu, J.-f., Shan, J.-f., Zhi, H.-l.: Research on Event-oriented Ontology Model. Computer Science 36, 189–192 (2009)
6. Liu, W., Liu, Z., Fu, J., Zhong, Z.: Extending OWL for Modeling Event Ontology. In: Proc. 4th International Conference on Complex, Intelligent and Software Intensive Systems (CISIS-2010), Krakow, Poland (2010)
7. Zhong, Z., Liu, Z., Liu, W., Guan, Y., Shan, J.: Event Ontology and its Evaluation. Journal of Information & Computational Science 7, 95–101 (2010)

Design and Optimization of PowerPC Instruction Set Simulator

Peng Shan, Jun-feng Han, Qi-wu Tong, and Jin-chang Liang

Dept. Electronic Information and Control Engineering, Guangxi University of Technology,
Liuzhou, 545006, China
shanpeng_china@163.com, hanjf@139.com,
tongqiwu2005@163.com, sqwer112@qq.com

Abstract. In many embedded systems simulators, instruction set simulator based software simulators become a research hot topic. This paper realized the PowerPC instruction set simulator based on the mode of interpretive simulation and used the optimization technology to realize the modes of dynamic translation. Finally, using two procedures to test the performance of the two modes, the results show that the performance of instruction set simulation is significantly improved.

Keywords: PowerPC instruction set simulation, Dynamic translation, Optimization.

1 Introduction

With the rapid development of embedded systems, embedded systems research and development has become an important branch of computer science. Because the software development and the hardware development are unable parallel to launch, on the one hand it costs too much time on research and development; on the other hand it makes the project work lack of flexibility. In order to solve above problems, software-based embedded system simulator has already become one of major technique methods in embedded system research and development. The present simulator can be divided into two categories, one kind is hardware simulator, in which the program is running on the target chip and the software only provides an interface for users to understand the program running situation. The other kind is the software simulator and this kind of simulator is the form of pure software simulation program running on the target chip. The current software simulators are generally simulated in the ISA (instruction set architecture) level.

The set of instructions simulator is the ISA level model of processor and the software simulator of ISA level model processor, it simulates the implementation processes on the target processor of the executable code. In the simulation process, the user can examine the target machine's internal state, such as processor's registers, program counter value and so on, and can also examine the memory's change situations. The instruction set simulation was aimed to test the correctness of the

Y. Yu, Z. Yu, and J. Zhao (Eds.): CSEEE 2011, Part II, CCIS 159, pp. 377–382, 2011.
© Springer-Verlag Berlin Heidelberg 2011

application algorithm, test interrupt handler, and make the behavior of the system and the external environment visual. PowerPC is RISC (Reduced Instruction Set Computer) architecture, which has outstanding performance and widespread applications in high-end server and the embedded domain, so it is necessity to simulate the PowerPC system.

2 Classification of Instruction Set Simulators

2.1 Interpretive Instruction Set Simulator

Interpretive instruction set simulator is a virtual machine basically realized by the software that execute the loaded target code on the host machine in the form of interpretation, as shown in Figure 1. Instruction execution is generally divided into three steps which is fetching, decoding and executing. Each time, one target program instruction is treated, by decoding and analysis, according to decoding results, and go to the corresponding functional simulation subroutine instructions to complete the instruction of logic functions, the entire process cycle is not completed until the end of the simulation.

Processing of all simulator instructions are completed at run time. Interpreted simulator mechanism makes it have advantages such as analog precision, simple realization, easy debugging. In the process of the target machine simulation, the decoding stage is a very time-consuming and inefficient operation.

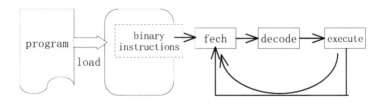

Fig. 1. Interpretive simulation

2.2 Compiled Instruction Set Simulator

The second technique is compiled simulation, also called static translation, as shown in Figure 2. The application program is decoded in compiling phase and has been translated into recognized procedures for the host machine. This technology makes the simulation speed greatly improved, but it is not as flexible as interpretive simulation. In the compilation phase, the application program can not be modified, so this method is not suitable for self-modifying code and dynamically new loaded code at running-time.

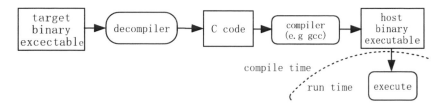

Fig. 2. Compiled simulation

3   Design and Optimization of Instruction Set Simulation

In order to compare the performance of different instruction set simulation, this article realizes the interpretive simulation mode, the dynamic translation model.

3.1   Interpreted Simulation Mode

This kind of mode, named D0, which process is simply described as to fetch instructions, decode and execute. It does not have translation pages and the concept of pseudo-instructions. The process of target instruction simulation is: according to the pc, fetch the target instructions, decode target instructions, then execute target instructions and update the state of the processor.

In order to improve extension and reuse, the simulation instructions is design for four types: arithmetic and logical instructions, jump instructions, addressing instructions, floating point instructions. When designing instruction execution functions, use data acquisition and data manipulation methods of separating, make the implementation function more versatility. When adding a new instruction, we only need to add the corresponding addressing mode and the instruction execution process which is not existed in the original system, the operation of adding the instruction is complete, so as to improve the extendibility for the system design.

The strategy of instruction dispatch simulation is to divide the instructions' format into fifteen categories: PPC_I_FORM, PPC_B_FORM, PPC_SC_FORM, PPC_D_FORM, PPC_DS_FORM, PPC_X_FORM, PPC_XL_FORM, PPC_XFX_FORM, PPC_XFL_FORM, PPC_XS_FORM, PPC_XO_FORM, PPC_A_FORM, PPC_M_FORM, PPC_MD_FORM, PPC_MDS_FORM. The same operation code is formed one class, in this class using a loop to locate the corresponding simulation instruction so as to complete the functions' corresponding operation. After using two cycle search, the speed of instruction scheduling is improved.

The cycle of instruction's Simulation and scheduling is as follows:

1.   Fetch instructions, read target instruction according to the current value of pc.
2.   Decode instructions, the instructions to be executed are decoded to be opcode, operands, condition codes and other information that needed in instruction simulation.
3.   Scheduler, jump to the corresponding instruction simulation function according to type of instructions.
4.   Execution, call the PPC_Instruction subclass exec method to finish functions of the target instruction.

5.   Interrupt detection and scheduling, in order to simulate the interrupt system, before running the simulator, the user simulate the external interrupt input by providing an interrupt configuration interface to set the variable of interrupt pins. By the end of each instruction simulation, must detect the interrupt pins' variable, if it detects an interrupt, then call the interrupt handler, pc is set to interrupt service routine entry address.

6.   Update pc value, if this cycle does not produce jump instruction, interrupt and load pc, then pc = pc +4. Remove the PPC_Instruction subclass object, jump to step 1.

In this mode, the decoding phase is time-consuming, the same instruction execute the second time, there will be no increase in speed, because the first decoding, the code will not be saved, it will affect the simulation speed.

3.2 Dynamic Translation Mode

The second mode (D1) is Dynamic translation with no specialization. For the embedded system usually contains a lot of code which need to repeat the execution, and instruction translation work will take a lot of system overhead, this model improve the speed of simulation by dynamic translation.

The main idea of dynamic translation is to fetch instructions from memory in the decoding stage; the process is of the same type with the interpretative simulation. A command, if it is the first time executed, it will be decoded, the simulator will put the instructions translated into intermediate code and stored in the Cache. When the instructions have been translated, the next application directly calls from the Cache. The process is shown in Figure 3. If the application is modified during operation, the simulator will make the code stored in the Cache as invalid. Dynamic compilation combines the advantages of Interpreted simulation and static compilation-based simulation, and supports dynamic loading and modifying code in the running, so the speed of this model is quicker than D0 model.

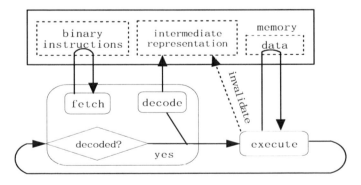

Fig. 3. Dynamic translation

D1 model based on D0 model uses paging and page cache decoding technology. Implementation of this model's function is compute_DT_with_instr. This technology has improved the performance of simulation speed.

4 The Results of Experiment

In this paper, we use two procedures to test the performance of instruction set simulator, use the function loop.c and crypto.c to test the speed of simulation instruction. loop.c is a simple loop program; Crypto is a complex testing procedures of the encryption and decryption, a lot of functions come from XYSSL library. We use the GCC cross compiler to compile the test program into different executable programs. 0.x as the suffix that is not optimized executable program , 3.x as the suffix that is optimized executable program, simulator uses GCC cross-compiler. no opt. (no optimaization) is not GCC optimal results, and with opt. (with optimization) is the GCC optimal results.

All experiments below are run on three different CPU, and then we get the average of values. The results of tests are shown as Table 1. Mips (Million instruction per second) is the unit to test the performance, that is, the number of millions of instructions per second. From the results of experiment, we can see that the speed of D1 mode is almost four times the D0 mode's.

Table 1. Results for the loop and crypto benchmark

Test Mode	loop.0.x no opt.	crypto.0.x no opt.	crypto.3.x with opt.
D0	15.45 s 9.77 Mips	183.60 s 9.38 Mips	60.91 s 9.28 Mips
D1	4.11 s 36.70 Mips	49.37 s 34.88 Mips	16.45 s 34.36 Mips

5 Summary

Instruction set simulator is simulated target machine on the target system's instruction set level, by simulating each instruction in the target processor to simulate the effect on the implementation of the target machine program. Users can view the target machine's internal state in simulation process, such as processor registers, program counter value and so on, can also view the memory changes.

In this paper, we realize the interpretive simulator, and then optimize it by dynamic translation. Experiment shows that the simulator has greater efficiency in the implementation. Currently, the simulator only implements the 32-bit PowerPC architecture version, the next step to achieve 64-bit PowerPC architecture version and use partial evaluation technology to optimize the instruction set simulator.

References

1. Keet, E.: A Personal Recollection of Software's Early Days (1960-1979): Part 1. IEEE Annals of the History of Computing 17(4), 24 (1995)
2. Voith, R.: The PowerPC 603 C++ Verilog Interface Model. In: Digest of Papers Spring CompCon, March 1994, pp. 337–340. IEEE Computer Society Press, Los Alamitos (1994)
3. Sutanuala, S., Paulin, P., Kumar, Y.: Insulin: An Instruction Set Simulation Environment. In: Proceedings of CHDL-1993, Ottawa, Canada (1993)
4. Faulh, A.: Beyond Tool-Specific Machine Descriptions. Code Generorion for Embedded Processors (1997)
5. Hadjiyiannis, G., Hanona, S., Devadas, S.: Lsdl: An Instruction Set Description Language for Retargetability. In: Proceeding of the Design Automation Conference (DAC), Anaheim, CA (June 1997)
6. Hartoog, M., Rowsan, A., Reddy, P.D., Desai, S., Dunlop, D.D., Harcoun, E.A., Khulla, N.: Generation of Software Tools From Processor Descriptions for Hardware/Software Codesign. In: Proceeding of the Design Automation Conference (DAC), Anaheim, CA (June 1997)
7. Scott, K., Davidson, J.: Strata: A Software Dynamic Translation Infastructure. In: Proceedings of the IEEE Workshop on Binary Translation (2001)
8. Grant, B., Philipose, M., Mock, M., Chambers, C., Eggers, S.L.: An Evaluation of Staged Run-Time Optimizations in Dyc. In: Proceedings of the ACM SIGPUN Conference on Programming Longuage Design and Implemanlalion (1999)
9. Hartoog, M., Rowson, J.A., Reddy, P.D., Desai, S., Dunlop, D., Harcourt, E.A., Khullar, N.: Generation of Software Tools from Processor Descriptions for Hardware/Software Codesign. In: Proc. of the Design Automation Conference (1997)
10. Leupers, R., Elste, J., Landwehr, B.: Generation of Interpretive and Compiled Instruction Set Simulators. In: Proc. Of the Asia South Pacific Design Automation Conference (1999)
11. Nohl, A., Braun, G., Schliebusch, O.A.: Universal Technique for Fast and Flexible Instruction-Set Architecture Simulation(A). In: Proceedings of the Design Automation Conference, vol. 39, pp. 138–141. IEEE Press, Design Automation Conference (2002)
12. Zhu, J., Gajski, D.D.: A Retargetable, Ultra-fast Instruction Set Simulator. In: Proceedings of Conference on Design Automation and Test in Europe, pp. 298–302. IEEE Press, Los Alamitos (1999)
13. Burtscher, M., Ganusov, I.: Automatic Synthesis of High-Speed Processor Simulators. In: Proceedings of the 37th annual International Symposium on Micro Architecture. IEEE Press, Los Alamitos (2004)
14. Bartholomeu, M., Azebedo, R., Rigo, S., Araujo, G.: Optimizations for Compiled Simulation Using Instruction Type Information. In: Proceedings of the 16th Symposium on Computer Architecture and High Performance Computing, pp. 74–81. IEEE Press, Los Alamitos (2004)
15. Cmelik, B., Keppel, D.: Shade: A Fast Instruction-Set Simulator for Execution Profiling. In: SIGMETRICS 1994, pp. 128–137. ACM, New York (1994)
16. Schnarr, E., Larus, J.R.: Fast Out-Of-Order Processor Simulation Using Memorization. SIGOPS Oper. Syst. Rev. 32(5), 283–294 (1998)
17. Hongwei, H., jiajia, S., Helmstetter, C., Joloboff, V.: Generation of Executable Representation for Processor Simulation with Dynamic Translation. In: Proceedings of the International Conference on Computer Science and Software Engineering. IEEE Press, Wuhan (2008)
18. Futamura, Y.: Partial Evaluation of Computation Process - An Approach to a Compiler-Compiler. Higher Order Symbolic Computation 12(4), 381–391 (1999)

Self-calibrating Sensitivity Based on Regularized Least Squares for Parallel Imaging

Xiaofang Liu[1,*], Yihong Zhu[2], Yongli Chen[1], and Wenlong Xu[1]

[1] Department of Biomedical Engineering, China Jiliang University, HangZhou310018, China
[2] No.118 hospital of PLA NO.15 JiaFuSi Lane WenZhou Zip 32500, China
liuxfang@cjlu.edu.cn, 511495621@qq.com,
{axuewuhen,wenlongxu}@cjlu.edu.cn

Abstract. In order to improve the accuracy of sensitivity estimate from small number of self-calibrating data, which is extracted from a fully sampled central region of a variable-density k-space acquisition in parallel images, a novel scheme for estimating the sensitivity profiles is proposed in the paper. On consideration of truncation error and measurement errors in self-calibrating data, the issue of calculating sensitivity would be formulated as a linear estimation problem, which is solved by regularized least squares algorithm. When applying the estimated coil sensitivity to reconstruct full field-of-view(FOV) image from the under-sampling simulated and in vivo data, the quality of reconstruction image is evidently improved, especially when a rather large accelerate factor is used.

Keywords: parallel magnetic resonance imaging (pMRI), self-calibrating technique, regularized least squares (RLS), conjugate gradients (CG).

1 Introduction

Parallel magnetic resonance imaging (pMRI) is a rapid acquisition technique and considered to be one of the modern revolutions in the field of MRI. In parallel imaging, since a certain amount of the spatial encoding, traditionally achieved by the phase-encoding gradients, is substituted by evaluating data from several coil elements with spatially different coil sensitivity profiles, the choice of sensitivity calibration strategy is at least as important as the choice of reconstruction strategy [1].

Unfortunately, the existing techniques for determining sensitivity functions are not satisfactory. The most common technique has been to derive sensitivities directly from a set of reference images obtained in a separate calibration scan before or after the accelerated scans. In order to eliminate the separate calibration scan, a more general method is the self-calibrating technique, which acquires variable-density k-space data during the accelerated scan [2,3]. In addition to the down-sampled lines at outer k-space, the variable-density acquisition includes a small number of fully sampled lines at the center of k-space, which is named self-calibrating lines. These

---

* Corresponding author.

Y. Yu, Z. Yu, and J. Zhao (Eds.): CSEEE 2011, Part II, CCIS 159, pp. 383–388, 2011.
© Springer-Verlag Berlin Heidelberg 2011

self-calibrating lines after Fourier transformation produce low-resolution reference images for any given component coil l. To derive the sensitivities, these low-resolution reference images are divided by their sum-of-squares(SoS) combination[4,5]:

$$\hat{c}_l(\vec{r}) \approx \frac{[f(\vec{r})c_l(\vec{r})]^{low-resolution}}{\sqrt{\sum_l |[f(\vec{r})c_l(\vec{r})]^{low-resolution}|^2}} \tag{1}$$

In general, the approximation in Eq.1 requires the range of spatial frequencies cover in the self-calibrating lines is much broader than the spatial frequency band of the coil sensitivity functions [6]. However, this increases the acquisition time associated with pMRI and contradicts the goal of pMRI. On the other hand, if a small number of self-calibrating lines are used, truncation of high spatial frequency components of transverse magnetization $f(\vec{r})$ results in Gibbs ringing artifacts in the extracted sensitivity reference images.

Towards quieter and faster imaging, the paper would use only the self-calibrating data from a variable-density acquisition, a novel method for estimating the sensitivity profiles would be proposed in the paper. On consideration of Gibbs ringing artifacts and noise in sensitivity reference images, which is generated from self-calibrating data, the method would view the issue of estimating $c_l(\vec{r})$ from these images as a linear estimation problem, and regularized least-squares methods are used to estimate the sensitivity profiles. In order to obtain the stable solution, the conjugate gradients algorithm will be used to solve this estimation problem. In combination with the generalized encoding matrix (GEM) reconstruction [7] method, the effectiveness of this self-calibration method would be demonstrated via phantom and in vivo brain imaging study.

2   Theory and Methods

To improve SNR and reduce the acquisition time, the use of multiple receive coils has become increasingly popular in MRI. Let c_l denote the sensitivity of the lth coil, for $l=1, \ldots, L$, where L denotes the number of coils. Let y_l denotes the recorded measurements associated with the lth coil. With only a modest loss of generality[8], the general forward model for the MR measurement signal associated with the lth coil may be approximated as the following matrix-vector equation:

$$y_l = c_l \rho + \varepsilon_l \tag{2}$$

Where ρ denotes the object's transverse magnetization; The measurement errors ε_l are modeled by additive, complex, zero-mean white gaussian noise.

For coil sensitivity varies slowly as a function of dominant spatial variations position, low-resolution images suffice to form sensitivity references data [9]. In order to self-calibrate the sensitivity profiles, the variable-density data acquisition schemes are usually adopted. Here, k-space is effectively split into two regions: a central region in which all phase-encode lines are fully sampled, and an outer region in which the lines are uniformly under-sampled. The central lines of k-space are extracted and Fourier transformed to yield low-resolution reference images as the measurement data of sensitivity reference $y_l^{reference}$. As a result, the low-resolution reconstruction image can be obtained by sum-of-squares (SoS) combination of these reference images, which is noted as G in the paper. According to Eq.2, the following equation would be proposed to calculate the sensitivity profiles in the paper:

$$y_l^{reference} = Gc_l + e_l \qquad (3)$$

Here, e_l denotes any error, such as truncation error and measurement errors. $y_l^{reference}$ is the low-resolution reference images.

Because c_l appears linear scaling in Eq. 3, the problem of estimating c_l would be a linear estimation problem, which can be resolved by minimizing the following regularized least-squares cost function:

$$\hat{c}_l = \arg\min_{c_l} \Phi(c_l) \quad \Phi(c_l) \cong \frac{1}{2} \| y_l^{reference} - Gc_l \|^2 + \frac{\beta}{2} R(c_l) \qquad (4)$$

Where $\hat{c}_l$ denotes the estimation values of c_l, and $y_l^{reference}$ denotes the sensitivity reference images; $R(c_l)$ is regularizing roughness penalty functions that encourages piecewise-smooth estimates. If regularization is not included, then the sensitivity estimate $\hat{c}_l$ will suffer from noise and Gibbs ringing artifacts. Here β is regularized parameters that control the smoothness of the estimates. Thus, our goal is to compute an estimate $\hat{c}_l$ of c_l by finding the minimum of the objective function $\Phi(c_l)$ as in Eq.4.

3 Data Acquisition and Analysis

Unaccelerated anatomical images of a standard resolution phantom were obtained from a 3T (Siemens Medical Solutions) human using an 8-channel head array coil. Imaging parameters: echo time (TE)=3.45ms, repetition time (TR)=2530ms, T1=1100 ms, Flip angle=7deg, slice thickness=1.33mm, FOV=256*256 mm. The B1 coil maps were calculated using Biot-Savart's law, and the fully sampled k-space data were obtained by inverse Fourier transforming the acquired unaccelerated images.

A in-vivo fully sampled brain dataset was obtained from PULSAR(a matlab toolbox for parallel MRI) [10], which was acquired using MR systems with eight-channel head array and multi-channel receiver from a healthy male volunteer with fast spoiled gradient-echo sequence, TR/TE =300/10 ms, matrix size = 256×256, tip angle=15° and FOV = 22*22 cm.

To simulate the under-sampled datasets in the manner of traditional variable-density data acquisition, the acquired full-sampled k-space data was decimated using reduction factors, namely accelerated factor R=2, 3, 4 and 5 for simulated and in vivo dataset. Meanwhile, the central k-space data were fully sampled to generate the sensitivity reference images.

In order to quantitatively analyze the quality of reconstruction image, so-called pixel-to-pixel normalized signal-to-noise ratio (SNR) and the normalized mean squared error (NMSE) of image would be calculated.

The normalized SNR can be calculated by the following equation:

$$SNR^{normalised}(\vec{r}) = \frac{1}{(\frac{N^{full}}{N^{acquired}})^{1/2}(\sum_l \mid f_l^{reference}(\vec{r})\mid^2)^{1/2}(\sum_{k_l} \mid \hat{c}_{\vec{r},k_l}^{-1}\mid^2)^{1/2}} \qquad (5)$$

Here, N^{full}, $N^{acquired}$ represent the un-accelerated and accelerated number of phase-encoded lines respectively. $f_l^{refrence}$ denotes a low-resolution reference image in a given component coil l. $\hat{c}_l$ represents the coil sensitivity profiles calculated as above.

NMSE is defined as the normalized difference square between the reconstructed image and the standard image, which was the SoS combination of unaccelerated images.

4   Results

4.1   Simulated Results

Fig.1 shows the B1 maps, the reference images, rough sensitivity maps and RLS sensitivity profiles of 8 receiver coils in simulated study.

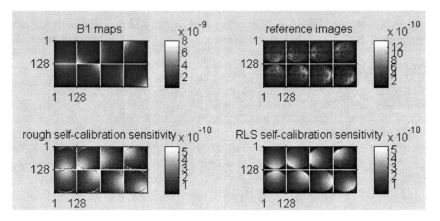

Fig. 1. Reference images and sensitivity profiles. Reference images extracted from central k-space data along 16 lines. From reference images, rough sensitivity profiles and RLS sensitivity profiles estimated respectively by Eq. 1 and Eq.4. For RLS sensitivity profiles, iterative order=128, β=4, φ(t)=t²/2. different order=2.

In Fig.1, the B1 maps are calculated by Biot-Savart's law; the reference images are extracted from 16 self-calibrating lines along phase encoding. From the reference images, the rough sensitivity maps are calculated by Eq.1, and RLS sensitivity profiles are estimated by Eq.4.

With the SoS of full images as the gold standard, Table.1 lists mean of normalized SNR and NMSE of reconstruction images respectively using B1 maps, rough sensitivity profiles and RLS sensitivity.

Table 1. Normalized SNR and NMSE of reconstruction images in phantom study

R	B1 coil maps		Rough sensitivity		RLS sensitivity	
	Mean of NSNR	NMSE	NSNR	NMSE	Mean of NSNR	NMSE
2	0.5312	1.0000	0.4981	1.0096	0.5865	0.4775
3	0.0837	4.2017	0.0859	4.3606	0.2128	2.6059
4	0.0455	9.0000	0.0400	9.1678	0.1368	5.9173
5	2.1865e-004	695.9739	2.6217e-004	214.1697	0.0315	15.9599

In table.1, R denotes the under-sampling rate, namely accelerate factor; NSNR is the pixel-to-pixel normalized SNR; From the fully-sampled data in 16 k-space central lines, Rough sensitivity were estimated by Eq.1, and RLS sensitivity estimated by Eq.4

4.2 In Vivo Study Results

Using rough sensitivity and RLS sensitivity profiles calculated as above simulated study, Table.2 lists mean of the normalized SNR and NMSE of reconstruction images respectively from the under-sampling data of accelerate factor R= 2, 3, 4 and 5.

Table 2. NSNR and NMSE of reconstruction images in in vivo study

R	Rough sensitivity			RLS sensitivity		
	Mean of NSNR	Std of NSNR	NMSE	Mean of NSNR	Std of NSNR	NMSE
2	0.5810	0.0939	1.0030	0.6376	0.0612	0.3730
3	0.1795	0.0907	4.7863	0.2448	0.1399	2.4113
4	0.1291	0.0646	10.5668	0.2035	0.0890	5.2800
5	0.0454	0.0498	58.7018	0.0761	0.0881	47.1165

5 Conclusions

The accuracy of coil sensitivity estimates is a major determinant of the quality of parallel magnetic resonance image reconstructions. Self-calibrating the coil sensitivity profiles can eliminate the need for an external sensitivity reference, and thereby reduces the total examination time. The paper proposed a novel self-calibrating sensitivity method, which viewed the issue of estimating the sensitivity profiles from self-calibrating data as a linear estimation problem [11,12]. On consideration of measurement errors and truncation error, the regularized least-squares method was used to estimate the sensitivities profiles. When the estimated sensitivity profiles were used to reconstruct full FOV image from under-sampling data, as seen in Table1 and Table 2, the quality of reconstruction images was remarkably improved, especially as a rather large accelerate factor was used.

References

1. Sodickson, D.K., McKenzie, C.A.M., Ohliger, A., et al.: Recent Advances in Image Reconstruction, Coil Sensitivity Calibration, and Coil Array Design for SMASH and Generalized Parallel MRI. Magnetic Resonance Materials in Physics, Biology and Medicine 13(3), 158–163 (2002)
2. McKenzie, C.A., Yeh, E.N., Ohliger, M.A., et al.: Self-Calibrating Parallel Imaging with Automatic Coil Sensitivity Extraction. Magnetic Resonance in Medicine 47(20), 529–538 (2002)
3. Schoenberg, S.O., Dietrich, O., Reiser, M.F.: Parallel Imaging in Clinical MR Applications. Springer, Heidelberg (2007)
4. Lin, F.H., Kwong, K.K., Belliveau, J.W., et al.: Parallel Imaging Reconstruction Using Automatic Regularization. Magnetic Resonance in Medicine 51(21), 559–567 (2004)
5. Ying, L., Sheng, J.H.: Joint Image Reconstruction and Sensitivity Estimation in SENSE (JSENSE). Magnetic Resonance in Medicine 57(30), 1196–1202 (2007)
6. Yuan, L., Ying, L., Xu, D., et al.: Truncation Effects in SENSE Reconstruction. Magn. Reson. Imaging 24(12), 1311–1318 (2006)
7. Pruessmann, K.P.: Encoding and Reconstruction in Parallel MRI. Magnetic Resonance in Medicine 19(10), 288–299 (2006)
8. Sodickson, D.K., McKenzie, C.A.: A generalized Approach to Parallel Magnetic Resonance Imaging. Medical Physics 28(8), 1629–1643 (2001)
9. Sodickson, D.K.: Tailored SMASH Image Reconstructions for Robust in Vivo Parallel MR Imaging. Magnetic Resonance in Medicine 44(2), 243–251 (2000)
10. Ji, J.X., Son, J.B., Rane, S.D.: PULSAR: A MATLAB Toolbox for Parallel Magnetic Resonance Imaging Using Array Coils and Multiple Channel Receivers. Concepts in Magnetic Resonance Part B 31B, 24–36 (2007)
11. Bauer, F., Kannengiesser, S.: An alternative Approach to the Image Reconstruction for Parallel Data Acquisition in MRI. Math. Meth.Appl.Sci. 30(21), 1437–1451 (2007)
12. Uecker, M., Hohage, T., Bloc k, K.T., et al.: Image Reconstruction by Regularized Nonlinear Inversion—Joint Estimation of Coil Sensitivities and Image Content. Magnetic Resonance in Medicine 60(13), 674–682 (2008)

Based on Human Position and Motion Trajectory for Behavior Detection

Hanqing Zhang[1], Yucai Wang[1], and Haitao Li[2,*]

[1] Hebei Vocational College of Foreign Language,
Qinhuangdao 066311, China
[2] College of Information Science and Engineering,
Yanshan University, Qinhuangdao 066004, China
{zhq_hello,ycwang,lht}@126.com

Abstract. For elderly people in regular daily activities, human position and motion trajectory can be used as a description of behavioral characteristics. Base on the improved HMM model and feedback sliding window, an efficient algorithm was proposed for abnormal detection. Identified the trajectory landmark points and HMM initial value with the fuzzy C-means clustering and given the improved formula revaluation, then carried on training and testing with feedback sliding window. Comparative experiments show that the method is effective and can ensure higher accuracy and lower false negative rate.

Keywords: Position; HMM; Abnormal detection; Trajectory.

1 Introduction

With the artificial intelligence and computer vision research in the field of increasing in-depth, target tracking and behavioral analysis technology continues to develop 1,2, the intelligence services and care for the elderly living alone in real-time way aroused wide attention and become a new research focus. When elderly in abnormal states, they may have many reactions, such as abnormal action, abnormal physiological parameters, etc. Because of the technical reasons, not all of the information is suitable for the real-time system, so the key problem of judging person's abnormal states by the data collected from sensors, such as camera and laser, is to find the appropriate feature information and represent the abnormal state pattern with it.

In many special locations, target behavior patterns can be extracted through the analysis of the target trajectory and position. For example, in intelligent traffic supervision , the motor vehicle generally move along a fixed path and the specified direction, through the study of the distribution pattern of these normal trajectory can automatically detect the reverse, crossing the road and other illegal acts 2.

In recent years, the behavior analysis method based on trajectory attracted extensive attention from scholars at home and abroad and a variety of algorithms have

---

* Corresponding author.

Y. Yu, Z. Yu, and J. Zhao (Eds.): CSEEE 2011, Part II, CCIS 159, pp. 389–394, 2011.
© Springer-Verlag Berlin Heidelberg 2011

been proposed. Some have more representative, such as the Euclidean distance 4, the longest common subsequence 5,6 and Hausdorff distance 7-9 and so on. They are usually used in the unsupervised clustering method to measure the differences between the different trajectories.

Service robot works in home environment and monitors the states of elderly human real-time. In case the elderly human is in abnormal state, the service robot should be able to find and give the judgment correctly. Considering the daily living habits of the elderly are more disciplinarian than the younger at home and a map of home environment will be used by the robot when it works, we use the location and trajectory as the feature information to find the abnormal state of the elderly human.

We propose an approach for the service robot to detect abnormal state of the elderly in a home environment in a real-time way with HMM (Hidden Markov Models) 10, and take correct measures according to the level and reliability of the abnormalities. We presents a trajectory distribution patterns extraction based on HMM and the method to detect abnomals, and improve the determination of the number of states in HMM model training, as well as some key issues, then ensured the accuracy and robustness in track testing through feedback sliding window. Finally, comparison experiments with the edit distance measurement proved the method is effectiveness and feasibility in the home environment.

2   The Preprocessing of Trajectory

2.1   Trajectory Generation and Quantization

Taking into account of the speed of the target is sometimes fast and sometimes slow in the actual monitoring scene as well in order to analyze the trajectory is much more accurate, the original trajectory needs to be re-quantize and re-sampled. Here the initial location information is pre-processed according to the time to ensure the robustness of the trajectory data. This paper selects the continuous trajectory in a fixed time as the basic information to deal with, referring to the literature 11.

Re-quantize and re-sample all the original trajectories that are tracked on the map $T' = \{(x'_1, y'_1), (x'_2, y'_2), ...(x'_n, y'_n)\}$: n is the original length of trajectory. Suppose the time interval contains k location information, considering the mean of location is a good expression of people's activity center in the period, and the variance is a very good description of the characteristic of activities from the scope of the activities, so the coordinate information of all the locations in each period is quantified by mean and variance. Quantified vector of each period t can be get $p = (\overline{x}, \overline{y}, \sigma_x, \sigma_y)$, $\overline{x}$, σ_x , $\overline{y}$, σ_y represent respectively the mean and the variance of location coordinates in current time interval t .After being handled, we can get the new post-sampling to quantify the trajectory sequence $T = \{(\overline{x_1}, \overline{y_1}, \sigma_{1_x}, \sigma_{1_y}), (\overline{x_2}, \overline{y_2}, \sigma_{2_x}, \sigma_{2_y}), ...(\overline{x_m}, \overline{y_m}, \sigma_{m_x}, \sigma_{m_y})\}$: m is the real trajectory length after quantization, that the total number of all period t .

Using this sampling method, we can quantify the trajectory better and can also filter some local points abnormal, ensuring the smoothness of trajectory.

2.2 Selection of the Marked-Points

In order to identify the trajectory better within the family environment, a number of marked points are often necessary in the environment which can be reference points of comparison between trajectory sequences. The difference of the marked-points selection is great for different methods. The existing method is to appoint some special location artificially where the activities are more frequently, such as sofas, dining, desk and so on, but its randomness and uncertainty is great and it can not well reflected the characteristics of trajectories, more importantly, it does not apply the HMM model to be adopted in this passage. Therefore, this article does some improvement on the selection of marked-points, adopt the method of environmental maps combined with FCM clustering algorithm, classifying the m vectors of the trajectory sequence T of a vector classification, assuming that the category that the vectors are divided into is c, then for all the vectors of the trajectory sequence, its membership grade $u_{ij} \in \{0,1\}$, that is to say the trajectory quantified vectors in the period j can be classified into the fuzzy membership grade of the ith category, meeting the condition $\sum_{i=1}^{c} u_{ij} = 1$, $i = 1...c$, $j = 1...m$. About the details of FCM algorithm, refer to the literature 12.

The Fig.1 shows the effect that all the quantified vectors of trajectories on the map are clustered, and each cluster center can be used as marked points of the trajectories. This kind of method avoids some errors prone to appear with manual intervention, and its results are closer to the characteristics of the target's actual trajectory, improving the robustness of the system as well.

3   HMM of the Trajectory

HMM is a double stochastic process, one of which is an implicit Markov chain with limited states, which describes the transfer of states; another random process describes statistical correlation between the states and the observations 10.

Usually model can be expressed with parameters $\lambda = (A \cdot B \cdot \pi)$. In the process of training, we divided the observed sequence T into sub-sequences whose length is C, $O = \{O^1, O^2, ..., O^L\}$, each O^z ($z = 1...L$, $L = \lfloor m/c \rfloor$) is the sub-sequence with the length of C.

After getting the model's new parameter λ, to determine whether it meet the Condition $P(O \mid \lambda) - P(O \mid \overline{\lambda}) < \varepsilon$, ε is the default threshold, and the selection of its value is determined through the experiment data. If it doesn't meet the condition, the model's parameter need to be re-estimated until the condition is met, then the loop iteration is over. It can be known that repeated assessment can improve the output probability of training sequence, the final outcome of this reassessment is known as the maximum likelihood estimate of HMM. After training, we can get the HMM' parameters describing the normal mode of the system.

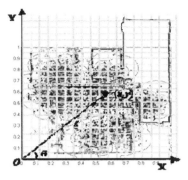

Fig. 1. sampling points clustering of the trajectory

4   Identification and Detection of the Trajectory with Feedback Sliding-Window Scheme

First, we first cut apart a section of the trajectory sequence tracked in real time with the sliding window whose length is Kt, the sliding window moves backward single order every time, that is a time interval t. And then quantify the window sequence that has been divided in the way referred at 2.1, suppose each time the number of the windows detected in real time is L, namely, the window sequence can be expressed with set $WT = \{T^w{}_1, T^w{}_2, ... T^w{}_l)\}$, each $T^w{}_i$ is a short sequence containing k quantified vectors of the trajectories.

As to each of the short sequence $T^w{}_i$, we get the output probability $P(T^w{}_i \mid \lambda)$, with which we can judge whether it is matching with the normal pattern. The abnormality of the trajectory δ can be calculated as follows:

$$\delta = 1 - \sum_{i=1}^{L} P(T^w{}_i \mid \lambda) \Big/ L \tag{1}$$

It can be seen that the length of sliding window is determined combining with the recognition accuracy of training samples, if the normal trajectory sequence of the sample can not reach the predetermined recognition accuracy, we need to consider to adjust the value of k. Here the threshold of the abnormality δ needs to be discussed according to the data of experiment.

5   Experiment

First, we record the specific location of the target in the family environment in some moments using the artificial method. The experiment data is obtained like this: collecting the behavior of the same serving object for several days continuously. We try our best to ensure that experiment data obtained in the ideal experimental environment is valid. Then, using the SLAM simulation system in the platform of Pioneer II , supplement the points of some specific locations generating the trajectory sequences needed by the experiment.

First of all, get the marked-points of the trajectories according to different values of C. Set a range (10-50) to deal with the data, and the experimental results is as follows Fig.2.

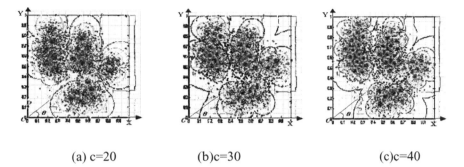

| (a) c=20 | (b)c=30 | (c)c=40 |

Fig. 2. Clustering results of the different values c

Here t is set 0.5 second, for the value of C is different according to different marked-points, we can train HMM respectively to get the value of the stable parameters. The experiment data of the abnormity δ that detect the trajectory sequences is as follows:

Table 1. analysis table of abnormity δ

trajectory measure k c	HMM			editorial distance
	20	30	40	1
3	0.125	0.084	0.091	0.135
5	0.054	0.076	0.041	0.112
7	0.101	0.097	0.078	0.099
9	0.114	0.101	0.093	0.157

The smaller the abnormity δ is, the better the matching degree of the detection result and the training model is. As can be seen, the value of C is not of great influence on the experiment results, but with the value of C increasing, the modeling time of the system will have a multiple increase. At the same time, we can know there is an optimum value while the value of k is in the vicinity of five, too large or too small value will both affect the final testing results.

6 Conclusion

About the detection of abnormal behavior for the elderly, by quantifying and sampling the daily trajectory of the elderly who move about regularly, the paper proposes a detection method of the abnormal behaviors based on improved HMM, and redefines the model's initial value as well as the training formula at the same time. Further more the paper tests the observations through the feedback sliding-window scheme. Through the experiment, it has proved that the method is effective to be used as anomaly detection of human behavior in family environment, and can ensure the higher accuracy and the lower false negative rate.

References

1. Hu, W.-M., Tan, T.-N., Wang, L.: A Survey on Visual Surveillance of Object Motion and Behaviors. IEEE Transactions on Systems, Man and Cybernetics, Part C: Applications and Reviews 34(3), 334–352 (2004)
2. Valera, M., Velastin, S.A.: Intelligent Dist Ributed Surveillance Systems: A Review. In: Proceedings of the Vision. Image and Signal Processing, pp. 192–204 (2004)
3. Yuan, H., Zhang, Y.: A Trajectory Pattern Learning Approach Based on the Normalized Edit Distance and Spectral Clustering Algorithm. Journal of Computer-Aided Design & Computer Graphic 6, 753–758 (2008)
4. Hu, W.M., Xie, D., Tan, T.N.: Learning Activity Patternsusing Fuzzy Self-Organizing Neural Network. Ieeetransactions on Systems, Man and Cybernetics -Part B: Cybernetics 34(3), 1618–1626 (2004)
5. Vlachos, M., Kollios, G., Gunopulos, D.: Discovering Similar Multidimensional Trajectories. In: Proceedings of the 18th Hinternational Conference on Data Engineering, San Jose, pp. 673–684 (2002)
6. Buzan, D., Sclaroff, S., Kollios, G.: Extraction and Clustering Ofmotion Trajectories in Video. In: Proceedings of T He 17th International Conference on Pattern Recognition, London, pp. 521–524 (2004)
7. Junejo, I.N., Aved, O., Shah, M.: Multi Feature Pat H Modelingfor Video Surveillance. In: Proceedings of T He 17th International Conference on Pattern Recognition, London, pp. 716–719 (2004)
8. Atev, S., Masoud, O., Papanikolopoulos, N.: Learning Trafficpatterns at Intersections by Spectral Clustering of Motiont Rajectories. In: Proceedings of the 2006 IEEE International Conference on Intelligent Robots and Systems, Beijing, pp. 4851–4856 (2006)
9. Qu, L., Zhou, F., Chen, Y.-W.: Trajectory Lcassification Based on Hausdorff Distance for Visual Surveillance System. Journal of Jilin University (Engineering and Technology Edition) 6, 1618–1624 (2009)
10. Wang, Q., Ni, G.: Anomaly Detection of System Calls Based on Improved Hidden Markov Model(HMM). Journal of Data Acquisition and Processing 4, 508–513 (2009)

Knowledge Representation for Fuzzy Systems Based on Linguistic Variable Ontology and RDF

Hongyang Bao[1], Jun Zhai[2], Mudan Bai[1], and Xiangpei Hu[1]

[1] Systems Engineering Institute, Dalian University of Technology,
Dalian 116024, China
[2] Transportation Management College, Dalian Maritime University,
Dalian 116026, China
baohongyang.dl@ccb.com, zhaijun_dlmu@yahoo.com.cn,
bmdhappy@gmail.com, drhxp@dlut.edu.cn

Abstract. The Semantic Web is turning into a new generation web, where ontology is adopted as a standard for knowledge representation, and Resource Description Framework (RDF) is used to add structure and meaning to web applications. In order to incorporate fuzzy systems into the Semantic Web, this paper utilizes fuzzy ontology to represent formally the fuzzy linguistic variables, considering the semantic relationships between fuzzy concepts. Then fuzzy rule is described as a RDF resource with properties: IF and THEN, and rule's antecedent and consequent is represented in RDF statement. Taking the fuzzy control system of industrial washing machine for example, the fuzzy system with ontology and RDF is built, which shows that this research enables distributed fuzzy applications on the Semantic Web.

Keywords: fuzzy knowledge, fuzzy system, linguistic variable, fuzzy ontology, RDF, the Semantic Web.

1 Introduction

The Semantic Web, the second-generation WWW, is based on a vision of Tim Berners-Lee, the inventor of the WWW. The great success of the current WWW leads to a new challenge: A huge amount of data is interpretable by humans only; machine support is limited. Berners-Lee suggests to enrich the Web by machine-processable information which supports the user in his tasks. Semantic Web application based on fuzzy logic greatly enhances user experience. It allows the user to enter queries using natural language with fuzzy concepts. Therefore, it is important to represent fuzzy knowledge on the Semantic Web [1]. To handle uncertainty of information and knowledge, one possible solution is to incorporate fuzzy theory into ontology. Then we can generate fuzzy ontologies, which contain fuzzy concepts and fuzzy memberships. Lee et al. [2] proposed an algorithm to create fuzzy ontology and applied it to news summarization. Tho et al. proposed a Fuzzy Ontology Generation Framework (FOGA) for fuzzy ontology generation on uncertainty information [3]. This framework is based on the idea of fuzzy theory and Formal Concept Analysis

Y. Yu, Z. Yu, and J. Zhao (Eds.): CSEEE 2011, Part II, CCIS 159, pp. 395–400, 2011.

(FCA). To enable representation and reasoning for fuzzy ontologies, Kang et al. [4] proposed a new fuzzy extension of description logics called the fuzzy description logics with comparison expressions (FCDLs). Calegari and Ciucci [5] presented the fuzzy OWL language. Lau presented a fuzzy domain ontology for business knowledge management [6]. Zhai et al. [7] presented a fuzzy ontology model using intuitionist fuzzy set to achieve fuzzy semantic retrieval for E-Commerce.

This paper applies ontology and Resource Description Framework (RDF) to represent fuzzy knowledge for fuzzy systems in the Semantic Web. The rest of this paper is organized as follows: Section 2 presents fuzzy linguistic variable ontology models and their RDF description. Section 3 introduces formal representation for linguistic variable using ontology and RDF. Section 4 presents formal representation for fuzzy rule in RDF. Finally, section 5 concludes the paper.

2   Fuzzy Linguistic Variable Ontology

The fuzzy linguistic variables proposed by Zadeh are the basic of fuzzy knowledge and fuzzy system. To achieve the knowledge share and reuse for fuzzy systems on the semantic web, it is necessary to represent the fuzzy linguistic variables with ontology.

Definition 1. (Fuzzy linguistic variable ontology) – A fuzzy linguistic variable ontology is a 6-tuple $O_F = (c_a, C_F, R, F, S, U)$, where:

(1) c_a is a concept on the abstract level, e.g. "price", "speed" etc.

(2) C_F is the set of fuzzy concepts which describes all values of c_a.

(3) $R = \{ r \mid r \subseteq C_F \times C_F \}$ is a set of binary relations between concepts in C_F. A kind of relation is set relation $R_S = \{$inclusion (i.e. $\subseteq$), intersection, disjointness, complement$\}$, and the other relations are the order relation and equivalence relation $R_O = \{\leq, \geq, =\}$. C_F and an order relation r compose the ordered structure $< C_F, r >$.

(4) F is the set of membership functions at U, which is isomorphic to C_F.

(5) $S = \{ s \mid s : C_F \times C_F \to C_F \}$ is a set of binary operators at C_F. These binary operators form the mechanism of generating new fuzzy concepts. Basic operators are the "union", "intersection" and "complement" etc., i.e. $S = \{ \vee, \wedge, \neg, \cdots \}$. C_F and S compose the algebra structure $< C_F, S >$.

(6) U is the universe of discourse.

To simplify the transform from fuzzy linguistic variables to fuzzy ontology, we introduce the basic fuzzy ontology model as follows.

Definition 2. (Basic fuzzy ontology) – A basic fuzzy ontology is a 4-tuple $O_F = (c_a, C_F, F, U)$, where c_a, C_F, F, U have same interpretations as defined in definition 1, which satisfy the following conditions:

(1) $C_F = \{c_1, c_2, \cdots, c_n\}$ is a limited set.

(2) Only one relation of set, the relation of disjointness, exists in C_F, and C_F is complete at U. In the other words, C_F is a fuzzy partition of U.

(3) C_F has an ordered relation $\leq$, and $< C_F, \leq>$ is a complete ordered set, i.e. all concepts in C_F constitute a chain $c_1 \leq c_2 \leq \cdots \leq c_n$.

(4) F is optional element of ontology.

An example of basic fuzzy ontology is $O_F = ($ $c_a =$ price of product, $C_F = \{$cheap, appropriate, expensive$\}$, $U = [0,100])$.

Another fuzzy ontology model can be derived from linguistic labels, which have been studied and applied in a wide variety of areas, including engineering, decision making, artificial intelligence, data mining, and soft computing [8].

Definition 3. (Linguistic labels ontology) – A linguistic labels ontology is a 4-tuple $O_F = (c_a, C_F, F, U)$, where c_a, C_F, F, U have same interpretations as defined in definition 1, which satisfy the following conditions:

(1) $C_F = \{c_0, c_1, c_2, \cdots, c_n\}$ is a limited set, whose cardinality value is odd, such as 7 and 9, where the mid term represents an assessment of approximately 0.5 and with the rest of the terms being placed symmetrically around it..

(2) C_F is ordered: $c_i \geq c_j$ if $i \geq j$.

(3) The reversion operator is defined: reversion $(c_i) = c_j$ such that $i + j = n$.

(4) $U = [0,1]$.

An example of linguistic labels ontology is $O_F = ($ $c_a =$ degree of loyalty, $C_F = \{c_0 =$ none, $c_1 =$ very low, $c_2 =$ low, $c_3 =$ medium, $c_4 =$ high, $c_5 =$ very high, $c_6 =$ perfect$\}$.

3 Formal Representation for Linguistic Variables in Fuzzy Systems

Fuzzy inference depends on fuzzy knowledge base, i.e. the set of fuzzy rules, which are summarized from human experiences expressed by fuzzy linguistic variables. Therefore, definition of fuzzy linguistic variables is the basic of construction of fuzzy rules. To share and reuse fuzzy knowledge, it is necessary to represent formally the fuzzy linguistic variables through ontology. Representing input variables in fuzzy control system for industrial washing machine, there are three basic fuzzy ontologies defined in definition 2 as followings:

(1) $O_1 = ($ $c_a =$ quality of cloth, $C_F = \{$chemical fiber, textile, cotton$\}$, $U = [0,100])$, where "chemical fiber" $\leq$ "textile" $\leq$ "cotton" from point of view of cotton content.

(2) $O_2 = ($ $c_a =$ quantity of cloth, $C_F = \{$little, middle, much, very much$\}$, $U = [0,25])$, where "little" $\leq$ "middle" $\leq$ "much" $\leq$ "very much" .

(3) $O_3 = ($ $c_a =$ degree of squalidity, $C_F = \{$clean, dirtish, dirty, filthy$\}$, $U = [0,100])$, where "clean" $\leq$ "dirtish" $\leq$ "dirty" $\leq$ "filthy" .

At the same time, there are several output variables, such as the quantity of scour, the quantity of water, the temperature of water and the washing time etc. The corresponding fuzzy ontology is as following:

(4) $O_4 = ($ $c_a =$ quantity of scour, $C_F = \{$little, middle, much, very much$\}$, $U = [0,500])$, where "little" $\leq$ "middle" $\leq$ "much" $\leq$ "very much" .

(5) $O_5 = ($ $c_a =$ quantity of water, $C_F = \{$little, middle, much, very much$\}$, $U = [0,20])$.

(6) $O_6 = ($ $c_a =$ temperature of water, $C_F = \{$low, middle, high$\}$, $U = [0,30])$.

(7) $O_7 = ($ $c_a =$ washing time, $C_F = \{$short, middle, long$\}$, $U = [0,60])$.

The Resource Description Framework (RDF) is the first layer where information becomes machine-understandable: According to the W3C recommendation [9], RDF "is a foundation for processing metadata; it provides interoperability between applications that exchange machine-understandable information on the Web." The RDF statements to represent these ontologies are as following:

```
<rdf: Description  ID= "quality of cloth">
<t: values> <rdf: Seq> <rdf: li  resource= "# chemical fiber"/>
<rdf: li  resource= "# textile"/>  <rdf: li  resource= "# cotton"/>
    </rdf: Seq> </t: values>
</rdf: Description>
<rdf: Description  ID= "quantity of cloth">
<t: values> <rdf: Seq> <rdf: li  resource= "# little"/>
<rdf: li  resource= "# middle"/> <rdf: li  resource= "# much"/>
    <rdf: li  resource= "# much"/></rdf: Seq></t: values>
</rdf: Description> ............
```

4 Formal Representation for Fuzzy Rules in RDF

Fuzzy systems use fuzzy IF-THEN rules. A Fuzzy IF-THEN rule is of the form: IF X1=A1 and X2=A2 … and Xn=An THEN Y=B, where Xi and Y are linguistic variables and Ai and B are linguistic values. The IF part of a rule is known as the antecedent, while the THEN part is known as the consequence or conclusion. The collection of Fuzzy IF-THEN rules is stored in the Fuzzy Rule Base, which is accessed by the Fuzzy Inference Engine when inputs are being processed.

The Fuzzy Rule Base for fuzzy control system of washing machine includes:

Rule 1: IF quantity of cloth is much and degree of squalidity is dirty THEN quantity of scour is much.

Rule 2: IF quantity of cloth is much and degree of squalidity is filthy THEN quantity of scour is very much.

Rule 3: IF quantity of cloth is little and degree of squalidity is clean THEN quantity of scour is little.

Rule 4: IF quantity of cloth is little and degree of squalidity is filthy THEN quantity of scour is middle.

Rule 5: IF quantity of cloth is middle and degree of squalidity is clean THEN quantity of scour is little.

Each rule has two properties, namely: "IF" that represents the antecedent part of the rule, "THEN" that represents the consequent part of the rule. The RDF statements to represent fuzzy rule are as following:

<rdf: Description ID= "Fuzzy rule">
<t: IF> <rdf: Bag> <rdf: li resource= "#antecedent 1"/>
<rdf: li resource= "#antecedent 2"/> </rdf: Bag> </t: IF>
<t: THEN> <rdf: Bag> <rdf: li resource= "#consequent 1"/>
<rdf: li resource= "#consequent 2"/> </rdf: Bag> </t: THEN>
</rdf: Description>

Each rule's antecedent and consequent can be represented in RDF Statement (rdf:Statement) with Subject (rdf:subject), Predicate (rdf:predicate), and Object (rdf:object) shown in Fig. 1.

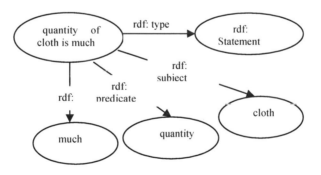

Fig. 1. RDF for antecedent

The RDF statements are as following:

<rdf: Description ID= "antecedent 1">
<rdf: Statement> <rdf: subject resource= "# cloth"/>
<rdf: predicate resource= "# quantity"/>
<rdf: object resource= "# much"/>
</rdf: Statement>
</rdf: Description> ....

5   Conclusions

Along with the development of WWW, it is useful to share and reuse fuzzy knowledge for fuzzy systems on the Semantic Web.

In this paper we have applied ontology and RDF to represent formally the fuzzy linguistic variables and fuzzy rule. Taking the fuzzy control system for washing machine for example, we have built the fuzzy system based on fuzzy ontology.

Our further researches lay on the automatic construction of fuzzy ontology from fuzzy systems and fuzzy ontology mapping.

Acknowledgments. This work is partially supported by the National Natural Science Funds for Distinguished Young Scholar (No. 70725004), and the Research Fund for the Ph.D. Programs Foundation of Ministry of Education of China under Grant No.20070151022.

References

1. Lukasiewicza, T., Stracciab, U.: Managing Uncertainty and Vagueness in Description Logics for the Semantic Web. J.Web Semantics 6(4), 291–308 (2008)
2. Lee, C.S., Jian, Z.W., Huang, L.K.: A Fuzzy Ontology and Its Application to News Summarization. IEEE Transactions on Systems, Man and Cybernetics (Part B) 35(5), 859–880 (2005)
3. Tho, Q.T., Hui, S.C., Fong, A.C.M., Cao, T.H.: Automatic Fuzzy Ontology Generation for Semantic Web. IEEE Transactions on Knowledge and Data Engineering 18(6), 842–856 (2006)
4. Kang, D., Xu, B., Lu, J., Li, Y.: Description Logics For Fuzzy Ontologies On Semantic Web. J. Southeast University (English Edition) 22(3), 343–347 (2006)
5. Silvia, C., Davide, C.: Fuzzy Ontology And Fuzzy-Owl In The Kaon Project. In: Proceedings Of 2007 IEEE International Conference On Fuzzy Systems Conference, London, UK, pp. 1–6 (2007)
6. Lau, R.Y.K.: Fuzzy: Domain Ontology Discovery for Business Knowledge Management. IEEE Intelligent Informatics Bulletin. 8(1), 29–41 (2007)
7. Zhai, J., Liang, Y.D., Yu, Y.: Semantic Information Retrieval Based On Fuzzy Ontology for Electronic Commerce. J. Software 3(9), 20–27 (2008)
8. Xu, Z.S.: Linguistic Aggregation Operators: an Overview. In: Bustince, H., et al. (eds.) Fuzzy Sets and Their Extensions: Representation, Aggregation and Models, pp. 163–181. Springer, Heidelberg (2008)
9. Lee, J.W.T., Wong, A.K.S.: Information Retrieval Based on Semantic Query on Rdf Annotated Resources. In: Proceedings of the 2004 IEEE International Conference On Systems, Man and Cybernetics, pp. 3220–3225 (2004)

A Mathematical Model for Finite Screen Assessment of Urban Noise

Baoxiang Huang, Yong Han[*], and Ge Chen

College of Information Science and Engineering Ocean University of China
266071 Qindao, China
hbx3726@163.com, chinahanyong@126.com, gechen@public.qd.sd.cn

Abstract. In order to obtain a good acoustical environment, the acoustic performance of screen needs to be well predicted. This paper presents a finite screen model for the assessment of airborne screen barrier, and it's application to an actual project. The main emphasis is placed on the diffraction from screen edges, incoherence between noise rays and sound pressure calculation. The proposed model is verified by comparsion with traditional model and measurement.

Keywords: Noise screen barrier, Noise impact assessment, Sound ray, VRGIS.

1 Introduction

Road traffic noise is one of the major environmental concerns of densely populate areas all over the world. The urban expressway volume makes the today's extensive road networks a very prominent sound source, the choice of using noise screen barriers has become available to road authorities [1]. Noise screen are walls built alongside expressways, railroads and at the end of airport runways in order to block the direct path from source to receiver [2]. In order to create shielded urban areas, a study of possible noise abatement schemes is an obvious issue [3]. The purpose of the present work is to report the development of a model that predicts the effect of noise screen in the presence of a directional noise source and to illustrate an application of the developed model by using data for urban expressway.

2 Traditional Infinite Model

The noise field around a screen, modeled as an infinitely half plane, generally sound will reach the receiver point by diffraction over the top of barrier or by direct transmission through the barrier [4]. When calculating the attenuation due to diffraction over a barrier, it is convenient to initially investigate if there is adequate diffraction to impede the noise propagation. This is achieved by calculating the difference in sound path length, δ i.e. the difference in path length that sounds would travel from source to receiver with and without the screen.

---

[*] Corresponding author.

Y. Yu, Z. Yu, and J. Zhao (Eds.): CSEEE 2011, Part II, CCIS 159, pp. 401–406, 2011.
© Springer-Verlag Berlin Heidelberg 2011

For pure diffraction, with the absence of ground effect, the attenuation may be given as Eq. 1.

$$A_{bar} = \begin{cases} 10\log_{10}(3+(40/\lambda)C''\delta) & (40/\lambda)C''\delta \geq -2 \\ 0 & (40/\lambda)C''\delta < -2 \end{cases} \tag{1}$$

Where λ is the wavelength of sound of the nominal central frequency for each considered octave band and δ is the difference in path length between the diffracted path and direct path, C'' is the coefficient accounting for multiple diffraction:

$$C'' = (1+(5\lambda e)^2)/(1/3+(5\lambda e)^2) \tag{2}$$

Where, e is the total distance between two extreme diffraction edges and direct paths. The calculated values for A_{bar} must lie between 0 and 25 dB.

3 Finite Screen Model

The sound propagation model described in GB17247.2-1998-7 assumes that contributions from interacting rays are added coherently, at the same time, traditional screen model assumes that the screens have an infinite length, in practice, screens will have a finite length and it is therefore possible that the screen effect will be reduced significantly when sound is diffracted around the vertical edges of the screen, so incoherent and averaging effects will smooth out the interference pattern, the effect of frequency band averaging and turbulence [5] are taken into account as incoherent and averaging effects.

The effect of frequency band averaging F_f cover the averaging within each one-third octave band, solution to F_f have been given in Eq. 3, k is the wave number, R_p and R_s are the length of the primary ray and secondary ray respectively.

$$F_f = \begin{cases} \dfrac{\sin(0.115k(R_s - R_p))}{0.115k(R_s - R_p)} & 0.115k(R_s - R_p) < \pi \\ 0 & 0.115k(R_s - R_p) \geq \pi \end{cases} \tag{3}$$

The effect of partial incoherence between rays [6] includes the reduction in coherence due to turbulence is defined by Eq. 4, C_W^2 and C_T^2 are the turbulence strength corresponding to wind and temperature, $\rho = 2h_s h_R/(h_s + h_R)$ is the transversal separation for flat ground where h_S and h_R are heights of source and receiver above the ground and d is the horizontal distance, T_0 is the mean temperature, c_0 is the mean sound speed, k is the wave number.

$$F_c = \exp\left(-\frac{3}{8}0.364\left(\frac{C_T^2}{T_0^2} + \frac{22}{3}\frac{C_W^2}{c_0^2}\right)k^2\rho^{5/3}d\right) \tag{4}$$

The contribution from diffraction around vertical edges is in an approximate manner taken into account by adding additional transmission paths as shown in Fig. 1. The solid line in the figure is the transmission path from the source S to the receiver R over the screen considered in above Section while the dashed lines are two subsidiary transmission paths introduced to account for diffraction around the vertical edges.

The sound pressure is calculated for of each path using the comprehensive propagation model with propagation parameters measured along the vertical propagation planes and added incoherently after a correction for the screening effect of the vertical edges. Plane SR is the direct propagation plane from the source to the receiver and Plane l and Plane r are the propagation planes around the left and the right edge of the screen. Plane l and r are passing just outside the screen so that the screen is not included in the propagation planes. The corresponding sound pressures are p_{SR}, p_l and p_r. The sound pressure can be calculated by Eq. 5, F_l and F_r are calculated by $F = F_f F_c$.

$$|p|^2 = |w_{SR}\, p_{SR} + F_l w_l\, p_l + F_r w_r\, p_r|^2 + \left(1 - F_l^{\,2}\right)|w_l\, p_l|^2 + \left(1 - F_r^{\,2}\right)|w_r\, p_r|^2 \tag{5}$$

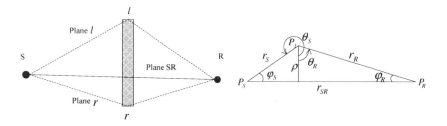

Fig. 1. Top view of vertical propagation planes   **Fig. 2.** Left vertical edge of the screen

The weight w_{SR}, w_l, w_r can be achieved by calculate the sound pressure reduction from diffraction around the vertical edges. The geometry used to calculate the contribution of sound diffracted around left vertical edge of the screen is shown in Fig. 2. P_S, P_R and P_l are the horizontal projections of the source, receiver and left edge. The weight w_l is calculated by Eq. 6. where the diffracted sound pressure p_{diffr} as a function of frequency is calculated by calculation equation in references [7], but with simplification that the screen faces are assumed to be non-reflection as mentioned above, $p_0 = 1/r_{SR}$ is the free-field pressure, r_S, r_R, θ_S, θ_R, β can be determined according to Eq. 7 to Eq. 12.

$$w_l = \left| \frac{p_{diffr,l}\left(r_S, r_R, \theta_S, \theta_R, \beta, f\right)}{p_0} \right| \tag{6}$$

$$\cos\varphi_S = \frac{\overrightarrow{P_S P_R} \cdot \overrightarrow{P_S P_l}}{\left|P_S P_R\right|\left|P_S P_l\right|} \tag{7}$$

$$\cos\varphi_R = \frac{\overrightarrow{P_R P_S} \cdot \overrightarrow{P_R P_l}}{\left|P_R P_S\right|\left|P_R P_l\right|} \tag{8}$$

$$\theta_R = \frac{\pi}{2} - \varphi_R \tag{9}$$

$$\theta_S = \frac{\pi}{2} + \varphi_S \tag{10}$$

$$r_S = \left|P_S P_l\right| \tag{11}$$

$$r_R = \left|P_R P_l\right| \tag{12}$$

The weight w_r corresponding to the contribution of sound diffracted around the right vertical edge of the screen can be calculated in a similar manner with the only alteration that P_l has to be replaced by P_R, w_{SR} has to be determined in a way where the w_{SR}, w_l, w_r and add up to a value of unity.

$$w_{SR} = -F_l w_l - F_r w_r + \sqrt{1 - w_l^2 - w_r^2 + \left(F_l w_l\right)^2 + \left(F_r w_r\right)^2} \tag{13}$$

When the screen length becomes 0 the method described above will produce a result identical to the result obtained when ignoring the screen.

4   Validation of the Model

To make a validation of the model presented, a part of Qingdao was chosen as a study area, as shown in Fig. 3. The area contains local streets with low and medium amounts of traffic volume. The east-west expressway is crossing the area, and is situated on a viaduct about 20m high, with noise barriers on part sides. To be able to realistically model the traffic noise [8], correct traffic data was necessary, the traffic noise model described in directive JTJ 005-96.

Fig. 4 presents noise simulation [9] around the gap between screen barriers created with a grid spacing of 10m. The noise values are comparable.

Fig. 3. The road network and screen of study area

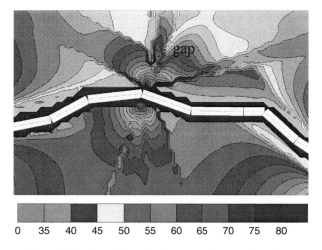

0   35   40   45   50   55   60   65   70   75   80

Fig. 4. The simulation of noise around screen junction

Using AWA6228 instruments mounted at a height of 1.2m, distance 5m,1/3 octave center values were measured, for a period of 20 minutes. The deviation between infinite model calculated values with infinite model and finite model and measurement is shown in Fig. 5.

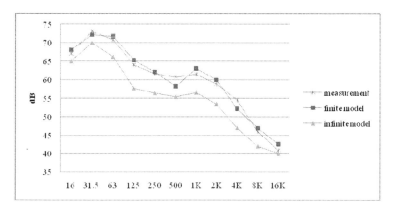

Fig. 5. 1/3 octave center values of position 5m

The mean differences between the predicted 1/3 octave band insertion loss values with infinite model behind the screen and the corresponding measured results are within 2dB.

5 Conclusion and Future Work

This paper provides an overview and information useful for noise screen assessment, although the new model provides only small average improvements relative to traditional model, it is easily argued that such small average improvement are important. It is note that adequate sound rays is important for optimization and improvements in the prediction model.

The model developed will be used as the basis of more extensive analysis of multi-screen structures and some improvements to the accurate of the model may be brought for more detailed evaluating noise mitigation activity.

Acknowledgement

This work is supported by the National Natural Science Foundation of China under Grant No.40971207; The Innovation Fund for Technology Based Firms of Hefei under Grant No.09C26213401458; Scientific and technical personnel services company Action Fund service of Beijing under Grant No. SQ2009GJA0002426.

References

1. Hamed, M., Effat, W.: A GIS-based Approach for the Screening Assessment of Noise and Vibration Impacts from Transit Projects. J. Environmental Management 84(6), 305–313 (2007)
2. Menounou, P., Papaefthymiou, E.S.: Shadowing of Directional Noise Sources by Finite Noise Barriers. J. Applied Acoustics 71(5), 351–367 (2010)
3. Salomons, E.M.: Noise Barriers in a Refracting Atmosphere. J. Applied Acoustics 47(3), 217–238 (1996)
4. Naish, D.: A method of Developing Regional Road Traffic Noise Management Strategies. J. Applied Acoustics 71(7), 640–652 (2010)
5. Han, N., Mak, C.M.: A Further Study of a Mathematical Model for a Screen in Open-plan Offices. J. Applied Acoustics 69(3), 1114–1119 (2008)
6. Hornikx, M., Forssén, J.: Noise Abatement Schemes for Shielded Canyons. J. Applied Acoustics 70(7), 267–283 (2009)
7. Hadden, J.W., Pierce, A.D.: Sound Diffraction Around Screens and Wedges for Arbitrary Point Source Locations, J. Acoust. Soc. Am. 69(2), 1266–1276 (1981)
8. Naish, D.: A Method of Developing Regional Road Traffic Noise Management Strategies. J. Applied Acoustics 71(7), 640–652 (2010)
9. Lee, S., Chang, S.I., Park, Y.: Utilizing Noise Mapping for Environmental Impact Assessment in a Downtown Redevelopment Area of Seoul, Korea. J. Applied Acoustics 69(8), 704–714 (2008)

Research and Implementation of Workflow Engine Based on MOSS

Yongshan Liu, Lele Wang, and Haihong Lv

College of Information Science and Engineering, Yanshan University,
Qinhuangdao, China
ysulys@sohu.com, wangLeVIP2010@163.com,
haixuanlanling@163.com

Abstract. Workflow engine is the core of the workflow management system. In order to make the business processes in MOSS achieve quickly and flexibly, this paper presented a kind of new workflow engine implementations based on MOSS. Analyzed various of business processes comprehensively, designed the common engine with WF state machine workflow, specific implemented the main function of the engine, and verified the feasibility of the common workflow engine through the application at actual project.

Keywords: MOSS, WF State Machine, Workflow Engine, XML Template.

1 Introduction

MOSS (Microsoft Office SharePoint Server 2007) is a web-based information services products published by Microsoft, it is a unified, secure enterprise application and management platform, compared to its previous version SPS 2003 (SharePoint Portal Server 2003), MOSS introduce workflow feature to provide a powerful support for the enterprise information collection and processing.

Workflow engine is the core of the workflow management system, the quality of the workflow engine will directly related to the execution efficiency and scalability of the workflow [1]. There are a number of studies on the workflow engine, through the discussion of the architecture of workflow engine, document [2] think that the workflow engine mainly comprises three models: agency model, information model and control model. Document [3] makes extended finite state machine as a model, reference WfMC standards, implemented a workflow engine based on extended finite state machine. There are other studies describe the workflow engine [4,5]. But as Microsoft's vacancy in workflow framework, in order to construct web-based workflow applications system on the Microsoft platform, the technology previously used mainly in Exchange and Biztalk. MOSS constructs its workflow function based on WF (Windows Workflow Foundation), how to implement workflow applications on MOSS platform is required to be studied in order to compensate the current academic gaps in this area, to provide better technical support for applications.

For the features of MOSS, this paper presented a new workflow engine implementation based on MOSS. Designed and implemented a common workflow

Y. Yu, Z. Yu, and J. Zhao (Eds.): CSEEE 2011, Part II, CCIS 159, pp. 407–411, 2011.

engine based on WF state machine workflow, by parsing the XML format workflow templates we can achieve business processes quickly and flexibly, simplify the work of administrator and improve work efficiency.

2   Related Concept

MOSS is Microsoft's enterprise application and management platform, which through an integrated platform rather than rely on decentralized systems to support all Intranet, Extranet and web applications across enterprises that provide features such as collaboration, portal, enterprise search, content management, electronic forms and business processes and business intelligence. The product builds on Microsoft. NET Framework, and is tightly integrated with Visual Studio 2008 and Office System 2007. MOSS introduced the concept of workflow, so that developers and IT staff can establish dynamic business processes cross-platforms, cross application, cross enterprise on the Internet very easily.

WF is the next generation of Windows workflow framework published by Microsoft, which can be used to create the interactive programs which need response to the signal from external entities. It can be used in both C/S and B/S frameworks, and support two workflow abstracts: sequential workflow and state machine workflow. WF give a great deal of support in the workflow persistence, transaction management, exception handling and activity compensation, role based access control, and communication based Web Service, and the dynamic update of process, etc.

3   The Implementation of Workflow Engine

3.1   Design of Common Workflow Engine

Analyzed and summarized the enterprise's various business processes, abstracted all of the states of common workflow engine based on WF state machine workflow, the workflow complete the transition of business management between these states. In order to meet enterprise's business needs, it contains seven different states, illustrate as follows:

WorkflowInitialState: The entry point of process, and the initial state of the process. In this state, workflow engine accept start event, accept initial information, such as the information of XML format process template, the project information associated;

TaskState: the main state of process. In this state, workflow engine will create tasks to the users, capture complete and update event for each task, and determine the next process state with the user's operation;

ToTaskState: After ending the current layer, except some special processes such as back, redistribute to others, etc, it is a intermediate state to the next level, back to taskState;

TaskResubmitState: This state is a state that the task was returned to the initiator. In this state, workflow engine create tasks to the initiator and deal with the operation results;

TaskReassignState: This state is a state that process will be redistributed to another person. In this state, the workflow engine will create tasks to approving people who has assigned task and monitor the approval of tasks;

ToFinalizationState: The intermediate state to end process. In this state, the workflow engine will deal with some information before ending the process, this state can only flow to the end state;

WorkflowFinalizationState: The end state, the process will end when it running to this state, without any code in this state, it only marks the end of a process.

3.2 Function Implementation of Common Workflow Engine

434. A business process must be defined as the process template that the workflow engine can parse and execute it. The workflow management system provide a graphical workflow template customizing interface, save the templates in the database as XML format after they are designed, the XML document include workflow template ID, the name of workflow, the approval level of process, the task title, the e-mail template ID, etc. When the engine needs to execute a process, the parser in engine parse the XML template to generate object which the engine can invoke, and obtain the initiator information, the user information (according to user No, LoginName or organization), agent information from them to assign tasks.

Process Instance Management. Workflow engine manage and monitor all of the workflow instances, modify workflow instances and the various states of task lists (including start, suspend, resume, stop, etc.). The life cycle of process instances described as follows: after creating a new process instance, the instance will be in Created state. Then start the instance, will migrate it to the Running state, and enter the Completed state after it running completed. A running instance can pause, and a pausing instance can resume execution. Incomplete instances can suspend, and move to Terminated state. When a process instance is idle as waiting for external data, it will be stored to external persistent storage media and destroy the process instance in the memory. When receive related signal from external entities, it will be activated automatically, load into memory from the persistent storage media and resume execution, this process may be cross process or cross machine.

Activity Executing Control. Activities are the basic building blocks of MOSS workflow, workflow engine parses the executing logic of each activity, interaction with external resources to complete tasks and parse routing rules for navigation of processes and activities execution.

Activity scheduling: When an activity started, it will take some CPU resources to complete a series of tasks quickly, then report its completion, or it will create one or more "bookmarks" (the logic fix point of program resume) first, conceded the resources and wait for external signals. The scheduler assigned a work item one time from the scheduler work queue by the way of FIFO. The essence of work items in scheduler work queue is a delegate, each work item corresponds to a approach of an active, it is the handle of the work item. Scheduler's behavior is strictly non-preemption, once the work items have been distributed, the engine will never intervene their execution.

Activity executing: The execution of activities are based on asynchronous fragmentary, its life cycle is a finite state machine, called the activity automaton. The initial state of an activity is initialized, once start work, the state shifts to the executing state, when the work is completed, activities will enter into the closed state. Activities will enter into the canceling state when the work is canceled, if an exception is thrown, activities will move to the Faulting state to handle errors. If a program has successfully executed a number of activities, but its execution has problems later, it can be migrated to the compensating state for compensate treatment, and cancel those effects which have been executed by activities.

3.3 The Core Control Algorithm of Common Workflow Engine

WF is a programming model to prepare and implement the interactive process, workflow flow downward according to the user's results. The transition between states can be described using the following algorithm, the main idea of the algorithm is: based on the principle of finite state machine, trigger the current state transited to the next state by the input and conditions. The current state of the engine receive task approval event, workflow engine obtain the information of next approval level after changing the task, and then determine the direction of the state according to the process set and the results of approver.

```
Algorithm: Workflow Engine State Transition Algorithm.
Input: current status; the event of task approving.
Output: result={taskResubmitState, taskState,
taskReassignState, toFinalizationState}
Procedure WfStateTransAlgorithm(Q,E)
Begin
SWFM•WE:e;
WE.receive(e);
WE.store(Aresult);
if(Aresult==backto)
nextlevel=backto;
if(Aresult==approved)
WE.require(Wflc){
if(!Isparallel&&Needapprove)
keepstate=1;
else
nextlevel==WE.get(Wft)}
WE•SWFM:result;
END
```

4   The Feasibility Verify

Qinhuangdao AGC automobile glass Co., Ltd. use MOSS as its enterprise information construction platform, the companies involved in a variety of business approval process in their office management, such as the human resources approval process, the contract approval process, the approval process of employee attendance, the business approval process of IT department, the corporate announcements, the press releases processes, etc. As the large enterprise organization structure, business approval processes are very complex accordingly. Make the position change between departments as an example, this process initiated by assistant director or attendance statisticians in recall sector, approved in proper order by the manager of application sectors, the director of application sectors, the manager of transferred to sectors, the director of transferred to sectors, HR manager of Human Resources, plant manager, general manager, after confirmed by the Human Resources Department the process will be ended. In many nodes of this process also exits parallel approval for several people, agency approval and other special circumstances, for example, the process doesn't need to be approved by senior leadership when the positions changing between base departments, process will end when approved by the HR manager.

This workflow engine implementations obtained the expected results in the enterprise's office platform, solved the problems that lack of flexibility and ease of use in MOSS workflow applications and other issues. Provided a flexible workflow management system for users, supported process branch, process rollback, multi-layer approval, more people approval (serial or parallel), load process template dynamic and other functions. And without modify the programs when the process changed, users can change the process only by creating, modifying the XML process configuration information, greatly met the complex business needs of enterprises.

5 Conclusion

This paper researched the design and implementation methods of workflow engine, designed and implemented a common workflow engine based on MOSS platform, can shorten the development cycle largely, guarantee the stability of the system effectively. Currently, the program has been successfully applied in the office platform of an enterprise, through the practical application finding that this scheme can satisfy the needs of enterprises complex business requirements, and it has some references value on the promotion of MOSS workflow application.

References

1. Workflow Management Coalition: The Workflow Reference Model. WFMC TC00-1003 (1996)
2. Qingfa, H., Guojie, L., Limei, J., Lili, L.: The Lightweight Workflow Engine Based on the Relationship Structure. Computer Research and Development 38, 137–139 (2001)
3. Yang, L., Baoxiang, C.: Design and Implementation of Workflow Engine Based on Extended Finite State Machine. Computer Engineering and Applications 32, 93–96 (2006)
4. Martinez, G., Heymann, E., Senar, M.: Integrating Scheduling Policies into Workflow Engines. Procedia Computer Science 1, 2737–2746 (2010)
5. Youxin, M., Feng, W., Ruiquan, Z.: Design of Workflow Engine Based on WCF. In: 2009 WRI World Congress on Software Engineering, pp. 100–103 (2009)

A Fully Automatic Image Segmentation Using an Extended Fuzzy Set

Ling Zhang[1] and Ming Zhang[2]

[1] School of Mathematics and System Sciences, Shandong University,
Jinan, Shandong 250100, China
[2] Department of Computer Science, Utah State University,
Logan, UT 84322, USA
zhanglingjn@sdu.edu.cn, ming.zhang@aggiemail.usu.edu

Abstract. Segmentation is a key step in image processing. Most existing segmentation algorithm doesn't work on blurry images. In medical CAD system, it is difficult to do segmentation fully automatically due to the property of images. In this paper, we employ a novel extended fuzzy watershed method for image segmentation and develop a fully automatic algorithm for breast ultrasound (BUS) image segmentation. A U domain is defined and employed to evaluate the uncertainty. The experiments show that our approach is fully automatic. It can get good results on blurry ultrasound images.

Keywords: Automatic Image Segmentation, Fuzzy Set, Uncertainty, Ultrasound Image.

1 Introduction

Image segmentation is a critical procedure for a CAD system. It has been widely used in many areas such as medical diagnosis [1, 2]. The purpose of image segmentation is to separate the objects and background into non-overlapping sets by tracing the edge between them. We know the edge pixels are the pixels separate objects from other objects or background. But these pixels may have small gray value difference with other pixels in many blurry images. Therefore, more precise segmentation algorithms are required for image analysis, evaluation and recognition techniques.

Fuzzy logic is a method in modeling vagueness and uncertainty problems for about 40 years [3]. It is widely used in digital image processing and pattern recognition. Many segmentation algorithms based on fuzzy set theory have been reported [4, 5, 6]. Fuzzy logic is a multi-value logic defined in [0, 1]. However, there are a lot of paradoxes and proposition cannot be described in Fuzzy Logic.

In this paper, an extended fuzzy logic set will be defined. We will combine Shannon's function and fuzzy entropy principle to perform the fuzzification. Then we define another uncertainty domain to help distinguish boundaries from background and objects. A new parameter U which stands for uncertainty factors is introduced.

Y. Yu, Z. Yu, and J. Zhao (Eds.): CSEEE 2011, Part II, CCIS 159, pp. 412–417, 2011.

Finally, watershed algorithm is applied to the extended fuzzy set to get segmentation result in BUS.

2 Proposed Method

In ultrasound images, the boundary of the object is unclear and hard to be distinguished. In order to measure with this uncertainty, we define a new domain U. In this paper, we define the object as T, the boundaries of the object as U and the background as F. A pixel in the image can be represented as $A\{T,U,F\}$, which means the pixel is $t\%$ true (object), $u\%$ uncertainty (boundary) and $f\%$ false (background), where $t \in T$, $u \in U$, and $f \in F$, $0 \le t,u, f \le 100$.

Map Image and Decide. $A\{T,F\}$, Given an image I, $p(x,y)$ is a pixel in the image,and (x,y) is its position. A 20x20 mean filter is applied to I for making the image uniform. Then the image is converted by using a S-function:

$$T(x, y) = S(g_{xy}, a, b, c) = \begin{cases} 0 & 0 \le g_{xy} \le a \\ \dfrac{(g_{xy} - a)^2}{(b-a)(c-a)} & a \le g_{xy} \le b \\ 1 - \dfrac{(g_{xy} - c)^2}{(c-b)(c-a)} & b \le g_{xy} \le c \\ 1 & g_{xy} \ge c \end{cases} \tag{1}$$

$$F(x, y) = 1 - T(x, y)$$

Where g_{xy} is the intensity value of pixel $P(i, j)$. Variables a, b and c are the parameters can be determined by using method in [7].

Enhancement. Use intensification transformation to enhance the image in the new domain:

$$\begin{aligned} T_E &= E(T(x, y)) = 2T^2(x, y) & 0 \le T(x, y) \le 0.5 \\ T_E &= E(T(x, y)) = 1 - 2(1 - T(x, y))^2 & 0.5 < T(x, y) \le 1 \\ F_E &= 1 - T_E \end{aligned} \tag{2}$$

Find the Thresholds in T_E and F_E. Two thresholds are needed to separate the new domains T_E and F_E. The histogram is applied on the fuzzy image obtained in last step. The number of N bins in histogram can be changed as a parameter. Two thresholds T_1 and T_2 are given for the first and second valley of the histogram (assume the background is dark, or we will use the complement of the image). The pixels with gray levels below T_1 are considered as the background, and those with the gray levels after T_2 are considered as objects and the ones with the gray levels between T_1 and T_2 are the blurred edges as shown in Fig. 1. The background and objects are set 0 and edges are set 1. Then the image is changed from the gray level to binary.

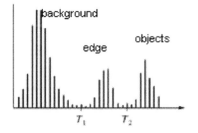

Fig. 1. Histogram of image

Define Uncertainty in $\{U\}$. Uncertainty is related to the local information, and plays an important role in image segmentation. We define uncertainty by using the standard deviation and edge information. Standard deviation describes the contrast within a local region, objects and background are more uniform, and blurry edges are gradually changing from objects to background. The uncertainty value of objects and background are smaller than that of the edges.

A size $d \times d$ window centered at (x, y) is used for computing the standard deviation of pixel $P(i, j)$:

$$sd(x, y) = \sqrt{\frac{\sum_{p=x-(d-1)/2}^{x+(d-1)/2} \sum_{q=y-(d-1)/2}^{y+(d-1)/2} (g_{pq} - \mu_{xy})^2}{d^2}} \tag{3}$$

where μ_{xy} is the mean of the intensity values within the window.

Edge information is defined as:

$$eg = \sqrt{G_x + G_y} \tag{4}$$

Where G_x and G_y are horizontal and vertical derivative approximations.

Normalize the standard deviation and define the uncertainty as

$$U(x, y) = \frac{sd(x, y)}{sd_{max}} \times \frac{eg}{eg_{max}} \tag{5}$$

Where $sd_{max} = \max\{sd(x, y)\}$ and $eg_{max} = \max\{eg(x, y)\}$.

The value of $I(x, y)$ has a range of $[0,1]$. The more uniform of the region surrounding a pixel is, the smaller the uncertainty value of the pixel is. The window size should be quite big to include enough local information, but it has to be less than the distance between two objects. We choose d=7 in our experiments.

Convert the Image to a Binary Image Based on $\{T,U,F\}$. In this step, we first divide the given image into 3 parts: Objects (O), Edges (E), and Background (B). $T(x,y)$ represents the degree of being an object pixel, $U(x,y)$ is the degree of being an edge pixel, and $F(x,y)$ is the degree of being a background pixel for pixel $P(x,y)$, respectively. The three parts are defined as follows: where t_1 and t_2 are the thresholds computed in T_E and F_E, and $\lambda = 0.01$

$$
\begin{aligned}
O(x, y) &= \begin{cases} true & T(x, y) \geq t_1, U(x, y) < \lambda \\ false & otherwise \end{cases} \\
E(x, y) &= \begin{cases} true & T(x, y) < t_1 \vee F(x, y) < t_2, U(x, y) \geq \lambda \\ false & otherwise \end{cases} \\
B(x, y) &= \begin{cases} true & F(x, y) \geq t_2, U(x, y) < \lambda \\ false & otherwise \end{cases}
\end{aligned} \tag{6}
$$

After O, E, and B are determined, the image is mapped into a binary image for further processing. We map the objects and background to 0 and map the edges to 1 in the binary image. The mapping function is as following. See Fig. 1 (e).

$$
Binary(x, y) = \begin{cases} 0 & O(x, y) \vee B(x, y) \vee \overline{E(x, y)} = true \\ 1 & otherwise \end{cases} \tag{7}
$$

Apply the Watershed to the Converted Binary Image. Watershed algorithm is good for finding the connected segmentation boundaries. The following is the watershed algorithm for processing the obtained binary image [8].

3 Experimental Results

One of the difficult problems in BUS image segmentation is how to fully and automatically find the tumor. Many existing methods need manually to select a ROI which includes the tumor as the initial step of segmentation. Active contour (AC) model is a 'region-based' method, which does not need manual region selection. It utilizes the means of different regions to segment image [9].

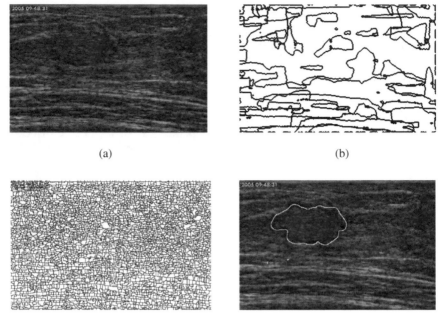

(a)   (b)

(c)   (d)

Fig. 2. (a) BUS image. (b) Segmentation result by applying AC model to (a). (c) Segmentation result by using normal watershed method. (d) Segmentation result by applying the proposed algorithm to (a).

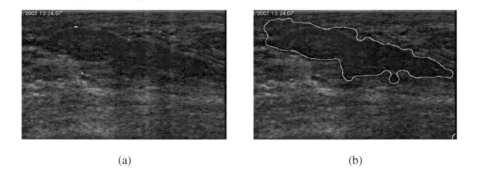

(a)   (b)

Fig. 3. (a)Bus Image. (b)segmentation result by the proposed algorithm to (a)

Fig.2 (b) is the result of applying AC model with 200 iterations to fig. 2 (a). The segmentation result is poor, and it cannot separate the tumor from BUS. Fig.2 (c) is the result by using watershed method in Matlab. We can see the normal watershed method is very sensitive to speckles in BUS image. Fig. 2 (d) is the result by applying the proposed method on fig. 2 (a), which can clearly segment tumor from BUS image. Fig. 3 gives another segmentation result by the proposed algorithm. (a) is the original BUS image and (b) is the segmentation result by our method.

4 Summary

Fuzzy logic is successfully used in many fields. But there are a lot of problems cannot be solved by fuzzy logic. In this paper, we extended fuzzy set into 3 domains T, F and U. By introducing U domain, it can handle the uncertainty better. Then we apply this method in BUS images. The results show that it locates and segment tumor from BUS image automatically and effectively. The proposed method works well for the image with blurred edges and noise. Another advantage of using this method is more noise-tolerant than the original watershed method.

References

1. Wells III, W.M., Grimson, W.E.L., Kikinis, R., Jolesz, F.A.: Adaptive Segmentation of MRI Data. J. IEEE Trans. Medical Imaging 15, 429–442 (1996)
2. Olabarriaga, S.D., Smeulders, A.W.M.: Interaction in the Segmentation of Medical Images: A Survey. J. Medical Image Analysis 5, 127–142 (2001)
3. Zadeh, L.A.: Fuzzy sets. J. Information and Control 8, 338–353 (1965)
4. Adlassnig, K.-P.: Fuzzy Set Theory in Medical Diagnosis. IEEE Trans. Systems, Man and Cybernetics 16, 260–265 (1986)
5. Tobias, O.J., Seara, R.: Image Segmentation by Histogram Thresholding Using Fuzzy Sets. IEEE Trans. Image Processing 11, 1457–1465 (2002)
6. Cai, W., Chen, S., Zhang, D.: Fast and Robust Fuzzy C-Means Clustering Algorithms Incorporating Local Information for Image Segmentation. Pattern Recognition 40, 825–838 (2007)
7. Cheng, H.D., Wang, J.L., Shi, X.J.: Microcalcification Detection Using Fuzzy Logic and Scale Space Approaches. Pattern Recognition 37, 363–375 (2004)
8. Gonzalez, R.C., Woods, R.E.: Digital Image Processing. Prentice Hall, San Francisco (2002)
9. Chan, T.F., Vese, L.A.: Active Contours without Edges. IEEE Trans. Image Processing 10, 266–277 (2001)

A Mathematical Model and Solution for Cost-Driven Job-Shop Scheduling Problem

Kai Sun[1], Gen-ke Yang[2], and Jie Bai[2]

[1] Shandong Provincial Key Laboratory of AM&MC Technology of Light Industry Equipment,
Shandong Institute of Light Industry, Shandong, China
[2] Department of Automation, Shanghai Jiao Tong University, Shanghai, China
{sunkai,gkyang,baijie}@sjtu.edu.cn

Abstract. This paper presents a cost-driven model of the job-shop scheduling problem and an effective hybrid optimization algorithm to solve it. The cost model is developed in terms of a combination of multi-dimensional costs generated from product transitions, revenue loss, earliness / tardiness penalty, and so on. The new hybrid optimization algorithm combines the strong global search ability of scatter search with the strong local search ability of simulated annealing. In order to illustrate the effectiveness of the hybrid method, several test problems are generated, and the performance of the proposed method is compared with other evolutionary algorithms. The experimental simulation tests show that the hybrid method is quite effective at solving the cost-driven job-shop scheduling problem.

Keywords: Cost-Driven, Job-Shop Scheduling Problem, Scatter Search.

1 Introduction

The job-shop scheduling problem (JSP) is one of the most well-known machine scheduling problems and one of the strongly NP-hard combinatorial optimization problems [1]. Historically, JSP was primarily solved by the branch-and-bound method and some heuristic procedures based on priority rules [2]. During the past decade, researches on meta-heuristic methods to solve the JSP have been widely studied, such as genetic algorithm [3], simulated annealing [4], tabu search [5] and particle swarm optimization [6]. The majority of studies on JSP, however, are driven by production criteria, such as total flowtime, maximum complete time (makespan), maximum tardiness and number of tardy jobs, etc.

With the globalization of the world economy and the revolutionary development of information technology, the critical challenge to manufacturing enterprise is to become more flexible and profitable. The development and application of an appropriate scheduling solution with seamless integration of business and manufacturing play a critical role in any modern manufacturer. Jiang et al. [7] present a cost driven objective function for job shop scheduling problem and solve it by using genetic algorithm, and the experimental results demonstrate the effectiveness of the algorithm. However, the study didn't provide precise mathematical model for the

Y. Yu, Z. Yu, and J. Zhao (Eds.): CSEEE 2011, Part II, CCIS 159, pp. 418–423, 2011.
© Springer-Verlag Berlin Heidelberg 2011

cost-driven job-shop scheduling problem. In the paper, a mathematical model of cost-driven JSP is presented. The cost of operational transitions between products, the revenue loss due to machine idle time during the phase of product transitions, and the penalty due to missing the required on-time delivery date, and so on are included in the model. And then, a new hybrid evolutionary algorithm, which combines the strong global search ability of scatter search (SS) with the strong local search ability of simulated annealing (SA), is developed to solve the cost-driven JSP.

The organization of remain contents is as follows. In section 2, the formulation and model of the cost-driven JSP is presented. Section 3 presents the conceptual introduction to SS and SA and proposes the hybrid scatter search (HSS) algorithm to solve the cost-driven JSP. Section 4 provides experimental results and performance analyses. Section 5 offers concluding remarks.

2 Problem Statement

The key ideas behind the developed scheduling system are constrained cost-driven optimization solution, namely, the cost is defined as an objective function subject to the relevant constraints. In the paper, the cost of the product transitions, revenue loss due to the machine idle time and earliness / tardiness penalty are included in the cost model for optimizing production scheduling solutions.

Firstly, we introduce a null job (job 0) whose processing time and transition cost with other job are zero and all sequences on each machine are started from job 0 and ended at job 0, and then the decision variable X_{ijk} is defined as:

$$X_{ijk} = \begin{cases} 1 & \text{if job } j \text{ is preceded by job } i \text{ on machine } k \\ 0 & \text{else} \end{cases} \quad (1)$$

The precedence coefficient Y_{ijk} is defined as:

$$Y_{ihk} = \begin{cases} 1 & \text{if machine } k \quad \text{process job } i \text{ right after machine } h \\ 0 & \text{else} \end{cases} \quad (2)$$

Based on the above discussion of the problem, the mathematical model of cost-driven JSP can be formulated as:

$$Min\{\sum_{k=1}^{m}\sum_{i=1}^{n}\sum_{j=1}^{n}t_{ijk} \cdot X_{ijk} + \sum_{k=1}^{m}\sum_{i=1}^{n}\sum_{j=1}^{n}(C_{jk} - C_{ik} - p_{ik}) \cdot W_k \cdot X_{ijk}$$
$$+ \sum_{i=1}^{n}[\alpha_i \cdot \max(0, e_i - C_i) + \beta_i \cdot \max(0, C_i - d_i)]\} \quad (3)$$

Subject to:

$$\sum_{j=0}^{n}X_{ijk} = 1 \quad i \neq j, i = 1,...n, k = 1,...,m \quad (4)$$

$$\sum_{i=0}^{n} X_{ijk} = 1 \quad j \neq i, \; j = 1,...n, \; k = 1,...,m \tag{5}$$

$$C_{jk} - p_{jk} + (1 - X_{ijk})M \geq C_{ik} \quad i, j = 1,...n, \; k = 1,...,m \tag{6}$$

$$C_{ik} - p_{ik} + (1 - Y_{ihk})M \geq C_{ih} \quad i = 1,...n, \; h, k = 1,...,m \tag{7}$$

$$X_{ijk} \in \{0,1\} \quad i, j = 0,1,...,n, \; i \neq j, \; k = 1,...,m \tag{8}$$

Constraint (4) defines that a job should be right before another job on each machine. Constraint (5) denotes that a job should be right after another job on each machine. Constraint (6), where M is a very large positive number, shows that each machine can process at most one job at any time. Constraint (7) shows that each job can be processed on at most one machine at any time. Constraint (8) ensures that the variable only takes the integer 0 or 1.

3 Hybrid Scatter Search for JSP

Scatter search (SS) is an evolutionary method that has been successfully applied to hard optimization problems [8]. Unlike genetic algorithm, a scatter search operates on a small set of solutions (called reference set) and makes only limited use of randomization as a proxy for diversification when searching for a globally optimal solution. The paper develop a new hybrid scatter search, which introduces the simulated annealing into the improvement method to enhance the local search ability of scatter search, to solve the cost driven job-shop scheduling problem. The detailed discussion of applying hybrid scatter search is shown as follows.

3.1 Encoding Scheme

One of the key issues in applying SS successfully to cost-driven JSP is how to encode a schedule of the problem to a search solution. We utilize an operation-based representation [3] that uses an unpartitioned permutation with m-repetitions of job numbers. A job is a set of operations that has to be scheduled on m machines. In this formulation, each job number occurs m times in the permutation, i.e. as often as there are operations associated with this job. By scanning the permutation from left to right, the *kth* occurrence of a job number refers to the *kth* operation in the technological sequence of this job. A permutation with repetition job numbers merely expressed the order in which the operations of jobs are scheduled.

3.2 Diversification Generation Method

This method was suggested by Glover [8], which generates diversified permutations in a systematic way without reference to the objective function. Assume that there are a $n \times m$ JSP, a given trial solution S used as a seed is representing by indexing its elements, so that they appears in consecutive order to yield $S = \{[1],[2],....,[l]\}$, where $l = m \times n$. Define the subsequence $S(h:t)$, where t is a positive integer between 1 and h, to be given by: $S(h:t) = ([t],[t+h],[t+2h],....,[t+rh])$ where r is the largest nonnegative integer such that $t + rh \leq l$. Then define the $S(h)$ for $h \leq n$, to be: $S(h) = \{S(h:h), S(h,h-1),...,S(h:1)\}$.

3.3   Improvement Method

Each of the new trial solutions which are obtained from the diversification generation method or solution combination method is subjected to the improvement method. This method aims to enhance the quality of these solutions. In the paper, we take two versions of local search meta-heuristics to improve trial solutions. A long-term SA-based improvement method is only applied to the best new trial solution, and a short-term swap-based local search is taken to enhance other new trial solutions. With the hybridization of these two local methods, we can get a compromise between solution quality and computational effort.

3.4   Reference Update Method

The reference set is a subset of both diverse and high-quality trial solutions. In the paper, the reference set, *RefSet*, consists of two subsets, *RefSet₁* and *RefSet₂*, of maximum size b_1 and b_2, respectively. The first subset contains the best quality solutions, while the second subset should be filled with solutions promoting diversity.

Assume that the population of improved trial solutions at each generation is *Pop*. The construction of the initial *RefSet₁* starts with the selection of best b_1 solutions from *Pop*. These solutions are added to the *RefSet₁* and deleted from *Pop*. And then the minimum distance from each improved solution to the solutions in *RefSet₁* is computed. Then b_2 solutions with the maximum of these minimum Euclidean distances are selected.

3.5   Subset Generation Method and Solution Combination Method

The subsets generation method consists of generating the subsets that will be used for creating new solutions with the solution combined method. In the paper, the method is organized to generate three different types of 2-elements subsets, which are shown as S_1, S_2, S_3. The implementation of the subsets is explained as follows:

S_1 : Generated by pairs of solutions from the *Refset₁*.
S_2: Generated by pairs of solutions from the *Refset₂*.
S_3: The first element of *ith* subset is the *ith* solution in the *Refset₁* and the second element is the solution of the *Refset₂* which has the maximum Euclidean distance to the first element.

Solution combination method uses each generated subset and combines the subset solutions, returning one or more trial solutions. An operation-based crossover operator is taken for the JSP in the paper [3].

4 Computational Experiments

To illustrate the effectiveness of the algorithm described in this paper, we consider several instances originated from two classes of standard JSP test problems: Fisher and Thompson instances (FT06, FT10, FT20) [9], and Lawrence instances (LA01, LA06,..., LA26) [10]. Originally, each instance only consists of the machine number and processing time (p_{ik}) for each step of the job. Furthermore, to apply these instances to the cost-driven JSP presented in this paper, some extra instance data should be generated.

The statistical performance of 20 independent runs of these algorithms are listed in table.1, including the optimum value known so far (*BKS*), the best objective value ($C*$), the percentage value of average objective value over *BKS* (%) and the average CPU time ($\bar{t}$) of the HSS. Table.1 shows that the results obtained by HSS are much better than those obtained by GA. The superiority of the best optimization quality demonstrates the effectiveness and the global search property of the hybrid search, and the superiority of the average performance over 20 random runs shows that the hybrid probabilistic search is more robust than these algorithms. It can be seen that the HSS algorithm can get desirable solutions in a reasonable computation time even for problems with 30 jobs and 10 machines. For more large-scale problems, we can trade off between computation time and solution quality by adjusting the number of generations or the parameters of SA-subprogram.

Table 1. Experiment Results

Inst.	Size	BKS	GA		HSS		
			$C*$	%	$C*$	%	$\bar{t}$
FT06	6×6	426	426	0	426	0	1.85
FT10	10×10	2732	2741	4.05	2732	1.61	45.21
FT20	20×5	2442	2495	2.32	2442	0.92	62.92
LA01	10×5	935	935	1.26	935	0	5.26
LA06	15×5	1670	1670	1.22	1670	0	10.83
LA11	20×5	1899	1906	3.85	1899	0.56	58.63
LA16	10×10	3061	3176	4.16	3061	1.28	38.25
LA21	15×10	3978	4122	3.83	3978	1.61	132.62
LA26	20×10	3781	3997	4.92	3781	2.37	367.17

5 Conclusions

Cost optimization is an attractive and critical research and development area for both academic and industrial societies. In the cost-driven JSP model we proposed, the solutions are driven by business inputs, such as market demand and the costs of inventory and machine idle time during the product transition phase. And then, a hybrid optimization algorithm that combines SS with SA is proposed to solve the problem. This hybrid method combines the advantages of these two algorithms and mitigates the disadvantages of them. The obtained results indicate that this hybrid method is superior to GA and is an effective approach for the cost-driven JSP.

Acknowledgments. The research is supported by the Natural Science Foundation of Shandong Province of China (No. ZR2010FQ009).

References

1. Garey, M.R., Johnson, D.S., Sethi, R.: The Complexity of Flowshop and Jobshop Scheduling. Mathematics of Operations Research 1(2), 117–129 (1976)
2. Adams, J., Balas, E., Zawack, D.: The Shifting Bottleneck Procedure for Job Shop Scheduling. Management Science 34(3), 391–401 (1988)
3. Park, B.J., Choi, H.R., Kim, H.S.: A hybrid Genetic Algorithm for the Job Shop Scheduling Problems. Computers & Industrial Engineering 45(4), 597–613 (2003)
4. Kolonko, M.: Some New Results on Simulated Annealing Applied to the Job Shop Scheduling Problem. European Journal of Operational Research 113(1), 123–136 (1999)
5. Eswaramurthy, V.P., Tamilarasi, A.: Tabu Search Strategies for Solving Job Shop Scheduling Problems. Journal of Advanced Manufacturing Systems, 59–75 (2007)
6. Sha, D.Y., Hsu, C.Y.: A Hybrid Particle Swarm Optimization for Job Shop Scheduling Problem. Computers & Industrial Engineering 51(4), 791–808 (2006)
7. Jiang, L.W., Lu, Y.Z., Chen, Y.W.: Cost Driven Solutions for Job-shop Scheduling with GA. Control Engineering of China S1(26), 72–74 (2007)
8. Glover, F.: Scatter Search and Path Relinking. In: Corne, D., Dorigo, M., Glover, F. (eds.) New Ideas in Optimization, pp. 297–316. McGraw-Hill, New York (1999)
9. Fisher, H., Thompson, G.L.: Probabilistic Learning Combinations of Local Job-shop Scheduling Rules. In: Industrial Scheduling. Prentice-Hall, Englewood Cliffs (1963)
10. Lawrence, S.: An Experimental Investigation of Heuristic Scheduling Techniques. In: Resource Constrained Project Scheduling. Carnegie Mellon University, Pittsburgh (1984)

Rolling Bearing Fault Diagnosis Using Neural Networks Based on Wavelet Packet-Characteristic Entropy

Weiguo Zhao[1], Lijuan Zhang[2], and Xujun Meng[3]

[1] Hebei University of Engineering, Handan, 056038, China
[2] Hohhot Vocational College, Huhehaote Neimenggu, 010051, China
[3] Jinyu Real Estate, Mongolia Hohhot, 010020, China
{zwg770123,zhlj1979,mxj2021}@163.com

Abstract. In this paper, a new fault diagnosis method of vibrating of hearings is proposed, in which three layers wavelet packet decomposition of the acquired vibrating signals of hearings is performed and the wavelet packet-characteristic entropy is extracted, then the eigenvector of wavelet packet is constructed and trained to implement the intelligent fault diagnosis. The simulation results show the proposed method is effective and feasible, so it is of great significance for improving the fault diagnosis and state identification.

Keywords: Wavelet Packet-characteristic Entropy, Rolling Bearing, Neural Networks, Fault Diagnosis.

1 Introduction

Machine condition monitoring is gaining importance in industry because of the need to increase reliability and to decrease the possibility of production loss due to machine breakdown. The use of vibration and acoustic emission (AE) signals is quite common in the field of condition monitoring of rotating machinery. By comparing the signals of a machine running in normal and faulty conditions, diagnosis and detection of faults like inner fault and outer fault is possible [1]. These signals can also be used to detect the incipient failures of the machine components, through the online monitoring system, reducing the possibility of catastrophic damage and the downtime. Although often the visual inspection of the frequency domain features of the measured signals is adequate to identify the faults, there is a need for a reliable, fast, and automated procedure of diagnostics [2].

Artificial neural networks (ANN) have potential applications in automated detection and diagnosis of machine. The applications of ANN are mainly in the areas of machine learning, computer vision, and pattern recognition because of their high accuracy and good generalization capability conditions. In this paper, wavelet packet features of entropy (Wavelet Packet-Characteristic Entropy, WP-EE) combined with ANN was used to diagnosis for the rolling bearing fault, the specific approach is to extract the characteristics entropy in various bands, and then establish three-layer BP neural

Y. Yu, Z. Yu, and J. Zhao (Eds.): CSEEE 2011, Part II, CCIS 159, pp. 424–429, 2011.
© Springer-Verlag Berlin Heidelberg 2011

networks classifier based on wavelet packet energy, the simulation results show that the method can effectively identify a variety of bearing fault.

2 Classifier Based on Neural Networks

The multi-layer neural networks (MNN) are the most commonly used network model for image classification in remote sensing. MNN is usually implemented using the Backpropagation (BP) learning algorithm [4]. The learning process requires a training data set, i.e., a set of training patterns with inputs and corresponding desired outputs. The essence of learning in MNN is to find a suitable set of parameters that approximate an unknown input-output relation. Learning in the network is achieved by minimizing the least square differences between the desired and the computed outputs to create an optimal network to best approximate the input-output relation on the restricted domain covered by the training set.

A typical MNN consists of one input layer, one or more hidden layers and one output layer. 10 hidden nodes in the hidden layer, and 5 output nodes in the output layer often noted as 4 -10 - 5. All nodes in different layers are connected by associated weights. For each input pattern presented to the network, the current network output of the input pattern is computed using the current weights. At the next step, the error or difference between the network output and desired output will be backprogated to adjust the weights between layers so as to move the network output closer to the desired output. The goal of the network training is to reduce the total error produced by the patterns in the training set. The mean square error J (MSE) is used as a classification performance criterion given by

$$J = \frac{1}{2N} \sum_{i=1}^{N} \varepsilon_i^2 \tag{1}$$

Where N is the number of training patterns. ε^2 is the Euclidean distance between the network output of the pattern and the desired output. This MSE minimization procedure via weigh adjusting is called learning or training. Once this learning or training process is completed, the MNN will be used to classify new patterns. MNN is known to be sensitive to many factors, such as the size and quality of training data set, network architecture, learning rate, overfitting problems, etc. To date, there are no explicit methods to determine most of these factors. Fortunately, based on many previous researches, there are many practical suggestions to help choose these factors.

The size and quality of the training data set have a considerable influence on the generalization capability of the resulted network classifier and the final classification accuracy. The selection of the training data set is often related to how many classes would be expected to derive. First of all, these classes must be determined carefully so that they would have enough spectral separability so that the classifier is able to discriminate them. Second, the training data set must contain sufficient representatives of each class. Third, the size of training set is related to the number of associated weights and the desired classification accuracy.

3 The Extraction of Wavelet Packet Feature Entropy for Rolling Bearing Vibration Signal

Wavelet Packet Decomposition. The following recursive equation (2) can be used to decompose wavelet pack for bearing vibration signals [5]:

$$
\begin{cases}
u_{2n}(t) = \sqrt{2} \sum_{k} h(k) u_n(2t - k) \\
u_{2n-1}(t) = \sqrt{2} \sum_{k} g(k) u_n(2t - k)
\end{cases}
\tag{2}
$$

$h(k)$ is the high-pass filter; $h(k)$ is the low-pass filter.

From the view of the multi-resolution analysis, the real of wavelet packet decomposition is that vibration signals can pass the high and low combination filter, for each decomposition, the original signal is always decomposed into two frequency channels, and then separately to the same decomposition up to meet the requirements.

Wavelet Packet Feature Entropy. The wavelet packet decomposition serial $S_{(j,k)}$ $(k = 0, \hbar, 2^j - 1)$ was obtained after j layer decomposition of signal, so wavelet packet decomposition of the signal can be as a division of the signal, the definition of this divided measure is as follows:

$$
\varepsilon_{(j,k)}(i) = \frac{S_{F(j,k)}(i)}{\sum_{i=1}^{N} S_{F(j,k)}(i)}
\tag{3}
$$

$S_{F(j,k)}(i)$ is the i value of Fourier transform for $S_{(j,k)}$, N is the length of the original signal. According to the basic theory of Shanno, wavelet packet features entropy is as follows:

$$
H_{j,k} = -\sum_{i=1}^{N} \varepsilon_{(j,k)}(i) \log \varepsilon_{(j,k)}(i) \ (k = 0, \hbar, 2^j - 1)
\tag{4}
$$

$H_{j,k}$ is the No. k Wavelet packet features entropy in j layer. When $\varepsilon_{(j,k)} = 0$, $\varepsilon_{(j,k)}(i) \log \varepsilon_{(j,k)}(i) = 0$, information entropy H is an information measure of the positioning system in a certain state, it is a measure of the unknown degree sequence and can be used to estimate the complexity of random signals.

4 Extraction of Vibration Signal with Wavelet Packet Entropy

Type 197726 Bearing was used to be experiment, the working speed is 1 250r/min, and the frequency is 10.24 kHz. The method for extraction of wavelet packet entropy is as follows [6]:

(1) Signal decomposition: The vibration signals were decomposed for the three-tier multi-wavelet packet decomposition based on the above equation (1).

(2) Signal reconstruction: after step (1), the sequences in eight bands were reconstructed.

(3) The characteristics of wavelet packet entropy vectors: from equation (2), (3), eight characteristics of wavelet packet entropy were obtained, and then constructed a feature vector T:

$$T = [H_{3,0}, H_{3,1}, H_{3,2}, H_{3,3}, H_{3,4}, H_{3,5}, H_{3,6}, H_{3,7}] \qquad (5)$$

When the characteristics of wavelet packet entropy are too large, it will be inconvenience for analysis, so this feature vector can be normalized:

$$H = (\sum_{j=0}^{7} \left\| H_{3,j} \right\|^2)^{1/2} \qquad (6)$$

T' is feature vectors after normalized:

$$T' = [\frac{H_{3,0}}{H}, \frac{H_{3,1}}{H}, \frac{H_{3,2}}{H}, \frac{H_{3,3}}{H}, \frac{H_{3,4}}{H}, \frac{H_{3,5}}{H}, \frac{H_{3,6}}{H}, \frac{H_{3,7}}{H}] \qquad (7)$$

Based on the above methods, vibration signal for bearing were analyzed in different operating conditions, wavelet packet features entropy are shown in Table 1:

Table 1. WP-CE of vibrating signals of hearing

Bearing state	$H_{3,0}$	$H_{3,1}$	$H_{3,2}$	$H_{3,3}$	$H_{3,4}$	$H_{3,5}$	$H_{3,6}$	$H_{3,7}$
Normal	9.3735	80.7256	2.7521	15.5489	0.5212	0.5423	0.4218	0.1475
Normal	6.1872	77.1542	4.2754	12.7452	0.4418	0.6542	0.2814	0.2463
Normal	6.4521	75.1845	3.4512	13.5742	0.5214	0.5579	0.3452	0.2019
Inner fault	22.4086	55.2163	2.7415	8.1542	0.1274	0.2019	5.1293	0.9246
Inner fault	13.4592	60.5283	2.8541	9.2019	0.1648	0.2948	4.9216	0.8952
Inner fault	15.6213	63.1275	3.1285	8.1282	0.2016	0.2413	5.1243	0.7596
Outer fault	11.2562	52.1354	2.1643	25.9216	0.0951	0.6318	1.2546	0.3549
Outer fault	14.2512	42.5986	5.2146	23.5486	0.1524	0.5869	1.8542	0.6318
Outer fault	10.2541	46.2351	3.4526	21.5623	0.2983	0.6875	1.2536	0.8192

5 Implementation and Experimental Results

Firstly, wavelet packet features entropy is extracted as a training network from vibration signals in different operating conditions, Part of the training samples are shown in Table 1, feature vectors were signed by fault codes to identify standard output.. The normal bearing is 20 groups, inner fault data is 16 groups and outer fault data is 16 groups, the other 30 groups is used to test [7, 8]. When the network training tends to be stable, freezing the network weights, then it can be diagnosed.

The ANN is one hidden-layer ANN, the ANN is trained using fast back-propagation method. Training parameters are set as follows: learning rate is 0.01, and Momentum constant is 0.8. The weights and biases are initialized randomly. The ANN network is trained with the same training samples, and the same testing samples are used too in testing. The target error is set as 0.001,the hidden layer is set 11,22,33,44,55 and 66.The results of the results of comparison are shown in Table 2, it is obvious that the result is the best when the number of hidden is 33,

Table 2. Comparison of BP network with different number of hidden neurons

Method	Number of hidden neurons	Training accuracy	Test accuracy
BP1	11	98.73%	99.72%
BP2	22	97.04%	97.61%
BP3	33	99.42%	98.36%
BP4	44	97.62%	94.85%
BP5	55	96.21%	94.11%
BP6	66	95.17%	92.82%

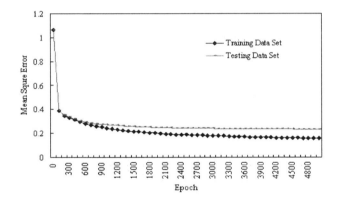

Fig. 1. Mean square error of BP3 training and test epochs

Via preliminary experiments, the network architecture with 8–33–3 proved to be optimal and was used as the network architecture for this particular application. The epoch training method was used to train the network. In epoch training, the weight update for each input training sample is computed and stored (without changing the weights) during one pass through the training set, which is called an epoch. At the end of the epoch, all the weight updates are added together, and only then will the weights be updated with the average weight value. The mean square error of neural network epoch training and testing is plotted as a function of the number of the training epochs for the 8-33-3 network, as shown in Fig. 1.

This well-trained network was then used as a feed-forward network to classify the whole vibration signal. From the results, the network classifier can accurately identify the bearing vibration, a large number of experiments shows that the recognition results of the network are in line with the actual situation, it has high accuracy and good practical value.

6 Conclusions

The MNN is a supervised classification method. Its accuracy highly depends on the quality of the selected training data. The training data for each class must be able to provide sufficient information about the class with which it is associated. In this paper, a diagnostic method was presented based on wavelet packet features entropy and neural network technology. In the method wavelet packet entropy was decomposed to bearing vibration signals and to extract features, then format entropy vector and enter into the BP network to train, when the network is stable, vibration signals can be identified effectively. The simulation results it is a better way to solve the practical difficulties.

Aknowledgement. Project supported by the Science Research Foundation of Hebei Education Department of China(No. 2009422) and Natural Science Foundation of Hebei Province of China (No. E2010001026).

References

1. Lei, Y., He, Z., Zi, Y.: Application of an Intelligent Fault Diagnosis Method to Rotating Machinery. Expert Systems with Applications 36(6), 9941–9948 (2009)
2. Cudina, M.: Detection of Cavitation Phenomenon in a Centrifugal Pump Using Audible Sound. Mechanical Systems and Signal Processing 17(6), 1335–1347 (2003)
3. Al-Zubaidy, S.N.: A Proposed Design Package for Centrifugal Impellers. Computers & Structures 55(2), 347–356 (1995)
4. Davidov, T.: General Idea of Generating Mechanism and its Application to Bevel Gear. Mechanism and Machine Theory 33(5), 505–515 (1998)
5. Zhonghua, G., Fengqin, H.: Neural Network Based on Wavelet Packet Characteristic Entropy for Fault Diagnosis of Draft Tube. In: Proceedings of the CSEE, vol. 25(4), pp. 99–102 (2005)
6. Chang, C.S.: Separation of Corona Using Wavelet Packet Transform and Neural Network for Detection of Partial Discharge in Gas-Insulated Substations. IEEE Transactions on Power Delivery 20(2), 1363–1369 (2005)
7. Sracl, I., Alguindigue, E.: Monitoring and Diagnosis of Rolling Element Bearings Using Artificial Neural Network. IEEE Tran. on Industrial Electronics 40(20), 623–628 (1993)
8. Ham, F.M., Kostanic, I.: Principles of Nero Computing for Science & Engineering. McGraw-Hill, New York (2001)

SAR Image Denoising Based on Dual Tree Complex Wavelet Transform

Huazhang Wang

Institute of Electrics and Information Engineering, Southwest
University for Nationalities 610041, chengdu, China
wanghuazhang@126.com

Abstract. This paper firstly applies DC-CWT filter into SAR image denoising. on the base of studying and analyzing the mathematical model of the DT-CWT, especially for multi-scale, multi-resolution performance. According to the merit of strong directory selection and shift invariance of DC-CWT, it proposes a novel noise reduction algorithm. Firstly, it transforms noise image into logarithmic form, then one operates high-frequency coefficients to reduce noise through selecting the optimal threshold. Finally, it reconstructs image. which reduces Computational complexity and improves the filtering effect. In contrast with traditional filter methods, the DT-CWT exhibits a good performance.

Keywords: Image denoising, DC-CWT, Computational complexity.

1 Introduction

Synthetic aperture radar (SAR) images are becoming more widely used in many applications such as high-resolution remote sensing for mapping, terrain classification, search-and-rescue, mine detection and so on[1].SAR images are generated by a coherent illumination system and are affected by the coherent interference of the signal backscatter by the elements on the terrain. This effect is called speckle noise, which seriously degrades the image quality and hampers the interpretation of image content. Thus, it appears sensible to reduce or remove speckle in SAR images, provided that the structural features and textural information are not lost.

Speckle filtering is an important pre-processing step to improve the overall performance of automatic target detection and recognition algorithms based on SAR images. In the past years, many filters for SAR denoising have been proposed. Kuan filter[2] and frost filter[3] are among the better denoising algorithms for SAR images. Kuan considered a multiplicative model and designed a linear filter based on the minimum mean square error (MMSE) criterion, but it relies on the ENL from a SAR image to determine a different weighting function W to perform the filtering. Frost filter adapts to the noise variance within the filter window by applying exponentially weighting factors M, the decrease as the variance within the filter windows reduces. These filters are thought to preserve the edges and detail well while efficiently reduced noise speckle in the image even if multiplicative noise is rather intensive. However, this only true when there are more homogeneous areas in the image and the

Y. Yu, Z. Yu, and J. Zhao (Eds.): CSEEE 2011, Part II, CCIS 159, pp. 430–435, 2011.

edges are within the defined filter window. The Lee MMSE filter was a particular case of the Kuan fliter based on a linear approximation made for the multiplicative noise model [4]. In recent years, wavelet-based denoising algorithm has been studied and applied successfully for speckle reduction in SAR image[5]. Some comparisons was made between the application of classical speckle reduction filters and the application of the denoising methods based on wavelets, which were proved the superiority of the aforementioned methods [6]. Because compact support of wavelet basis functions allows wavelet transformation to efficiently represent functions or signals. And wavelet provides an appropriate basis for separating the noise signal from the image signal. So we can select appropriate threshold to separate the noise without affecting the feature information of image. There has been a mount of research on wave threshold selection for signal denoising.[7,8]. In most research, the DWT was most widely used, but it is shift variant, because the transform coefficients behave unpredictably under shifts of input signal, so it requires introducing large mount of redundancy into the transform to make it shift invariant, and the computing is much complex. The problems mentioned above can be overcome by dual-tree complex wavelet transform (DT-CWT) [9][10], the DT-CWT has been found to be particularly suitable for image decomposition and representation. Because it provides texture information strongly oriented in six different directions at some scales, which is able to distinguish positive and negative frequencies. So this paper proposes a novel dual tree complex wavelet transform method to reduce the noise of SAR image. Which has not only the properties of shift invariance, but also has translation invariance for image.

The paper is organized as follows: Section 2 introduces the basic principles of 2-D dual-tree complex wavelet transform. Section 3 gives introduction and elaboration of Sar image denoising algorithm based DT-CWT. Section 4 analyses and explains the experimental results in detail. Section 5 draws a short conclusion about this paper.

2   2-D Dual-Tree Complex Wavelet Transform

Similarly to the DWT, the DT-CWT is a multiresolution transform with decimated subbands provding perfect reconstruction of the input. DT-CWT is an enhancement to the discrete wavelet transform (DWT), with important wavelet properties. In contrast, it uses analytic filters instead of real ones and thus overcomes problems of the DWT at the expense of moderate redundancy. The DT-CWT is implemented using two real DWT, DT-CWT can be informed by the existing theory and practice of real wavelet transforms. The DT-CWT gives much better directional selectivity when filtering multi-dimensional signals, and there is approximate shift invariance, what is more, perfect reconstruction using short linear-phase filters and limited redundancy can improve the performance of image processing greatly. Shift invariance can be achieved in a dyadic wavelet transform by doubling the sampling rate at each level of the tree, to get uniform intervals between samples from the two trees below level one, the filters in one tree must provide delays that are half a sample from those in the other tree. For the linear phase, it requires odd-length filters in one tree and even-length filters in the other. So each tree uses odd and even filters alternately from level to level (except one level).

Due to its good performance of DT-CWT, especially for its shift invariance and improved directional selectivity, so the DT-CWT outperforms the critically decimated DWT in a range of applications, such as, image retrieval, edge detection, texture discrimination, image fusion and denoising.

3   SAR Image Denoising Based on DT-CWT

3.1   Speckle Noise Model

In general, the denoising framework described is based on the assumption that the distribution of the noise is additive zero mean Gaussian, but for SAR images, the noise content is multiplicative and non-Gaussian, such noise is generally more difficult to remove than additive noise because of the noise varies with the image intensity. Let I(x,y) represent a noise observation of 2-D function S(x,y). $N_m(x, y)$ and $N_a(x, y)$ are the corrupting multiplicative and additive speckle noise components, respectively. One can write

$$I(x, y) = S(x, y) \cdot N_m(x, y) + N_a(x, y) \tag{1}$$

Where the indices x,y represent the spatial position over the image, generally, the effect of additive noise in SAR images is less significant than the one of multiplicative noise., in order to simplify the model, it ignores term $N_a(x, y)$, one can rewrite (1) as

$$I(x, y) = S(x, y) \cdot N_m(x, y) \tag{2}$$

In most applications involving multiplicative noise, the noise content is assumed to be stationary with unitary mean and unknown noise variance, so we have to apply a logarithmic transformation on the speckle image I(x, y). Some researchers have shown that when the image intensity is logarithmically transformed, the speckle noise is approximately Gaussian additive noise, which tends to a normal probability much faster than the intensity distribution. So the equation (2) is logarithmically transformed, The expression is denoted as follow:

$$LnI(x, y) = LnS(x, y) + LnN_m(x, y) \tag{3}$$

At this stage, one can consider $LnN_m(x, y)$ to be white noise and subsequently apply any conventional additive noise reduction technique, such as wiener filtering, Kuan filtering, Frost filtering and so on. In fact, we can rewrite expression (3) as:

$$I'(x, y) = S'(x, y) + N'(x, y) \tag{4}$$

Where I'(x, y), S(x, y), N'(x, y) are the logarithms of I(x, y), S(x, y), N(x, y), respectively. Now, we assume that the signal and noise components are independent random variables.

3.2 Proposed Method

In general speaking, single-scale represents of signals are often inadequate when attempting to separate signals from noisy data, so the paper uses DT-CWT to separate noise through multi-scale analysis. The block diagram shown in Fig 1 depicts the basic step of proposed method.

Fig. 1. Flow chart of DT-CWT filitering

The first step is to transform noise image into logarithmic form, which is decomposed up to three levels, then in each subband, except the lowest band, one operates high-frequency coefficients through selecting the optimal threshold, as the noise mainly distributes in the high-frequency coefficients. Then one reconstructs image to get the image which noise is reduced. Currently, there are mainly two methods: soft threshold and hard threshold, the soft threshold method can be denoted as

$$\omega I(x,y) \begin{cases} \text{sgn}(\omega I(x,y))(\omega I(x,y)-T) & \text{if } |\omega I(x,y)| \ge T \\ 0 & \text{if } |\omega I(x,y)|T < T \end{cases} \tag{5}$$

This paper adopts the hard threshold method to adjust the coefficient of DT-CWT, the selection of threshold can be referenced literature [6].

$$T = \hat{\sigma}\sqrt{2\log m} \; . \tag{6}$$

Where m is the number of samples for data processing, T is the global threshold, $\hat{\sigma}$ is the estimation value of noise, the expression can be denoted as:

$$\hat{\sigma} = median(|w_{j,i}| : i = 0,1,\cdots 2^j - 1)/0.6745 \tag{7}$$

Where j represents the scale of the wavelet coefficient, and i represents the number of wavelet coefficients, since when a image is decomposed, noise distributes in each scale, but the energy mainly concentrate at the smallest scale, so we can use the square of error at this scale as the estimation for one of noise.

4   Experimental Results and Analysis

This section is dedicated to show the result of applying DT-CWT filtering method to real SAR images. It compares the results of our approach with current speckle filtering methods. For different methods, we select the most appropriate parameters to gain optimal result.

Original SAR image   Frost filter   Kuan filter

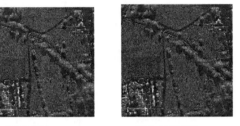

DWT filter   DT-CWT filter

Fig. 2. Visual evaluation

In order to quantify the achieved performance improvement, three different measures were computed based on the original and the denoised SAR images, mainly including the ability to save information, the ability of speckle reduction and computation efficiency. To decide whether or not holding average backscattering coefficient σ^0, it can be evaluate by using normalized mean(NM), NM is a criterion to evaluate that filter is whether of not unbiased estimation, if the NM approaches 1, it shows the estimation is unbiased, which illuminates that the filtering method saves good original image information. The NM is defined as

$$NM = M_{filtered} / M_{original} .$$ (8)

Where $M_{filtered}$ is the mean value of post-filter, $M_{original}$ is the mean value of original noise image. Also, in order to quantify the speckle reduction performance, it computed the standard-deviation- to-Mean ration(S/M), the quantity is a measure of image speckle in homogeneous regions. For remote sensing image data is very large, so the efficiency of different filtering algorithms is very important indicator. Efficiency mostly consider the operation time of all kinds of filter methods.

Table 1. Filtering Measure by Six Denoising Methods

method	NM	S/M	Runtimes(s)
Kuan filter	0.9968	0.4722	7.6899
Frost filter	1.0007	0.4857	3.4061
DWT	0.9954	0.4213	1.3542
DT-CWT	0.9969	0..4026	2.2558

Aimed at the three indicators above, it makes a lot of experiments, and some statistics is shown in table.1. The measures are calculated on an average of four noise realizations. The NM of the image filtered by kuan filter, DWT filter and DT-CWT is near the original image. What is more, the NM by DT-CWT is Closer to the original image, which shows the ability to save information of DT-CWT is much better, but the frost filter brings some small enhancement. It is evident from the table (1) that DT-CWT method is more successful in the noise reduction than other filters. The runtime of both mean filter and DWT filter is less than other filter, the kuan filter requires the longest time. In contrast, the time required for the DT-CWT filter is followed , execution time of the frost filter is middle. From the overall performance, the effect of the DT-CWT is the best relatively.

5 Conclusion

This paper firstly applies DC-CWT filter into SAR image denoising. on the base of studying and analyzing the mathematical model of the DT-CWT, especially for multi-scale, multi-resolution performance, according to the merit of DC-CWT for decomposition characteristics of image, it proposes a novel noise reduction algorithm, which reduces Computational complexity and improves the filtering effect. In contrast with traditional filter methods, the DT-CWT exhibits a good performance.

Acknowledgments. This paper is supported by the Fundamental Research Funds for the Central Universities（09NZYZJ05）, Southwest University for Nationalities.

References

1. Soumekh, M.: Synthetic Aperture Radar Signal Processing. John Wiley and Sons, New York (1999)
2. Kuan, D.T., Sawchuk, A.A., Stand, T.C., Chavel, P.: Adaptive Noise Smoothing Filtering for Image with Signal-Dependent Noise. IEEE Transaction of Pattern Analysis and Machine Intelligence 7, 165–177 (1985)
3. Frost, V.S., Stiles, J.A., Shanmugan, K.S., Holtzman, J.C.: A Model For Radar Images and its Application to Adaptive Digital Filtering of Multiplicative Noise. IEEE Transaction of Pattern Analysis and Machine Intelligence 4, 157–166 (1982)
4. Lee, J.S.: Digital Image Enhancement and Noise Filtering By Use Of Local Statistics. IEEE Transaction on Geoscience and Remote Sensing. 2, 165–168 (1980)
5. Achinm, A., Tsakalides, P., Bezerianons, A.: SAR Image Denoising Via Bayesian Wavelet Shrinkage Based on Heavy-Tailed Modeling. IEEE Transaction on Geoscience and Remote Sensing 41(8), 1773–1784 (2003)
6. Donoho, D.L., Johnstone, I.M.: Ideal Spatial Adaptation Via Wavelet Shrinkage. Biometrika 81, 425–455 (1994)
7. Grance Change, S., Yu, B., Vottererli, M.: Adaptive Wavlet Thresholding for Image Denoising and Compression. IEEE Transaction of Image Processing. 9, 1532–1546 (2000)
8. Mastriani, M., Giraldez, A.E.: Kalman's Shrinkage for Wavelet-Based Despeckling of SAR images. International Journal of Intelligent Technology 3(1), 190–196 (2006)
9. Kingsbury, N.: Complex Wavelets for Shift Invariant Analysis and Filtering of Singals. Journal of Applied and Computational Harmonic Analysis 10(3), 234–253 (2001)
10. Selesnick, I.W., Baraniuk, R.G., Kingsbury, N.G.: The Dual-Tree Complex Wavelet Transform. IEEE signal processing magazine, 123–151 (November 2005)

Demosaicking Authentication Codes VIA Adaptive Color Channel Fusion

Guorui Feng[*] and Qian Zeng

School of Communication and Information Engineering
Shanghai University Shanghai, China
{fgr2082,zxx1001}@gmail.com

Abstract. Data hiding is used for raw images instead of the full-resolution colour version. Among various color filter array patterns, we choose the most popular Bayer pattern and cover authentication codes in the pseudo host from original sampling in terms of high correlation within red, blue and green channels. Pseudo host is created by interpolating in this pattern according to the key. This procedure confirms the security, in the meantime, and less brings the artifact of embedding codes. The simulations test the robustness and transparency after some demosaicking methods and JPEG compression. These results imply that hiding data accompany with less demosaicking traces and the better robustness.

Keywords: Color filter array, data hiding, pseudo host, raw image.

1 Introduction

Techniques for hiding messages into a host artwork for the certain purposes of copyright protection, content authentication and traitor tracing have been in popular way for a long time. Nowadays, most valuable results in digital images pay attention to the full-resolution image, less concern theirs raw versions from natural scenes. In many consumer electronics systems, digital imaging devices have widely been embedded instead of film cameras for capturing and processing the digital image to provide the user with a viewable image [1]. As a popular digital camera, most natural scenes captured by a single sensor array require at least three color samples each pixel location, which are red(R), green(G), and blue(B). The majority of these digital devices use monochromatic sensors covered with a color filter array (CFA), to capture the image scene for reducing cost and complexity of the system. Fig.1 shows the most widely used Bayer CFA pattern. It can be found that the G component is mosaic on a quincunx grid and the R and B components are on rectangular grids. The sensor allows only one color to be measured each pixel. It is necessary to estimate two missing color values each pixel, and the process was called CFA interpolation or demosaicking [1]. Recently, multiple demosaicking algorithms were reported in refs. [6-10].

---

[*] Corresponding author.

Y. Yu, Z. Yu, and J. Zhao (Eds.): CSEEE 2011, Part II, CCIS 159, pp. 436–441, 2011.
© Springer-Verlag Berlin Heidelberg 2011

As mentioned before, the authentication security of raw pictures plays an important role in images released. It can tolerate acceptable data manipulations by authentication codes. An early result was suggested by embedding visible watermarks at CFA data acquisition level [3]. This approach embedded in CFA is considered with direct spread spectrum (DSSS), derived from enhanced secure features of raw images. Post-processing images are easily available for the storage or distribution. After down-filtering, the watermark embedded in the low-resolution raw data led to better performance better DSSS at the cost of the complex implementation [4]. Despite a variety of applications, the essential attributes were relatively featured as: transparency, payload, robustness, security, detectability [5]. The aforementioned methods are absent of some key attributes to fit for certain applications.

This paper provides a type of mechanism that content authentication can be detected. Simultaneously, the received multimedia signal is slightly different from original version. We propose a set of secure pseudo host scheme using in the non-overlapped block of the host after the local map. It is robust to the common CFA demosaicking processing and can survive JPEG compression. The principal part of the scheme belongs to the transform domain algorithm, while we focus on the security and robustness in simultaneity. The proposed technique is blind, i.e., neither the original image nor any side information extracted is used during the detection process. Data embedding is done using quantization techniques. The security of whole scheme is supported by the high correlation of sampled levels, which is controlled by the key.

2   CFA Data Hiding

By far, information hiding may be mainly divided into spatial and transform domain algorithms. The mechanism devised over the pseudo host by the key, but not the CFA sample pattern directly, confirms the security. Because attackers have not known the main body of schemes, this algorithm satisfies the Kerckhoffs principle of the cryptography.

2.1   Security Mechanism

In order to recover missing samples of the CFA, a color demosaicking algorithm has to rely on some additional statistical properties about input color signals. A commonly exploited property is the correlation between the sampled primary color channels: red, green, and blue. Some results had been used to examine the high relationships between the green and red channels, and between the green and blue channels [2,8].

In our approach, we propose a pseudo host based embedding algorithm to enable the platform to cover hiding information. First of all, we assume that I_g, I_r, I_b are the corresponding the green, red and blue components of the CFA pattern. The gray image from Bayer pattern can be generated using $I_{CFA} = I_g + I_r + I_b$. The output after embedding authentication codes can be represented by

$$I_{ph}(i, j) = F(I_g^p, I_r^p, I_b^p, k_{ij}) \tag{1}$$

where k_{ij} is the key at pixel (i,j) and I_g^p, I_r^p, I_b^p are corresponding the green, red and blue components of pseudo host are given respectively as following $I_g^p(i,j) = I_g(i,j)$,

$$I_r^p(i,j) = \begin{cases} I_r(i,j), & I_r(i,j) \in \{\min(f_{ij}^r(k_{ij}), f_{ij}^r(1-k_{ij})), \max(f_{ij}^r(k_{ij}), f_{ij}^r(1-k_{ij}))\} \\ f_{ij}^r(\frac{1}{2}), & otherwise \end{cases},$$

$$I_b^p(i,j) = \begin{cases} I_b(i,j), & I_b(i,j) \in \{\min(f_{ij}^b(k_{ij}), f_{ij}^b(1-k_{ij})), \max(f_{ij}^b(k_{ij}), f_{ij}^b(1-k_{ij}))\} \\ f_{ij}^b(\frac{1}{2}), & otherwise \end{cases}, \tag{2}$$

where

$$f_{ij}^r(k_{ij}) = k_{ij} I_g(i,j-1) + (1-k_{ij}) I_g(i+1,j)$$
$$f_{ij}^b(k_{ij}) = k_{ij} I_g(i-1,j) + (1-k_{ij}) I_g(i,j+1)$$.

The pseudo host makes the system workload more securely and effectively. It aims to avoid potential malicious process and improve the system scalability.

2.2 Data Hiding Algorithm

One basic modality for data hiding is available here via the spatial frequency-domain technique. First of all, we describe a modulation method that reply on DCT of pseudo host. There are still a key issue that wait for the further research. Which image processing is considered as non-malicious way? It demands that we can distinguish malicious and non-maliciously image processing. In following section, we assume JPEG compression as non-malicious image processing. This algorithm treats this point as same as the ref. [4].

Suppose I_{ph} is the pseudo host image and I_w is the stego-image, then watermarking scheme is defined as: $I_w = G(I_{ph})$, where operator $G(\bullet)$ denotes the embedding procedure. The host image is first divided into the equally non-overlapped square block of size $n \times n$, which is dealt in a way similar to JPEG standard. Data embedding is mainly done using quantization techniques, i.e., the transform coefficients are quantized into an integer multiple of a small quantization step Δ to accommodate the watermark. To be convenient for realize, a typical technology denoted as quantization index modulation (QIM) is used to the DCT transform of cover image. Due to lack of involving the key, security of the algorithm will face to the crucial challenge. So we cover information using pseudo host, stead of the original. A scalar, the uniform quantizer is

$$I_w(s) = Q(s) + \frac{1}{4} w_i \Delta, \tag{3}$$

where $Q(s) = \Delta \left\lfloor \frac{s}{\Delta} \right\rfloor + \frac{\Delta}{2}$. Note that we take value w_i in $\{-1,1\}$ with pseudo-random sequence and the polarity sequence of DCT coefficients. We modify the DC coefficient in order to seek the highest anti-compressibility of DCT algorithm.

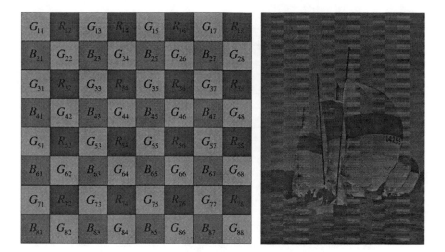

Fig. 1. Bayer CFA pattern

3 Simulations

In this section, simulation results are presented. We simulated a color mosaicking image by sampling a typical color images drawn from the Kodak set (768×512) for various testing purposes and benchmarking. This example is sampled in Fig. 1(b), and interleaving the samples according to the Bayer pattern. We evaluate this algorithm based on their ability to survive common demosaicking methods and incidental image distortions of JPEG compression. In addition, the quality of stego-image is measured by color-peak signal-to-noise ratio (CPSNR) given by $CPSNR=10\log_{10}(255^2/CMSE)$, where $CMSE=\dfrac{1}{3MN}\sum_{i=r,g,b}\sum_{x=1}^{M}\sum_{y=1}^{N}(I(x,y,i)-I_d(x,y,i))^2$, I and I_d represent the original and demosaicking image of size $M\times N$ each. In the implementation, equation (1) is set as $I_{ph}(i,j)=I_g^p+I_r^p+I_b^p$.

Test results are carried out to illustrate the effect of the proposed authentication method. According to the previous section, Δ will determine the entire bit error rate (BER) of extracted bits and image distortion. We can adjust it so as to seek the tradeoff of imperceptibility and robustness. To the JPEG compression, we know that there is error bits produced at the higher compression quality.

Table 1. BER of different demosaicking methods by tuning Δ (Q=95)

	Bilinear	ACPI[7]	Chang[8]	Zhang[6]	Menon[9]	Lian[10]
$\Delta=16$	0.3413	0.3216	0.3247	0.3226	0.3122	0.3185
$\Delta=24$	0.1631	0.1377	0.1387	0.1375	0.1292	0.1247
$\Delta=36$	0.0627	0.0381	0.0384	0.0366	0.0286	0.0332
$\Delta=60$	0.0133	0.0042	0.0034	0.0033	0.0021	0.0015

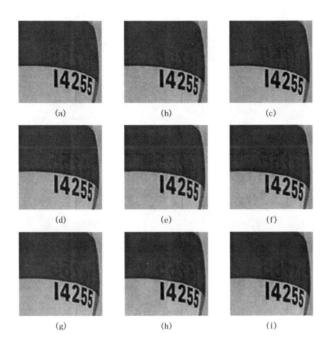

Fig. 2. Part of the demosaicking results of the benchmark image. (a) Orignal. (b) Zhang's method without hiding data. (c) Menon's method without hiding data. (c)-(f) Stego-images ($\triangle$ =20,36,60) with Zhang's method. (g)-(i) Stego-images ($\triangle$=20,36,60) with Menon's method.

Table 2. BER of different demosaicking methods by various JPEG compression factors (Δ=60)

	Bilinear	ACPI[7]	Chang[8]	Zhang[6]	Menon[9]	Lian[10]
Q=90	0.0133	0.0042	0.0034	0.0033	0.0021	0.0015
Q=70	0.0584	0.0392	0.0547	0.0480	0.0396	0.0449
Q=50	0.1551	0.1188	0.1382	0.1289	0.1276	0.1313
Q=30	0.3024	0.2839	0.2878	0.2822	0.2830	0.2869
Q=10	0.4696	0.4660	0.4644	0.4642	0.4627	0.4632

Table 1 shows BER performance comparison of different demosaicking methods on various quantizer steps. It can be observed that better demasaicking algorithms imply lower the BER under the same compression quality. Note that, Zhang [6], Menon [9] and Lian [10] are alternative better algorithms. When Δ=16, CPSNR(dB) of stego-image after their demasaicking are 39.87, 39.66 and 40.01. It appears to outperform others by the noticeable margin. Table 2 shows BER performance comparison of different demosaicking methods on various compression factors. Resemble the results shown in Table 1, as the better demosaicking algorithms, Zhang, Menon and Lian appear to outperform others and achieve the lower BER with same compression factors and quantizer step. For example, we designate the Zhang method to full-resolution. When Δ=16, 24, 36, 60, the CPSNR(dB) of stego-images are respectively 42.66, 42.22, 41.58, 39.87. These results are similar to those regular demosaicking without authentication codes [6]. Figure 2 shows some local demosaicking results of the benchmark image. They show that the proposed algorithm can preserve fine texture patterns and produce less color artifacts even if higher quantizer steps.

4 Conclusion

An effective authentication code in this paper is suggested to apply to raw images from photoelectric sensor. Based on the observation that region across color channels are highly correlated, we firstly set an interpolating pattern as the host in place of the original CFA sample. After blocking, QIM of DCT coefficient is adopted as identification codes. Pseudo host decided by the key can enhance the security of the data hiding scheme. As compared with various demosaicking methods, this proposed algorithm produces the less stego-artifact and confirms the benefits of authentication code. Experimental results show that stego-image is highly similar to the original version without the authentication code.

Acknowledgements. This work was supported by the Natural Science Foundation of Shanghai, China (09ZR1412400), Innovation Program of Shanghai Municipal Education Commission (10YZ11, 11YZ10).

References

1. Lukac, R., Plataniotis, K.N.: Color Filter Arrays: Design and Performance Analysis. IEEE Trans. Consum. Electron 51(21), 1260–1267 (2005)
2. Harro, S., Theo, G.: Selection and Fusion of Color Models for Image Feature Detection. IEEE Trans. Pattern Anal. Mach. Intell. 29(12), 371–381 (2007)
3. Lukac, R., Plataniotis, K.N.: Secure Single-Sensor Digital Camera. IEE Electron Lett. 42(24), 627–629 (2006)
4. Peter, M., Andreas, U.: Additive Spread-Spectrum Watermark Detection in Demosaicked Images. In: 11th ACM Workshop on Multimedia and Security, pp. 25–32. ACM Press, NY (2009)
5. Moulin, P., Koetter, R.: Data-Hiding Codes. Proc. of the IEEE 93, 2083–2126 (2005)
6. Zhang, L., Wu, X.: Color Demosaicking Via Directional Linear Minimum Mean Square-Error Estimation. IEEE Trans. Image Process. 14, 2167–2178 (2005)
7. Adams, J.E., Hamilton, J.F.: Adaptive Color Plane Interpolation in Single Color Electronic Camera. U.S. Patent 5, 506–619 (1996)
8. Chang, L., Tan, Y.P.: Effective Use Of Spatial and Spectral Correlations for Color Filter Array Demosaicking. IEEE Trans. Consum. Electron 50, 355–365 (2004)
9. Menon, D., Andriani, S., Calvagno, G.: Demosaicing with Directional Filtering and A Posteriori Decision. IEEE Trans. Image Process. 16, 132–141 (2007)
10. Lian, N.X., Chang, L., Tan, Y.P., Zagorodnov, V.: Adaptive Filtering for Color Filter Array Demosaicking. IEEE Trans. Image Process. 16, 2515–2525 (2007)

Detecting Algorithm for Object Based-Double Haar Transform

Yingxia Liu[1,2] and Faliang Chang[1]

[1] School of Control Science and Engineering,
Shandong University, Jinan, China
[2] Shandong Communication and Media College,
Jinan, China
84881756@qq.com, flchang@sdu.edu.cn

Abstract. According to the feature of double Haar transform, a new moving object detecting algorithm is proposed. Firstly, the gray difference image is pre-processed based on double Haar transform in order to realize de-noising. Secondly, make use of the characteristic of wavelet domain energy, an optimal dynamic threshold is obtained, and the gray difference image is judged by the threshold, so, a binary image only with object is got. The experimental result is showed that this method can conquer the contradiction between background noise and integrity of object and it can get better object detecting result.

Keywords: Object detecting, Double Haar transform, De-noising, Dynamic threshold.

1 Introduction

Object detecting plays an important role in the smart surveillance system and pattern recognition. In surveillance system, object detecting is a key step and it decides the tracking result directly. A dynamic threshold is demanded by the tracking system. As we know, the algorithm is simple and the system is real time with a fixed threshold. But, the circumstance is complex and the object is moving, the fixed threshold can't adapt the change of all the circumstances. So, a fast, exact and efficient detecting algorithm is still a challenging topic, more and more detecting algorithm is proposed[1]-[3].

Wavelet transform is a new mathematical tool appeared in recent years. It has better de-noising and edge detecting ability, so, it has been widely used in image processing domain. In computer vision field, new detecting algorithm based on wavelet transform are proposed[4][5]. Paper[6] proposes a wavelet based adaptive tracking controller for uncertain nonlinear systems.

In this paper, the contradiction between background noise and integrity of object is considered. According to analyzing the characteristic of wavelet transform energy, a dynamic threshold is obtained to separate object from background. This method can improve detecting result and make the object detection more accurately.

Y. Yu, Z. Yu, and J. Zhao (Eds.): CSEEE 2011, Part II, CCIS 159, pp. 442–446, 2011.
© Springer-Verlag Berlin Heidelberg 2011

2 Pre-processing for Gray Difference Image

Double Haar wavelet is a three channel filter group based on two Haar wavelet in which the input signal $x(n)$ is split by a low-pass filter and two high-pass filters. For reconstruction, the output signal $y(n)$ is obtained. In order to reconstruct the signal, the output signal should equal input signal, so, we have $y(n) = x(n)$.

According to Lee filter theory, the wavelet shrinkage coefficient K is defined as[7]:

$$K_j(n) = \begin{cases} 1 - \dfrac{\sigma_w^2}{x_k^2(n)}, & x_k^2(n) > \sigma_w^2 \\ 0, & x_k^2(n) \le \sigma_w^2 \end{cases} \qquad j = 1,2 \qquad (1)$$

Here, σ_w^2 is variance of the noise signal. The case of $x_k^2(n) \le \sigma_w^2$ implies that it is a flat part of signal. Thus, by letting $K_j(n) = 0, k = 1,2$, the wavelet coefficients of noise are cleaned out entirely.

3 Object Detecting Based on Double Haar Wavelet

According to wavelet transform theory, energy distribution on each subspace is a good description of the image feature. The splitting structure image under one scale with double Haar is showed in fig.1.

Here, (a) is gray difference image, (b) is the splitting image with double Haar transform, (c) is the position of different frequency component after wavelet transform. We can see that the energy concentrates on the low frequency part mainly. On the other hand, the edge and noise concentrate on high frequency part.

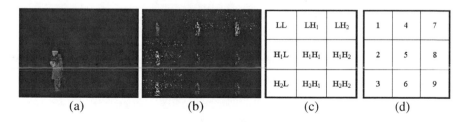

Fig. 1. Result of double Haar wavelet transform

Since double Haar wavelet is composed of one low-pass filter and two high-pass filter, the structure image after wavelet splitting has 9 passages, showed in (d), and passage 1 is low frequency part. The energy of each passage is calculated by:

$$E_n = \frac{1}{M} \sum_{i,j} x_n^2(i, j) \qquad n = 1,2,3,\cdots,9 \tag{2}$$

Here, M is number of pixel in the passage, $x_n(i, j)$ is gray value of each pixels.

Since wavelet coeffients represent the gray change of the original image, the all enery information of the image is contained in coeffients. So, the pixels with bigger energy stand for the apparent feature of the original image. As we know, the energy concentrates on low freqncy part(passage 1) mainly, the energy of high frequency parts (passage2,3,4,5,6,7,8,9) is much less, relatively. The object can be separated from background noise according to the ratio of energy. For background image, the energy difference between low frequency part and high frequency part is small, but for object, low frequency part is much greater than sum of high frequency parts. So, the ratio of low frequency and sum of high frequency can be used as the threshold to separate object from background. The ratio r is defined as:

$$r = \frac{E_1}{E_2 + E_3 + E_4 + E_5 + E_6 + E_7 + E_8 + E_9} \tag{3}$$

By letting r as the threshold, separating object from background based on double Haar wavelet, we can get better detecting result.

The process of the algorithm can be described as:

(1) Obtain the gray difference image $D(i, j)$ with background subtraction.

(2) Conduct double Haar wavelet transform. We can obtain double Haar wavelet split result $W(i, j)$ in a small window (9 pixels) and obtain low frequency part and high frequency part.

(3)Clean up noise from image $W(i, j)$ and we can get image without noise.

(4) Calculate the optimal judging threshold T. The optimal threshold $T = r$ can be obtained.

(5)Separate object from background.

$$E(i, j) = \begin{cases} 1, & |C(i, j)| > T \\ 0, & otherwise \end{cases} \tag{4}$$

Here, $E(i, j)$ is the binary image with the object only. $C(i, j)$ is the center pixel of the small window with 9 pixels. T is the optimal threshold.

4 Experimental Results

The comparative detecting results between the traditional fixed threshold method and the new algorithm proposed in this paper under outdoor and indoor circumstance are shown in fig.2.

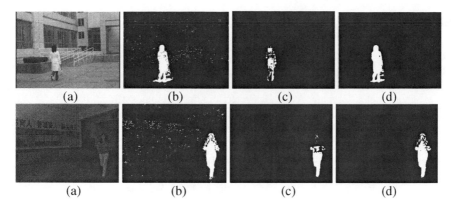

Fig. 2. Comparison between fixed threshold and optimal threshold

The original images are showed in fig.2 (a). Figure (b) is the detecting result when threshold is 10, there are noises in image. Figure (c) gives the result when threshold equals 20 indoor and 30 outdoor. We can see that there are no noises in images, but the object is incomplete. So, the fixed threshold is limited. We can hardly get ideal result with fixed threshold. The optimal threshold is used to detect object which proposed in this paper, we can obtain the result in (d). There are no noises in the images and the detecting object is complete.

So, in the surveillance system, because the influence the circumstance, the fixed threshold can't realize the detecting task accurately. The algorithm proposed in this paper can overcome the contradiction between background noise and integrity of object and separate the object from the background. The threshold which determined by the parameter of wavelet energy is an adaptive value, so, it can conquer the shortcoming that the threshold is too big to ensure the object region integrated and the threshold is too small to remove noise from the binary image. The new method can give a better detecting result.

5 Conclusion

When the object is detected in a video sequence, because of the influence of the circumstance, a fixed threshold is limited, and the accuracy of the object detection will be affected greatly. In this paper, combined with double Haar wavelet transform and energy characteristic, a new dynamic threshold is obtained and the gray difference image is processed to detect object. The value of the threshold is determined by ratio of the energy between low frequency and sum of high frequency. It is proved by the experiment that the threshold is a optimal threshold and it can restrain the noise effectively and finish the object detection accurately.

Acknowledgement. This paper is supported by China Natural Science Foundation Committee (60775023, 60975025). The authors would like to thank the anonymous reviewers and the Associate Editor, for their helpful comments and suggestions.

References

1. Huang, K., Wang, L., Tan, T., Maybank, S.: A Real-time Object Detecting and Tracking System for Outdoor Night Surveillance. Pattern Recognition 41(1), 432–444 (2008)
2. Qing, L., Jianhui, X., Shitong, W., Yongzhao, Z.: Automatic Detection of Regions of Interest in Moving People. Computer Aided Design &Computer Graphics 21(10), 1446–1450 (2009)
3. Shiping, Z., Xi, X., Qingrongp, Z.: An Edge Detection Algorithm in Image Processing Based On Point By Point Threshold Segmentation. Optoelectronics Laser 19(10), 1383–1387 (2008)
4. Liu, D., Zhang, G.: Special Target Detection of the SAR Image Via Exponential Wavelet Fractal, Xidian University, vol. 37(2), pp. 366–373 (2010)
5. Hsuan Cheng, F., Chen, Y.L.: Real Time Multiple Object Tracking and Identification Based on Discrete Wavelet Transform. Pattern Recognition 39(6), 1126–1139 (2006)
6. Sharma, M., Kulkarni, A., Puntambekar, S.: Wavelet Based Adaptive Tracking Control for Uncertain Nonlinear Systems with Input Constraints. In: 2009 International Conference on Advances in Recent Technologies in Communication and Computing, pp. 694–698 (2009)
7. Wang, X.: Moving: Window-Based Double Haar Wavelet Transform for Image Processing. IEEE Trans on Image Processing 15(9), 2771–2779 (2006)

Optimal Trajectory Generation for Soccer Robot Based on Genetic Algorithms

Songhao Piao[1], Qiubo Zhong[2], Xianfeng Wang[1], and Chao Gao[1]

[1] School of Computer Science and Technology, Harbin Institute of Technology, China
[2] College of Electronic and Information Engineering, Ningbo University of Technology, China
piaosh@hit.edu.cn, zhongqiubo@yahoo.com.cn,
{wangxianfeng211,gaochaohit}@hit.edu.cn

Abstract. According to the feature of humanoid soccer robot, a model of kicking motion is built. Kinematics and dynamics analysis is achieved by this model and optimal trajectory for kicking motion is generated using genetic algorithms. The object of optimization is energy consumed during the processing of kicking under the guaranty of stabilization of robot. Simulations and experiment testify the efficiency of the method.

Keywords: Soccer robot, Genetic algorithms, Motion planning, Optimal trajectory.

1 Introduction

Soccer robot is very popular recently and the humanoid soccer robot is being a hot research all over the world. There are many kinds of humanoid robots in the world, and their control and mechanical structure very are different[1]. RoboCup and FIRA(Federation of international robot soccer association) also organized humanoid robot football match to provide a new way and platform for research on humanoid robot. Philipp [2] implemented this method mentioned above on the robot ASIMO, and achieved good experimental results. Cupec [3]studied the loop iteration between the biped walking and visible observation, they developed a single vision system, which can deal with the wider visual margin in the environment within obstacles. Analysis for scene is achieved after every step of walking, and the location of the obstacle is determined by the distance between the visible edges relative to foot reference frame. The edge detection of the obstacle is present by analyzing the scenes on line. Kanehiro[4] present a method, which can generate the 3D model environment of humanoid robot by stereo vision (three cameras). Through this model, a feasible path can be found and planned on line in the moveable space. The motion planning is achieved by a pose generator for whole body on line. It can update the height of waist and pose of the upper body according to the size of the moveable space. Nagasaka studied the local increment of PD control in adjusting the biped robot gaits through GA[5]. Capi[6] developed a method based on genetic algorithm to generate a human-like motion. Humanoid robot gait was generated using two different cost functions: minimum consumed energy (CE) and minimum torque change (TC). Since the GA

Y. Yu, Z. Yu, and J. Zhao (Eds.): CSEEE 2011, Part II, CCIS 159, pp. 447–451, 2011.
© Springer-Verlag Berlin Heidelberg 2011

generally requires longer time in the process of evolution, if the problem of evolution speed is not solved, it is only just suitable in off-line planning for humanoid robot.

2 Kicking Model and Kinematics and Dynamics Equation

2.1 Building the Kicking Model

In this paper, model of kicking played in the sagittal plane for the soccer robot through kinematics and dynamics. There are two joints include knee and pelvis of the swing leg which play a key role during the motion of kicking. The rotation for these joints consumed the major energy during the processing of kicking. Although the rotation of ankle in the swing leg and any joints in the supporting leg can also affect the stabilization of the robot, however, for the simplicity, these joints excepted the knee and pelvis joints are not considered. As shown in the Fig. 1, a model of kicking motion for soccer robot is built during its single supporting period.

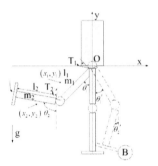

Fig. 1. Model diagram of kicking process for soccer robot

Where m_1, m_2 are the mass of link. For simplicity, we supposed that the mass is focused on the end of link. l_1, l_2 are the length of link, while T_1, T_2 are the torque of link, and θ_1, θ_2 are the ankles of between rotation of link and the vertical.

2.2 Kinematics and Dynamics Equation for Kicking Motion

According to the model of kicking motion for soccer robot, kinematics and dynamics equation for kicking is built by the Lagrangian dynamic equation. The process of kicking is supposed be finished during the single supporting period. The process consists of two subprocess, which are rising the swing leg and falling to touch the ball on the level ground. From the Fig. 1, the center of torso is supposed to be the original coordinate, and the displacements of (x_2, y_2) and speeds of $(\dot{x}_2, \dot{y}_2)$ can be obtained from the Eq. (1) and (2).

$$x_2 = l_1 \sin \theta_1 + l_2 \sin \theta_2; \quad y_2 = -(l_1 \cos \theta_1 + l_2 \cos \theta_2) \tag{1}$$

$$\dot{x}_2 = l_1 \cos\theta_1 \dot{\theta}_1 + l_2 \cos\theta_2 \dot{\theta}_2; \quad \dot{y}_2 = l_1 \sin\theta_1 \dot{\theta}_1 + l_2 \sin\theta_2 \dot{\theta}_2 \tag{2}$$

The kinetic energy K_1 of link 1 and potential energy P_1 can be get by Eq. (3)

$$K_1 = 1/2 m_1 l_1^2 \dot{\theta}_1^2; \quad P_1 = -m_1 g l_1 \cos\theta_1 \tag{3}$$

Also, the kinetic energy K_2 of link 2 and potential energy P_2 can be get by Eq. (4)

$$K_2 = 1/2 m_2 \left(\dot{x}_2^2 + \dot{y}_2^2 \right); \quad P_2 = m_2 g y_2 \tag{4}$$

Eq. (5) can be get by substituting Eq. (1) and (2)

$$K_2 = 1/2 m_2 l_1^2 \dot{\theta}_1^2 + 1/2 m_2 l_2^2 \dot{\theta}_2^2 + m_2 l_1 l_2 \dot{\theta}_1 \dot{\theta}_2 \cos\left(\theta_1 - \theta_2\right) \tag{5}$$

$$P_2 = -m_2 g l_1 \cos\theta_1 - m_2 g l_2 \cos\theta_2 \tag{6}$$

As $L = K - P$ and $K = K_1 + K_2$, $P = P_1 + P_2$. The Lagrangian dynamic equation is described as followed:

$$T = \frac{d}{dt}\frac{\partial L}{\partial \dot{\theta}} - \frac{\partial L}{\partial \theta} \tag{7}$$

The state equation of the problem is $w_1(t) = \dot{\theta}_1(t)$ and $w_2(t) = \dot{\theta}_2(t)$, the constraint condition of kinetics equation is $|T_1(t)|, |T_2(t)| \leq T_{max}$; $-\theta_{1s} \leq \theta_1(t) \leq \theta_{1f}$; $\theta_1(t) - \pi \leq \theta_2(t) \leq \theta_1(t)$, where T_{max} is the max torque, θ_{1s} and θ_{1f} are the initial and final angle of θ_1 respectively. The initial state is $\theta_1(0) = \theta_{1s}; \theta_2(0) = \theta_{2s}; w_1(0) = w_1(0) = 0$. The object function of optimization is described as in Eq. (8)

$$J = J_1 + \int_0^{tf} J_2 dt \tag{8}$$

Where $J_1 = \Delta w(\theta_{1s}, \theta_{2s}) = (m_1 + m_2) g l_1 (1 - \cos\theta_{1s}) + m_2 g l_2 (1 - \cos\theta_{2s})$ is the energy of lifting the leg, $J_2 = \int_0^{tf} \left[T_1(t) w_1(t) + T_2(t) w_2(t) \right] dt$ is the energy of kicking.

3 Optimizing Using Genetic Algorithm

Comparing to the traditional optimal methods, genetic algorithm has the advantages of global optimization, less of dependence of problem, needless of the information of gradient and continuous for function. Therefore, genetic algorithm is also introduced into this paper to optimal the trajectory of robot joints during the motion of kicking. The steps of algorithm can be shown bellow:

Step1: Initialization and set the parameters of genetic algorithm.
Step2: Coding for the variables to optimize.
Step3: Generating the initial population.
Step4: Decomposition and decoding the chromosome
Step5: Inverse dynamics computation
Step6: Computing the value of fitness
Step7: Judging whether the evolution generation is reached to the requirement, if yes, optimal parameters are achieved, and stop the algorithm.
Step8: Chromosome replication, Crossover and mutation and go to Step 4.

4   Simulation and Experiment

Joints trajectories for soccer robot during the motion of kicking is run by Matlab. The result of optimization by genetic algorithm is shown in Fig. 2 (left), where, the iterations is 100, the number of population is 200 and the crossover probability p_c is 0.7, mutation probability p_m is 0.03. the optimal trajectories of knee and pelvic is described as in Fig. 2 (right). Simulation for kicking motion by Mablab is shown in Fig. 3.

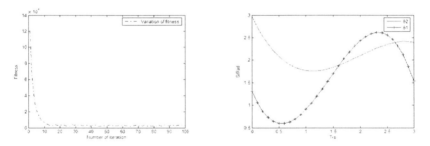

Fig. 2. Optimizing by genetic algorithms (left) and joint trajectories simulation for soccer robot during the motion of kicking (right)

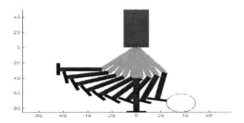

Fig. 3. Simulation for kicking motion by Matlab

Fig. 4 describes the simulation pictures of kicking motion based on Robonova robot. The simulation shows that the robot can do the kicking motion successfully. Snapshots of kicking motion by real robot is shown in Fig. 5, which shows that the robot can kick effectively using the trajectories resulted by the optimal algorithm.

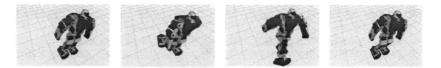

Fig. 4. Simulation of kicking by right foot based on Robonova robot

Fig. 5. Snapshots of experiment for kicking motion based on Robonova robot

5 Conclusion

According to the characteristics of mechanical structure for humanoid soccer robot, a kicking motion is designed under the requirement of soccer match for humanoid robot held in FIRA. Kinemics and dynamic equation of kicking motion is derived and object function of least energy consumed is optimized by genetic algorithm. Simulations and experiment testify the efficiency of the method and achieve a good result.

Acknowledgement. Project 61075077 supported by National Natural Science Foundation of China.

References

1. Iishida, T.: A Small Biped EntertainmentRobotSDR-4X. In: Proceedings 2003 IEEE International Symposium on Computational Intelligence in Robotics and Automation, pp. 1024–1030. IEEE Press, Los Alamitos (2003)
2. Philipp, M., Chestnutt, J., Chuffer, J., et al.: Vision-guided Humanoid Footstep Planning For Dynamic Environments. In: Proceedings of IEEE/RAS International Conference on Humanoid Robots, pp. 13–18. IEEE Press, Los Alamitos (2005)
3. Cupec, R., Schmidt, G., Lorch, O.: Vision-Guided Walking in a Structured Indoor Scenario. Automatika 4, 1–2, 49–57 (2005)
4. Kanehiro, F., Yoshimi, T., Kajita, S., Morisawa, M.: Whole Body Locomotion Planning of Humanoid Robots Based on a 3D Grid Map. In: Proceedings of the 2005 IEEE International Conference on Robotics and Automation, Barcelona, Spain, April 2005, pp. 1072–1078. IEEE Press, Los Alamitos (2005)
5. Nagasaka, K.: Acquisition of Visually Guided Swing Motion Based on Genetic Algorithms and Neural Networks in Two-Armed Bipedal Robot. In: Proceeding of IEEE International Conference of Robotics and Automation, pp. 2944–2948. IEEE Press, Korea (2001)
6. Capi, Y., Nasu, L., Barolli, K.: Real Time Gait Generation for Autonomous Humanoid Robots: A Case Study For Walking. Robotics and Autonomous Systems 42, 107–116 (2003)

Novel Method for Measuring Water Holdup of Oil and Water Mixture in High Temperature

Guozhong Wu*, Jiadong Sun, Xianzhi Yang, and Dong Li

College of Architecture and Civil Engineering, Northeast Petroleum University,
Hci Longjiang Daqing 163318, China
{WGZDQ,lidonglvyan}@126.com,
{sunjiadong0211,yangxianzhi}@163.com

Abstract. The experimental method of testing the water content of oil-water mixture in high temperature by microwave was proposed in this paper, and the experimental device of indoor microwave test was established. The device can be used under the temperature of 20℃-200℃ and the pressure of 0MPa-2.5MPa. This experiment was accomplished under the temperature of 40℃-150℃ and the pressure of 1.0MPa-1.5MPa. The feasibility of experimental device was verified by dynamic and static experiment; the impact of the impact of temperature on microwave test of water content of crude oil was further analyzed. The results show that: (1) the largest error between dynamic and static test which appeared in the water content range of 30%-50% is 76.45%. (2) The impact of temperature on the test of water content in high temperature is great and has no regularity. When the water content is less than 10% and greater than 30%, the microwave signal values will increase steadily with the increasing temperature. When the water content is within the range of 10% to 30%, the microwave signal values fluctuate greatly with the change of temperature. When the water content is greater than 40%, the microwave signal values get slower with water content. The impact of low temperature on the microwave signal values of water content is smaller and largest microwave signal values of water content error between high temperature and room temperature is 65.7%.

Keywords: oil water mixture, water holdup, microwave, high temperature.

1 Introduction

The on-line detecting of water cut of crude oil is significant to determine the oil-water reservoir, estimate the oil production and predict the cycle life of oil well [1]. The main traditional measurement method of water cut of crude oil is manual detection method with timing sampling distillation. This measurement method with long time duration of sampling, no property of on-line, great randomness of sampling and big artificial error can not satisfy the requirement of oilfield production automation management. The methods of on-line testing include ray method, electrical capacitance method, shortwave

---

* Corresponding author.

Y. Yu, Z. Yu, and J. Zhao (Eds.): CSEEE 2011, Part II, CCIS 159, pp. 452–456, 2011.
© Springer-Verlag Berlin Heidelberg 2011

method, and so on [2,5]. Compared with other test methods, penetration ability of microwave is powerful, and it detects not only the surface moisture of substance, but also the interior moisture. It has remarkable characteristics, such as such as fast speed, short time, high efficiency and wide measuring range (0%-100%), so the application prospect is good. At present, the testing error of using this technology in high water-content area is between 5%-10%[5] in China. But most water cut of microwave tests are going on when the water-oil mixture under normal temperature. At present, many oilfields have entered into high-temperature crude oil exploitation phase [6]. It is important to study the impact of temperature on the water cut test of oil-water mixture.

The experimental method of testing the water content of oil-water mixture in high temperature by microwave is proposed in this paper, and the experimental device of indoor microwave test was established. This experiment was under the temperature of 40°C-150°C and the pressure of 1.0MPa-1.5MPa. The feasibility of experimental device was verified by dynamic and static experiment.

2 Experimental Set

The experimental device which consists of the electric heating system, power cycle system and the signal collection system are shown in Fig.1.

The heating tubes length of electric heating system is 1.5 m. One flange which links heating device is at each end of hating tube, and each heating device consists of three roots of 75 cm length, the heating power of two heating tubes are 2kW, another's is 1kW. The six heat tubes are controlled by three groups of switches, the temperature of oil-water mixture is regulated by controlling heating power.

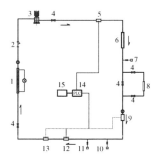

1-electric heater 2-Y type filter 3-circulating pump 4- valve 5-turbine flow meter 6-oil-water mixture 7- drain port 8- observation section 9-microwave water cut analyzer 10- outlet 11-infusion port 12- temperature sensor 13-pressure sensor 14-PLCsystem 15- computer

Fig. 1. Schematic diagram of the experiment system

Power circulate system mainly include circular pipe and circulating pump (maximum flow is 5m³/h), while the circulation pipe includes the Y-type filter (filter rust and other impurities in oil-water mixture, prevent damaging the measuring instrument and effecting measurement) and oil-water mixture (make oil and water to mix fully). The total length of the circular pipe is about 1100 cm, the diameter of which is DN40.

Signal acquisition system include temperature sensor (the temperature range is 0℃ to 300℃, the test accuracy is ± 1%), pressure sensor (the temperature range is 0℃ to 300℃ and the test accuracy is ± 1%), the flow meter (the pressure range is 0 MPa to 2.5 MPa, the test accuracy is ± 0.5%), microwave water cut analyzer, PLC system and the computer. The signal values of each transducer, flow meter and moisture analyzer is inputted to PLC system in current form, Then it'll be connected with the computer through the PC/PPI cables and signal values will be displayed by the treatment of internal program of the computer.

3   Results and Discussion

3.1   Static and Dynamic Contrast Test

In order to verify the feasibility of experimental device, static and dynamic contrast experiments are preceded under 30℃ and atmospheric pressure. The result was shown in Fig 2.

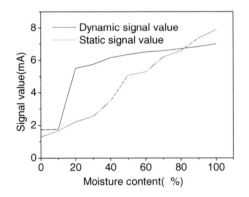

Fig. 2. The electric current signal values of dynamic and static test of water cut of oil-water mixture

Fig.2 shows that the electric current signal value increases with the increasing of water cut in static experiment, although there is some fluctuation in the process, it is smaller than that in dynamic experiment, so it can be considered as a linear change. And the electric current signal value increases with the increasing of water cut in dynamic experiment, but it is a nonlinear change. When the water cut is 0-10% and 30%-100%, the electric current signal value changes slowly and when it is 10%-30%, the electric current signal values change rapidly. Meanwhile, by the result of the static and dynamic contrast test, the largest and smallest absolute error is respectively 76.45% and 5.2%. The reason of the large error is that in the static experiment, the industrial white oil and the water was stratified and in the dynamic experiment, the stratified phenomenon was weaker.

3.2 Impact of Temperature

In order to analyze the impact of temperature on the test of water content, microwave test of the water content of oil-water mixture are proceeded under 40℃-150℃ and 1Mpa-1.5Mpa, the result of test is showed in Fig 3.

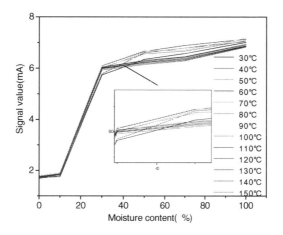

Fig. 3. The water cut under different temperatures

Fig.3 shows that the electric current signal values under high temperature are similar to the values under 30℃, and the electric current signal values change obviously with different water cuts. When the water cut of the oil-water mixture was less than 10% and greater than 30%, the flow pattern of the mixture is respectively water-in-oil and oil-in-water and the two flow states can keep static states, this moment the flow state of the fluid has little influence on the sensitivity of the equipment. When the water cut is 10%-30%, the flow state is not stable, and the flow pattern changes between water-in-oil and oil-in-water continually, so the electric current signal values change greatly. When the water cut is 0-40%, under the same water cut, the signal value increases with the increasing of the temperature. When the water cut is greater than 40%, under the same water cut, the signal value decreases with the increasing of the temperature. Meanwhile, the change of the signal values under 40-90℃ is similar to the one under 30℃. And when the temperature is greater than 90℃, the change trends are quite different from the one under 30℃. It shows that when the temperature of the mixture is greater than 90℃, it has great influence on the test of the electric current signal values of the water cut.

The temperature is 100℃, the error gets the maximum, it is 65.7%. The signal values curve chart shows that the change trend of the signal values of the water cut, which is changes with the changing of temperature, has no uniform regularity.

4 Conclusion

(1) Experimental device of microwave test of water cut of oil-water under high temperature is established in this paper, it can meet the microwave test of water cut of oil-water mixture with the temperature range of 20℃-200℃ and the pressure range of 0 MPa-2.5 MPa.

(3) The largest error between dynamic and static test appeared in the water cut range of 30%-50% is 76.45%. When the water cut is less than 10% and greater than 30%, the microwave signal values increase steadily with the increasing temperature. When the water cut is between 10% and 30%, the microwave signal values fluctuate greater with the change of temperature. When the water cut is greater than 40%, the change gradient of the microwave signal values with the changing water cut begin to be slow. When it is in high temperature, the largest error compared with room temperature test is 65.7%.

References

1. Nailu, Z., Chaomei, X., Jingtian, X.: Measuring Technique for Water Content of Crude and Its Development. Technology Supervision in Petroleum Industry 21(11), 68–72 (2005) (in Chinese)
2. Fahong, L., Junkuan, H., Shuping, W.: Ray-based Automatic Monitor of Void Fraction and Water Content of Crude Oil and Its Application in Ansai Oilfield. Technology Supervision in Petroleum Industry 19(10), 9–12 (2003) (in Chinese)
3. Xiuxin, J.: Principle and Applications of Measuring Water cut through Capacitance. Well Logging Technology 9(6), 73–80 (1985) (in Chinese)
4. Hongzhang, J.: Monitors of Water cut of Oil-Water Mixture with short Wave Absorption. Foreign Oil Fifid Fngineering 12(10), 54–58 (1985) (in Chinese)
5. Li, L., Daoqing, H., Xiaoling, Z.: Microwave Detection of Crude Oil. China Instrumentation 21(4), 60–62 (2009) (in Chinese)
6. Ning, X.: Measurement Technology Investigate of High Water Cut Well. Petroleum Planning and Engineering 12(4), 11–13 (2004) (in Chinese)

Research of Group Communication Method on Multi-Robot System

Songhao Piao[1], Qiubo Zhong[2], Yaqi Liu[1], and Qi Li[1]

[1] School of Computer Science and Technology,
Harbin Institute of Technology, China
[2] College of Electronic and Information Engineering,
Ningbo University of Technology, China
piaosh@hit.edu.cn, zhongqiubo@yahoo.com.cn,
liuyaqi@163.com, liqi22@hit.edu.cn

Abstract. Based on the characteristics of multi-agent communication behaviors, firstly this article proposes the channel allocation model which is based on Q learning so that the limited communication resources can be used effectively, secondly it designs the Communication protocol of multi-agent system for ensuring all parties of communication could exchange and understand its intention, in the end, it reflects the characteristics of communication mechanism comprehensively and proves this method by Simulation test under group strategy.

Keywords: Group strategy, Q learning, Communication protocol, Packet communication.

1 Introduction

The communication mechanism of multi-agent system is the important expression of agents sociality and the core issues of multi-agent system researching [1-4]. For the multi-agent systems, they must solve the limitation of experience, ability, resources and information through the communication, its communication is an important means of intelligence, while it is also the key way to solve the interdependent and conflict and competition among agents [5]. Whatever the purpose of collaboration and competition is, coordinate between agents need each other, but the foundation of coordination is communication [6-8].

2 Method of Channel Allocation

2.1 Dynamic Allocation

Agent communication can be abstracted as discrete event systems。 The new arrival events of agent i subjects to average incidence λ Poisson distribution:

$$P_n = \frac{(\lambda t)^n}{n!} e^{-\lambda t} \tag{1}$$

Y. Yu, Z. Yu, and J. Zhao (Eds.): CSEEE 2011, Part II, CCIS 159, pp. 457–461, 2011.
© Springer-Verlag Berlin Heidelberg 2011

Time interval $\tau_{arrival}$ has exponential density which is defined as:

$$f\left(\tau_{arrival}\right) = \lambda e^{-\lambda \tau_{arrival}} \tag{2}$$

Assuming the duration of event $\tau_{holding}$ with average events time $1/\mu$ fits exponential distribution, then:

$$f\left(\tau_{holding}\right) = \lambda e^{-\mu \tau_{holding}} \tag{3}$$

For using the Q learning method, the dynamic allocation problem must be described as follows:

Step 1: Current status: Assumptions there are N agents and M channels in a mobile system. The state x_t is defined time t:

$$x_t = \left(i, A(i)\right)_t \tag{4}$$

Where $i \in \{1, 2, \ldots, N\}$ is the marks of agents, show there is event happen and event over for agent i, $A(i) \in \{1, 2, \ldots, M\}$ is the number of available channels for agent i on time t. Defining channel state for agent q, $q = 1, 2, \ldots, N$ is M dimension vector:

$$u(q) = \{u_1(q), u_2(q), \ldots, u_M(q)\} \tag{5}$$

If the agent uses this channel, its members for value is 1, or it is 0. From the channel state of agent i and its interference agents $j \in I(i)$, efficiency vector s(i) can be expressed as:

$$s_k(i) = \bigvee_{q=1, 2, \ldots, P, i} u_k(q) \tag{6}$$

Where $k = 1, 2, \ldots, M$, P is the number of agents which happen interference with agent I, V is logic or operation.

Step 2: Using action: Using action a is defined as distributing channel k from the available channel A(i) to the agent i which raises a request at the moment.

$$a = k, \quad k \in \{1, 2, \ldots, M\}, \quad s_k(i) = 0 \tag{7}$$

2.2 Algorithm Realizing

The proposed method is an online learning method. In the cases of researching, channel can be distributed to agent as studying the dynamic allocation strategies. This article considers lookup table and neural network at the same time. Corresponding to the algorithm learning and allocation steps is as below:

Step 1: State action: Determining the number of current agents I and channel Using information to establish current status $x_t = (i, A(i))$. Searching available channel m_x sequence is made sure by set $L(m_x)$.

Step 2: Restoring of Q value: from independent xa=(xt,k), $k \in L(m_x)$ forms m_x set, then gets Q (xt,k) value of m_x.

Step 3: Channel allocation:The optimal channel has the smallest Q value:

$$k^* = \min_{k \in L(m_x)} \{Q(x_t, k)\} \tag{8}$$

3 Multi-agent Communication Protocol

3.1 Design of Communication Protocol

Make A express the set made by all intelligence collection, S express all possible states of agents, B express all possible actions of agents, And assumption power set of any sets Q to be p(Q), then a communication protocol can be defined $InP =< Ag, St, Ob, R, Bh, Act >$ Where $Ag \in p(A)$ express object involved in communication protocol, $St \in p(S)$ express each communication state of the communication process, $Ob \in p(S)$ express achieving purpose of the communication protocol, $R \subseteq St \times St$ express the direct relationship among communications, $Bh \in p(B)$ express the agent actions in the communication process, $Act : R \rightarrow Bh$ express the relationship between the changing of communication state and one certain sure action.

3.2 Strategy of Group Communication

Communication strategies involve in the analysis of related problems, the analysis of ability of related agents and the analysis of communication protocol. The purpose of building multi-agent system is to solve problem of some aspects, it must take corresponding communication strategy facing some specific issues.

Calling the differential equation p(t) as information transmission equation, then:

$$\frac{dp(t)}{dt} = \beta \{1 - p(t)\} \{1 - e^{-\rho A p(t)} e^{-\rho_n A}\} \tag{9}$$

This formula is the expansion form of logic equation, then the time required for communication can be calculated easily. The communication area Ag and the density of agent ρ_g are calculated as below:

$$A_g = R^2 (2c_r (k-1) + \pi), \rho_g = \frac{k}{A_g} \tag{10}$$

Where cr is the coefficient of communication radius R, k is the number of agents in group.

4 Communication Simulation Experiment

Simulation experiment on performance of dynamic channel allocation which is based on Q learning.

Measure the channel allocation methods by the communication obstruction probability P_n, if P_n reach 25%, the system can't guarantee the communication smooth, the result of this experiment is figure 1:

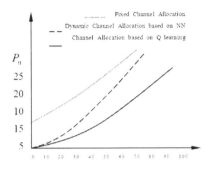

Fig. 1. The Number of Intelligent Communication performance compare test

It can be found through the comparison that the information among agents can be made spread quickly and accurately by grouping communication strategy in multi-agent system. The results of experiment show that the strategy can accelerate the information transferring among agents. Especially when there are many agents, the property of whole system will be greatly improved. The results of simulation experiment are Figure 2 as below:

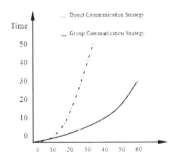

Fig. 2. The Number of Intelligent Diffuse time contrast test

5 Conclusion

This chapter proposes the hierarchical communication mechanism of multi-agent system. Firstly, it studies the channel allocation methods based on Q learning, which is the basis of dynamic communication resource allocation with events. Then it formulates the communication protocol according to the characteristics of multi-agent communication, which can express the intention of agents completely and clearly. Finally, this article achieves a stable and efficient multi-agent mechanism by using the time of communication system of the packet communications strategy. In computer simulation experiment, it proves that the constructed communication system mechanism can replies the distributing, time-varying, heterogeneous and large-scale multi-agent system perfectly.

Acknowledgement

Project 61075077 supported by National Natural Science Foundation of China.

References

1. Attila, P., Tamas, K., Zoltan, I.: Compass and Odometry Based Navigation of a Mobile Robot Swarm Equipped by Bluetooth. In: International Joint Conference on Communication Computational Cybernetics and Technical Informatics (ICCC-CONTI 2010), pp. 565–570. IEEE Press, New York (2010)
2. Zhao, Z., Guo, S.: An FHMA Acoustic Communication System for Multiple Underwater Robots. In: IEEE International Conference on Information and Automation, pp. 1223–1228. IEEE Press, New York (2010)
3. Jung, J.H., Park, S., Kim, S.-L.: Multi-Robot Path Finding with Wireless Multihop Communications. Communications Magazine 48(7), 126–132 (2010)
4. Carlone, L., Ng, M.K., Du, J., Bona, B.: Rao-Blackwellized Particle Filters Multi Robot SLAM with Unknown Initial Correspondences and Limited. In: IEEE International Conference on Communication Robotics and Automation, pp. 243–249 (2010)
5. Trawny, N., Roumeliotis, S.I., Giannakis, G.B.: Cooperative Multi-Robot Localization Under Communication Constraints. In: IEEE International Conference on Robotics and Automation, pp. 4394–4400 (2009)
6. Man, B., Pei-Yuan, G.: Realization Of Robot Communication System Based on Linux. Intelligent Systems and Applications, 1–4 (2009)
7. Fukuda, T., Taguri, J., Arai, F., Nakashima, M., Tachibana, D., Hasegawa, Y.: Facial Expression of Robot Face for Human-Robot Mutual Communication. In: Proceedings of Robotics and Automation, pp. 46–51 (2002)
8. Zhijun, G., Guozheng, Y., Guoqing, D., Heng, H.: Research Of Communication Mechanism Of Multi-Agent Robot Systems. Micromechatronics and Human Science, 75–79 (2001)

Optimization Control Techniques for Environment Pollution of Cokery

Bin Zhou[1], Yong-liang Chen[2], Yue Hui[1], Shu-mei Lan[1],
Hua-jun Yu[1], Ping Zhang[1], and Yao-fu Cao[1]

[1] Computer Institute of Science and Technology, Jilin University,
Changchun, 130026, China
[2] Research Institute of Mineral Resources Prognosis on Synthetic Information,
Jilin University, Changchun, 130026, China
zhoubin@jlu.edu.cn, chenyongliang2009@hotmail.com,
{huiyue,lansm}@jlu.edu.cn, 309562885@qq.com,
zhangping87hap@163.com

Abstract. Since fuzzy control technology provides a convenient method for constructing nonlinear controllers via the use of heuristic information, it is a practical alternative for the nonlinear, big inertia and big hysteresis plant. In industrial production, the pressure of cokery is nonlinear and has more strong couple property change parameter essentially. It is highly impossible to describe the course that it changes with general mathematics model. Relying on the traditional PID technology is not effective. This paper gives a kind of cokery gas collector pressure control scheme based on fuzzy control theory. With the use of fuzzy control system, we are not only obtaining improvement pressure steady, but also alleviate the pollution level in certain scope.

Keywords: Cokery Control Techniques, Fuzzy Algorithm, Environment Pollution.

1   Introduction

Fuzzy cybernetics is the theoretical foundation of fuzzy control technology. Fuzzy control provides a user-friendly formalism for representing and implementing the ideas we have about how to achieve good control results.

In this paper, we talk about two cokery and use one set air-cooling system. Gas transported pipeline is interlinked, it exists serious couple disturbing. The gas collector pressure of two cokeries is hard to reach balance. Typical problem is: when putting coal into cokeries, gas collector pressure will be down greatly and more big change in the dish valve of gas collector begins to regulate. Along with the switching of the control valve, another gas collector pressure can arouse change, since pipeline is interlinked. Then, the valve of gas collector which pressure is steady also regulate at the same time, this has formed couple disturbing. The imbalance of pressure has caused the serious air pollution and waste of gas resource. Relying on traditional PID instrument is highly impossible to reach ideal level. Therefore, we want to solve the problem that exists in coke production with fuzzy control technology. It has realized

Y. Yu, Z. Yu, and J. Zhao (Eds.): CSEEE 2011, Part II, CCIS 159, pp. 462–466, 2011.

the automatic control in the gas collector pressure and cooling machine suction, and makes the pressure steady in certain scope.

2 Paper Preparation

The control system is composed of industry personal computer (IPC), PC-7484 multi-function comprehensive board, fuzzy algorithm control module, object to be controlled, valve actuator, pressure sensor and monitoring warn system etc. What the inputs signal of control system is the gas collector pressure of two cookeries, negative pressure of main suction. The outputs control is the dish valve actuator of gas collector and main. The 4mA -- 20mA pressure current signals of each channel are sent into computer via PC-7484'A/D converter, the IPC gets outputs quantity through the fuzzy algorithm and then is sent into dish valve actuator via PC-7484'D/A.

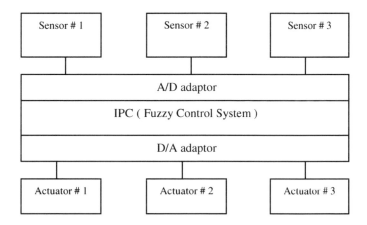

Fig. 1. Control Systematic Composition

2.1 Fuzzy Controller Inputs and Outputs

According to eventually object to be controlled and the signal of measure, the systematic structures have be determined. The signals that the fuzzy control system need are: the pressure values of two gas collectors, the negative pressure value of main suction and the pressure value before the fan.

We adopt PC-7484 multi-function card for each detection pressure signal and export control current. PC-7484 interface board has the 16(12bit) input channel of A/D, 4(12bit) independent output of D/A, 24 switching channels for input/output. Meet the needs of data measurement and industrial control system expressly. It has highly integrated degree and good reliability. PC-7484 has offered standard Dynamic-Link Library (pc7484.dll), copy the pc7484.dll to the system catalog of WINDOWS first, then using LoadLibrary() dynamic loading, and define a kind of type indicator and directional every function, get the entrance address of each function with GetProcAddress(). Through using the function of Dynamic-Link Library in library you can carry out the operation of I/O for hardware equipment. Fallow is two main function of PC-7484, the collection function of A/D : Int pc7484_ad (int ChannelNo, unsigned short Port, long Delay) are the function of I/O of the most important operation, it returns 0~4095 number collection data of the channel-number. In data collection course, get each gas collector data of corresponding pressure through using this function. The export function of D/A: Void pc7484_da(int ChannelNo, int DaData, unsigned short Port, long Delay); through using this function, the data of calculation outputs changes simulated current to control corresponding dish valve actuator.

2.2 Fuzzy Control Algorithm

Fuzzy control is a kind of language controller. The fuzzy controller is to be designed to automate how a human expert who is successful at this task would control the system. The human expert provides a description of how best to control the plant in some natural language. We seek to take this "linguistic" description and load it into the fuzzy controller. It can reflect the experience of the expert. It has very strong robust and control stability.

Since fuzzy controller has the property that trailed fast, therefore, applies to the nonlinear object to be controlled of big inertia and big hysteresis especially. The property of fuzzy control again depends on the definite and its adjustability of fuzzy control rule on great level. We adopt parameter adjustable fuzzy controller and using the controller into the automatic control system of cokeries gas collector pressure. Its control rule can be adjusted from correction factor. Replacing instrument PID controller with fuzzy controller, have formed the fuzzy control system of cokeries gas collector pressure. What fuzzy controller modular adopted normally is that digital increment type of control algorithm. The structure of fuzzy control modular shows as Figure 2.

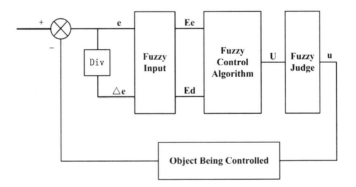

Fig. 2. Schematic for Fuzzy Control System

Where e(n) is the systematic error between gas collector pressure value and ideal value at certain momentary; $\Delta e(n)$ is the change-in-error. The reciprocalt of error value is the sampling time of control system. u(n) is the outputs value of control system at certain momentary, $\Delta u(n)$ is the outputs change value of the control system.

After systematic error Ee and change-in-error Ed be defuzzied, projection goes to vague view region, Ee and Ed are e(n) and Δe(n) the fuzzy value on fuzzy view region. On fuzzy view region we get the fuzzy value U by fuzzy control algorithm. Through fuzzy control judge we get the control outputs value u of the controller. This value u is the plant (object to be controlled) inputs value. In order to control the plant accurately, it is needed to be change fuzzy control value u into accurate value.

The control rule of fuzzy controller can be described as a analysis expression:

$$U=- [\alpha \, Ee+ (\, 1-\alpha \,) \, Ed] \quad \alpha \, \{0,1\} \tag{1}$$

In the expression, Ee, Ed and U are the fuzzy variable via quantification, its corresponding view region is systematic error, change-in-error and control value respectively. The α is correction factor ,and also called as adding right factor. Through adjust the value, it can change different add right level of the systematic error and change-in-error.

3 Implement of Fuzzy Control System

The establishment of fuzzy control rule is based on the control strategy by hand. Through summarizing and gathering we can get the strategy of fuzzy control. For guarantee the safe of cokeries operation, it must control the pressure of main suction and gas collector at the same time. According to expert operating experience, we can summarize some following regulation rules of opening dish valves angle:

(1) When two gas collector pressures are higher than ideal, two gas collector dish valve need not be regulated, contrary main suction set-point should be modified, so main suction pressure value will increase;

(2) The pressure of two gas collectors reflected the gas of two cokeries to occur quantity directly. Its dish valve angle affects gas output directly. If the pressure is too high, the angle of valve should increases, otherwise, should reduce the angle of the valve;

(3) If main pipeline suction is lower than set-point, open degree of the main suction valve should increase and increase air volume;

(4) The best location of each valves is about %50, regulation is most sensitive in such place;

(5) Limit the angle of each valve to %0 (as valves close), in order to avoid store gas jug break up;

The control rule of fuzzy controller is a group of fuzzy condition sentence. It is some sets of vocabulary of inputs and outputs variable state in sentence to be called as variable word set. The choice of the inputs to the controller and outputs of the controller can be more difficult. Once the fuzzy controller inputs and outputs are chosen, we must determine what the reference inputs are. There will be "linguistic variables" that describe each of the fuzzy controller inputs and outputs.

Set up linguistic description of error Ee is:

{ NB, NM, NS, NZ, PZ and PS, PM, PB } in which { NB, NM, NS, NZ, PZ and PS PM, PB } show { negative big, negative medium, negative small, negative zero and positive zero , positive small, positive medium, positive big} 8 level variable word set.

The linguistic descriptions of change-in-error Ed is:

{ NB, NM, NS, ZE and PS, PM, PB } in which { NB, NM, NS, ZE and PS, PM, PB } show { negative big, negative medium, negative small, zero, positive small, positive medium, positive big } 7 level variables word set.

Outputs of fuzzy controller view region of u, is:

{ NB, NM, NS, ZE and PS ,PM, PB } in which { NB, NM, NS, ZE and PS, PM, PB } show { negative big, negative medium, negative small, zero, positive small, positive medium, positive big } 7 level variables word set.

According to the word set of definition fuzzy subclass. Definite fuzzy subclass subjection function curve gets the subjection degree of limited point.

The establishment of fuzzy rule is based on the control strategy by hand. Adding the knowledge set that man accumulate through study and test, long-time experience.

4 Conclusion

Using the fuzzy control system, the phenomenon of gas collector pressure disequilibrium can be gotten great improvement. The air quality parameter of city has been improved. At the same time, it prolonged the service life of coke oven, raised the recovery of coke and chemical products. Not only get obvious economic benefits, but also get obvious social benefits.

Acknowledgments. This Work was Supported by National Natural Seience Foundation of China (No.41072244).

References

1. Passino, K.M.: Fuzzy Control. The Ohio State University, Columbus (1997)
2. Huaguang, Z.: Fuzzy Adapte Theoretical from Adaptive Control and Its Application. Beijing Aviation Space flight University, Beijing (2002)
3. Junhua, L.: Virtual Instruments Design Based on Lab View, Electric Industrial, Beijing (2003)
4. Zengliang, L.: Florilegium of Fuzzy Technology and Neural Networks, Beihang University, Beijing (2001)

Analysis of Indoor Thermal Environment and Air-Conditioning Mode of Ventilation in Large Space Building

Guozhong Wu, Kang Xue, Jing Wang, and Dong Li

The College of Architecture and Civil Engineering, Northeast Petroleum University,
Hei Longjiang Daqing 163318, China
{WGZDQ,lidonglvyan,xuekang8478,wangjing1108}@126.com

Abstract. Heat transfer and flow characteristics of air in large space building were analyzed, the boundary conditions of which were simplified, the normal k-ε turbulence model was used to simulate and calculate the air distribution of air-conditioning system and the influence of two schemes on air quality of large space. The results show that: (1) adopting the scheme of air-conditioning system in whole space, the wind speed nearby air return open is larger, and the temperature around ground, wall, people and lampshade is higher. (2) adopting the scheme of straticulate air-conditioning system, the wind speed along the jet axis and in the recirculation zone is larger, the temperature between 18℃ and 19℃ in air-conditioning working area is lower, while the temperature in non-working area is higher, which is over 20℃. (3) adopting straticulate air-conditioning, the reasonable air distribution and thermal comfort of working area is attained, the initial investment is decreased and the operation cost is saved, which have great importance to building energy saving.

Keywords: Large Space, Thermal Environment, Air-conditioning Mode, Ventilation.

1 Introduction

Using CFD technology to study indoor air quality problem is that wind speed, temperature, relative humidity, pollutant concentration in each position of room and so on, which are obtained by numerical simulation, to evaluate ventilation and air exchange efficiency, thermal comfort and pollutant removing efficiency. Yan simulated and analyzed indoor natural ventilation situation in summer typical climate, also discussed the influence of wind direction, wind speed and temperature on velocity field and thermal comfort index of indoor natural ventilation. Zhang used Airpak 2.0 software to numerically simulate the temperature field and velocity field of a certain gymnasium and experimentally demonstrated the simulation results. Cen made a thermal analysis on summer simplified double glass curtain by means of FLUENT

Y. Yu, Z. Yu, and J. Zhao (Eds.): CSEEE 2011, Part II, CCIS 159, pp. 467–471, 2011.
© Springer-Verlag Berlin Heidelberg 2011

software, revealed the law of airflow in thermal channel and compared the difference of several different design schemes. Hirnikel established mixing ventilation system and room model of hot source, verified CFD model according to ASTM standard and considered that CFD model is effective to the simulation of temperature and particle concentration of mixing ventilation. Lai and Chen established motion model of small particles that suspended in indoor air and simulated transmission and deposition process of the small particles.

In this paper, the air supply characteristics of large space air-conditioning under working condition of summer were analyzed, and the air supply model of large space air-conditioning was established and the boundary conditions were determined according to the working condition of design. Using CFD software to simulate and analyze the air-conditioning system in different air distributions, the temperature, wind speed and typical section flow filed of the two situations were obtained, the comfort and energy consumption were also obtained.

2 Model Establishment

The length of exhibition hall is 140m (On one side of the computer room is a room of 14m long, 9m wide), the width is 70m and the height is 20m. This large space design scheme adopts the whole space air-conditioning system form, selects three air conditioning units, the nominal airflow rate of which is 85000m3/h. The total nominal supply air volume is 255000m3/h, the return air volume is 204000m3/h, the fresh air volume is 51000m3/h and the exhaust air volume is 51000m3/h. The supply air dusts are relative to air conditioning units, that is, each air conditioning unit has its own supply air dust. Three supply air dusts are symmetrically set at the 13m high of exhibition hall. Each supply air dust has its symmetric branch pipeline. And swirl supply air outlet is set up in each end of the branch pipeline, the total number of swirl supply air outlet of Φ630mm is 18, every service area is 18m×18m and the swirl supply air outlet is 13m away from ground. Supply air outlet work under the condition of environmental stress and temperature, after the mixture of fresh air and supply air treated by surface air cooler, they will be transported into exhibition hall.

CFD method is used to simulate the problem of air flow and heat transfer within the tall space buildings, namely research of temperature field, velocity field and thermal comfort problems [8]. The assumption of indoor airs belong to steady incompressible gas, the unsteady term is zero in the differential equation. Based on the analysis of the characteristics of air turbulence, using an average turbulent energy model that is k-ε two-equation model supposed by Spalding and Launder to solve equations is the main method. The control equations including continuity equation, the momentum equations, the energy equation, k–ε equation and turbulent viscosity coefficient (turbulent viscosity) formulas [9].

3   Results and Discussion

3.1   Original Scheme

In the ducts air-conditioning system which applies without return air, air inlets are located on the side walls, air supply outlet below. Air inlets return air in the environmental pressure and temperature conditions. Numerical simulation results are as shown in Fig.1.

As shown in Fig.1, wind is smaller on the location which is below 4 meters height, while the wind is bigger near air inlet position and speed to the middle slowly decrease. Also we can see that the temperature is high around the ground, wall, people and lampshade. It is compared with human thermal comfort below 5 meters in air diffuser, and 5 meters above is low temperature position.

3.2   Optimization Scheme

In the two side walls which X=0m, X=60m have vents, in each side sets one which is Φ400mm, center height is 7m, blowing distance for 30m. When the average speed in the central air is 1m/s, air output of each outlet is 0.69m3/s when cooling, effective air diffuser speed for 7.8 m/s, supply air temperature difference is 8 ℃ and vents have beyond 10 ℃. Air inlets are located below two jets; the size is 0.3m×0.85m. Maintenance structure of cold load boundary conditions are setting in air-conditioning zone, according to the constant heat flux are still set, while in the non air conditioning area, constant wall temperature is changed. Numerical simulation results are shown in Fig.2.

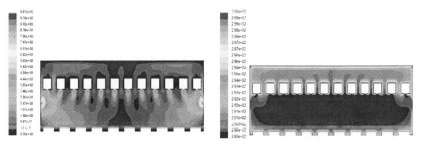

(a) Velocity distribution   (b) temperature distribution

Fig. 1. Simulation results of the original plan

As shown in Fig.2, layered air-conditioning can be seen in the low temperature, working between 18 and 19 ℃, according with the working temperature request. While the temperature is higher, in the non-work is beyond 20 ℃. Considering the comparison and analysis, it embodies energy-saving characteristics of layered air-conditioning. At the same time, the wind is bigger in jet axis and the backflow area and local winds of backflow area can reach 3.2 m/s.

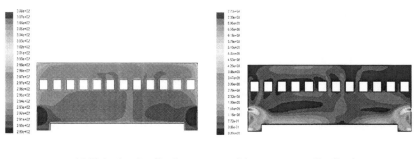

(a) Velocity distribution   (b) temperature distribution

Fig. 2. Simulation results of the Optimization plan

3.3 Schemes Compared

The two schemes are compared. The summer cooling load of the space air conditioning is 100% while layered air conditioning of summer cold load is 50% ~ 85% for all compared with the space of air condition cold load. Through the improvement scheme of simulation calculation, this building layered air condition cold load is 71% in full space air condition cold load and air outputs reduce 1/3, namely reduce an air conditioning unit, and reduce the initial investment and also reduce the operating cost. From the comparative whole temperature, we know that air temperature is lower than working status besides within working 2m; the rest of height air temperature in layered air conditioning is above current situation. As temperature is below 19°C, from the ground to 14m before stratification, air conditioning ensure working area is 19°C within 10m, reached the goal. The highest temperature can reach 26°C for layered air conditioning, and the highest temperature in the current situation is less than 23°C. From both the temperature, the layered air conditioning not only can provide relatively low temperature of workspace, but obvious energy saving effect.

4   Conclusion

(1) Adopting the scheme of air-conditioning system in whole space, wind is smaller on the location which is below 4 m, while the wind is bigger near air inlet position; the temperature is high around the ground, wall, people and lampshade. It is compared with human thermal comfort below 5 m in air diffuser, and above 5 m is low temperature position.

(2) Adopting the scheme of staticulate air-conditioning system, the wind speed along the jet axis and in the recirculation zone is larger, the temperature between 18°C and 19°C in air-conditioning working area is lower, while the temperature in non-working area is higher, which is beyond 20°C.

(3) With staticulate air-conditioning, the reasonable air distribution and thermal comfort of working area can be ensured, the initial investment is decreased and the operation cost is saved, which have great importance to building energy saving.

References

1. Zhao, B., Lin, B., Li, X., Yan, Q.: Comparison of Methods for Predicting Indoor Air Distribution. Heating Ventilating & Air Conditioning 31(4), 82–85 (2001) (in Chinese)
2. Qingyan, C.: Using Computational Tools to Factor Wind into Architectural Environment Design. Energy and Buildings 36(12), 1197–1209 (2004)
3. Yan, F., Wang, X., Wu, Y.: Simulation of Interior Natural Ventilation and Thermal Comfort Based on CFD. Journal of Tianjin University 42(5), 407–412 (2009) (in Chinese)
4. Zhang, H., Yang, S., You, S., Zhou, G.: CFD Simulation for the Air Distribution in a Gymnasium. Heating Ventilating & Air Conditioning 38(3), 87–90 (2008) (in Chinese)
5. Cen, X., Zhan, J., Yang, S., Ma, Y.: CFD Simulation and Optimization Design of Double Skin Facade. Acta Scientiarum Naturalium Universitatis Sunyatseni 47(2), 18–21 (2008) (in Chinese)
6. Hirnikel, D.J., Lipowicz, P.J., Lau, R.W.: Predicting Contaminant Removal Effectiveness of Three Air Distribution Systems by CFD Modeling. ASHRAE Transactions 109(1), 350–359 (2003)
7. Lai, A.C.K., Chen, F.Z.: Comparison of a New Eulerian Model with a Modified Lagrangian Approach for Particle Distribution and Deposition Indoors. Atmospheric Environment 41, 5249–5256 (2007)
8. Li, Z.: Review of the Biennial Meeting of China s HVAC & R 2002. Heating Ventilating & Air Conditioning 33(2), 1–14 (2003) (in Chinese)
9. Wang, Z., Zhang, Z., Lian, L.: Discussion of Thermal Comfort Indices and Design Indoor Temperature for Winter Heating. Heating Ventilating & Air Conditioning 32(2), 46–57 (2002) (in Chinese)

Numerical Simulation on Aerodynamic Characteristic of an Air-to-Air Missile

Chun-guo Yue[1], Xin-Long Chang[1], Shu-jun Yang[2], and You-hong Zhang[1]

[1] Xi'an Research Inst. of Hi-Tech Hongqing Town, 710025 Xi'an, China
[2] 203 Institute of China North Industries Group Corporation, 710065 Xi'an, China
wsgangzi802@qq.com, xinlongch@sina.com.cn, zyhnpu@hotmail.com,
chuangshawudi@etang.com

Abstract. In order to compute the aerodynamic characteristic of an Air-to-air Missile in different mach numbers and different attack angles were carried though making use of CFD software Fluent, the 3-D Reynolds-averaged incompressible Navier-Stokes equations were solved numerically by using a finite volume method. The movement trend of pitching moment coefficient, lift coefficient, and drag coefficient with variety of mach numbers and attack angles were gained, meanwhile, distributing trends of pressure, temperature and airflow velocity were also obtained. The results indicated that the basis and references could be offered by numerical computation results for shape design of missile and definite preponderances were showed than traditionary numerical computation methods.

Keywords: Aerodynamic characteristics, Lift coefficient, Drag coefficient, Pitching moment coefficient, Attack angle.

1 Introduction

Aerodynamic momentums acting on the missile were made up of two parts of forces when compressible air flows detoured the surfaces of all missile parts, one was arosed by pressure difference because of intensity of asymmetrical pressure distribution on the missile surface, the other was viscous friction brought by air acting on the missile surface [1]. It would be interfered by other parts due to the existence of other parts when air flowed of any missile components. The flow instance was different from the case rounding the part singly, thus, different role in the parts of the air force was also happened to a certain extent change accordingly. An air power increment was formed comparing with the separate parts. The traditional calculation method was that force stress instance of alone component was calculated separately in advance, then components of force were added after proper amendments. Therefore, the traditional method had low-rise accuracy [2].

The aerodynamics in this paper was calculated by commercial CFD software FLUENT. The whole process was made up of three main parts. First, appearance of the missile on CAD geometry model was put up and solving area was determined according to the geometrical model established. Secondly, the solving area was

Y. Yu, Z. Yu, and J. Zhao (Eds.): CSEEE 2011, Part II, CCIS 159, pp. 472–476, 2011.

determined not only conforming to the actual area but also making for effective grid discrete of calculation of boundary conditions. Finally the iterations were not stopped until a minimum residual difference of whole field numerical solution converged to target residuals [3,4].

2 Physical Model

Missiles and calculation mesh areas were shown in figure 1 and figure 2. Because the missile was axisymmetric, half of the missile was calculated in order to reduce the calculation time and improve the calculation efficiency. This calculated area was a cylinder with a diameter 3000mm, high 6000mm. Another 600×6×3000mm cuboid area in the cylindrical was set up (the missile located in the center of rectangular) in order to plotted more dense gridding and the area within the cylinder area and outside the cuboid could be plotted sparseness grid slightly. In this way, the accuracy of the calculation was assured and the computation time was reduced at the same time. Grid number of the whole area was 297024.

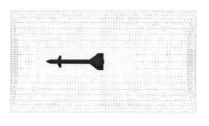

Fig. 1. Gridding plot of calculational area   **Fig. 2.** Gridding plot of missile exterior

3 Mathematical Model

The calculation of the plotted grid was done by FLUENT software. The $k - \varepsilon$ model was adopted as the turbulence model. Continuity equation was shown as below.

$$\frac{\partial(\rho u_i)}{\partial x_i} = 0 \tag{1}$$

Momentum equations viz Reynolds average Navier-Stockes equation was shown as below.

$$\frac{\partial(\rho u_j u_i)}{\partial x_j} = -\frac{\partial p}{\partial x_j} + \frac{\partial\left[\mu_e\left(\frac{\partial u_i}{\partial x_j} + \frac{\partial u_j}{\partial x_i}\right)\right]}{\partial x_j} \tag{2}$$

The standard $k - \varepsilon$ double model equation was shown as below

$$\left.\begin{aligned} \mu_e &= \mu + \mu_t \\ \mu_t &= \rho C_\mu \frac{k^2}{\varepsilon} \end{aligned}\right\} \tag{3}$$

$$\left.\begin{aligned} \frac{\partial(\rho k u_i)}{\partial x_i} &= \frac{\partial\left[\left(\mu + \frac{\mu_t}{\sigma_k}\right)\frac{\partial k}{\partial x_j}\right]}{\partial x_j} + G_k - \rho\varepsilon \\ \frac{\partial(\rho \varepsilon u_i)}{\partial x_i} &= \frac{\partial\left[\left(\mu + \frac{\mu_t}{\sigma_\varepsilon}\right)\frac{\partial \varepsilon}{\partial x_j}\right]}{\partial x_j} + C_{1e}\frac{\varepsilon}{k}G_k - C_{2e}\rho\frac{\varepsilon^2}{k} \end{aligned}\right\} \tag{4}$$

There into,

$$G_k = -\rho u_i' u_j' \frac{\partial u_j}{\partial x_i} \tag{5}$$

The constants in above formula were chosen as below.

$C_{1e} = 1.44$, $C_{2e} = 1.92$, $C_\mu = 0.09$, $\sigma_k = 1.0$, $\sigma_\varepsilon = 1.3$。

4   Calculation Method

1) Coupling implicit method was used and the standard $k - \varepsilon$ model was chosen for viscous model which was solved using conduction equation;
2) The liquid material attribute was set as below: material item was set as air and density item was set as the ideal gas. Sutherland was chosen as "viscosity" item because gas viscosity sutherland law was very suitable for compressible flow speed;
3) The surface condition without sliding was chosen for surface condition and surface roughness was chosen to its default value 0.5.Impermeable surface conditions were adopted by all other scalar;
4) The difference format selections of numerical calculation process were shown as follow: a) choosing the acquiescent standard method as pressure interpolation; b) adopting SIMPLEC for pressure-speed coupled; c) momentum, turbulent kinetic energy and dissipation rate were chosen for turbulence scheme, namely second order upwind;
5) Relaxation factors were set as follow: relaxation factor of pressure item was set for 0.3, density item and quality item were set for 1, momentum was set for 0.5, turbulent kinetic energy, turbulent dissipation rate of and turbulence viscosity were all set for 0.6;

6) The convergence criteria was that both sides calculating differences of continuous equation denoted with differential equation were less than 0.001.

5 Calculated Results and Analysis

Figure 3 was drag coefficient of the missile with mach and attack angle Figure 4 was Lift coefficient of the missile with mach and attack angle. Figure 5 was pitching moment coefficient of the missile with mach and attack angle. From Figure 3 to figure 5, drag coefficient and lift coefficient with the increscence of angle attacks were increscent and pitching moment coefficient decreased with the increscence of attack angles, the calculation results agreed with many reports in the literature. Surface pressure distribution and temperature distribution during missile flight was shown in Figure 6. The pressure of head and the front rudder fin of missile body was the most, and the most big head pressure reached to 34979.76Pa. Surface temperature distribution during missile flight was as shown in figure 7, high temperature parts of the missile were head and rudder and the maximal temperature of the missile was 250.7981K. Surface flow velocity distribution during missile flight was as shown in figure 8. The part with higher flow speed by missile surface was rudder fin and maximum speed could be reached to 1.0974Ma.

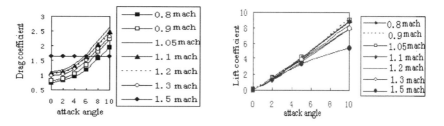

Fig. 3. Drag coefficient change   **Fig. 4.** Lift coefficient change

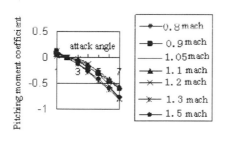

Fig. 5. Pitching moment coefficient change   **Fig. 6.** Static pressure contours of missile

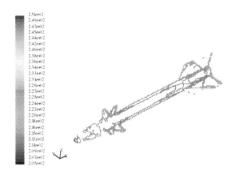
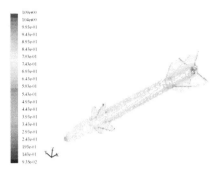

Fig. 7. Temperature contours of the missile under Ma=0.8, attack angle=10°

Fig. 8. Mach number contours of the missile Under Ma=0.8, attack angle=10°

6 Conclusion

Numerical computations of whole missile aerodynamics were carried though in virtue of Fluent of CFD software and traditional programming calculation mode was changed. The explanation was needed. The different place of FLUENT compared with and traditional calculation method was that it was to calculate the missile force and moment, then calculation results were gained according to the given reference condition. The calculation results may have some deviations, so the comparison of the calculation results and the data gained by wind tunnel test was needed [5], then improvement would be processed according to the insufficient places found.

References

1. Heng-yuan, Y.: The Engineering Calculation and Analysis Aerodynamic Characteristics of Vehicle. Northwestern Polytechnical University Press, Beijing (1990)
2. Zai-Xin, C.: Aerodynamics. Aviation Industry Press, Beijing (1993)
3. Jin-Yu, G.: Winged Missiles Structure Design Principle. National Defense Industry Press, Beijing (1988)
4. Carr, L.W.: Progress in Analysis and Prediction of Dynamic Stall. AIAA Journal of Aircraft 5(1), 6–17 (1988)
5. Bartlett, G.E., Vidal, R.J.: Experiment Investigation of Influence of Edge Sharp on the Aerodynamic Characteristic of Low Aspect Ration Wings at Low Speed. Aero Sci. 22(3), 517–533 (1955)

Numerical Simulation of a Pintle Variable Thrust Rocket Engine

Chun-guo Yue[1,*], Xin-Long Chang[1], Shu-jun Yang[2], and You-hong Zhang[1]

[1] Xi'an Research Inst. of Hi-Tech Hongqing Town, 710025 Xi'an, China
[2] 203 Institute of China North Industries Group Corporation, 710065 Xi'an, China
wsgangzi802@qq.com, xinlongch@sina.com.cn, zyhnpu@hotmail.com,
chuangshawudi@etang.com

Abstract. The most prominent feature of variable thrust rocket engine were a large scale changes in working conditions and the adaptive control of the thrust. With proper design and manufacturing pintle injector, high performance and inherent combustion stability could be typically delivered. So, the pintle injector became perfect select of variable thrust engine. With the support of powerful calculational ability of CFD software Fluent, integrative simulation research on the flux-orientation variable thrust liquid propellant rocket engine was developed. Numerical simulation of interior flow field of a variable thrust rocket engine with flux-oriented injector was done. The distributions of pressure, temperature, molar fraction of product and flow mach numbers were attained. By the contrast of the calculational results, the effects of structure parameters and working condition etc. on total whole performance of variable thrust rocket engine were analysed. The results also provided theoretic references for design and optimization of variable thrust rocket engine.

Keywords: Pintle injector, Variable thrust rocket engine, Numerical simulation, Fluent.

1 Introduction

The most prominent features of variable thrust engine were adaptability on wide working conditions and controllable characteristic of thrust. Due to the large scope change of engine thrust needed to adjust to changes in the scope of working conditions, the flow field structure of variable thrust engine had significant differences form fixed thrust liquid rocket engines.

Now special procedures which could carried through numerical simulation of spontaneous combustion process of propellant rocket engine had been emerged at France, Germany and China. One PHEDRE[1] was put forward by the French institute of space, the droplet size distribution model and droplet rate of evaporation model were the simpler and one-step chemical reaction model was adopted. ROCFLAM program[2] developed by German aerospace research considered the fuel of MMH

---

* Corresponding author.

Y. Yu, Z. Yu, and J. Zhao (Eds.): CSEEE 2011, Part II, CCIS 159, pp. 477–481, 2011.
© Springer-Verlag Berlin Heidelberg 2011

decomposition and antioxidant NTO the dissociation in chemical reaction model and reflected the spontaneous combustion characteristic of propellant in a certain extent. HPRECSA procedures[3,4] were studied by equipment command technical institute and national university of defense technology of China, the EBU turbulent combustion model was adopted and its characteristic was that high propellant combustion theory of solving the evaporation process of propellant droplet was adopted, considering the dissociation of propellant decomposition and evaporation process.

Based on the previous research, combustion flow integrated simulations of variable thrust engine were developed. In-depth analysis of structure parameters and operating parameters on the overall performance of the engine were processed, which could provide theory support to design and optimization of variable thrust engine.

2   Mathematical Model and Physical Model

2.1   Physical Model

The model and mesh of the variable thrust liquid rocket engine studied in this paper was shown in figure 1. The combustion chamber parameters were shown as below. The length was 135 mm, diameter was 75.5 mm and nozzle throat diameter was 15 mm. The pintle injector was used by the engine and its structure principle was shown in figure 2. The vertical angle scheme was adopted in this article. The fuel was in the center and the injection angle was 90°, injection angle of antioxidant road was 40°. The lengths of pintles embed to combustion chamber were 10 mm, 15 mm and 20 mm respectively.

   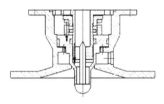

Fig. 1. The model and mesh of the engine   **Fig. 2.** Structure principle of pintle injector

2.2   Mathematical Model

Eulerian-Lagrangian approach[4]was adopted in the simulation of liquid rocket engine spray combustion process. The control equations of two phase reactant flow control were made up of Navier-Stokes equations set in Eulerian coordinate system and the control equations set describing the motion law of spray in Lagrangian coordinate system. The coupling between two phases used source items of liquid and gas interaction for consideration. The concrete forms of control equations could be seen in reference[5]. The $k - \varepsilon$ double model equations of Renaull average were used for turbulence model. EBU model with similar finite rate/vortex dissipative model (EDM) was used for turbulent combustion model, the model took into account both the chemical control, and considered the reaction rate by turbulence pulsation of the decay rate of reaction rate, and decided that the rate of smaller was burning control rate.

The flow field of variable thrust liquid rocket engine thrust was very complex, the assumption as follow were done to facilitate the flow field.

1) The combustion flow field of liquid engine combustion chamber was 3d multiphase flow turbulence;

2) The reaction of fuel and antioxidant was one step reaction, namely did not exist intermediate product at any point;

3) Each gas phase component of the combustion chamber was all ideal gas according with the perfect gas state equation $p = \rho RT$;

4) The effect of gravity on flow field was ignored.

According to the assumptions, the combustion chamber reaction was shown as below.

$$CH_3NHNHCH_3 + 2N_2O_4 \rightarrow 2CO_2 + 4H_2O + 3N_2$$

3 Boundary Conditions and the Solution

The engine working conditions in vacuum were simulated. Fuel inlet was flux entry condition, its parameters were total temperature 300 K and total mass flow rate 0.0658 kg/s, oxidant inlet was also flux entry condition, its parameters were total temperature 300 K and total mass flow rate 0.0329 kg/s. Export boundary was dealt with supersonic speed, parameters were obtained according to external value-inserting method of neighboring upstream parameters. In the border of symmetry, vertical velocity component was 0, all variables in the direction of vertical gradient were 0. Solid wall boundary was adiabatic boundary sliding take.

The control equations were dispersed by the finite volume method. Discrete control equations were solved using SIMPLE algorithm.

4 Calculated Results and Analysis

So far, 3 (models) x 5 (working conditions) = 15 conditions had been the simulated., the flow field of combustion obtained had similarity, so simulation results of the 1# model were only analyzed (seeing figure ~ figure 10).

As the figure 3 showed, because of the free jet of fuel and oxidants, the high pressure area came forth near injection surface and appeared larger changes. In the combustion process, the combustion chamber pressure appeared small changes. The pressure along the axial dropped rapidly when the air flow passed through throat part and expand section area. As the figure 4 showed, the maximum temperature of combustion chamber was reached 3340 K. The temperature distribution uniformity was not so good, it showed mixing effect of antioxidant and fuel was not good enough. Combustion temperature peak located high concentrations of antioxidant basically. The high temperature area was apart from the combustion chamber wall relatively close in the front of chamber. The temperature peak of the convergence part located near the wall region of engine[3].

As the figure 6 showed, the evaporation rate of UDMH (unsymmetric dimethyl Hydrazine) was slower and evaporation began in a certain distance away from injection face, a majority of UDMH finished evaporation almost to the nozzle throat. Combining with mass fraction distribution figure of reaction products, the rate of evaporation of UDMH was slower than the combustion rate. Therefore, this model engine combustion process was controlled by evaporation of UDMH. As the figure $7\sim9$ showed, maximum mass concentration of CO_2 was reached 0.335, maximum mass concentration of N_2 reached 0.320 and maximum mass concentration of H_2O reached 0.442. From the distribution maps of the various components, components distribution areas were quite obvious that mixing effect of propellant in combustion chamber was bad and reactions focused on the border of fuel and antioxidant fully.

By the figure 10, injector after jet fuel spray came under the effect of sector momentum and gas, two main backflow areas were come into being in the combustion chamber. One was the outer race backflow area, which was mainly rich, propellant combustion gas, liquid propellant droplets collision. Second was the center backflow area, which was mainly backflow of propellant combustion toward the middle of needle axial backflow. The center backflow area made the effect of baffle and mixer for those droplets who had movement trend to nozzle throat movement trend, but unburned. Figure 11 was the isoclines distribution of mach numbers, as could be seen from the graph that velocity of antioxidant and fuel to begin ejecting from injector was lesser, the gas in the combustion chamber happened the rapid expansion and speedup after a chemical reaction and the flow velocity in throat part 1.0 Ma, nozzle exit was reached 2.77 Ma. A group from one of the fuel droplets (65 μm) of the trajectory map (shown in Figure 12) shows that the droplets in the area near the pintle evaporated quickly completed, and the droplets in the region away from the pintle evaporated relatively slow.

5 Main Conclusions

1) The distributions of pressure, velocity, mach number, temperature and the components mass fraction and so on of three combustion chamber models were attained, which could provide supports and foundations to further analysis and simulation. 2) Simulation results showed that, pintle embedding length had an effect on combustion flow field of engine. The reasonable injectors should be designed in order to get high efficiency and better stability and must make technology to meet the design requirements. 3) The pintle injector combustion chamber had two different backflow areas comparing with flow fields of ordinary liquid rocket engines, namely outer race backflow area and center backflow area. 4) The rate of evaporation of fuel was slower than the combustion rate. Therefore, this model engine combustion process was controlled by evaporation of fuel.

Fig. 3. Pressure distribution   **Fig. 4.** Temperature distribution   **Fig. 5.** Oxidant distribution

Fig. 6. Fuel distribution   **Fig. 7.** CO$_2$distribution   **Fig. 8.** H$_2$O distribution

Fig. 9. N$_2$ distribution   **Fig. 10.** Part flown line   **Fig. 11.** Mach number distribution

Fig. 12. Trajectory map of 65 μm fuel droplet

References

1. Yu-Lin, Z.: Change of Liquid Rocket Engine Thrust and Control Technology. National Defense Industry Press, Beijing (2001)
2. Fu-Yuan, X.: Injector Design Improvement of Bipropellant Variable Thrust Rocket Engine. Propulsion Technology 10(2), 40–43 (1990)
3. Feng-Chen, Z.: Liquid Rocket Engine Spray Combustion Theory, Model and Application. National Defense Science and Technology University Press, Beijing (1995)
4. Jian-xing, Z.: Numerical Simulation of Combustion. Science Press, Beijing (2002)
5. Ning-Chang, Z.: The Design of Liquid Rocket Engine. Aerospace Press, Beijing (1994)

Author Index